*PROGRESS IN*

# Molecular Biology and Translational Science

*Volume 106*

*PROGRESS IN*

# Molecular Biology and Translational Science

## Protein Phosphorylation in Health and Disease

*edited by*

Shirish Shenolikar, Ph.D.

*Professor & Senior Associate Dean of Research*
*Signature Research Programs in Cardiovascular and*
*Metabolic Disorders & Neuroscience and Behavioral Disorders*
*Duke-NUS Graduate Medical School Singapore, Singapore*
*and*
*Professor of Psychiatry and Behavioral Sciences and Professor of Pharmacology and Cancer Biology, Duke University Medical Center, Durham, USA*

*Volume 106*

ELSEVIER

AMSTERDAM • BOSTON • HEIDELBERG • LONDON
NEW YORK • OXFORD • PARIS • SAN DIEGO
SAN FRANCISCO • SINGAPORE • SYDNEY • TOKYO
Academic Press is an imprint of Elsevier

Academic Press is an imprint of Elsevier
32 Jamestown Road, London, NW1 7BY, UK
Radarweg 29, PO Box 211, 1000 AE Amsterdam, The Netherlands
225 Wyman Street, Waltham, MA 02451, USA
525 B Street, Suite 1900, San Diego, CA 92101-4495, USA

This book is printed on acid-free paper. ∞



**Library of Congress Cataloging-in-Publication Data**
A catalog record for this book is available from the Library of Congress

**British Library Cataloguing in Publication Data**
A catalogue record for this book is available from the British Library

ISBN: 978-0-12-396456-4
ISSN: 1877-1173

For information on all Academic Press publications
visit our website at elsevierdirect.com

Printed and bound by CPI Group (UK) Ltd, Croydon, CR0 4YY

# Contents

# Contributors

*Numbers in parentheses indicate the pages on which the authors' contributions begin.*

**Steven P. Braithwaite,** Signum Biosciences, Monmouth Junction, New Jersey, USA (343)

**Dmitry V. Bulavin,** Cell Cycle Control and Tumorigenesis Group, Institute of Molecular and Cell Biology, Proteos, Singapore (307)

**Perry M. Chan,** sGSK Group, A-Star Neuroscience Research Partnership, Singapore, Singapore (171)

**John H. Connor,** Department of Microbiology, Boston University School of Medicine, Boston, Massachusetts, USA (223)

**Lucy E. Dalton,** Division of Respiratory Medicine, Department of Medicine, University of Cambridge, Cambridge Institute for Medical Research, Cambridge, United Kingdom (189)

**Ewan F. Dunn,** Department of Microbiology, Boston University School of Medicine, Boston, Massachusetts, USA (223)

**Shogo Endo,** Aging Regulation Research Team, Tokyo Metropolitan Institute of Gerontology, Tokyo, Japan (381)

**Xin-Yuan Fu,** Department of Biochemistry, and Cancer Science Institute of Singapore, National University of Singapore, Singapore; and Department of Microbiology and Immunology, Indiana University School of Medicine, Indiana, USA (63)

**Melissa J. Fullwood,** Program in Neuroscience and Behavioral Disorders, Duke-NUS Graduate Medical School Singapore, Singapore (75)

**Anne-Claude Gingras,** Samuel Lunenfeld Research Institute at Mount Sinai Hospital; and Department of Molecular Genetics, U Toronto, 1 King's College Circle, Toronto, Ontario, Canada (3)

**Jean-Antoine Girault,** Institut du Fer à Moulin, and Inserm UMR-S 839; and Université Pierre et Marie Curie, Paris, France (33)

**Serge Hardy,** Goodman Cancer Research Centre, McGill University, Montréal, Québec, Canada (253)

**Eleanor Healey,** Division of Respiratory Medicine, Department of Medicine, University of Cambridge, Cambridge Institute for Medical Research, Cambridge, United Kingdom (189)

**James Irving,** Division of Respiratory Medicine, Department of Medicine, University of Cambridge, Cambridge Institute for Medical Research, Cambridge, United Kingdom (189)

**Erika Segear Johnson,** Department of Pharmacology and Cancer Biology, Box 3813, Duke University School of Medicine, Durham, North Carolina, USA (327)

**Wen Chun Juan,** Program in Cancer and Stem Cell Biology, Duke-NUS Graduate Medical School, Singapore, Singapore (107)

**Sally Kornbluth,** Department of Pharmacology and Cancer Biology, Box 3813, Duke University School of Medicine, Durham, North Carolina, USA (327)

**David P. Labbé,** Goodman Cancer Research Centre; and Department of Medicine, Division of Experimental Medicine, McGill University, Montréal, Québec, Canada (253)

**Cheh Peng Lim,** Department of Biochemistry, National University of Singapore, Singapore (63)

**Paul J. Lombroso,** Child Study Center, Yale University School of Medicine, New Haven, Connecticut, USA (343)

**Ed Manser,** sGSK Group, A-Star Neuroscience Research Partnership, Singapore, Singapore (171)

**Stefan J. Marciniak,** Division of Respiratory Medicine, Department of Medicine, University of Cambridge, Cambridge Institute for Medical Research, Cambridge, United Kingdom (189)

**Angus C. Nairn,** Department of Psychiatry, Yale University School of Medicine, New Haven, Connecticut, USA (343)

**S. Tiong Ong,** Program in Cancer and Stem Cell Biology, Duke-NUS Graduate Medical School, Singapore, Singapore; and Division of Medical Oncology, Department of Medicine, Duke University Medical Center, Durham, North Carolina, USA (107)

**Kanaga Sabapathy,** Division of Cellular & Molecular Research, Humphrey Oei Institute of Cancer Research, National Cancer Centre, and Cancer and Stem Cell Biology Program, Duke-NUS Graduate Medical School; and Department of Biochemistry, National University of Singapore, Singapore (145)

**Shirish Shenolikar,** Program in Neuroscience and Behavioral Disorders; and Program in Cardiovascular and Metabolic Disorders, Duke-NUS Graduate Medical School Singapore, Singapore (75)

**Nicole St-Denis,** Samuel Lunenfeld Research Institute at Mount Sinai Hospital, Toronto, Ontario, Canada (3)

**Jeffry B. Stock,** Department of Molecular Biology, Princeton University, Princeton, New Jersey, USA (343)

**Michel L. Tremblay,** Goodman Cancer Research Centre, and Department of Medicine, Division of Experimental Medicine; and Department of Biochemistry and Oncology, McGill University, Montréal, Québec, Canada (253)

**Wei Zhou,** Program in Neuroscience and Behavioral Disorders; and Program in Cardiovascular and Metabolic Disorders, Duke-NUS Graduate Medical School Singapore, Singapore (75)

**Yun-Hua Zhu,** Cell Cycle Control and Tumorigenesis Group, Institute of Molecular and Cell Biology, Proteos, Singapore (307)

# Preface

Reversible protein phosphorylation plays a critical role in controlling virtually all aspects of mammalian cell physiology. Conservative estimates of cellular phosphoproteins suggest that more than one-third of proteins in the mammalian cell undergo covalent modifications on serine, threonine, and tyrosine residues. Given that the vast majority of the known phosphoproteins are modified at multiple sites—examples of proteins phosphorylated at a single amino acid by a single protein kinase are extremely rare—this has led many to speculate that up to 250,000 distinct phosphorylation events occur in most human cells. When considered in the wider context of the cellular content of phosphorylatable amino acids (serines, threonines, and tyrosines), estimated at 700,000, this suggests that a substantial portion of these residues are substrates for protein kinases and phosphatases. However, to date, 60,000 phosphorylation events have been clearly documented. While indicating the tremendous progress that has been made in identifying phosphoproteins and their phosphorylated amino acids, particularly over the past decade, there are still some ways to go before we have a full understanding of all phosphorylations that control mammalian cell physiology.

With this in mind, readers should recognize that no volume aimed at highlighting the role of protein phosphorylation in human health and disease can aim to be comprehensive, covering all relevant phosphoproteins or phosphorylation events or all diseases influenced by aberrant protein phosphorylation. Thus, this volume is focused on selected topics that together exemplify the different ways that phosphoproteins and/or phosphorylations impact human disease, illustrate the role of protein kinases and phosphatases in these processes, and provide the views of international experts recognized for their significant contributions to the current research in protein phosphorylation. I hope that this volume will provide the reader with a broad overview of the ongoing research in the selected disease arenas, an appreciation for how this research has improved our understanding of the disease mechanisms and the potential of these studies to guide the future development of novel therapeutics.

The contents of this volume are separated into three broad sections solely for the purpose of communicating simple concepts. In this regard, it is very difficult to distinguish phosphoproteins, often defined as substrates of protein kinases and phosphatases, from the kinases and phosphatases themselves, as many of these are also phosphoproteins and therefore modulated by reversible phosphorylation. The first chapter in the phosphoproteins section is by St-Dennis and Gingras and outlines the current methods for analyzing phosphoproteins and identifying individual phosphorylation sites. This is followed by chapters by Girault, Lim and Fu, and Fullwood and colleagues that describe the phosphorylation of distinct classes of proteins, from ion channels, transcription, and translational factors, that may represent valuable biomarkers for various human diseases. Finally, the chapter by Juan and Ong describes how the analysis of signaling pathways may be used to overcome drug resistance frequently seen with currently approved therapies targeting the oncogenic kinase, BCR-Abl, to treat chronic myelogenous leukemia. This then provides the transition to the next section on protein kinases, a family of drug targets that constitutes a major fraction of the discovery portfolio of most major pharmaceutical companies. The chapters by Sabapathy, Chan and Manser, and Dalton and colleagues highlight families of protein kinases the contributions of which to many human diseases have only recently been fully realized. While in some cases, the aberrant activity of kinases exacerbates the pathophysiology of the disease such that their inhibition by small molecules would be anticipated to ameliorate clinical symptoms, other kinases play a key role in cell survival and are ideally the targets of pharmacological activators, a significant new challenge for drug discovery. Together, these chapters highlight that the improved understanding of their physiological contributions combined with the three-dimension structures of the protein kinases will likely play a critical role in the design of specific kinase inhibitors that can achieve the desired therapeutic goal without unexpected toxicity. Finally, the chapter by Dunn and Connor highlight a current strategy, namely, elucidating virus–host interactions, with the goal of potentially targeting host mechanisms to treat serious viral infections. It is hoped that this experimental approach will reduce or overcome the rapid resistance seen with many existing antiviral drugs but also raises the possibility that targeting host mechanisms may carry an increased risk of side effects or drug toxicity.

The equilibrium between protein kinases and phosphatases dictates the overall phosphorylation state of cellular phosphoproteins. Thus, the pharmacological inhibition of protein phosphatases to bolster or prolong kinase action has been frequently proposed as a mechanism for correcting defects on hormone or neurotransmitter activity. Despite considerable effects from both industry and academic scientists to develop compounds that inhibit tyrosine phosphatases, these efforts have yet to provide us with effective therapeutic agents. Nevertheless, as described by Labbé, Hardy, and Tremblay, aberrant function of protein tyrosine phosphatases contributes to many human diseases and these enzymes remain very attractive if challenging targets for drug discovery. Chapters by Zhu and Bulavin, Johnson and Kornbluth, and Stock and colleagues highlight several areas in which serine/threonine phosphatases can play a role in biological processes that impinge on human disease. Finally, Endo describes the physiological role of an endogenous serine/threonine phosphatase regulator in propagating cGMP signals in the brain. The existence of selective and regulated or tunable phosphatase inhibitors as well as the identification of numerous xenobiotics that inhibit cellular protein phosphatases suggests that targeting serine/threonine phosphatases will require an innovative approach or out-of-the box thinking that potentially exploits the presence of numerous cellular phosphatase complexes that are dedicated regulators of specific physiological processes. While at first blush, this appears to be a major hurdle for future drug development, recent discoveries of small molecules that disrupt specific protein phosphatase-1 complexes highlight a potential winning strategy for the development of therapeutically effective and largely safe serine/threonine phosphatase inhibitors.

In summary, emerging studies have highlighted that changes in protein phosphorylation play key roles in the physiology and pathophysiology of human disease. The growing understanding of the kinases and phosphatases involved bodes well for the future development of biomarkers to monitor disease progression and drugs to treat a wide variety of human diseases. I will end by thanking all my colleagues for their very thoughtful contributions that will hopefully help the reader to appreciate and understand the role of protein phosphorylation in human health and disease.

Shirish Shenolikar

# Section I

## Phosphoproteins

# Mass Spectrometric Tools for Systematic Analysis of Protein Phosphorylation

Nicole St-Denis[*] and Anne-Claude Gingras[*,†]

[*]*Samuel Lunenfeld Research Institute at Mount Sinai Hospital, Toronto, Ontario, Canada*

[†]*Department of Molecular Genetics, U Toronto, 1 King's College Circle, Toronto, Ontario, Canada*



The study of protein phosphorylation has long been performed one protein and one modification site at a time, with a major emphasis placed on the functional relevance of one or a small number of given phosphorylation event (s). This has enabled exquisitely detailed views of many intracellular signaling pathways but has left a large portion of the phosphoproteome relatively uncharted. The past several years have seen an explosion in the development of generic and global approaches to study protein phosphorylation, thanks in large part to advances in robotics, mass spectrometry, and computational biology. As of July 2011, there were more than 60,000 nonredundant phosphorylation sites for human proteins annotated in a large repository. This explosion of data has, however, highlighted important issues regarding both the reliability of these types of identifications and the problem of assigning function to each of these phosphorylation events. Parallel advances in the identification of


DOI: 10.1016/B978-0-12-396456-4.00014-6

1877-1173/12 $35.00

consensus sites for kinases, and systematic mapping of protein–protein interactions in signaling pathways, have provided complementary information that should help in obtaining a more holistic view of signaling. Here, we provide a perspective on system-wide approaches based on mass spectrometry to understand phosphoregulation.

## I. Introduction

Protein phosphorylation as a regulatory mechanism was first described in 1955.[1] Since then, phosphorylation has emerged as a crucial component of cellular signal transduction, regulating pathways controlling everything from cell division to cell death.[2] Reversible phosphorylation can lead to changes in protein structure and stability, protein–protein interactions, enzyme activation, or subcellular localization. The study of phosphorylated proteins and the enzymes that modulate them—kinases and phosphatases, which catalyze the addition and removal of phosphate, respectively—has provided a wealth of information on signal transduction pathways in both normal and disease states. It is clear that proper regulation of phosphorylation is crucial to human health, as several human diseases are caused by aberrant protein phosphorylation, including cancer, neurodegeneration, inflammatory disease, and diabetes.[3] The development of therapies targeting phosphoregulation is thus an area of tremendous interest. Following the success of imatinib, a kinase inhibitor targeting the Bcr-Abl fusion protein found in chronic myelogenous leukemia,[4] several additional kinase inhibitors have received regulatory approval, and many more are undergoing clinical trials. Despite these successes, there remains much to be learned about the effects of phosphorylation on cellular signaling.

In the postgenomic era, our knowledge of the influence and extent of protein phosphorylation have expanded dramatically, and large-scale proteomic approaches allow the study of phosphorylation on a global scale, including the integration and cross talk between cellular pathways. The human genome contains 518 kinases, including 428 that phosphorylate serine and threonine and 90 that phosphorylate tyrosine.[5] The number of phosphatases encoded in the genome is lower, at 147,[6] and consists of 40 serine/threonine phosphatases and 107 tyrosine phosphatases (not all of which are active enzymes). In a typical eukaryotic cell, there is an estimated 700,000 phosphorylatable residues.[7] Phosphoamino acid analysis and, more recently, global phosphoproteomic analysis indicate that protein phosphorylation occurs primarily at serine residues—~85% of sites—while threonine and tyrosine phosphorylation are considerably rarer (~15% and ~2%, respectively).[8–10] Regulated phosphorylation events have now been characterized on a global scale for a variety of cellular stimuli, cell lines, and disease states. System-wide analysis of phosphorylation pathways is necessary to gain a full understanding of the roles of phosphorylation but presents several challenges that need to be overcome in order to develop a fully integrated model of protein phosphorylation in response to various stimuli. In this chapter, we review several strategies for mass spectrometric analysis of regulatory phosphorylation sites.

## II. Mass-Spectrometry-Based Identification of Phosphorylation Sites

Long before global phosphoproteomics, researchers employed mass spectrometry to identify phosphorylation sites in their protein(s) of interest. Mass spectrometry can be harnessed for phosphorylation site mapping because the addition of a phosphate group changes the mass of the protein. Figure 1 depicts a

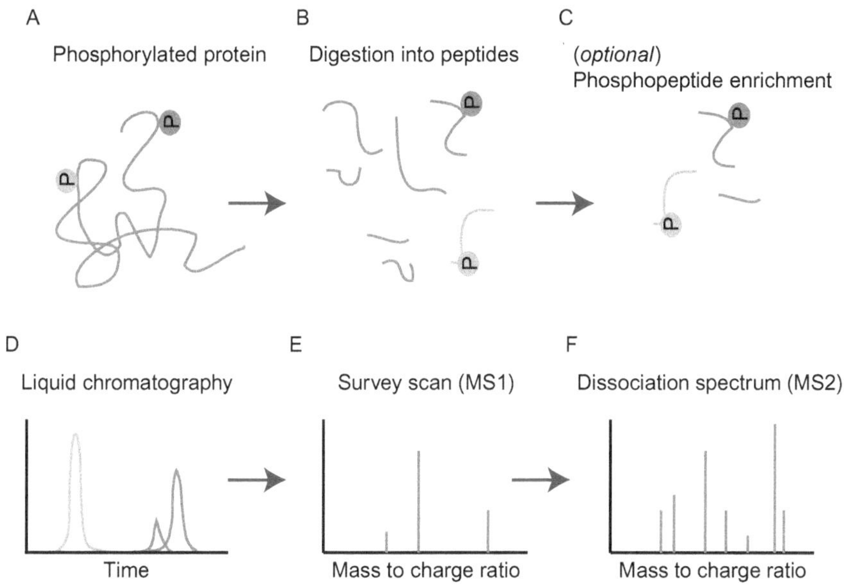


FIG. 1. Workflow for the analysis of phosphorylation using mass spectrometry. The Phosphorylated proteins (A) are digested into peptides (B). Phosphopeptides can be enriched by a variety of methods (C) before separation by liquid chromatography (D) and analysis by mass spectrometry. The mass spectrometer first performs a survey (MS1) scan to measure the intensity of each peptide (E). High intensity peptide ions are then selected for fragmentation, and the fragmentation pattern is used to determine the sequence of the peptide and identify phosphorylation events (F). (See Color Insert.)

typical modern workflow for phosphorylation site mapping. The protein (or more often a protein mixture) is first digested with a proteolytic enzyme (trypsin is preferred in most cases). Phosphopeptides are then either enriched (see Section III), or the entire mixture is subjected to separation on a liquid chromatography column placed inline with a mass spectrometer (for an excellent and accessible review of this process, refer to Ref. 11). The mass spectrometer first performs a survey scan, a step in which the mass-to-charge ratio and the intensity of every eluting peptide are monitored. In most cases, the *n* most abundant peptides will next be sequentially selected for fragmentation. Specific fragmentation conditions are used to produce breaks primarily at peptide bonds. The mass of the resulting fragments is then measured in what is often referred to as an MS/MS or MS2 spectrum. In the fragmentation type known as collision-induced dissociation, two primary ion series are formed, depending upon whether the amino- or carboxy-terminal end of the peptide has maintained a charge; the pattern of fragments can be used in a manner similar to a fingerprint to identify the initial peptide. However, not all possible fragments are detected above background in each spectrum; this becomes important when analyzing posttranslational modifications. Following the acquisition of thousands of MS/MS spectra by the mass spectrometer, software tools are used to deduce sequence information for the peptide from which they are derived. Database search tools match the pattern of peaks observed in the MS/MS spectrum to theoretical spectra calculated on the basis of DNA or protein sequences in databases. The peptide with the best match to the observed spectrum (providing that a certain threshold is achieved) is assigned as the identification. There are issues (especially with earlier data) with false-positive identifications at the peptide level. In recent years, statistical approaches have been employed to limit (or at least report) the false discovery rate. Note that, while on a global scale, these statistical tools perform very well, it remains possible that a given peptide is wrongly assigned, and closer inspection of spectra is important if the identification of a specific peptide is crucial to the experiment.

A phosphopeptide can be differentiated from a nonphosphorylated peptide by its mass/charge change; in ideal cases (Fig. 2), there will be only one phosphorylatable residue, and the phosphate group will remain conjugated during fragmentation. Ions that maintain these phosphate groups will be shifted from the unmodified peptide by the mass/charge of the phosphate group. The situation becomes more complex, however, if the phosphate group is lost during fragmentation (this is observed as a neutral loss of 98 Da—corresponding to phosphoric acid, and occurs frequently, especially with serine and threonine phosphorylation). In this case, positional information is lost (note that this phosphate release is not all bad news, as it is possible to harness this neutral loss as a tool for identifying phosphopeptides; see e.g., Ref. 12). There are intermediate cases, with both a neutral loss and residual—albeit

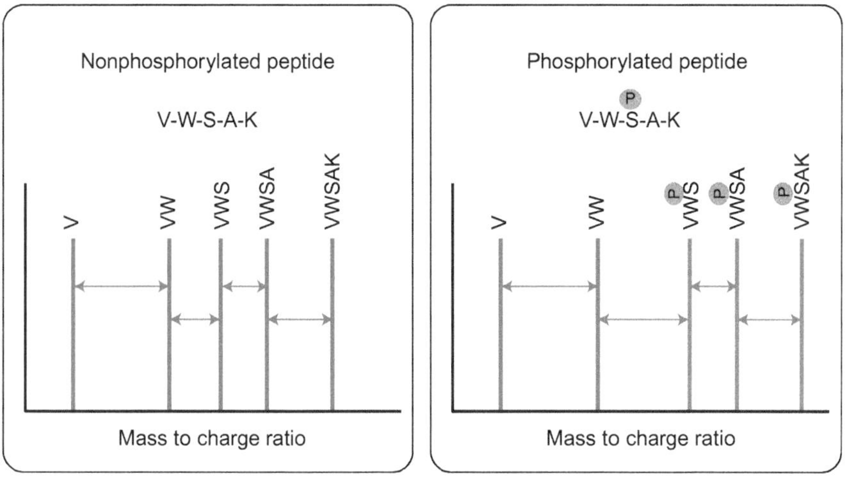


FIG. 2. Simplified view of mass spectrometric identification of a phosphorylation site. Left: Fragmentation of a nonphosphorylated peptide. The distance between the peaks is proportional to the mass-to-charge ratio (*m/z*) of each amino acid; for simplicity, only one ion series is depicted. Right: Fragmentation of a phosphopeptide. Ions retaining the phosphate group (orange) will be shifted in *m/z* as compared to their nonphosphorylated counterparts (green). (See Color Insert.)

smaller—ion peaks still bearing the phosphate; in these cases, phosphorylation site identification may or may not be possible. If there is more than one phosphorylatable residue in the peptide, a challenge is to use the information available to identify the position of the modified residues. Again, software tools, including Ascore,[13] have been designed for this operation. These tools provide a nice improvement upon previous approaches where such positional effects were not taken into consideration. As with the peptide scoring methods, however, the positional scoring tools are imperfect, and caution should be applied when interpreting automatically generated phosphopeptide lists.

## III. Enrichment Approaches for Phosphorylation and Large-Scale Phosphoproteomics

A key element that enabled the burgeoning of the global phosphoproteomics field has been the development of a number of enrichment methods to isolate phosphopeptides from unphosphorylated peptides. The most commonly used enrichment approach is immobilized metal affinity chromatography (IMAC), either using metals such as iron, nickel, or cobalt, or a column made of titanium dioxide (reviewed in Ref. 14; note that most groups now employ the term IMAC to refer to the nontitanium version of the enrichment). Depending upon the conditions employed for phosphopeptide capture, more or less specificity for the phosphopeptides may be achieved. These methods are easy to implement in the laboratory, and several companies offer phosphopeptide enrichment kits. In addition to these straightforward approaches, alternative strategies, including covalent capture of phosphopeptides, for example, on dendrimers, are also employed.[15] So which of the approaches is best? The answer is complex, especially as a recent report by Bodenmiller *et al.* suggested that there is a large nonoverlapping population of phosphopeptides identified by each method.[16] This suggests that the nature of the phosphopeptides identified will depend upon the selected method, and that combining multiple approaches will result in the identification of more phosphorylation sites than either method alone. This observation also indicates that many sites are likely to be missed by current enrichment/detection methods. In addition to a single enrichment step, IMAC or $TiO_2$ enrichment methods may be coupled with additional fractionation steps, with strong cation exchange (SCX) being frequently used in high-throughput studies.[17] On an SCX column, phosphorylated peptides tend to elute much earlier than their nonphosphorylated counterparts, allowing for a rapid enrichment orthogonal to affinity chromatography.

The methods highlighted above have the potential to enrich any phosphorylated peptide. However, other recently implemented affinity approaches are able to identify specific phosphorylation sites. The prime example is the use of antibodies recognizing phosphorylated tyrosine to immunoprecipitate peptides.[18] Often coupled with another enrichment technique (such as $TiO_2$ or IMAC), these antibodies enable the identification of hundreds of tyrosine-phosphorylated peptides per experiment.[19] This is in sharp contrast with the number of tyrosine-phosphorylated peptides that are identified in standard IMAC experiments (tyrosine-phosphorylated peptides represent only ~0.5–2% of all phosphopeptides).

Following the impressive success of anti-Tyr peptide immunoprecipitation, other antibodies have also been used for enriching specific phosphorylated sequences. While generic antiphosphoserine or antiphosphothreonine antibodies have, in general, worked poorly for immunoprecipitation, antibodies generated against specific sequences (e.g., the consensus site for a specific serine/threonine kinase) have been more powerful. For example, a survey of the sites phosphorylated on serine/threonine-glutamine sites after DNA damage treatment has enabled the identification of a network of ~700 substrates for the ATM and ATR kinases. The identified substrates were highly enriched for proteins involved in the DNA damage response, demonstrating the robustness of the approach, and also identified a number of networks not previously linked to DNA damage signaling.[20]

## IV. Phosphorylation Databases

Given the wealth of data generated concerning protein phosphorylation in both systems-wide and focused studies, several computational resources have been developed to facilitate analysis. These resources may be useful for a researcher trying to determine whether phosphorylation sites have been detected on their protein(s) of interest, but they can also be used for making general conclusions regarding phosphorylation sites. Several resources are currently available (see Table I), but for the purpose of this discussion, we will only review a few of them. PhosphoSitePlus was created by Cell Signaling Technology and represents an extensive curation effort of posttranslational modifications on mammalian proteins, supplemented with mass spectrometry efforts at the company.[9] As of July 2011, the database included more than 12,000 curated publications reporting nearly 100,000 nonredundant sites on more than 13,000 proteins (including more than 60,000 nonredundant sites on human proteins). Key features of the database include a straightforward search interface supported by an appealing graphical user interface, and the possibility to download datasets of interest for further analysis. Importantly, while the raw source data for the mass spectrometry-based detection of phosphorylation sites are not included (nor are the data from the original publication researched in house), PhosphoSitePlus lists the number of instances of detection of a given phosphorylation site in small studies and in mass spectrometry screens. This is very useful to the individual researcher, as confidence in identification increases with the number of observations. Other databases such as Phospho. ELM[21,22] and PhosphoGRID[27] are similar in reporting the results of the phosphopeptide identifications without a direct link to the annotated spectra. Specialized mass spectrometry databases for posttranslational modification data have also been created, however, that provide links to the mass spectra, thereby enabling experienced users to make a judgment call regarding the particular phosphorylation event. These specialized mass spectrometry databases include PhosphoPep[25] and PHOSIDA.[23,24] Lastly, other databases exhibit specific features (e.g., the Phospho3D database attempts to place the phosphorylation site in a structural context[29] and the PhosphoNet database combines experimentally validated and predicted phosphorylation sites [Unpublished, www.phosphonet.ca]).

In combination with other systems biology resources, phosphopeptide identification data can contribute to a better characterization of signaling pathways and networks. In fact, having a large repository of phosphorylation sites as a reference is critical for a global view of phosphorylation. However, there remain problems associated with data analysis and curation. As described above, confidence in the phosphopeptide identification, but more critically in the phosphorylation site assignment, is often difficult to establish. Databases

TABLE I

DATABASES FOR THE ANNOTATION AND ANALYSIS OF PHOSPHORYLATION SITES

| Phosphorylation database | Website | Description | References |
|---|---|---|---|
| PhosphoSitePlus | http://phosphosite.org | Comprehensive and well-designed repository of posttranslational modifications for human and mouse | 9 |
| Phospho.ELM | http://phospho.elm.eu.org | Stores a large number of phosphorylation sites and allows analysis of disordered regions and structural accessibility; sites can be viewed on multiple alignments | 21,22 |
| PHOSIDA | http://www.phosida.com | Specialized for mass spectrometry (MS) data; contains both quantitative information and prediction of kinase sites | 23,24 |
| PhosphoPep | http://www.phosphopep.org | Contains data from MS-based large-scale experiments from four species (fly, yeast, worm, and human). The database is coherent, in that the raw data was all searched with the same sets of parameters, and the identified spectra can be viewed. This database also specializes in the generation of transitions (or specific peptide identifiers) for targeted proteomics approaches | 25,26 |
| PhosphoGRID | http://phosphogrid.org | Encompasses multiple phosphorylation sites for the budding yeast *S. cerevisiae* in a nice simple visual interface with enhanced search functions. Kinase motifs are highlighted, and download functions are built in | 27 |
| PhosPhAt | http://phosphat.mpimp-golm.mpg.de | Specialized phosphorylation database for *Arabidopsis thaliana* with links to the spectra and quantitative mass spectrometric information | 28 |
| Phospho3D | http://www.phospho3D.org | This database integrates phosphorylation data from phospho.ELM and places it in a structural context using structures from the Protein Data Bank | 29 |
| SwissProt | http://www.uniprot.org | Curated protein database that provides a high level of annotation, including posttranslational modifications, to each protein | 30 |
| HPRD | http://www.hprd.org | Contains information about human proteins, including on phosphorylation | 31,32 |
| PhosphoNet | http://www.phosphonet.ca | Includes both experimentally defined and predicted phosphorylation sites on human proteins; most of the sites are predicted by Kinexus, and 14% have been experimentally validated. Also includes prediction on the kinases responsible for phosphorylation | |

that simply import phosphorylation sites directly from publications without reanalyzing the mass spectra are more susceptible to inconsistency in the quality of the data. Problems with data analysis (given the same raw mass spectrometry data) were already reported for general protein identification data.[33] Not surprisingly, the same type of issues also applied to phosphopeptide identification (study group from the Association of Biomolecular Resource Facilities; ABRF, poster presentation). These problems are somewhat reduced with databases that reanalyze all mass spectrometry data using a consistent set of rules and for which the annotated spectra are also provided. As mentioned above, repeatability in the detection of a given phosphorylation site is useful for assessing its reliability (this information is easily accessible on PhosphoSitePlus), but other parameters (including protein abundance and phosphorylation site stoichiometry) also influence the ease of detection of a phosphopeptide. As with any other systems-wide resource, the quality should be high overall (to enable global analyses), but there will be some level of false-positive identifications. This is important to take into consideration before attempting labor-intensive experiments based on nonvalidated high-throughput data.

## V. Are All Sites Functional?

A major caveat in the use of large-scale phosphorylation data is that just because a particular site is phosphorylated *in vivo* does not necessarily indicate that phosphorylation at that site has any measureable biological function, especially if phosphorylation at this site is taken in isolation. While very detailed mechanistic studies are available for several phosphorylation sites (take, e.g., phosphorylation of tyrosine residues that create docking sites for specific SH2 containing proteins[34]), functional data is currently available for only a small fraction of the sites identified by mass spectrometry. Classically, determination that a site is functional has involved point mutation of the phosphorylation site, and some kind of a functional readout (either *in vivo* or *in vitro*). Clearly, this will not be possible to generate for all reported sites, and alternative approaches have therefore been taken to assess whether phosphorylation sites are likely to be biologically important.

Now that several large-scale phosphoproteomic studies have been conducted in diverse organisms, computational groups have begun asking the question: are phosphorylation sites conserved throughout evolution? Conservation in the position of the phosphorylation site could suggest that the function of the phosphorylation event is important for the organism, and a conservation of the flanking residues could indicate that an orthologous kinase is responsible for the phosphorylation. A lack of conservation could therefore potentially point to a lack of function for the site, or, alternatively, that there is a plasticity in signaling that is afforded by mutations in the phosphorylation region. While several studies have highlighted that phosphorylation introduces constraints on evolution,[35–38] there are a great number of sites that do not display conservation.[10] How many of the sites are deemed conserved also depends on the methodology used for the detection of the conservation (e.g., selecting only well-characterized sites or all sites identified by mass spectrometry, or requiring a phosphorylation event to have been detected in both species or just that the site be positionally conserved, etc.), and the evolutionary distances between species tested. Clearly, much work remains to be performed before functional relevance of a given site can be assigned solely on the basis of evolution.

Another confounding factor in the use of conservation as a proxy for biological relevance comes from studies of multiply phosphorylated proteins (these proteins tend to contain disordered regions that are prone to phosphorylation[39]). For instance, a phosphoproteomic screen for *Saccharomyces cerevisiae* Cdk1 substrates found that only a small fraction of the 547 sites identified were evolutionarily conserved in the strict sense (these tended to occur in metabolic enzymes), but that many more displayed localized conservation, in which the same region of a given orthologous protein was phosphorylated, albeit on different residues, across different species.[40] Regions with conserved phosphorylation tended to contain multiple Cdk1 consensus sequences, suggesting that the precise position of the phosphate may be somewhat flexible, allowing for adaptation of signaling. The allowance for flexibility in multiphosphorylated regions may also be important for functionality. For example, multiple phosphorylation sites in the N-terminal region of the yeast Cdk-inhibitory protein Sic1, which is intrinsically disordered, allow Sic1 to complex with the ubiquitin ligase subunit Cdc4. Multiple phosphorylation events within the region provide enough electrostatic energy to form a binding site in the disordered protein, enabling Sic1 degradation and passage through the G1/S transition of the cell cycle.[41,42] Interestingly, precise phosphorylation events at specific positions were not as important as the number of sites phosphorylated, indicating that multiple phosphorylation sites may act together to reach a required threshold for signaling.[43] How widespread this phenomenon is remains to be determined.

On a similar note, there also remains the possibility that a phosphorylation site may have an important functional role, but only in the context of other phosphorylation sites. Indeed, there are several kinases that can use previously phosphorylated residues as consensus determinants for phosphorylation (GSK3, CK1, CK2), resulting in hierarchical phosphorylation catalyzed by either a single kinase or a combination of kinases.[44–46] Hierarchical phosphorylation may be widespread in the phosphoproteome, as a striking 54% of phosphoserine or phosphothreonine residues in the phosphoproteome (70,000 sites from different species were analyzed) were no farther than four amino acids away from another phosphoserine or phosphothreonine in multiply phosphorylated proteins.[47] An independent study of known mouse and human phosphorylation sites also revealed an overabundance of phosphorylation sites that not only clustered to specific regions of substrates but were also activated simultaneously.[48] This adds yet another layer of complexity to determining the functionality of a phosphorylation site, as some sites may be functional only as part of a combinatorial phosphorylation module.

Lastly, there remains the likely possibility that a certain number of sites are the result of off-target phosphorylation, from a chance encounter with an active kinase. These would be expected to be present in substoichiometric amounts (i.e., less abundant than functional sites), yet still detectable by modern mass spectrometry. By looking at the frequency of detection of one site in databases, or better yet at the stoichiometry of phosphorylation at that site (see below), it may be possible to increase the accuracy of the functional predictions.

## VI. Analysis of Regulated Phosphorylation

With current global phosphoproteomics surveys indicating that a majority of proteins are phosphorylated,[49] it becomes important to look at the regulation of these phosphorylation sites and not their mere detection. Quantitative mass spectrometric approaches are increasingly being used to monitor phosphorylation dynamics. Quantitative approaches can help to unravel, for example, the timing of phosphorylation events after growth factor treatment. They are also useful to determine enzyme–substrate relationships (as detailed below). Lastly, they can help in monitoring the stoichiometry (or at least relative abundance) of phosphorylation sites across cell types, or again following treatment. Here, we briefly review the general approaches for mass spectrometry-based quantification and provide a few examples as applied to phosphorylation.

### A. Isotope-Based Quantification

Stable isotopes, such as $^{13}C$ or $^{15}N$, are ideally suited for use in mass spectrometry-based quantification because they change the mass of a peptide without affecting its chemical properties. For example, a peptide labeled with six heavy carbons ($^{13}C$) will cofractionate with the cognate peptide possessing six light carbons ($^{12}C$), but their mass difference will be easily detected in the mass spectrometer. The intensity of the signal for each of the peaks in the mass spectrometer can then be used to calculate the relative amounts of the light and heavy peptides in the sample. In a typical workflow, the quantification occurs at the level of the survey scan (MS1), and the peptides are identified after fragmentation in the MS2 spectrum. While, as described, this only provides relative information regarding the abundance of the peptides, absolute information can also be obtained if a labeled peptide is spiked into the sample in a known amount (this is the principle behind the absolute quantitation – or AQUA – technique[50]). For relative quantification, the labeling of a peptide is affected by chemical modification, an enzymatic reaction, or the incorporation of stable isotopes metabolically. Each approach has its advantages (and drawbacks).

Among the metabolic labeling approaches, SILAC (stable isotope labeling by amino acids in cell culture) is the most widely used.[51] In SILAC, pairs (or more) of cell lines are grown in parallel in culture medium in which one or more amino acids have been replaced by heavy isotopic variants. After several generations in culture, nearly all proteins will be labeled. Cells can be treated as desired and combined from the lysis step onwards for downstream analysis. The mixture of the cell lysate early on has clear advantages, as it helps maintaining consistency between samples. Even after extensive separation (by standard chromatography, affinity purification, or both), cognate peptides will cofractionate and quantification will be possible. Specialized software (such as MaxQuant[52]) can be used to perform quantification. SILAC has been extensively used for comparative phosphoproteomics studies.[53–55] Importantly, using an approach such as SILAC is essential for phosphoproteomics studies of bidirectional signaling (signaling induced in one cell from the contact with another cell, and vice versa), as was demonstrated by Jorgensen *et al.*[56] In this study, the authors were interested in analyzing the signaling networks resulting from ephrin signaling within the context of self-sorting. Cells expressing the Eph receptor or its ephrin ligand were independently labeled with SILAC prior to a short coculture (10 min) to initiate signaling (a reference point, labeled with a different SILAC amino acid, was used as a noninduced control). Following antiphosphotyrosine enrichment, the samples were analyzed by mass spectrometry; in this case, the isotopic variant of the SILAC reveals the origin of the cell (expressing the Eph receptor or the ephrin ligand), which allowed the authors to conclude that the reciprocal effect of signaling in the two cell populations was asymmetrical.

While SILAC is easily amenable to differential phosphoproteomics, a limitation is that only a few samples can be analyzed in parallel (up to five[57]; but most often two or three). To overcome this limitation, one of the isotopic channels can be set as the reference point to normalize the ratios across experiments. This was indeed done in 2006 by Olsen *et al.*,[8] who were able to apply quantitative phosphoproteomics to the epidermal growth factor signaling pathway by using triple-encoded SILAC, and a 5-min stimulation as a reference point. This early study demonstrated that—as predicted from small-scale studies—phosphorylation dynamics were different not only for different proteins but also for different sites within the same protein.

A limitation to the use of SILAC (in addition to its cost, which is nonnegligible, and the low multiplexing) is that the sample has to be grown in SILAC medium, something that is not always possible. This said, several groups have now used the SILAC approach to create fully labeled animals. In these cases, the animals, for example, mice[58] or flies,[59] are fed diets consisting of labeled material. Tissues can then be harvested from the animals kept under SILAC-light and SILAC-heavy conditions, enabling a tissue-level analysis. A variation on the SILAC approach, called Super-SILAC, was also recently introduced, which essentially bypasses this requirement for growth in SILAC medium.[60] Super-SILAC consists of a spike-in of a labeled sample (pool of five tumor cell lines) into a nonlabeled sample (e.g., patient tumor). Each patient sample is independently compared to the Super-SILAC, and normalization (ratio of ratios) could provide an after-the-fact comparison between the different patient samples. A precursor of the Super-SILAC approach has already been successfully used for the determination of the insulin-induced changes in phosphorylation in mouse liver: each of the PBS-treated and insulin-treated samples was independently mixed with a SILAC-label spike-in as an internal reference standard. This study enabled the quantification of ~10,000 phosphopeptides between the insulin-treated and untreated samples, and 1000 of the sites were found to be responsive to insulin, including several on not only known insulin pathway components but also new putative effectors.[61]

Metabolic labeling is not the only way to introduce isotopes in a peptide, and several different chemical labeling approaches have also been used for mass spectrometry-based quantification. In 1999, Gygi *et al.* introduced the ICAT (isotope-coded affinity tag) reagent, which labels cysteine residues in peptides, while also providing a means for the isolation of modified peptides.[62] In a standard ICAT workflow, two populations of proteins or peptides are reacted in parallel with the cysteine-reactive compound available in two isotopic forms. After quenching the reaction, the samples can be mixed and the affinity tag present on the ICAT reagent is used to isolate the ICAT-modified peptides. As with the SILAC strategy, quantification occurs at the MS1 level while identification occurs during MS2 (note that the mass of the ICAT reagent, either isotopically heavy or light, will be added to the mass of the cysteine-containing peptide). While the cysteine reactivity of the ICAT reagent clearly is not ideal for use in phosphoproteomics (only phosphopeptides that contain a cysteine can be analyzed), the introduction of ICAT reagent has spearheaded a number of other chemical labeling strategies.[63] One that is particularly well suited to the study of dynamic phosphorylation is known as iTRAQ (isobaric tag for relative and absolute quantification).[64] iTRAQ reagents are amine reactive, meaning that each peptide can be labeled via its amino group, and are available in a series of eight different reagents, enabling multiplexing. Quantification with iTRAQ reagents differs from that of ICAT or SILAC in that the eight iTRAQ reagents are isobaric when they are reacted with the peptides. As the mass of the peptides is not changed initially, quantification cannot be performed in the survey scan. Instead, the iTRAQ reagents contain an internal fragile bond that readily dissociates during fragmentation to produce a low-molecular-weight reporter ion and a balancing ion that remains attached to the peptide. The masses of the reporter ions differ for each of the eight reagents, and quantitation can therefore be based on the relative intensity of each of these eight reporter ions at the MS/MS level (concomitant with the identification). If a phosphopeptide enrichment protocol is implemented in the iTRAQ workflow, phosphorylation dynamics can be studied.[65–67]

## B. Label-Free Quantification

While stable isotopes represent a powerful approach to compare peptide levels across samples, several approaches have also been designed that enable quantification in the absence of labeling. These can broadly be separated into "counting," "MS1 intensity," and "MS2 intensity" approaches (a thorough review describes these approaches, their respective advantages, and their pitfalls[68]).

Counting approaches are the simplest and easiest to perform; for example, spectral counting is the number of times a specific peptide was selected for fragmentation and identified in the mass spectrometer. Because of the stochastic nature of the most common form of mass spectrometry acquisition, and the fact that peptides elute over many seconds or even minutes in the mass spectrometer, the same peptide can often be sequenced multiple times through a mass spectrometry run. How often the peptide is sequenced can provide a rough estimate of abundance. Importantly, this information is available from any mass spectrometric analysis, without modifications to the protocol; variants of the approach include using different scoring metrics and the implementation of normalization factors (reviewed in Ref. 69). Spectral counting is particularly useful at the protein level, when data for many individual peptides are available. While simple in theory, this approach has not been widely employed for comparative phosphopeptide analysis, as it is not believed to be particularly precise, especially if only a few instances of the phosphopeptide are detected (statistically speaking, spectral counting is effective only when the given species have been observed multiple times, and with significant counts[70]). Regardless of this drawback, counting approaches can serve as a proxy for phosphopeptide abundance and have been used in a global comparative study of tissue-specific phosphoproteome.[71]

Quantification at the MS1 level consists of monitoring the intensity of the eluting peak (for a given phosphopeptide) across multiple samples. In theory, if the samples are fairly similar to each other, and the chromatography is robust, the MS1 intensity of a peptide in one sample should be equal to the MS1 intensity in the other sample, if the peptides are present in the same amounts initially. Technically, however, this approach is difficult; even small changes in sample preparation or chromatography between the samples will introduce drastic changes in the quantification (this difficulty increases with each fractionation step). Quantification is performed using software tools that attempt to align chromatography across runs, to find features (here, phosphopeptides) and to extract the total ion current (intensity) for each of the peptides. Confidence in the quantification is much improved from the analysis of biological replicates (the logic here is that, if the peak finding software is picking up noise, it is likely not reproducible). When the approach works, however, it can be very effective, as described below for the systematic analysis of yeast kinase and phosphatase deletions.[72] Other success stories of MS1 quantification include, for example, determination of ATM-dependent and -independent signaling following induction of DNA damage breaks,[73] a survey of changes in phosphorylation imparted by phosphatase inhibition,[74] a global analysis of hormone-mediated changes in signaling in isolated renal cells,[75] and a study of fibroblast growth factor 3-mediated tyrosine signaling in multiple myeloma.[76]

Last but not least, label-free quantification can also be performed at the level of the MS2 spectra. This is most often done within the context of a type of mass spectrometry protocol known as selective reaction monitoring (or SRM; multiple reaction monitoring, MRM, is another term employed). SRM has not only been employed for more than 30 years as an accurate method to quantify small molecules (e.g., metabolites, pesticides, and hormones[77]) but is also increasingly used for accurate peptide quantification (reviewed in Ref. 78). SRM experiments differ from the other workflows discussed in this review, in that the list of species to be measured is selected by the experimentalist at the onset of the experiment (e.g., a phosphopeptide of a given mass and its unphosphorylated counterpart). Selection of the precursor ions (intact peptides) is followed by fragmentation, but in this second step, only selected daughter ions are recorded. The intensity of the daughter ions (also called "transitions") is used not only for confirming the identity of the peptide but also for abundance determination. In the case of phosphopeptide identification, the selected transitions need to provide positional information on the modification. Described first for phosphopeptide quantification in 2003[50] and implemented for several small-scale studies (reviewed in Ref. 78; also see Ref. 79), SRM-based approaches have not yet been widely applied to phosphopeptide quantification on a global scale, in large part because of initial limitations on the number of peptides that could be analyzed simultaneously. This is, however, likely to change, as scheduled MRM enables the analysis of much larger numbers of peptides, and resources for the design and analysis of SRM transitions are becoming widespread.[80,81]

## C. Stoichiometry and Other Considerations

As mentioned above, phosphorylation site stoichiometry is of critical importance, and the quantitative proteomics approaches mentioned above can be harnessed for a stoichiometry determination (see e.g., Refs. 82–87). A more generalized application of stoichiometry-focused quantification would greatly help in the modeling of cellular signaling. It is also important to keep in mind that the apparent "phosphorylation dynamics" measured in quantitative phosphoproteomics experiments may not be phosphorylation changes at all, but instead due to changes in gene expression. For this reason, calibration of phosphorylation levels to the level of the protein is crucial to avoid spurious results. This can be difficult to accomplish within one experiment, especially with the requirement for phosphoenrichment, but improved techniques and instrumentation are beginning to make tandem quantification a reality. In a recent study in yeast, Wu *et al.* measured both phosphorylation and expression changes within the same experiment and demonstrated that a full 25% of detected phosphorylation changes were false positives caused by changes in gene expression.[88]

## VII. Consequence of Kinase or Phosphatase Depletion on the Phosphoproteome

One of the major challenges in the characterization of a novel phosphorylation site is the identification of the enzymes regulating the site. In 1979, Krebs and Beavo proposed a set of criteria that should be satisfied to demonstrate that phosphorylation of a substrate by a kinase has physiological relevance.[89] The field has expanded rapidly since the criteria were suggested (at the time, only 21 phosphoproteins had been identified), but they remain the gold standard for evaluating kinase–substrate–phosphatase relationships. The criteria, along with traditional and mass spectrometry-based methods for their investigation, are summarized in Table II. The ability to monitor phosphorylation by mass spectrometry allows both *in vitro* and *in vivo* validation studies to be performed in a quantitative matter without the traditional use of radiolabeled ATP. Phosphospecific antibodies can also be very useful in characterizing phosphorylation sites, but these reagents are often not available, and can be time consuming and expensive to produce. In reality, a well-designed mass spectrometry experiment can be used in almost all of the ways a phosphospecific antibody could (with the obvious exception of immunofluorescence studies), with higher sensitivity and much more precise quantitation. There are several quantitative mass spectrometry techniques that can be used to validate phosphorylation sites (see Section IV).

To investigate whether a particular kinase or phosphatase is involved in the regulation of a site, phosphoproteomic analysis of phosphorylation after depletion of a kinase or phosphatase by RNA interference (RNAi) may be employed (inhibitory small molecules may also be used, with the caveat that specificity is generally much lower than RNAi; also note that gene knockout can also be

TABLE II

KREBS–BEAVO CRITERIA FOR FUNCTIONALITY OF A PHOSPHORYLATION SITE

| Criteria[89] | Traditional approaches | Mass spectrometry approaches |
|---|---|---|
| 1. Demonstrate phosphorylation of substrate at significant stoichiometric levels by kinase and dephosphorylation by phosphatase *in vitro* | • *In vitro* kinase/phosphatase assays with $\gamma$-$^{32}$P-ATP and autoradiography<br>• Phosphospecific antibodies (if available) | • Cold kinase assay followed by quantitative proteomics with stoichiometry measurements |
| 2. Demonstrate that functional properties of substrate undergo meaningful changes that correlate with the degree of phosphorylation | • Phosphorylation site mutations<br>• Depends on function of substrate, for example:<br>○ Enzyme activity<br>○ Binding assays<br>○ Localization | • AP-MS with phosphorylation-site mutants<br>• Quantitative measurement of enzyme activity |
| 3. Demonstrate that substrate is phosphorylated *in vivo* or in an intact cell system with accompanying functional changes | • $^{32}$P labeling and autoradiography<br>• 2D electrophoresis<br>• Phosphospecific antibodies (if available) | • Isolation of the substrate from intact cells and analysis by mass spectrometry<br>• Global phosphoproteomics |
| 4. Demonstrate correlation between substrate phosphorylation and the cellular levels of the kinase/phosphatase | • RNAi knockdown or drug inhibition of kinase or phosphatase combined with:<br>○ $^{32}$P labeling and autoradiography<br>○ Phosphospecific antibodies (if available) | • RNAi knockdown or drug inhibition of kinase or phosphatase combined with quantitative proteomics, for example, selected reaction monitoring |

used, especially in the case of *S. cerevisiae*, where large collections of strains each lacking a kinase or phosphatase are readily available). In the simplest of cases, knockdown of the responsible kinase will decrease phosphorylation and knockdown of the responsible phosphatase will increase phosphorylation. However, these genetics experiments result in global changes in the phosphorylation patterns that extend beyond the effects on the direct substrates of the enzyme depleted. Kinase/phosphatase depletion will affect the phosphorylation of entire signaling cascades, potentially activating regulatory feedback mechanisms, and the complex interplay between different phosphoregulatory enzymes can lead to profound effects on the global phosphoproteome. Not only are kinases themselves highly phosphorylated,[90] but also our study of the yeast interactome found significantly more interactions between kinases than would be expected by chance.[91] Kinases and phosphatases also show a significant increase in genetic interactions.[92] Owing to the vast functional dependencies between kinases and phosphatases, knockdown of a specific enzyme can dramatically alter global cellular phosphorylation patterns.[40,72]

## VIII. Peptide Arrays to Systematically Map Phosphorylation Sites

Many of the phosphorylation sites identified in large-scale phosphoproteomic screens are not assigned to a particular kinase, and even more have not been assigned to a particular phosphatase. In these cases, the consensus sequence—the primary amino acid motif surrounding a phosphorylation site that primarily dictates kinase specificity—can facilitate the assignment of a particular kinase family to an identified phosphorylation site.[93] Peptide arrays range in complexity from oriented peptide arrays used to determine optimal consensus motifs[94] to proteome-wide peptide microarrays used to identify new phosphorylation sites for particular kinases,[95] but the general principle is the same. The kinase (along with radiolabeled ATP) is incubated with the peptides, which are covalently attached to a solid matrix. After autoradiography, the precise consensus determinants for a given kinase can be read off the filter by comparing the relative phosphorylation of peptides of different sequences. A systematic peptide array study of half of the *S. cerevisiae* kinome successfully determined consensus motifs for many kinases and predicted thousands of new phosphorylation motifs in the yeast proteome.[96] Compendiums of motif analysis for human kinases and phosphobinding domains (PBDs) have also been created, providing a useful resource for researchers hoping to match an identified phosphorylation site to a particular kinase or binding protein.[93]

As with other methods, a few caveats exist in the use of consensus determinants to assign a kinase to a site. First, authentic *in vivo* phosphorylation sites do not always conform to the optimal consensus sequence of the phosphorylating kinase.[97] In addition, the presence of a consensus sequence for a given kinase in an amino acid sequence does not guarantee that the site is phosphorylated *in vivo*. Phosphorylation at a site depends on many factors, including the structure and solvent accessibility of the phosphorylation site, temporal activation state of the kinase, protein complex formation, and subcellular localization.[3] Proper kinase–substrate–phosphatase interactions can require additional sequence requirements outside the immediate consensus, such as the D domains contained in MAP kinases,[98] which can also affect the likelihood of *in vivo* phosphorylation. This emphasizes the need for additional validation studies to assign a kinase to a phosphorylation site. Other *in vitro* phosphorylation-based approaches, for example, using protein array strategies[99,100] or modified kinases leaving a mark on the substrates recognizable by radioactivity incorporation[101] or by the appearance of a novel epitope (Refs. 102,103; also see below), can provide complementary information.

While the determination of the consensus sequence of a kinase is relatively straightforward, the same cannot be said for phosphatases. For PP1 and PP2A, the bulk of phosphatase specificity is conferred by combinatorial use of a large number of regulatory subunits that complex with the relatively few catalytic subunits.[104] Tyrosine phosphatases have highly conserved catalytic domains and gain substrate specificity through nonenzymatic modular domains.[105] The reliance on domains outside the phosphatase active site for specificity makes peptide-based approaches less useful for the determination of substrate specificity. However, motif analysis of the full substrate sequence can sometimes lead to the identification of phosphatase substrates. Some phosphatases, including PP1 and calcineurin, contain docking motifs that confer specificity. Using the rapidly expanding interactomes for each phosphatase, several novel motifs have been identified, which can be used to mine the proteome for additional substrates.[106] Additionally, as our knowledge of human phosphorylation patterns expands, phosphatase consensus sequence determination will likely benefit greatly from motif analysis of large-scale phosphoproteomics data.

## IX. Interactome Studies Help to Place Signaling Molecules in a Physiological Context

Interaction studies can complement and add meaning to identified kinase–substrate–phosphatase relationships by putting these catalytic events in a relevant biological context. Proteomic approaches to identify kinase/phosphatase interactors typically rely on affinity purification coupled to mass spectrometry (AP-MS).[107] These experiments can involve immunoprecipitation of the bait kinase/phosphatase with a single affinity tag, or make use of several tandem affinity tags that are now available.[108] Note that purification of endogenous proteins is also possible, though harder to control.[109] The choice of method will depend greatly on whether the aim of the experiment is to identify interacting proteins or substrates, as, in general, enzyme–substrate relationships tend to be transient. The interactions made by a kinase or phosphatase are crucial to understand, as complex formation can affect phosphorylation, and phosphorylation can affect complex formation.

Many kinases and phosphatases are regulated through protein–protein interactions. These interactions can regulate the enzyme in a variety of ways. For example, association between a kinase and a regulatory subunit can be a requirement for enzyme activity, as is the case for cyclin-dependent kinases. The serine/threonine phosphatases PP1 and PP2A (which together are believed to account for 90% of cellular serine/threonine dephosphorylation[6]) rely on a plethora of different regulatory subunits in order to target the catalytic subunits toward appropriate substrates.[104] Large proteins such as AKAPs can act as scaffolds, anchoring a kinase and/or a phosphatase in close proximity to appropriate targets or changing its subcellular localization.[110] Protein–protein interactions therefore act in multiple ways to ensure that kinase/phosphatase activity is appropriate, given the current cellular situation.

Protein phosphorylation can also have dramatic effects on complex formation, and this is reflected in the variety of modular PBDs present in the genome. Nonenzymatic PBD-containing proteins often contain multiple PBDs and act as linkers between components of cellular signaling pathways. The modularity of PBDs makes them quite amenable to proteomic analysis, and identification of phospho-dependent protein–protein interactions can offer significant insight into the organization of cellular signaling networks. For example, a conserved binding module in polo-like kinases, namely, the polo box domain, binds specifically to phosphorylated serine/threonine residues followed by a proline.[111] Plk1 is an important kinase for mitotic progression, and this consensus matches the consensus sequence for phosphorylation by Cdk1, which is the master mitotic regulatory kinase. A subsequent proteomic investigation of polo box domain interactors identified 622 mitosis-specific interactors for Plk1 and uncovered new biological functions for Plk1 kinase activity.[112] While kinases and phosphatases contain highly conserved catalytic domains, the remainder of the genes can be quite divergent, and often (particularly with tyrosine kinases and phosphatases) contain modular domains, PBDs included. These additional domains can act to target the enzyme to particular substrates and therefore can have a significant impact on the *in vivo* activity of a particular kinase or phosphatase. Clearly, the mapping of PDB interactions can provide a deeper understanding of the complexities of phosphorylation-based signaling.

A fascinating aspect of phosphoregulation is the surprising interconnectivity among kinases and phosphatases revealed in large-scale proteomic studies. Kinases and phosphatases interact with higher frequency than expected by chance,[91] and the regulation of phosphorylation does not occur in traditional isolated pathways, but rather as a large, complex network.[113] Proteomic studies have also identified intriguing complexes containing both kinases and phosphatases. For instance, during the mapping of the static PP2A interactome, specific subunits of PP2A were found in a conserved complex with a family of GCK-III kinases.[114,115] Several other examples have been reported to date in addition to the AKAP-mediated interactions described above. For example, tyrosine phosphatases and MAP kinase phosphatases associate with the MAP kinases, to regulate their activity in a spatiotemporal manner.[98,116] It may well be that the kinases and phosphatases in these protein complexes are working together, either synergistically or antagonistically, to regulate the function of the protein complex. Importantly, methods have been developed to identify complex components and their phosphorylation sites in the same experiment,[82] which will be useful in linking complex formation to component phosphorylation status.

## X. Getting It All Together—One of the Next Big Challenges

After almost six decades of studying protein phosphorylation, the field has amassed a tremendous body of knowledge regarding both protein phosphorylation as a regulatory mechanism and the enzymes responsible for changes in phosphorylation. However, full understanding of the impact of phosphorylation on cellular signaling pathways will require global approaches to identify and characterize the cellular phosphoproteome, including temporal and spatial cues. Mass spectrometry is a powerful method for identifying and tracking phosphorylation, and large-scale proteomic investigations of protein phosphorylation have uncovered a plethora of *in vivo* phosphorylation sites, allowing the field to begin to make conclusions about the underlying mechanisms regulating cellular signal transduction. However, they have also provided a glut of largely unvalidated data of differing quality.

In an age where we are capable of identifying tens of thousands of phosphorylation sites in a single experiment, a huge bottleneck has formed. Many of the thousands of identified sites have not been validated or investigated for physiological significance. While the annotation of identified sites in large databases provides a valuable tool for researchers, caution should be taken in basing a line of investigation on a false-positive result. However, the mass spectrometer also allows highly sensitive validation techniques without the need for radioactive material or phosphospecific antibodies, and techniques are currently being developed for large-scale validation of phosphorylation sites.[83,117]

Much progress has been made in developing bioinformatic solutions for the organization, analysis, and mining of high-throughput mass spectrometry data. However, the complexity and sheer size of phosphoproteomics experiments require a constant evolution of bioinformatic tools. In particular, the ability to link phosphoproteomic data with phenotypic outcomes from large-scale screens will be helpful in assigning biological functions to both phosphorylation sites and the kinases and phosphatases that control them.

Another considerable challenge for the field will be in elucidating the effects of combinatorial posttranslational modification. With respect to phosphorylation, the close proximity of most phosphorylation sites indicates extreme interconnectivity,[47,48] which, no doubt, stems from the interconnectivity between different kinases and phosphatases.[91] Phosphorylation can also combine with additional forms of posttranslational modifications, such as ubiquitylation,[118] acetylation,[119] and sumoylation.[120] When additional modifications are taken into account, the possible combinations can be endless. Indeed, this analysis has already begun for histones,[121] which are highly posttranslationally modified in a combinatorial manner, in which different combinations of PTMs lead to different transcriptional outcomes.[122] Mass spectrometric analysis paired with peptide array techniques will help illuminate the effects of combinatorial PTMs on substrate regulation.

By far, the greatest challenge facing the field is in linking the massive amount of phosphoproteomic data being produced with the actual biological significance of individual phosphorylation events. In addition to the methods mentioned above (RNAi depletion of enzymes, peptide-based approaches, etc.), several methods have been developed to aid in the identification of direct kinase/phosphatase substrates which are amenable to proteomics. Direct kinase substrates can be identified through the use of "gatekeeper" mutations that alter the catalytic site of the kinase. The enlarging of the active site enables the kinase to use ATP analogs to phosphorylate their substrates and makes the kinase sensitive to gatekeeper-specific inhibitors.[123] For example, these strategies have been used in yeast and human cells to catalog CDK1 substrates.[40,124,125] To identify direct substrates of tyrosine and dual-specificity phosphatases, AP-MS can be conducted using phosphatases containing substrate-trapping mutations that render the phosphatase inactive and, therefore, unable to dephosphorylate and release substrates.[126] This strategy was recently used in yeast to identify substrates of the dual-specificity phosphatase Cdc14p.[127] Unfortunately, serine/threonine phosphatases are not amenable to substrate-trapping mutant strategies, as they catalyze dephosphorylation through a different mechanism involving the coordination of metal ions.[105]

Another important aspect of phosphoregulation is determining the spatial and temporal cues that regulate substrate phosphorylation and dephosphorylation. Several proteomic studies have focused on determining the phosphoproteome of distinct subcellular compartments, such as the centrosome,[128] mitochondria,[129] and the mitotic spindle[130,131] during different phases of the cell cycle or after a particular stimulus. These studies give us snapshots of the cellular phosphorylation changes occurring dynamically in response to various stimuli. Key to understanding dynamic protein phosphorylation will be linking this dynamic phosphoproteomic data to information about the activation status and cellular localization of the regulatory enzymes. Peptide-based MS strategies can shed light on the temporal changes in phosphorylation patterns in response to various stimuli. A mass spectrometry approach named KAYAK (kinase activity assay for kinome profiling) uses a number of peptides, each representing a known phosphorylation site. Cell lysates are incubated with the peptides, and kinase activity against each peptide is quantified against a known concentration of stable isotope-labeled phosphopeptide of the same sequence.[132]

Additionally, Forster resonance energy transmission (FRET)-based biosensors have been developed for a number of kinases, facilitating the spatiotemporal localization of kinase activity in living cells. These biosensors contain a consensus motif for the kinase of interest and a PBD recognizing the site fused in tandem between two fluorophores (typically CFP and YFP). Phosphorylation of the consensus motif leads to PBD binding, bringing the CFP and YFP in close enough proximity for FRET to occur.[133] In living cells, these sensors can illustrate both the spatial and the temporal activations of the kinase in response to a given stimuli. For example, FRET-based biosensors have been used to define the localization and level of CDK1 activation required for the onset of mitosis,[134,135] correlate calcineurin phosphatase activity with intracellular calcium dynamics,[136] and measure BCR-ABL response to drugs in CML.[137]

Finally, large-scale RNAi screens have been quite useful in identifying kinases and phosphatases important for a variety of cellular processes. Kinase/phosphatase screens have been performed to identify phosphoregulatory enzymes important for biological processes such as mitotic exit,[138] cell invasion,[139] and apoptosis.[140,141] These studies have also been used to identify potential therapeutic targets for diseases such as cancer[142] and viral infection.[143] The combination of high-throughput mass spectrometry approaches with high-throughput RNAi screens can link the regulation of phosphorylation with specific phenotypic outcomes.

The advent of large-scale phosphoproteomics has changed not only how we study the process of regulatory phosphorylation but also our understanding of how phosphorylation achieves signaling homeostasis. We now know that kinases and phosphatases are intricately interlinked, contributing redundancy and stability to cellular signaling. Using phosphoproteomic techniques, it is now possible to measure the effects of cellular stimuli, mutations, and disease states on phosphoregulation at a global scale.

## Acknowledgments

We thank Brian Raught for a critical reading of this chapter. This work was supported by a grant from the Canadian Institutes of Health Research (CIHR) to ACG (MOP-84314). ACG holds a Canada Research Chair in Functional Proteomics and the Lea Reichmann chair in Cancer Proteomics.

## References

1. Krebs EG, Fischer EH. Phosphorylase activity of skeletal muscle extracts. *J Biol Chem* 1955;**216**(1):113–20.
2. Manning G, Plowman GD, Hunter T, Sudarsanam S. Evolution of protein kinase signaling from yeast to man. *Trends Biochem Sci* 2002;**27**(10):514–20.
3. Turk BE. Understanding and exploiting substrate recognition by protein kinases. *Curr Opin Chem Biol* 2008;**12**(1):4–10.
4. O'Dwyer ME, Mauro MJ, Druker BJ. STI571 as a targeted therapy for CML. *Cancer Invest* 2003;**21**(3):429–38.
5. Manning G, Whyte DB, Martinez R, Hunter T, Sudarsanam S. The protein kinase complement of the human genome. *Science* 2002;**298**(5600):1912–34.
6. Moorhead GB, Trinkle-Mulcahy L, Ulke-Lemee A. Emerging roles of nuclear protein phosphatases. *Nat Rev Mol Cell Biol* 2007;**8**(3):234–44.
7. Ubersax JA, Ferrell Jr. JE. Mechanisms of specificity in protein phosphorylation. *Nat Rev Mol Cell Biol* 2007;**8**(7):530–41.
8. Olsen JV, Blagoev B, Gnad F, Macek B, Kumar C, Mortensen P, et al. Global, *in vivo*, and site-specific phosphorylation dynamics in signaling networks. *Cell* 2006;**127**(3):635–48.
9. Hornbeck PV, Chabra I, Kornhauser JM, Skrzypek E, Zhang B. PhosphoSite: a bioinformatics resource dedicated to physiological protein phosphorylation. *Proteomics* 2004;**4**(6):1551–61.
10. Tan CS, Bodenmiller B, Pasculescu A, Jovanovic M, Hengartner MO, Jorgensen C, et al. Comparative analysis reveals conserved protein phosphorylation networks implicated in multiple diseases. *Sci Signal* 2009;**2**(81):ra39.
11. Steen H, Mann M. The ABC's (and XYZ's) of peptide sequencing. *Nat Rev Mol Cell Biol* 2004;**5**(9):699–711.
12. Old WM, Shabb JB, Houel S, Wang H, Couts KL, Yen CY, et al. Functional proteomics identifies targets of phosphorylation by B-Raf signaling in melanoma. *Mol Cell* 2009;**34**(1):115–31.
13. Beausoleil SA, Villen J, Gerber SA, Rush J, Gygi SP. A probability-based approach for high-throughput protein phosphorylation analysis and site localization. *Nat Biotechnol* 2006;**24**(10):1285–92.
14. Grimsrud PA, Swaney DL, Wenger CD, Beauchene NA, Coon JJ. Phosphoproteomics for the masses. *ACS Chem Biol* 2010;**5**(1):105–19.
15. Iliuk A, Tao WA. Quantitative phospho-proteomics based on soluble nanopolymers. *Methods Mol Biol* 2009;**527**:117–29. [ix].
16. Bodenmiller B, Mueller LN, Mueller M, Domon B, Aebersold R. Reproducible isolation of distinct, overlapping segments of the phosphoproteome. *Nat Methods* 2007;**4**(3):231–7.
17. Villen J, Gygi SP. The SCX/IMAC enrichment approach for global phosphorylation analysis by mass spectrometry. *Nat Protoc* 2008;**3**(10):1630–8.
18. Rush J, Moritz A, Lee KA, Guo A, Goss VL, Spek EJ, et al. Immunoaffinity profiling of tyrosine phosphorylation in cancer cells. *Nat Biotechnol* 2005;**23**(1):94–101.

19. Rikova K, Guo A, Zeng Q, Possemato A, Yu J, Haack H, et al. Global survey of phosphotyrosine signaling identifies oncogenic kinases in lung cancer. *Cell* 2007;**131**(6):1190–203.
20. Matsuoka S, Ballif BA, Smogorzewska A, McDonald 3rd ER, Hurov KE, Luo J, et al. ATM and ATR substrate analysis reveals extensive protein networks responsive to DNA damage. *Science* 2007;**316**(5828):1160–6.
21. Dinkel H, Chica C, Via A, Gould CM, Jensen LJ, Gibson TJ, et al. Phospho.ELM: a database of phosphorylation sites–update 2011. *Nucleic Acids Res* 2011;**39**(Database issue):D261–7.
22. Diella F, Cameron S, Gemund C, Linding R, Via A, Kuster B, et al. Phospho.ELM: a database of experimentally verified phosphorylation sites in eukaryotic proteins. *BMC Bioinformatics* 2004;**5**:79.
23. Gnad F, Gunawardena J, Mann M. PHOSIDA 2011: the posttranslational modification database. *Nucleic Acids Res* 2011;**39**(Database issue):D253–60.
24. Gnad F, Ren S, Cox J, Olsen JV, Macek B, Oroshi M, et al. PHOSIDA (phosphorylation site database): management, structural and evolutionary investigation, and prediction of phosphosites. *Genome Biol* 2007;**8**(11):R250.
25. Bodenmiller B, Campbell D, Gerrits B, Lam H, Jovanovic M, Picotti P, et al. PhosphoPep—a database of protein phosphorylation sites in model organisms. *Nat Biotechnol* 2008;**26**(12):1339–40.
26. Bodenmiller B, Aebersold R. Phosphoproteome resource for systems biology research. *Methods Mol Biol* 2011;**694**:307–22.
27. Stark C, Su TC, Breitkreutz A, Lourenco P, Dahabieh M, Breitkreutz BJ, et al. PhosphoGRID: a database of experimentally verified *in vivo* protein phosphorylation sites from the budding yeast *Saccharomyces cerevisiae*. *Database (Oxford)* 2010;**2010**:bap026.
28. Durek P, Schmidt R, Heazlewood JL, Jones A, MacLean D, Nagel A, et al. PhosPhAt: the Arabidopsis thaliana phosphorylation site database. An update. *Nucleic Acids Res* 2010;**38**(Database issue):D828–34.
29. Zanzoni A, Carbajo D, Diella F, Gherardini PF, Tramontano A, Helmer-Citterich M, et al. Phospho3D 2.0: an enhanced database of three-dimensional structures of phosphorylation sites. *Nucleic Acids Res* 2011;**39**(Database issue):D268–71.
30. Boeckmann B, Bairoch A, Apweiler R, Blatter MC, Estreicher A, Gasteiger E, et al. The SWISS-PROT protein knowledgebase and its supplement TrEMBL in 2003. *Nucleic Acids Res* 2003;**31**(1):365–70.
31. Peri S, Navarro JD, Amanchy R, Kristiansen TZ, Jonnalagadda CK, Surendranath V, et al. Development of human protein reference database as an initial platform for approaching systems biology in humans. *Genome Res* 2003;**13**(10):2363–71.
32. Keshava Prasad TS, Goel R, Kandasamy K, Keerthikumar S, Kumar S, Mathivanan S, et al. Human Protein Reference Database—2009 update. *Nucleic Acids Res* 2009;**37**(Database issue):D767–72.
33. Bell AW, Deutsch EW, Au CE, Kearney RE, Beavis R, Sechi S, et al. A HUPO test sample study reveals common problems in mass spectrometry-based proteomics. *Nat Methods* 2009;**6**(6):423–30.
34. Hong F, Root DD. Downscaling functional bioassays by single-molecule techniques. *Drug Discov Today* 2006;**11**(13–14):640–5.
35. Landry CR, Levy ED, Michnick SW. Weak functional constraints on phosphoproteomes. *Trends Genet* 2009;**25**(5):193–7.
36. Gnad F, Forner F, Zielinska DF, Birney E, Gunawardena J, Mann M. Evolutionary constraints of phosphorylation in eukaryotes, prokaryotes, and mitochondria. *Mol Cell Proteomics* 2010;**9**(12):2642–53.
37. Macek B, Gnad F, Soufi B, Kumar C, Olsen JV, Mijakovic I, et al. Phosphoproteome analysis of *E. coli* reveals evolutionary conservation of bacterial Ser/Thr/Tyr phosphorylation. *Mol Cell Proteomics* 2008;**7**(2):299–307.

38. Boekhorst J, van Breukelen B, Heck Jr. A, Snel B. Comparative phosphoproteomics reveals evolutionary and functional conservation of phosphorylation across eukaryotes. *Genome Biol* 2008;**9**(10):R144.
39. Tompa P. The interplay between structure and function in intrinsically unstructured proteins. *FEBS Lett* 2005;**579**(15):3346–54.
40. Holt LJ, Tuch BB, Villen J, Johnson AD, Gygi SP, Morgan DO. Global analysis of Cdk1 substrate phosphorylation sites provides insights into evolution. *Science* 2009;**325** (5948):1682–6.
41. Nash P, Tang X, Orlicky S, Chen Q, Gertler FB, Mendenhall MD, et al. Multisite phosphorylation of a CDK inhibitor sets a threshold for the onset of DNA replication. *Nature* 2001;**414** (6863):514–21.
42. Borg M, Mittag T, Pawson T, Tyers M, Forman-Kay JD, Chan HS. Polyelectrostatic interactions of disordered ligands suggest a physical basis for ultrasensitivity. *Proc Natl Acad Sci USA* 2007;**104**(23):9650–5.
43. Mittag T, Marsh J, Grishaev A, Orlicky S, Lin H, Sicheri F, et al. Structure/function implications in a dynamic complex of the intrinsically disordered Sic1 with the Cdc4 subunit of an SCF ubiquitin ligase. *Structure* 2010;**18**(4):494–506.
44. Litchfield DW, Arendt A, Lozeman FJ, Krebs EG, Hargrave PA, Palczewski K. Synthetic phosphopeptides are substrates for casein kinase II. *FEBS Lett* 1990;**261**(1):117–20.
45. Flotow H, Graves PR, Wang AQ, Fiol CJ, Roeske RW, Roach PJ. Phosphate groups as substrate determinants for casein kinase I action. *J Biol Chem* 1990;**265**(24):14264–9.
46. Fiol CJ, Wang A, Roeske RW, Roach PJ. Ordered multisite protein phosphorylation. Analysis of glycogen synthase kinase 3 action using model peptide substrates. *J Biol Chem* 1990;**265** (11):6061–5.
47. Schweiger R, Linial M. Cooperativity within proximal phosphorylation sites is revealed from large-scale proteomics data. *Biol Direct* 2010;**5**:6.
48. Yachie N, Saito R, Sugahara J, Tomita M, Ishihama Y. In silico analysis of phosphoproteome data suggests a rich-get-richer process of phosphosite accumulation over evolution. *Mol Cell Proteomics* 2009;**8**(5):1061–71.
49. Yachie N, Saito R, Sugiyama N, Tomita M, Ishihama Y. Integrative features of the yeast phosphoproteome and protein-protein interaction map. *PLoS Comput Biol* 2011;**7**(1): e1001064.
50. Gerber SA, Rush J, Stemman O, Kirschner MW, Gygi SP. Absolute quantification of proteins and phosphoproteins from cell lysates by tandem MS. *Proc Natl Acad Sci USA* 2003;**100** (12):6940–5.
51. Ong SE, Blagoev B, Kratchmarova I, Kristensen DB, Steen H, Pandey A, et al. Stable isotope labeling by amino acids in cell culture, SILAC, as a simple and accurate approach to expression proteomics. *Mol Cell Proteomics* 2002;**1**(5):376–86.
52. Cox J, Mann M. MaxQuant enables high peptide identification rates, individualized p.p.b.-range mass accuracies and proteome-wide protein quantification. *Nat Biotechnol* 2008;**26**(12):1367–72.
53. Blagoev B, Ong SE, Kratchmarova I, Mann M. Temporal analysis of phosphotyrosine-dependent signaling networks by quantitative proteomics. *Nat Biotechnol* 2004;**22**(9):1139–45.
54. Bose R, Molina H, Patterson AS, Bitok JK, Periaswamy B, Bader JS, et al. Phosphoproteomic analysis of Her2/neu signaling and inhibition. *Proc Natl Acad Sci USA* 2006;**103**(26):9773–8.
55. Munoz J, Heck AJ. Quantitative proteome and phosphoproteome analysis of human pluripotent stem cells. *Methods Mol Biol* 2011;**767**:297–312.
56. Jorgensen C, Sherman A, Chen GI, Pasculescu A, Poliakov A, Hsiung M, et al. Cell-specific information processing in segregating populations of Eph receptor ephrin-expressing cells. *Science* 2009;**326**(5959):1502–9.

57. Pimienta G, Chaerkady R, Pandey A. SILAC for global phosphoproteomic analysis. *Methods Mol Biol* 2009;**527**:107–16. [x].
58. Kruger M, Moser M, Ussar S, Thievessen I, Luber CA, Forner F, et al. SILAC mouse for quantitative proteomics uncovers kindlin-3 as an essential factor for red blood cell function. *Cell* 2008;**134**(2):353–64.
59. Sury MD, Chen JX, Selbach M. The SILAC fly allows for accurate protein quantification *in vivo*. *Mol Cell Proteomics* 2010;**9**(10):2173–83.
60. Geiger T, Cox J, Ostasiewicz P, Wisniewski JR, Mann M. Super-SILAC mix for quantitative proteomics of human tumor tissue. *Nat Methods* 2010;**7**(5):383–5.
61. Monetti M, Nagaraj N, Sharma K, Mann M. Large-scale phosphosite quantification in tissues by a spike-in SILAC method. *Nat Methods* 2011;**8**(8):655–8.
62. Gygi SP, Rist B, Gerber SA, Turecek F, Gelb MH, Aebersold R. Quantitative analysis of complex protein mixtures using isotope-coded affinity tags. *Nat Biotechnol* 1999;**17**(10):994–9.
63. Boersema PJ, Foong LY, Ding VM, Lemeer S, van Breukelen B, Philp R, et al. In-depth qualitative and quantitative profiling of tyrosine phosphorylation using a combination of phosphopeptide immunoaffinity purification and stable isotope dimethyl labeling. *Mol Cell Proteomics* 2011;**9**(1):84–99.
64. Ross PL, Huang YN, Marchese JN, Williamson B, Parker K, Hattan S, et al. Multiplexed protein quantitation in Saccharomyces cerevisiae using amine-reactive isobaric tagging reagents. *Mol Cell Proteomics* 2004;**3**(12):1154–69.
65. Zhang Y, Wolf-Yadlin A, Ross PL, Pappin DJ, Rush J, Lauffenburger DA, et al. Time-resolved mass spectrometry of tyrosine phosphorylation sites in the epidermal growth factor receptor signaling network reveals dynamic modules. *Mol Cell Proteomics* 2005;**4**(9):1240–50.
66. Wang Z, Gucek M, Hart GW. Cross-talk between GlcNAcylation and phosphorylation: site-specific phosphorylation dynamics in response to globally elevated O-GlcNAc. *Proc Natl Acad Sci USA* 2008;**105**(37):13793–8.
67. Iwai LK, Benoist C, Mathis D, White FM. Quantitative phosphoproteomic analysis of T cell receptor signaling in diabetes prone and resistant mice. *J Proteome Res* 2010;**9**(6):3135–45.
68. Neilson KA, Ali NA, Muralidharan S, Mirzaei M, Mariani M, Assadourian G, et al. Less label, more free: approaches in label-free quantitative mass spectrometry. *Proteomics* 2011;**11**(4):535–53.
69. Lundgren DH, Hwang SI, Wu L, Han DK. Role of spectral counting in quantitative proteomics. *Expert Rev Proteomics* 2010;**7**(1):39–53.
70. Choi H, Fermin D, Nesvizhskii AI. Significance analysis of spectral count data in label-free shotgun proteomics. *Mol Cell Proteomics* 2008;**7**(12):2373–85.
71. Huttlin EL, Jedrychowski MP, Elias JE, Goswami T, Rad R, Beausoleil SA, et al. A tissue-specific atlas of mouse protein phosphorylation and expression. *Cell* 2010;**143**(7):1174–89.
72. Bodenmiller B, Wanka S, Kraft C, Urban J, Campbell D, Pedrioli PG, et al. Phosphoproteomic analysis reveals interconnected system-wide responses to perturbations of kinases and phosphatases in yeast. *Sci Signal* 2010;**3**(153):rs4.
73. Bensimon A, Schmidt A, Ziv Y, Elkon R, Wang SY, Chen DJ, et al. ATM-dependent and -independent dynamics of the nuclear phosphoproteome after DNA damage. *Sci Signal* 2010;**3**(151):rs3.
74. Yang F, Jaitly N, Jayachandran H, Luo Q, Monroe ME, Du X, et al. Applying a targeted label-free approach using LC-MS AMT tags to evaluate changes in protein phosphorylation following phosphatase inhibition. *J Proteome Res* 2007;**6**(11):4489–97.
75. Gunaratne R, Braucht DW, Rinschen MM, Chou CL, Hoffert JD, Pisitkun T, et al. Quantitative phosphoproteomic analysis reveals cAMP/vasopressin-dependent signaling pathways in native renal thick ascending limb cells. *Proc Natl Acad Sci USA* 2010;**107**(35):15653–8.

76. St-Germain JR, Taylor P, Tong J, Jin LL, Nikolic A, Stewart II, et al. Multiple myeloma phosphotyrosine proteomic profile associated with FGFR3 expression, ligand activation, and drug inhibition. *Proc Natl Acad Sci USA* 2009;**106**(47):20127–32.
77. Baty JD, Robinson PR. Single and multiple ion recording techniques for the analysis of diphenylhydantoin and its major metabolite in plasma. *Biomed Mass Spectrom* 1977;**4**(1):36–41.
78. Elschenbroich S, Kislinger T. Targeted proteomics by selected reaction monitoring mass spectrometry: applications to systems biology and biomarker discovery. *Mol Biosyst* 2011;**7**(2):292–303.
79. Unwin RD, Griffiths JR, Whetton AD. A sensitive mass spectrometric method for hypothesis-driven detection of peptide post-translational modifications: multiple reaction monitoring-initiated detection and sequencing (MIDAS). *Nat Protoc* 2009;**4**(6):870–7.
80. MacLean B, Tomazela DM, Shulman N, Chambers M, Finney GL, Frewen B, et al. Skyline: an open source document editor for creating and analyzing targeted proteomics experiments. *Bioinformatics* 2010;**26**(7):966–8.
81. Reiter L, Rinner O, Picotti P, Huttenhain R, Beck M, Brusniak MY, et al. mProphet: automated data processing and statistical validation for large-scale SRM experiments. *Nat Methods* 2011;**8**(5):430–5.
82. Pflieger D, Junger MA, Muller M, Rinner O, Lee H, Gehrig PM, et al. Quantitative proteomic analysis of protein complexes: concurrent identification of interactors and their state of phosphorylation. *Mol Cell Proteomics* 2008;**7**(2):326–46.
83. Wu R, Haas W, Dephoure N, Huttlin EL, Zhai B, Sowa ME, et al. A large-scale method to measure absolute protein phosphorylation stoichiometries. *Nat Methods* 2011;**8**(8):677–83.
84. Steen H, Jebanathirajah JA, Springer M, Kirschner MW. Stable isotope-free relative and absolute quantitation of protein phosphorylation stoichiometry by MS. *Proc Natl Acad Sci USA* 2005;**102**(11):3948–53.
85. Jin LL, Tong J, Prakash A, Peterman SM. Measurement of protein phosphorylation stoichiometry by selected reaction monitoring mass spectrometry. *J Proteome Res* 2010;**9**(5):2752–61.
86. Kettenbach AN, Rush J, Gerber SA. Absolute quantification of protein and post-translational modification abundance with stable isotope-labeled synthetic peptides. *Nat Protoc* 2011;**6**(2):175–86.
87. Olsen JV, et al. Quantitative phosphoproteomics reveals widespread full phosphorylation site occupancy during mitosis. *Sci Signal* 2010;**3**(104):ra3.
88. Wu R, et al. Correct interpretation of comprehensive phosphorylation dynamics requires normalization by protein expression changes. *Mol Cell Proteomics* 2011;**10**(8) M111009654.
89. Krebs EG, Beavo JA. Phosphorylation-dephosphorylation of enzymes. *Annu Rev Biochem* 1979;**48**:923–59.
90. Chi A, et al. Analysis of phosphorylation sites on proteins from Saccharomyces cerevisiae by electron transfer dissociation (ETD) mass spectrometry. *Proc Natl Acad Sci USA* 2007;**104**(7):2193–8.
91. Breitkreutz A, et al. A global protein kinase and phosphatase interaction network in yeast. *Science* 2010;**328**(5981):1043–6.
92. Fiedler D, et al. Functional organization of the *S. cerevisiae* phosphorylation network. *Cell* 2009;**136**(5):952–63.
93. Miller ML, et al. Linear motif atlas for phosphorylation-dependent signaling. *Sci Signal* 2008;**1**(35):ra2.
94. Leung GC, et al. Characterization of kinase target phosphorylation consensus motifs using peptide SPOT arrays. *Methods Mol Biol* 2009;**570**:187–95.
95. Thiele A, et al. High-density peptide microarrays for reliable identification of phosphorylation sites and upstream kinases. *Methods Mol Biol* 2009;**570**:203–19.

96. Mok J, et al. Deciphering protein kinase specificity through large-scale analysis of yeast phosphorylation site motifs. *Sci Signal* 2010;**3**(109):ra12.
97. Meggio F, Pinna LA. One-thousand-and-one substrates of protein kinase CK2? *FASEB J* 2003;**17**(3):349–68.
98. Tanoue T, et al. A conserved docking motif in MAP kinases common to substrates, activators and regulators. *Nat Cell Biol* 2000;**2**(2):110–6.
99. Ptacek J, et al. Global analysis of protein phosphorylation in yeast. *Nature* 2005;**438**(7068):679–84.
100. Lin J, et al. Understanding protein phosphorylation on a systems level. *Brief Funct Genomics* 2010;**9**(1):32–42.
101. Shah K, et al. Engineering unnatural nucleotide specificity for *Rous sarcoma* virus tyrosine kinase to uniquely label its direct substrates. *Proc Natl Acad Sci USA* 1997;**94**(8):3565–70.
102. Allen JJ, et al. A semisynthetic epitope for kinase substrates. *Nat Methods* 2007;**4**(6):511–6.
103. Allen JJ, Lazerwith SE, Shokat KM. Bio-orthogonal affinity purification of direct kinase substrates. *J Am Chem Soc* 2005;**127**(15):5288–9.
104. Virshup DM, Shenolikar S. From promiscuity to precision: protein phosphatases get a makeover. *Mol Cell* 2009;**33**(5):537–45.
105. Moorhead GB, et al. Evolution of protein phosphatases in plants and animals. *Biochem J* 2009;**417**(2):401–9.
106. Roy J, Cyert MS. Cracking the phosphatase code: docking interactions determine substrate specificity. *Sci Signal* 2009;**2**(100):re9.
107. Chen GI, Gingras AC. Affinity-purification mass spectrometry (AP-MS) of serine/threonine phosphatases. *Methods* 2007;**42**(3):298–305.
108. Gingras AC, et al. Analysis of protein complexes using mass spectrometry. *Nat Rev Mol Cell Biol* 2007;**8**(8):645–54.
109. Malovannaya A, et al. Analysis of the human endogenous coregulator complexome. *Cell* 2011;**145**(5):787–99.
110. Welch EJ, Jones BW, Scott JD. Networking with AKAPs: context-dependent regulation of anchored enzymes. *Mol Interv* 2010;**10**(2):86–97.
111. Elia AE, Cantley LC, Yaffe MB. Proteomic screen finds pSer/pThr-binding domain localizing Plk1 to mitotic substrates. *Science* 2003;**299**(5610):1228–31.
112. Lowery DM, et al. Proteomic screen defines the Polo-box domain interactome and identifies Rock2 as a Plk1 substrate. *EMBO J* 2007;**26**(9):2262–73.
113. Bhalla US, Iyengar R. Emergent properties of networks of biological signaling pathways. *Science* 1999;**283**(5400):381–7.
114. Goudreault M, et al. A PP2A phosphatase high density interaction network identifies a novel striatin-interacting phosphatase and kinase complex linked to the cerebral cavernous malformation 3 (CCM3) protein. *Mol Cell Proteomics* 2009;**8**(1):157–71.
115. Glatter T, et al. An integrated workflow for charting the human interaction proteome: insights into the PP2A system. *Mol Syst Biol* 2009;**5**:237.
116. Pulido R, Zuniga A, Ullrich A. PTP-SL and STEP protein tyrosine phosphatases regulate the activation of the extracellular signal-regulated kinases ERK1 and ERK2 by association through a kinase interaction motif. *EMBO J* 1998;**17**(24):7337–50.
117. Kiyonami R, et al. Increased selectivity, analytical precision, and throughput in targeted proteomics. *Mol Cell Proteomics* 2011;**10**(2): M110.002931.
118. Hunter T. The age of crosstalk: phosphorylation, ubiquitination, and beyond. *Mol Cell* 2007;**28**(5):730–8.
119. Yang XJ, Seto E. Lysine acetylation: codified crosstalk with other posttranslational modifications. *Mol Cell* 2008;**31**(4):449–61.
120. Liu B, Shuai K. Regulation of the sumoylation system in gene expression. *Curr Opin Cell Biol* 2008;**20**(3):288–93.

121. Young NL, Dimaggio PA, Garcia BA. The significance, development and progress of high-throughput combinatorial histone code analysis. *Cell Mol Life Sci* 2010;**67**(23):3983–4000.
122. Lee JS, Smith E, Shilatifard A. The language of histone crosstalk. *Cell* 2010;**142**(5):682–5.
123. Blethrow J, et al. Design and use of analog-sensitive protein kinases. *Curr Protoc Mol Biol* 2004; [Chapter 18: p. Unit 18 11].
124. Blethrow JD, et al. Covalent capture of kinase-specific phosphopeptides reveals Cdk1-cyclin B substrates. *Proc Natl Acad Sci USA* 2008;**105**(5):1442–7.
125. Ubersax JA, et al. Targets of the cyclin-dependent kinase Cdk1. *Nature* 2003;**425**(6960):859–64.
126. Flint AJ, et al. Development of "substrate-trapping" mutants to identify physiological substrates of protein tyrosine phosphatases. *Proc Natl Acad Sci USA* 1997;**94**(5):1680–5.
127. Bloom J, et al. Global analysis of Cdc14 phosphatase reveals diverse roles in mitotic processes. *J Biol Chem* 2011;**286**(7):5434–45.
128. Keck JM, et al. A cell cycle phosphoproteome of the yeast centrosome. *Science* 2011;**332**(6037):1557–61.
129. Zhao X, et al. Phosphoproteome analysis of functional mitochondria isolated from resting human muscle reveals extensive phosphorylation of inner membrane protein complexes and enzymes. *Mol Cell Proteomics* 2011;**10**(1): M110.000299.
130. Santamaria A, et al. The Plk1-dependent phosphoproteome of the early mitotic spindle. *Mol Cell Proteomics* 2011;**10**(1): M110.004457.
131. Malik R, et al. Quantitative analysis of the human spindle phosphoproteome at distinct mitotic stages. *J Proteome Res* 2009;**8**(10):4553–63.
132. Kubota K, et al. Sensitive multiplexed analysis of kinase activities and activity-based kinase identification. *Nat Biotechnol* 2009;**27**(10):933–40.
133. Zhang J, Allen MD. FRET-based biosensors for protein kinases: illuminating the kinome. *Mol Biosyst* 2007;**3**(11):759–65.
134. Gavet O, Pines J. Progressive activation of cyclinB1-Cdk1 coordinates entry to mitosis. *Dev Cell* 2010;**18**(4):533–43.
135. Gavet O, Pines J. Activation of cyclin B1-Cdk1 synchronizes events in the nucleus and the cytoplasm at mitosis. *J Cell Biol* 2010;**189**(2):247–59.
136. Newman RH, Zhang J. Visualization of phosphatase activity in living cells with a FRET-based calcineurin activity sensor. *Mol Biosyst* 2008;**4**(6):496–501.
137. Mizutani T, et al. A novel FRET-based biosensor for the measurement of BCR-ABL activity and its response to drugs in living cells. *Clin Cancer Res* 2010;**16**(15):3964–75.
138. Schmitz MH, et al. Live-cell imaging RNAi screen identifies PP2A-B55alpha and importin-beta1 as key mitotic exit regulators in human cells. *Nat Cell Biol* 2010;**12**(9):886–93.
139. Lin G, et al. Identification of PTPN23 as a novel regulator of cell invasion in mammary epithelial cells from a loss-of-function screen of the 'PTP-ome'. *Genes Dev* 2011;**25**(13):1412–25.
140. MacKeigan JP, Murphy LO, Blenis J. Sensitized RNAi screen of human kinases and phosphatases identifies new regulators of apoptosis and chemoresistance. *Nat Cell Biol* 2005;**7**(6):591–600.
141. Niemi NM, et al. MK-STYX, a catalytically inactive phosphatase regulating mitochondrially dependent apoptosis. *Mol Cell Biol* 2011;**31**(7):1357–68.
142. Cole KA, et al. RNAi screen of the protein kinome identifies checkpoint kinase 1 (CHK1) as a therapeutic target in neuroblastoma. *Proc Natl Acad Sci USA* 2011;**108**(8):3336–41.
143. Lupberger J, et al. EGFR and EphA2 are host factors for hepatitis C virus entry and possible targets for antiviral therapy. *Nat Med* 2011;**17**(5):589–95.

# Signaling in Striatal Neurons: The Phosphoproteins of Reward, Addiction, and Dyskinesia

Jean-Antoine Girault[*,†,‡]

[*]*Institut du Fer à Moulin, Paris, France*
[†]*Inserm UMR-S 839, Paris, France*
[‡]*Université Pierre et Marie Curie, Paris, France*



The striatum is a deep region of the forebrain involved in action selection, control of movement, and motivation. It receives a convergent excitatory glutamate input from the cerebral cortex and the thalamus, controlled by dopamine (DA) released in response to unexpected rewards and other salient stimuli. Striatal function and its dysfunction in drug addiction or Parkinson's disease depend on the interplay between these neurotransmitters. Signaling cascades in striatal medium-sized spiny neurons (MSNs) involve multiple kinases, phosphatases, and phosphoproteins, some of which are highly enriched in these neurons. They control the properties of ion channels and the plasticity of MSNs, in part through their effects on gene transcription. This chapter summarizes signaling in MSNs and focuses on the regulation of multiple protein phosphatases through DA and glutamate receptors and the role of ERK. It is hypothesized that these pathways are particularly adapted to the specific computing properties of MSNs and the function of the basal ganglia circuits in which they participate.

Progress in Molecular Biology
and Translational Science, Vol. 106
DOI: 10.1016/B978-0-12-396456-4.00006-7




1877-1173/12 $35.00

# I. Introduction

Reward and punishment are two major incentives for learning. Highly sophisticated mechanisms underlying these basic behavioral regulations have been selected during evolution, allowing adaptation and survival of individuals in fairly unpredictable environments. These two powerful mechanisms teach the brain to avoid repeating potentially harmful situations or behaviors, and to look for or repeat those that are potentially beneficial. Arguably, these two deeply engrained mechanisms control to various degrees all types of learning and memory. In mammals, they depend to a large extent on deep brain regions, which regulate cortical functions and control behavior. Over the past decades, progress in understanding the underlying neural circuits and their molecular and cellular bases has been considerable. This progress sheds light on neurological and psychiatric disorders whose pathophysiology and symptoms depend in part on alterations in these learning mechanisms. This chapter is devoted to some of the phosphorylation reactions that are important for reward-controlled or incentive learning, and its pathology, mostly addiction. Importantly, similar brain circuits are involved in the control of movements and their alterations are important in the course of Parkinson's disease (PD).

Reward-controlled learning has been extensively studied from a behavioral standpoint since the pioneering work of Pavlov, Thorndike, and others at the beginning of the twentieth century. The existence of brain reward circuits was then demonstrated by the intracranial self-stimulation experiments of Olds and Milner in the 1950s, and the role of dopamine (DA) in these circuits was rapidly suspected. However, it is only recently that the neural mechanisms underlying these complex processes have been clarified by the work of many laboratories (see Refs. 1–4 for reviews). DA neurons in the substantia nigra and the ventral tegmental area (VTA) are proposed to code for errors in reward prediction[3]: they fire in response to unexpected rewards, or, following associative learning, in response to the conditioned stimulus that predicts the reward but not to the primary reward. Conversely, the absence of an expected reward inhibits their firing. Thus, the release of DA appears to be a signal for controlling learning in relation to reward-related events. Recent work shows that some DA neurons also fire for nonrewarding stimuli which nevertheless have motivational salience.[5] Schematically, current hypotheses imply that DA released in response to unexpected rewards (or reward-conditioned stimuli) facilitates synaptic plasticity in forebrain circuits and, thus, reinforces associations between specific environmental context and/or cues and a particular behavior. Much experimental and theoretical work has been devoted to the elucidation of the role of DA, with implications for animal and human behavior and pathology, and even "neuroeconomics." One of the major targets of DA is the striatum, which is described below in more details. Its dorsal region is involved in the control of movements, and DA was first characterized because of the dramatic consequences of its absence in the dorsal striatum, which is responsible for PD.

The study of DA function has progressed in parallel with that of the biological bases of drug addiction. Most, if not all, addictive drugs share the ability to increase extracellular DA in the ventral striatum (nucleus accumbens, NAc), a region of the basal ganglia important for goal-directed behaviors and motivation.[1] Addiction is thought to involve a chemically induced abnormal learning process due to the ability of drugs of abuse to directly enhance DA transmission.[6,7] Thus, drugs of abuse divert the mechanisms of reward-controlled learning, progressively driving the behavior toward a compulsive urge for drug intake. It should be kept in mind that "real" addiction results from repeated exposure to drugs in a vulnerable background, in humans as well as in rodents.[8] It is remarkable that similar mechanisms of DA-controlled neuronal plasticity take place in the dorsal striatum and are involved in learning motor patterns.[9] Recent work suggests a progressive ventro-dorsal shift of the striatal regions critically involved in the action of drugs responsible for the formation of drug-related habits, which is central to addiction.[10] In a different context, imbalance of DA control of plasticity in these motor circuits is a key factor in the occurrence of abnormal movements such as dyskinesia in L-DOPA-treated patients with PD. Thus, understanding the mechanisms of action of DA on striatal neurons is important for various aspects of physiology and pathology.

## II. Overview of the Anatomo-Functional Organization of the Striatum and Basal Ganglia

The striatum, which consists of a dorsal region, the caudate–putamen (CP), and a ventral region, the NAc, is the major entry station of the basal ganglia. It is comprised of medium-sized spiny neurons (MSNs) which are GABAergic efferent neurons making up to 95% of striatal neurons in rodents, and of several types of interneurons, including large cholinergic and diverse medium-size GABAergic neurons.[11] MSNs receive massive glutamatergic inputs from virtually all cortical areas and the thalamus, and a dense DA innervation from the *substantia nigra pars compacta* (dorsolateral CP) and the VTA (dorsomedial CP and NAc).[12] MSNs are characterized by the presence of spines on their dendrites and each of them receives several thousands corticostriatal synapses, with only very few originating from any individual corticostriatal axon.[13] Major synaptic targets of the DA terminals are the necks of spines that receive synaptic contacts from cortical terminals,[14] an arrangement ideally suited for the modulation of MSNs response to the cortical input. MSNs are the output neurons of the striatum and are divided in several subpopulations on the basis of their targets and the proteins they express: they project to the *substantia nigra pars reticulata* either directly (striatonigral MSNs) or indirectly through relays in the lateral *globus pallidus* and subthalamic nucleus (striatopallidal MSNs) (Fig. 1). D1 DA receptors (D1R) are preferentially expressed in striatonigral MSNs of the direct pathway and D2R in striatopallidal MSNs of the indirect pathway[15] (see also Refs. 16,17 for recent reviews).

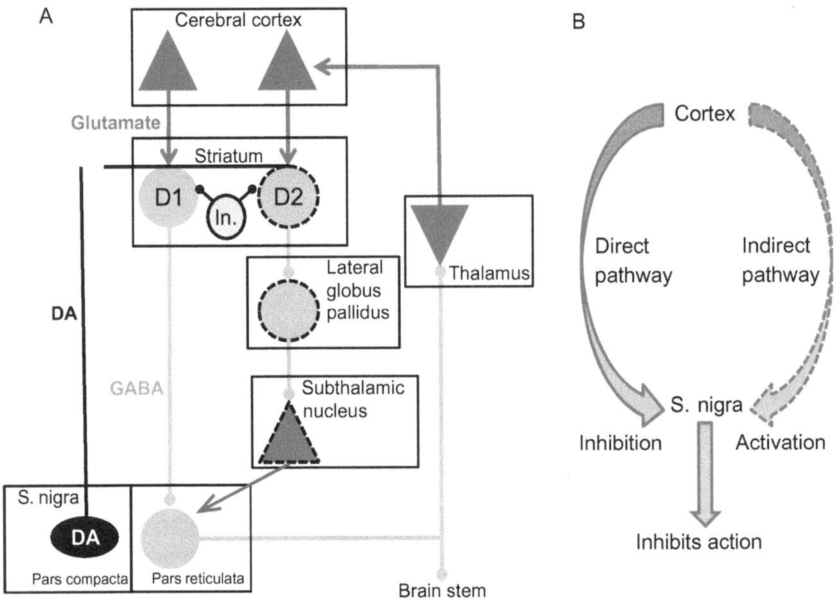


FIG. 1. Schematic anatomo-functional organization of the striatum. (A) The major connections of the basal ganglia circuits are indicated. Glutamatergic (excitatory) neurons are symbolized by triangular cell bodies and arrowed axons in dark gray. GABAergic (inhibitory) neurons are symbolized by round cell bodies and axon terminals in light gray. The main striatal neurons are medium-sized spiny neurons which belong to two populations: striatonigral which express D1 receptors and striatopallidal neurons which express D2 receptors. Striatopallidal neurons are part of the indirect pathway to the substantia nigra (S. nigra); neurons in this pathway are indicated by dashed cell body contours. The striatum also contains interneurons (In.). Many important anatomical connections are omitted for the sake of clarity. (B) Schematic representation of the function of the basal ganglia circuits shown in A. The substantia nigra pars reticulate neurons are tonically active and maintain a strong inhibition of their targets including the thalamus. Thus they inhibit action. Activation of the direct pathway inhibits S. nigra neurons and thus facilitates action (e.g., a movement due to the selective activation of appropriate muscle groups). Activation of the indirect pathway results in increased activity of subthalamic neurons and further activates S. nigra neurons, thereby reinforcing action inhibition. The function of the basal ganglia depends on the balance between the two pathways. Tonic release of DA inhibits the indirect pathway. Phasic release of DA activates the direct pathway.

DA controls the flow of information through corticostriatal synapses and their plasticity.[18] MSNs are in a basal state of low firing, whereas their targets in the substantia nigra or globus pallidus have a high discharge rate. Activation of GABAergic MSNs inhibits their targets.[19] This inhibition mechanism is thought to facilitate the function of neuronal circuits in the cerebral cortex and other brain regions and provides the basis for the proposed role of the basal ganglia in action selection.[20–22] This theory proposes that the basal ganglia carry out the important task of devoting most brain resources to a single movement or behavior at a time, in the context. These studies in mice show that the complex and changing environmental and internal stimuli. MSNs possess a rich repertoire of short- and long-term plasticity mechanisms which provide the bases for DA-controlled learning mechanisms.[23–26] In addition, dendritic spines are known to undergo constant rearrangement[27] and, in MSNs, to be modified by chronic alteration in DA transmission.[28–30] The resulting modifications in the efficacy of corticostriatal synapses could alter the "selection of action" in response to a given pattern of striatal input, underlying learning of complex motor sequences (dorsal striatum) or behavioral orientation ("motivation," ventral striatum). The role of the striatum in the control of many aspects of movements and motivation depends on the balance between the two trans-striatal circuits, involving the direct and indirect pathways (Fig. 1). Their disequilibrium accounts for the consequences of striatum dysfunction in pathological conditions, including PD due to the lack of DA following degeneration of DA neurons and addiction, related to the common capacity of abused substances to increase extracellular DA. Although the simple model of the two pathways is clearly an oversimplification, it is remarkable that it explains many of the functions and dysfunctions of the basal ganglia. This has been recently supported by several studies in which sophisticated gene-targeting techniques allowed the selective destruction[31] or stimulation using optogenetics[32] of one population or the other.

## III. DA Receptors in the Striatum

Paul Greengard and his collaborators showed that cascades of biochemical reactions similar to those identified in the control of liver and muscle glycogen metabolism were critical in the action of neurotransmitters in the nervous system.[33,34] They identified the first DA receptor by its positive coupling to adenylyl cyclase[35] and the downstream protein phosphorylation.[36] Later, a second type of DA receptor, D2, which is negatively coupled to adenylyl cyclase, was identified, first in the pituitary and then in the brain.[37] All DA receptors are G protein-coupled receptors (GPCRs, see Ref. 38 for a recent review). Molecular cloning identified five different receptors D1 and D5 positively coupled to adenylyl cyclase (D1-type) and D2–D4 negatively coupled to this enzyme (D2-type). D1R and D2R are the most highly expressed in the striatum. The D3R are also present mostly in the ventral regions of the striatum.[39] Importantly, D1R are expressed in striatonigral MSNs of the direct pathway, whereas D2R are found in striatopallidal neurons of the indirect pathway.[15] This dichotomy, which was originally proposed on the basis of careful anatomical studies, has been extensively confirmed recently using BAC transgenic mice in which various reporter genes are expressed under the control of D1R or D2R promoters (see Ref. 17 for a review). These studies in mice indicate that the segregation is high, with coexpression in only 5% of the neurons in the dorsal striatum and 15% in the NAc.[40]

D1R in the striatum are coupled to adenylyl cyclase by a heterotrimeric G protein with a specific composition, containing the $\alpha_{olf}$, $\beta_2$, and $\gamma_7$ subunits.[41,42] These proteins are expressed in the two types of MSNs which contain little if any "classical" and ubiquitous $\alpha_S$ subunit. In striatopallidal neurons, these heterotrimeric G proteins are associated mostly to adenosine A2a receptors (A2aR). In mutant mice devoid of $\alpha_{olf}$ or $\gamma_7$, the activation of striatal cyclase by DA or adenosine is severely impaired and the effects of DA profoundly altered.[43–46] The main adenylyl cyclase isoform in MSNs is AC5, which is not activated by $Ca^{2+}$ and whose deletion has severe functional consequences.[47] In striatal neurons, cAMP is degraded by several phosphodiesterases, including PDE1B and PDE10A, which are highly enriched in MSNs[48,49] where they play a key role in regulating cAMP signaling.[50,51]

D2R are coupled to Gi/o heterotrimeric G proteins. They inhibit AC, opposing the effects of adenosine in striatopallidal neurons and are also able to inhibit $Ca^{2+}$ channels and activate $K^+$ channels (see Ref. 38 for a recent review of D2R signaling). In addition, once they undergo endocytosis, they are able to activate a number of signaling pathways through the recruitment of β-arrestins and the formation of endocytic signaling vesicles, including regulation of Akt through activation of PP2A.[52]

In addition to these direct coupling, DA receptors can interact with other receptors. It has been shown that D1R and D2R can interact with each other and then activate phospholipase C.[53] This novel signaling would presumably be restricted to the small number of neurons that express significant levels of both receptors. In addition, the formation of heteromers between D2 and A2a and CB1, also highly expressed in MSNs, has been reported. D1 receptors may interact directly with glutamate receptors (see below). MSNs also express a large number of other GPCRs including for other monoamines, acetylcholine, endocannabinoids, neuropeptides, and orphan GPCRs for which the ligand is not yet known. In spite of their importance, these receptors are outside of the focus of this review and not discussed here.

# IV. Glutamate Receptors

Glutamate is the major excitatory neurotransmitter in the vertebrate central nervous system. It is responsible for the fast excitatory transmission mediated by binding of glutamate to extracellular regions of ligand-gated ion channels, the so-called ionotropic glutamate receptors (iGluR). Glutamate also acts on GPCRs, the metabotropic glutamate receptors (mGluR) that have important modulatory roles. The iGluR comprise two major groups in the context of this discussion, termed after their selective synthetic ligands the AMPA (2-amino-3-(5-methyl-3-oxo-1,2-oxazol-4-yl)propanoic acid) receptors (AMPAR) and NMDA (*N*-methyl-D-aspartate) receptors (NMDAR). These receptors are heterotetramers and their subunit composition controls their properties (see[54,55] for recent reviews). AMPAR are responsible for a large part of the fast transmission. They are permeable to cations and, at normal resting membrane potential, depolarize the cell by letting $Na^{+}$ flow in. AMPAR lacking the GluA2 subunit can also let $Ca^{2+}$ enter the cell (GluA2 subunits lose their capacity to allow $Ca^{2+}$ permeability due to mRNA edition[56]). The three-dimensional structure of AMPAR has been determined by X ray crystallography, which comprises two external amino terminal domains, one of which binds the ligand, a transmembrane ion channel forming domain, and an intracellular C-terminal domain.[57] NMDAR are related in structure but have unique properties that are critical for synaptic plasticity. They are heterotetramers composed of GluN1 subunits and GluN2 subunits which are responsible for glutamate binding and exist as four different isoforms (mostly GluN2A and GluN2B in the striatum, formerly known as NR2A and NR2B, respectively). To open NMDAR require binding not only of glutamate but also of glycine which is a coagonist.[58] Importantly, at negative intracellular membrane potential (i.e., resting potential or partial depolarization), NMDA channels are blocked by extracellular $Mg^{2+}$ ions.[59] This obstruction is lifted at positive intracellular potential and the channels then become permeable to $Na^{+}$ and $Ca^{2+}$ ions. As a consequence, NMDA channels are coincidence detectors that are activated when the presynaptic terminal is depolarized and releases glutamate and when the postsynaptic element is depolarized. They implement the rules theoretically defined by Donald Hebb more than 60years ago for synaptic plasticity: "When an axon of cell A is near enough to excite cell B and repeatedly or persistently takes part in firing it, some growth process or metabolic change takes place in one or both cells such that A's efficiency, as one of the cells firing B, is increased."[60] Not surprisingly, these receptors have been shown to be critical for plasticity in many synapses (albeit not all), including in the striatum. AMPA and NMDAR have large intracellular domains that interact with the dense cytoskeletal protein network, which underlies the postsynaptic membrane and forms the postsynaptic density. These domains contain many phosphorylation sites which regulate channel properties, protein–protein interactions, intracellular trafficking, and degradation.

## V. Synaptic Plasticity

Synaptic plasticity corresponds to changes in synaptic efficiency (i.e., increased or decreased postsynaptic response for a constant stimulation of the presynaptic element) of various durations. Long-term plasticity (several hours, days, or longer) is thought to be particularly important for learning and memory. The work of many laboratories has allowed the dissection of mechanisms of long-term potentiation (LTP) and long-term depression (LTD), including in the striatum.[25,61,62] These apparently simple modifications correspond in fact to a complex succession of multiple changes with convergent effects but whose molecular details can vary from one synapse to the other. Put in very simple terms, LTP usually corresponds to an increased function of AMPAR, with first an increased permeability of existing channels followed by an increased number of channels at the postsynaptic side, with possible changes in subunit composition. LTP is usually triggered by a massive postsynaptic $Ca^{2+}$ influx resulting from the opening of NMDAR. LTD, in contrast, results from a decreased permeability and mostly a decreased number of AMPAR in the synaptic region and is triggered by a small increase in postsynaptic $Ca^{2+}$. LTP and LTD are not only due to changes at the postsynaptic level, but they can also include changes in presynaptic properties triggered by retrograde messengers generated at the postsynaptic site. Important retrograde messengers include nitric oxide (NO) and endocannabinoids (ECs) anandamide, and 2-arachidonoylglycerol. ECs act on presynaptic cannabinoid receptors 1 (CB1R), which decrease neurotransmitter release and are highly expressed in MSNs and in their terminals in the substantia nigra and the globus pallidus.[63] It should be kept in mind that the process of synaptic plasticity implies a reversibility, allowing synaptic efficiency to evolve in both directions. In addition, synaptic plasticity is accompanied by important morphological changes including alteration in the morphology of existing spines, growth of new spines, and disappearance of existing ones.[64,65]

## VI. The cAMP Cascade

The action of DA on cAMP-production-coupled D1R was one of the first neurotransmitters effects to be identified in the brain and has been extensively studied (Fig. 2). cAMP major action is to bind to the regulatory subunit (R) of cAMP-dependent protein kinase (PKA), a heterotetramer containing two R and two catalytic subunits (C).[66] This binding releases the C subunits, which are fully active, phosphorylate membrane-bound and cytosolic substrates, and can penetrate the nucleus to phosphorylate nuclear targets. R subunits interact with a proteiform family of partners called A-kinase-associated proteins (AKAPs), which

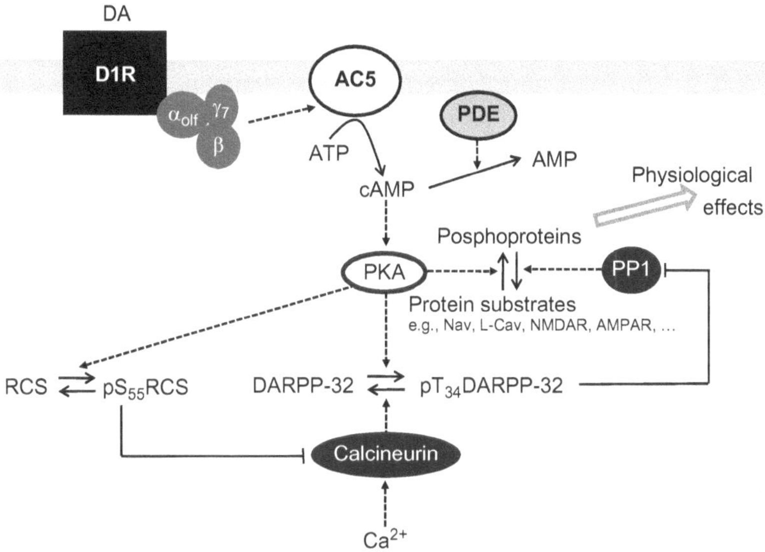


FIG. 2. Schematic representation of D1R signaling in striatonigral neurons. In the striatum, D1 receptors are mostly coupled to adenylyl cyclase 5 (AC5) through a heterotrimeric G protein comprising an $\alpha_{olf}$ and a $\gamma_7$ subunit. The regulatory roles of DARPP-32 and RCS are indicated (see text). Dashed arrows symbolize influences (e.g., catalysis); solid arrows, chemical reactions. Inhibition is symbolized by solid lines ending with a perpendicular segment.

enrich PKA at specific subcellular locations (e.g., postsynaptic sites or perinuclear region).[67] In addition, cAMP can activate directly cyclic nucleotide-gated cation channels and guanine nucleotide exchange factors (cAMP-GEF or EPACs).[68] The importance of these targets is increasingly recognized, but little information is yet available on their role in the striatum.

In the early 1980s, attempts were made to better understand the action of neurotransmitters by identifying the substrates of protein kinases enriched in specific neuronal populations.[69,70] This approach was very successful in the striatum, where it allowed the identification of several protein substrates highly enriched in DA-innervated regions and termed 32-kDa DA and cAMP-regulated phosphoprotein (DARPP-32),[71] 16-kDa cAMP-regulated phosphoprotein (ARPP-16),[72,73] and 21-kDa cAMP-regulated phosphoprotein.[74,75] These proteins are by no means the only targets of cAMP signaling in MSNs but have raised special interest due to their specific enrichment in particular cell types and, therefore, have been hypothesized to underlie some particular properties of these neurons. DARPP-32, which is by far the most extensively studied among these proteins, and the others, which have recently started to be better understood, are discussed in the following paragraphs.

## VII. DARPP-32: A Signaling Hub Important in MSNs

DARPP-32 is a small unstructured protein of about 202 amino acids (its length varies among species).[76] Its N-terminal region has sequence identities with protein phosphatase 1 (PP1 also termed phosphoprotein phosphatase 1 PPP1) inhibitors 1A and 1C (see Ref. 77 for a recent review). These three proteins form a family of cAMP/PKA-regulated PP1 inhibitors (PPP1-regulatory subunits 1A–C, or PPP1R1A/B/C, respectively). The motifs involved in interaction with PP1 catalytic subunit (PP1c) comprise a highly conserved docking motif found in most PP1c partners, K/R-K/R-V/I-x-F/W,[78,79] which is RKIQF in DARPP-32 (Fig. 3). In addition, a Thr residue at position 34 in DARPP-32 is phosphorylated by PKA- and cGMP-dependent protein kinase.[80] When Thr-34 is phosphorylated, DARPP-32 becomes a powerful inhibitor of PP1 ($IC_{50}$ in the nM range) with mixed competitive and uncompetitive kinetics.[81] It is assumed that this inhibition results from a bidentate interaction of DARPP-32 with binding of pThr-34 to the catalytic site of the phosphatase and binding of the docking motif to a hydrophobic groove on the other side of PP1c.[82] Thus, a major function of DARPP-32 is to provide a feed-forward amplification of PKA signaling by inhibiting dephosphorylation of substrates of this kinase that are dephosphorylated by PP1 (Fig. 2). It is also a potential source of cross talk with other signaling pathways by preventing dephosphorylation by PP1 of residues that are

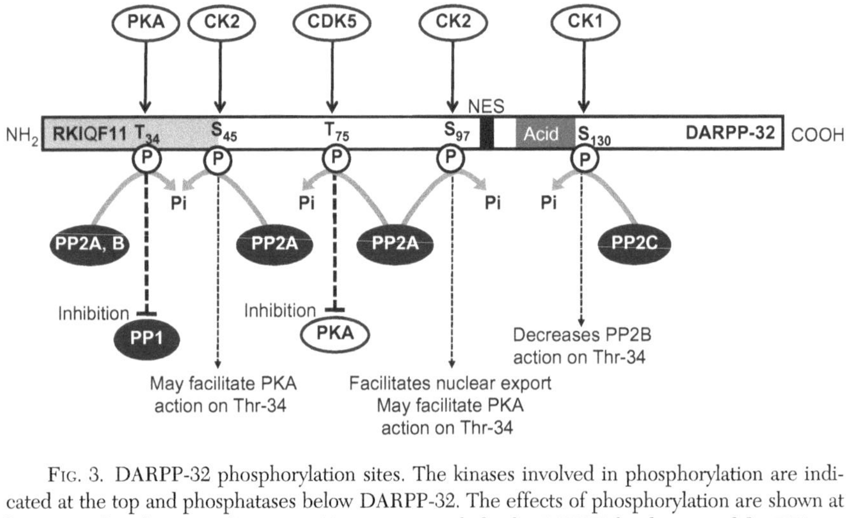


FIG. 3. DARPP-32 phosphorylation sites. The kinases involved in phosphorylation are indicated at the top and phosphatases below DARPP-32. The effects of phosphorylation are shown at bottom. Other features of DARPP-32 sequence include the RKIQF binding motif for PP1, an incomplete nuclear export sequence (NES), and an acidic stretch (acid). The region conserved in Inhibitor 1 and PPP1R1C is shaded light gray. Residue numbers are from mouse sequence.

substrates of other kinases. pThr-34 is dephosphorylated by two phosphatases, namely, $Ca^{2+}$/calmodulin-activated calcineurin (also known as PP2B or PPP3) and PP2A (or PPP2), *in vitro* and *in vivo*.[83–85] DARPP-32 is also phosphorylated by cyclin-dependent protein kinase 5 (CDK5) on Thr-75[86] (Fig. 3). CDK5 is activated in neurons by its p35 subunit. Importantly, when DARPP-32 is phosphorylated on Thr-75, it becomes a powerful inhibitor of PKA.[86] This endows DARPP-32 with the remarkable and complex property to be both a PKA-regulated inhibitor of PP1 and a CDK5-regulated inhibitor of PKA, providing CDK5 with the theoretical possibility to switch off the PKA pathway in striatal neurons. DARPP-32 is also phosphorylated by casein kinases 1 and 2 (CK1 and CK2)[87,88] (Fig. 3). CK1 phosphorylates Ser-130 (in mouse sequence, Ser-137 in rat), and this phosphorylation inhibits dephosphorylation of pThr-34 by calcineurin *in vitro* and in neurons.[88,89] CK2 phosphorylates Ser-45 and Ser-97 (Ser-102 in rat).[87] Phosphorylation by CK2 facilitates phosphorylation of Thr-34 by PKA *in vitro*, but it is not known whether this results from pSer-45, pSer-97, or both, and whether this effect is important in intact cells. In addition, Ser-97 phosphorylation facilitates nuclear export of DARPP-32 (Ref. 90 see below). Both pThr-75 and pSer-97 are dephosphorylated by PP2A.[86,87] PP2A is a heterotrimeric enzyme comprising a scaffolding A subunit, a regulatory B subunit, and a catalytic C subunit.[91] There are multiple isoforms of the B subunit. Two of them, B56d (PPP2R5D) and PR72 (PPP2R3A), which are expressed in MSNs, provide interesting regulatory properties, as the B56δ subunit is activated by PKA phosphorylation on Ser-566,[92] while the PR72 subunit is activated by $Ca^{2+}$ binding to its two EF-hands motifs[93] (Fig. 3). These isoforms of PP2A can account for the cAMP- and $Ca^{2+}$-induced dephosphorylation of pThr-75 and p-Ser-97 in striatal neurons. The CK1 site, Ser-130, is mostly dephosphorylated by PP2C (PPP2)[94] (Fig. 3).

Although most studies of DARPP-32 have addressed its properties in the cytoplasm, recent work uncovered that DARPP-32 undergoes a constant cyto-nuclear trafficking, regulated by phosphorylation of Ser-97 which increases nuclear export of DARPP-32.[90] Activation of the cAMP/PKA pathway, and presumably other signals, can increase nuclear DARPP-32 concentration by activating dephosphorylation of Ser-97 by PP2A. When it is pThr-34-DARPP-32 that accumulates in the nucleus, it is expected to inhibit nuclear PP1 and regulate multiple responses in the nucleus, as indicated by its role in the control of histone H3 Ser-10 phosphorylation.[90]

The multiple phosphorylation sites of DARPP-32 and their intrinsic and extrinsic regulatory effects result in complex signaling properties which have been investigated by various modeling approaches.[95–100] Classical kinetic models using differential equations showed that DARPP-32 is not only a robust signal integrator, but that its response also depends on the delay between cAMP and calcium signals[99] and that Thr-34 phosphorylation is potentiated by the coincidence of the two stimuli.[97] They also showed that DARPP-32 can translate the various states of glutamate and DA input into distinct steady states and/or dynamic responses[95] and that the positive feedback loop consisting of PKA, PP2A, and Thr75 phosphorylation may serve as a major switch for inducing LTD and LTP.[96] Interestingly, this loop displayed robust bistable responses but was disrupted by high basal levels of DA. Other authors, using large-scale numerical simulation and sophisticated mathematical approaches, have shown that the global topology of the DARPP-32 network provides a very high resistance to perturbations.[100] Thus, modeling studies indicate that DARPP-32 is a robust sensor of glutamate and DA inputs with an intrinsic ability to stably orient the cell responses in the presence of noise. They suggest that a major function of DARPP-32 is to increase the reliability of signal processing in MSNs.

A large number of studies have been devoted to the investigation of DARPP-32 regulation and of its potential role *in vivo*, mostly in the striatum (see Refs. 77,101–103 for reviews). Describing all these findings is beyond the scope of this chapter and we will only provide a short synopsis. DARPP-32 phosphorylation on its various sites can be regulated in MSNs by a wide diversity of neurotransmitters and drugs. DARPP-32 is expressed in the two classes of MSNs: the striatonigral and striatopallidal neurons.[40] A recent elegant work used BAC transgenic-mediated expression of differentially tagged DARPP-32 in these two populations of neurons, under the control of the D1R and D2R promoters, respectively, to study the regulation of phosphorylation sites.[104] The results confirmed and expanded previous work showing that DARPP-32 phosphorylation on its various sites can be modified in opposite directions in the two cell populations. For example, cocaine, which increases extracellular DA by preventing its reuptake, increased pThr-34 and decreased pThr-75 in D1R-expressing neurons, while it had the opposite effects in the D2R-expressing neurons.[104] In contrast, haloperidol, an antipsychotic drug, which is a D2R antagonist, increased specifically pThr-34 in neurons expressing these receptors.[104] This strong effect of D2R antagonist in behaving animals, also observed in many other experiments, is functionally revealing. It shows that there is a strong spontaneous permanent DA tone that inhibits adenylyl cyclase in striatopallidal neurons through stimulation of high-affinity D2R. The simple blockade of this tone by a pharmacological antagonist is sufficient to activate signaling pathways—as a result of another permanent stimulatory tone exerted mostly by A2aRs—that activate adenylyl cyclase. In contrast, activation of lower-affinity D1R appears to require a high level of DA release, following a bursting activity of DA neurons triggered by an unexpected reward, for example.[105,106]

The function of DARPP-32 has been investigated at the cellular level as well as in behaving mutant mice.[107,108] Several studies have shown that DARPP-32 plays an important role in the regulation of ion channels activity by controlling directly or indirectly their phosphorylation state through inhibition of PP1. Thus, DARPP-32 is strongly involved in DA-enhanced function of AMPAR, NMDAR, voltage-gated Na, and L-type $Ca^{2+}$ channels, as well as in DA-decreased function of GABA-A channels and N and P/Q-type $Ca^{2+}$ channels.[108–112] As a result, DARPP-32 appears to be important for regulating synaptic plasticity in striatal neurons, since in its absence LTP or LTD can be completely blocked in specific experimental conditions.[113] The functional importance of DARPP-32 *in vivo* has been investigated using knockout mutant mice and knock-in mice in which one or the other phosphorylation sites was replaced by Ala.[107,108] These mutant mice are viable and fertile and do not show major abnormalities in laboratory conditions. However, when they are challenged by drugs, their responses are profoundly altered, with decreased responses to many therapeutic and abused drugs.[90,107,108,114–117] A recent study has investigated the specific function of DARPP-32 in D1R- or D2R-expressing MSNs using a conditional knockout strategy.[118] Interestingly, in some tests, the behavior of these mice was more overtly altered than that of complete knockout mice, showing that any imbalance between the striatonigral direct and the striatopallidal indirect pathway has important functional consequences. Beyond the striatum, DARPP-32 is involved in other functions and DARPP-32-deficient female mice display impaired sexual behavior.[119] Genetic association studies suggest that DARPP-32 expression is positively correlated with a general cognitive performance in mice.[120] In humans DARPP-32 polymorphism, presumably reflecting expression levels, is correlated with performance in several cognitive tests that depend on frontostriatal function and with increased frontostriatal connectivity in multimodal brain imaging.[121] Some studies but not others have found an association between single nucleotide polymorphism in DARPP-32 gene and schizophrenia, and a decreased expression in postmortem cortex of patients with schizophrenia has been repeatedly reported (see Ref. 77 for a discussion). It should also been mentioned that DARPP-32 overexpression has been observed in several human cancers as well as expression of a truncated isoform, t-DARPP, devoid of the N-terminal region including Thr-34,[122] whose physiological function if any is not known.

## VIII. The Function of Other PKA Substrates Enriched in Striatal Neurons

ARPP-21 and ARPP-16 were described 25years ago and, although they were clearly regulated by DA and other neurotransmitters,[123–125] their function remained unknown until recently.[126] ARPP-21, when phosphorylated by

PKA, binds to calmodulin in the presence of $Ca^{2+}$ and thus inhibits CaM-dependent enzymes.[127] Therefore, ARPP-21 has been renamed "regulator of calmodulin signaling" (RCS). RCS is involved in the regulation of PKA substrates that are dephosphorylated by calcineurin and amplifies the response mediated by these proteins (Figs. 2 and 4). For example, it has been shown to increase phosphorylation of DARPP-32 and function of L-type $Ca^{2+}$ channels.[127]

The function of ARPP-16 has been identified very recently in a completely different context. ARPP-16 belongs to a family of several related gene products termed ARPP-16, ARPP-19, and endosulfine.[125] ARPP-16 is enriched in striatal neurons,[72] whereas the other two are more ubiquitously expressed. These proteins are major players in the control of the cell cycle at the level of initiation and maintenance of mitosis by inhibiting PP2A containing the B55δ isoform.[128,129] This effect requires phosphorylation by Greatwall, a protein kinase that has several orthologues in mammals (microtubule associated serine/threonine kinase like, MASTL1-4).[126] The function of the conserved phosphorylation of these proteins by PKA has yet to be identified. At any rate, they provide a potential and novel pathway for regulating PP2A in striatal neurons, as well as in other cell types.

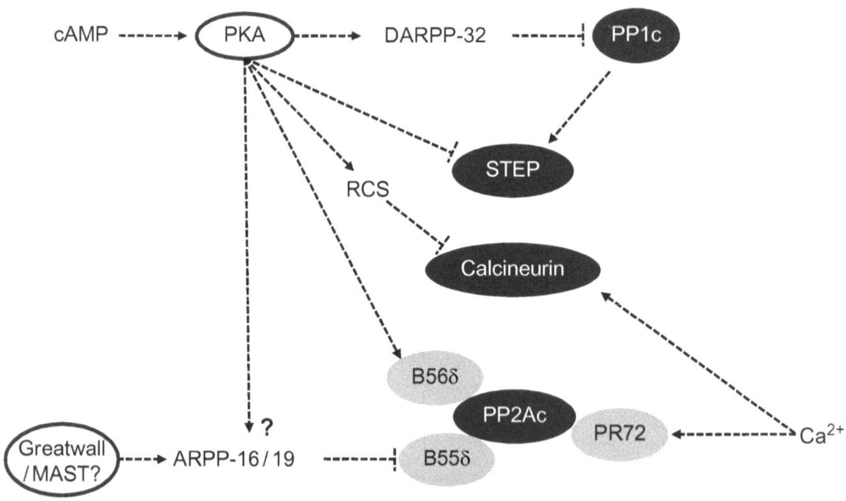


FIG. 4. Summary of known regulatory pathways of protein phosphatases in striatal neurons. Phosphatases are indicated in dark gray. PP2A regulatory subunits in light gray (note that only one of them interacts with the catalytic subunit at a time). Dashed arrows symbolize positive influences. Inhibition is symbolized by solid lines ending with a perpendicular segment. Striatal enriched tyrosine phosphatase (STEP).

## IX. The ERK Cascade

Mitogen-activated protein kinases (MAP kinases) are regulated by two upstream kinases, a MAP-kinase-kinase and a MAP-kinase-kinase-kinase, which form a highly conserved 3-kinase signaling modules. One type of MAP kinases, namely, the extracellular signal-regulated kinases 1/2 (ERK1/2), has been shown to play an important role in striatal signaling.[130] ERKs are ubiquitous enzymes activated by many extracellular signals including growth factors and neurotransmitters and regulate a variety of cellular functions including cell growth, differentiation, and synaptic plasticity.[131] In MSNs, ERK1/2 are activated by cocaine[132] and other drugs of abuse.[133–136] In fact, all tested drugs of abuse that have varied mechanisms of action were able to activate ERK in the ventral striatum,[134] inasmuch as they share the ability to increase DA release in this brain region.[1] In contrast, non-addictive psychoactive drugs did not have this property, although some of them activated ERK in other brain regions.[134] The importance of ERK in the long-term effects of drugs of abuse is supported by the prevention of conditioned place preference and locomotor sensitization by pharmacological inhibition of MEK1/2, the kinases that activate ERK.[132,133,137] Blockade of ERK activation during reexposure also suppresses previously acquired cocaine-conditioned place preference.[138,139] A recent report showed that ERK is necessary for corticostriatal LTP that is also induced by exposure to cocaine.[140] Importantly depotentiation of these synapses reversed locomotor sensitization, strongly supporting the causal relationship between the LTP and behavioral sensitization.

The regulation of ERK1/2 by drugs of abuse in MSNs involves a cross-talk between D1R and NMDAR, which are both necessary for this response.[117,132,133] This feature is especially important, as it provides a mechanism by which MSNs can detect the simultaneous occurrence of a raise in DA, for example, due to an unexpected reward, and the activity of specific cortico- and thalamostriatal glutamatergic inputs, linked to the external and internal context.[130] As ERK is important for long-term changes in behavior, this mechanism of coincidence detection or logical AND gate has a potential to play an important functional role in learning specific behaviors leading to reward. Its molecular bases have been particularly studied in the case of psychostimulants (cocaine and amphetamine)[117,140] (Fig. 5). As in other neurons, a strong stimulation of NMDAR is sufficient to activate ERK in striatal neurons in culture.[140] This activation can also be achieved *in vivo* by electrical stimulation of the corticostriatal pathway.[141] However, stimulation of D1R potentiates this effect and is necessary for activation of ERK by drugs of abuse *in vivo*. There are multiple levels of interaction between D1R and NMDAR. These two receptors can directly interact, and this interaction may be involved in these responses.[142,143] In addition, D1R can activate a tyrosine phosphorylation

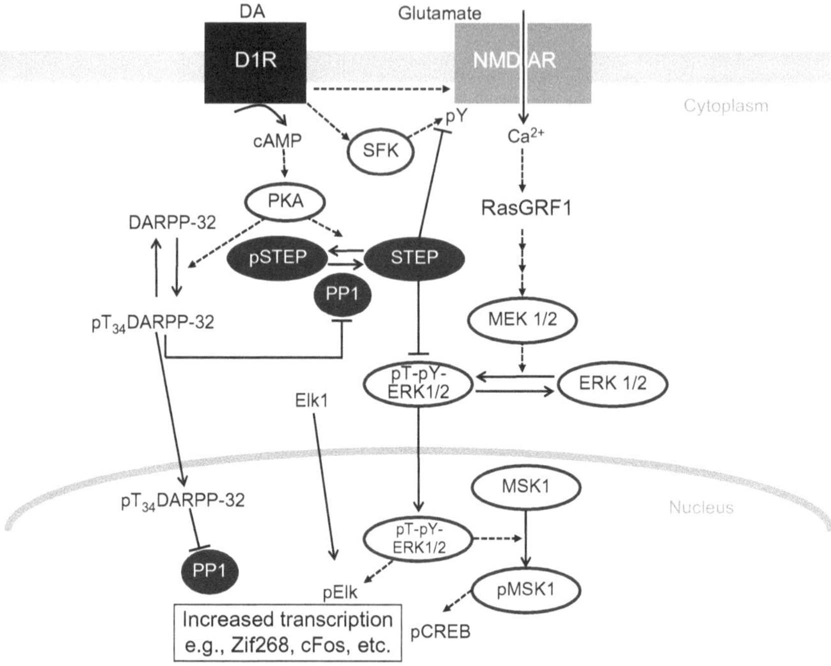


FIG. 5. Schematic representation of the activation of ERK1/2 by interaction between DA and glutamate. The Src-family kinases, presumably Fyn, are involved in the regulation of NMDAR by D1R. Elk1 can translocate to the nucleus with activated ERK. DARPP-32 can translocate to the nucleus following dephosphorylation of Ser-97 by PP2A (e.g., activated by PKA through B56δ subunit). In the nucleus, it can inhibit PP1 and thereby facilitate the action of protein kinases. PKA and PP1 are also directly active on NMDAR (not shown).

pathway that results in phosphorylation of GluN2B subunit of NMDAR by an Src-family kinase, presumably Fyn.[144,145] The mechanism of activation of SFK is not clear but may involve a tyrosine phosphatase.[146] Tyrosine phosphorylation of GluN2B and possibly GluN2A increases NMDAR permeability,[147] and D1R stimulation enhances the NMDAR-mediated $Ca^{2+}$ influx.[147] Although this mechanism appears sufficient in neurons in culture, DARPP-32 plays a critical additional role *in vivo*, as in DARPP-32 knockout or Thr-34-Ala knock-in mutant mice, the phosphorylation of ERK by various drugs of abuse was blocked in the striatum.[117] DARPP-32 exerts its effects in part by regulating striatal enriched tyrosine phosphatase (STEP), a tyrosine phosphatase that dephosphorylates the Tyr residue of ERK activation loop (Fig. 5). STEP is phosphorylated by PKA and dephosphorylated by PP1. This phosphorylation inactivates the enzyme. DA enhances STEP phosphorylation directly through

PKA and indirectly by blocking PP1 through pThr-34–DARPP-32. In addition, DARPP-32 is active upstream from MEK at level(s) yet to be identified, which may include the known regulation of NMDAR through regulation of NR1 phosphorylation.[148]

How does NMDAR activate the ERK pathway in the striatum? Although little evidence is available in the striatum *in vivo*, experiments in neurons in culture show that it is a $Ca^{2+}$-mediated response.[140] It has been shown that NMDAR associates with $Ca^{2+}$-activated Ras guanine nucleotide exchange factor-1 (Ras-GRF1) to activate ERK.[149] Indeed, in Ras-GRF1 knockout mice, activation of ERK was impaired in the striatum, whereas it was increased in mice overexpressing this factor.[150] The role of additional pathways through other types of guanine nucleotide exchange factors including CalDAG-GEFI or II is possible.[151]

What are the important targets of ERK1/2 in striatal neurons? Interestingly, the two isoforms of ERK (ERK1 and 2) appear to play a different role in the striatum and possibly other brain regions, as several ERK-mediated physiological and behavioral responses were paradoxically enhanced in ERK1 knockout mice.[152] It is possible that this unexpected effect results from a less potent effect of ERK1 than ERK2 on some of its targets, with a competition between the two isoforms. Although ERK can phosphorylate proteins in the cytoplasm of neurons that are important for plasticity, including ion channels,[131,153] their role has not been studied in the striatum. In the context of striatal signaling, the nuclear targets of ERK have been more thoroughly investigated. In most cell types, when ERK is activated, it can accumulate in the nucleus and some of its nuclear targets in striatal neurons have been characterized. The actions of ERK in the nucleus include direct phosphorylation of the transcription factor Elk1 on Ser383/389.[154–157] In striatal neurons, Elk1 can traffic in and out of the nucleus and its phosphorylation by ERK increases its nuclear relocalization and its transcriptional activity in association with serum response factor to form a ternary complex factor (TCF). Elk1 appears to be associated with endocytic complexes containing AMPAR and ERK, which contribute to the transport of ERK from the plasma membrane to the perinuclear region.[155] In the nucleus, TCF increases the transcription of immediate early genes such as Zif268 (also known as NGFI-A or Krox24). Zif268 is itself a transcription factor that appears to be critical for long-lasting behavioral response such as cocaine conditioned place preference.[158] ERK in the nucleus can also activate protein kinases, among which mitogen- and stress-activated protein kinase 1 (MSK1) is enriched in MSNs.[157] MSK1 plays a critical role in the phosphorylation of CREB and histone H3 and induction of cFos in response to cocaine. All these regulations are important for long-lasting effects that require regulation of gene transcription.

ERK activation by cocaine, as well as the phosphorylation of MSK1 and histone H3 Ser-10, occurs selectively in D1R-expressing neurons of the direct pathway.[40] However, ERK can be activated in D2R-expressing neurons of the indirect pathway in response to other types of stimuli. The simple blockade of D2R by haloperidol or raclopride triggers ERK phosphorylation, at least in part by removing the brake exerted by D2R on A2aR[40,159] (see above). However, in D2R-expressing striatopallidal neurons, the phosphorylation of some of the proteins that are targets of ERK in striatonigral neurons appears to occur through other mechanisms, including direct phosphorylation by PKA.[159] The origin of this difference between the two populations of striatal neurons is not clear. In addition, a powerful synergistic interaction of A2aR with the fibroblast growth factor receptor to activate ERK has been reported and may have a significant role in the physiology of the striatum.[160] Thus, the ERK pathway appears to be an important player in the striatum, with different regulations in striatonigral and striatopallidal MSNs.

## X. Protein Phosphorylation and Signaling in L-DOPA-Induced Dyskinesia in PD

PD is due to the progressive death of DA neurons in the substantia nigra, resulting in a loss of DA in the dorsal striatum. The lack of basal levels of DA impairs signaling by D2R in striatopallidal neurons and contributes to an imbalance between the indirect and the direct pathway.[161] The predominance of the indirect pathway results in a general inhibition of movement initiation (akinesia) and rigidity. The most common treatment of PD is L-DOPA, the metabolic precursor of DA, which alleviates the symptoms very efficiently at the early stage of the disease. L-DOPA is prescribed in combination with inhibitors of DOPA decarboxylase, the enzyme that transforms L-DOPA into DA, that do not cross the blood–brain barrier to avoid side effects due to excess peripheral L-DOPA. Unfortunately, after several years of efficient treatment of PD with L-DOPA, the "honeymoon phase," treatment-induced side effects start to appear including abnormal involuntary movements (AIMs) or L-DOPA-induced dyskinesia (LID).[162] These can be related to a transient excess in DA function, with a predominance of the direct over the indirect pathway and clinical consequences schematically opposed to those of the lack of DA. Much effort has been devoted to understanding the mechanisms leading to the appearance of LID, mostly using rat and mouse models of PD. These animal models are based on the selective destruction of DA neurons with 6-OH-DA a selective toxin. Interestingly, the signaling mechanisms involved in the long-term effects of L-DOPA in the DA-denervated striatum appear to be in part similar to those taking place in response to drugs of abuse. This is not surprising, as both conditions involve a transient drug-induced excess of DA transmission. Of course, there are also many differences including a predominance of PD in the dorsal striatum, whereas drugs of abuse critically involve the ventral striatum or NAc. L-DOPA has no effect in the normal striatum but exerts a powerful effect in DA-denervated tissue. This hypersensitivity is in part due to increased D2R and increased coupling of D1R through increased levels of Gαolf.[163] In part as a consequence of increased D1R signaling, L-DOPA induces a strong phosphorylation of ERK, MSK1, and H3, which is restricted to D1R-expressing neurons of the direct pathway.[164,165] The ERK pathway appears to be important for LID, as the occurrence of AIMs is prevented by a drug that blocks the activation of ERK.[164] DARPP-32 phosphorylation is also increased and seems to be important for ERK activation, as the latter is attenuated in DARPP-32 knockout mice,[164] although other authors have observed a persistence of ERK phosphorylation in these mutant mice.[166] At any rate, the importance of the D1/DARPP-32 signaling is shown by the absence of L-DOPA-induced AIMs in D1R knockout mice[167] and their attenuation in DARPP-32 knockout mice.[164] Moreover, AIMs are prevented in mice bearing a conditional knockout of DARPP-32 only in D1R-expressing neurons.[118] The importance of the ERK pathway is supported by the genetic- or viral-induced reduction of Ras-GRF1 in the striatum (which activates the ERK pathway, see above) which also prevents LID in mice and monkeys treated with a different toxin (MPTP).[168]

A signaling pathway that may mediate the role of ERK in LID in 6-OH-DA-lesioned mice results from the activation of the mammalian target of rapamycin (mTOR) complex 1 (mTORC1).[169] This pathway is known to be involved in several forms of synaptic plasticity.[170] Treatment of 6-OH-DA-lesioned mice with L-DOPA increased phosphorylation of S6-kinase on Thr-389 and its substrates ribosomal protein S6 on Ser-235, 236, 240, and 244, selectively in D1R-expressing neurons in an ERK-dependent manner.[169] Activation of mTORC1 also resulted in phosphorylation of the eukaryotic translation initiation factor (eIF) 4E-binding protein (4E-BP) on Ser-65, and indirectly in phosphorylation of eIF4E on Ser-209. Importantly, blockade of mTORC1 with rapamycin prevented the occurrence of LID in L-DOPA-treated 6-OH-DA-treated mice.[169] Although the changes in phosphorylation of important factors for protein synthesis reported in this study are expected to promote 5′ cap-dependent initiation of mRNA translation, the nature of the protein(s) responsible for the clinical effects is not known. Thus, study of signaling pathways has provided important insights into the mechanisms of LID and suggests novel therapeutic approaches.

## XI. Functional Considerations and Conclusions

Signaling mechanisms involved in the action of DA and other neurotransmitters in striatal neurons are among the most extensively studied in neurons. Although they share many similarities with other neurons, MSNs are endowed with a set of specific signaling proteins, which are much less abundant or absent in most other cell types (e.g., Gαolf, Gγ7, AC5, PDE10A, DARPP-32, RCS, ARPP-16, etc.). These proteins are likely to provide specific signaling properties suited for the particular physiological characteristics of MSNs. Although all MSNs share many properties and intracellular signaling components, the two populations, forming the direct and indirect pathways, respectively, are distinguished by the expression levels of a number of genes.[171] Recent work taking advantage of novel genetic tools in the mouse has shown that these two populations have very distinct and contrasted responses, depending on their membrane receptors and anatomical connections (see refs16,17 for recent reviews). How can the specific striatal signaling components participate both in the various common features of MSNs and to the characteristic responses of each population? A common electrophysiological property of striatal MSNs is that they are relatively silent and that only a small number can be activated at any given time. This is in part due to the combination of ion channels they express. In neural network models various mechanisms of lateral inhibition also play an important role, allowing only a few "winner neuronss to take all" in response to activation of corticostriatal inputs[22]. At the level of signaling pathways it is striking that several DA and glutamate actions are antagonistic, for example the dephosphorylation by calcineurin of the PKA site on DARPP-32.[84,172] This antagonism may possibly contribute to maintain low levels of basal PKA signaling in both MSNs populations. In this context, it will be interesting to determine whether phosphorylation of DARPP-32 on Thr-75 by Cdk5, which inhibits PKA[86], switches off PKA signaling and puts some MSNs in a state more resistant to activation of the cAMP pathway. Once the stimulation of D1 receptors by the transient increase in DA, triggered for example by an unexpected reward, reaches a certain threshold, things are likely to change since several signaling components have a strong potential for synergism. Phosphorylation of RCS by PKA would prevent dephosphorylation of DARPP-32 by calcineurin[127] and facilitate its feed-forward amplifier role of PKA signaling. PKA activation and DARPP-32 mediated inhibition of PP1 would in turn reinforce glutamate signaling by increasing the function of NMDAR[173], AMPAR[112,174] and voltage-gated $Ca^{2+}$-channels[110] and by inhibiting GABAR[111]. The synergistic activation of the ERK pathway by DA and glutamate, including through interaction between D1 and NMDA receptors[140] and inhibition of the tyrosine phosphatase STEP[117], may then become a predominant feature of the response. The strong contribution of the ERK pathway in long-term events, including in behavioral responses [132,137,164,168,175] and corticostriatal LTP[175], underlines its importance in striatal learning mechanisms. Thus the existence of antagonistic and synergistic aspects of the signaling pathways activated by DA and glutamate may have relevance for both the relative resistance of MSNs to activation and to the sharp contrast of the responses in D1R- and D2R-expressing neurons. The combination of modeling studies and of live monitoring of signaling pathways using biosensors should help testing these hypotheses. The dissection intracellular signaling pathways in striatal neurons, is already clinically relevant in the context of the action of addictive drugs or in the adverse side effects of L-DOPA in the DA-depleted striatum. It allows better understanding the pathophysiological mechanisms and may help to find novel therapeutic approaches in these major medical conditions. It is likely that, in combination with other approaches, it will also shed light in the future on other neurological and psychiatric diseases that are known to involve the basal ganglia.

## References


1. Di Chiara G. Drug addiction as dopamine-dependent associative learning disorder. *Eur J Pharmacol* 1999;**375**:13–30.
2. Robbins TW, Ersche KD, Everitt BJ. Drug addiction and the memory systems of the brain. *Ann NY Acad Sci* 2008;**1141**:1–21.
3. Schultz W, Dickinson A. Neuronal coding of prediction errors. *Annu Rev Neurosci* 2000;**23**:473–500.
4. Wise RA. Addictive drugs and brain stimulation reward. *Annu Rev Neurosci* 1996;**19**:319–40.
5. Bromberg-Martin ES, Matsumoto M, Hikosaka O. Dopamine in motivational control: rewarding, aversive, and alerting. *Neuron* 2010;**68**:815–34.
6. Berke JD, Hyman SE. Addiction, dopamine, and the molecular mechanisms of memory. *Neuron* 2000;**25**:515–32.
7. Everitt BJ, Dickinson A, Robbins TW. The neuropsychological basis of addictive behaviour. *Brain Res Brain Res Rev* 2001;**36**:129–38.
8. Deroche-Gamonet V, Belin D, Piazza PV. Evidence for addiction-like behavior in the rat. *Science* 2004;**305**:1014–7.
9. Graybiel AM, Canales JJ, Capper-Loup C. Levodopa-induced dyskinesias and dopamine-dependent stereotypies: a new hypothesis. *Trends Neurosci* 2000;**23**:S71–7.
10. Belin D, Jonkman S, Dickinson A, Robbins TW, Everitt BJ. Parallel and interactive learning processes within the basal ganglia: relevance for the understanding of addiction. *Behav Brain Res* 2009;**199**:89–102.
11. Tepper JM, Bolam JP. Functional diversity and specificity of neostriatal interneurons. *Curr Opin Neurobiol* 2004;**14**:685–92.
12. Wilson CJ. Basal ganglia. In: Shepherd GM, editor. *The synaptic organization of the brain*. Oxford: Oxford University Press; 2004. pp. 361–414.
13. Kincaid AE, Zheng T, Wilson CJ. Connectivity and convergence of single corticostriatal axons. *J Neurosci* 1998;**18**:4722–31.

14. Freund TF, Powell JF, Smith AD. Tyrosine hydroxylase-immunoreactive boutons in synaptic contact with identified striatonigral neurons, with particular reference to dendritic spines. *Neuroscience* 1984;**13**:1189–215.
15. Gerfen CR, Engber TM, Mahan LC, Susel Z, Chase TN, Monsma FJ, Jr., et al. $D_1$ and $D_2$ dopamine receptor-regulated gene expression of striatonigral and striatopallidal neurons. *Science* 1990;**250**:1429–32.
16. Bertran-Gonzalez J, Herve D, Girault JA, Valjent E. What is the degree of segregation between striatonigral and striatopallidal projections? *Front Neuroanat* 2010;**4**:136.
17. Valjent E, Bertran-Gonzalez J, Herve D, Fisone G, Girault JA. Looking BAC at striatal signaling: cell-specific analysis in new transgenic mice. *Trends Neurosci* 2009;**32**:538–47.
18. Reynolds JNJ, Wickens JR. Dopamine-dependent plasticity of corticostriatal synapses. *Neural Netw* 2002;**15**:507–21.
19. Chevalier G, Deniau JM. Disinhibition as a basic process in the expression of striatal functions. *Trends Neurosci* 1990;**13**:277–80.
20. Grillner S, Hellgren J, Menard A, Saitoh K, Wikstrom MA. Mechanisms for selection of basic motor programs—roles for the striatum and pallidum. *Trends Neurosci* 2005;**28**:364–70.
21. Mink JW. The basal ganglia: focused selection and inhibition of competing motor programs. *Prog Neurobiol* 1996;**50**:381–425.
22. Redgrave P, Prescott TJ, Gurney K. The basal ganglia: a vertebrate solution to the selection problem? *Neuroscience* 1999;**89**:1009–23.
23. Calabresi P, Picconi B, Tozzi A, Di Filippo M. Dopamine-mediated regulation of corticostriatal synaptic plasticity. *Trends Neurosci* 2007;**30**:211–9.
24. Kauer JA, Malenka RC. Synaptic plasticity and addiction. *Nat Rev Neurosci* 2007;**8**:844–58.
25. Kreitzer AC, Malenka RC. Striatal plasticity and basal ganglia circuit function. *Neuron* 2008;**60**:543–54.
26. Surmeier DJ, Ding J, Day M, Wang Z, Shen W. D1 and D2 dopamine-receptor modulation of striatal glutamatergic signaling in striatal medium spiny neurons. *Trends Neurosci* 2007;**30**:228–35.
27. Yuste R, Bonhoeffer T. Genesis of dendritic spines: insights from ultrastructural and imaging studies. *Nat Rev Neurosci* 2004;**5**:24–34.
28. Ingham CA, Hood SH, Taggart P, Arbuthnott GW. Plasticity of synapses in the rat neostriatum after unilateral lesion of the nigrostriatal dopaminergic pathway. *J Neurosci* 1998;**18**:4732–43.
29. Lee KW, Kim Y, Kim AM, Helmin K, Nairn AC, Greengard P. Cocaine-induced dendritic spine formation in D1 and D2 dopamine receptor-containing medium spiny neurons in nucleus accumbens. *Proc Natl Acad Sci USA* 2006;**103**:3399–404.
30. Robinson TE, Gorny G, Mitton E, Kolb B. Cocaine self-administration alters the morphology of dendrites and dendritic spines in the nucleus accumbens and neocortex. *Synapse* 2001;**39**:257–66.
31. Durieux PF, Bearzatto B, Guiducci S, Buch T, Waisman A, Zoli M, et al. D2R striatopallidal neurons inhibit both locomotor and drug reward processes. *Nat Neurosci* 2009;**12**:393–5.
32. Kravitz AV, Freeze BS, Parker PR, Kay K, Thwin MT, Deisseroth K, et al. Regulation of parkinsonian motor behaviours by optogenetic control of basal ganglia circuitry. *Nature* 2010;**466**:622–6.
33. Greengard P. The neurobiology of slow synaptic transmission. *Science* 2001;**294**:1024–30.
34. Nestler EJ, Greengard P. Protein phosphorylation in the brain. *Nature* 1983;**305**:583–8.
35. Kebabian JW, Greengard P. Dopamine-sensitive adenyl cyclase: possible role in synaptic transmission. *Science* 1971;**174**:1346–9.
36. Krueger BK, Forn J, Greengard P. Dopamine-sensitive adenylate cyclase and protein phosphorylation in the rat caudate nucleus. *Psychopharmacol Bull* 1975;**11**:10–1.

37. De Camilli P, Macconi D, Spada A. Dopamine inhibits adenylate cyclase in human prolactin-secreting pituitary adenomas. *Nature* 1979;**278**:252–4.
38. Beaulieu JM, Gainetdinov RR. The physiology, signaling, and pharmacology of dopamine receptors. *Pharmacol Rev* 2011;**63**:182–217.
39. Sokoloff P, Giros B, Martres M-P, Bouthenet M-L, Schwartz J-C. Molecular cloning and characterization of a novel dopamine receptor ($D_3$) as a target for neuroleptics. *Nature* 1990;**347**:146–51.
40. Bertran-Gonzalez J, Bosch C, Maroteaux M, Matamales M, Herve D, Valjent E, et al. Opposing patterns of signaling activation in dopamine D1 and D2 receptor-expressing striatal neurons in response to cocaine and haloperidol. *J Neurosci* 2008;**28**:5671–85.
41. Herve D, Levi-Strauss M, Marey-Semper I, Verney C, Tassin JP, Glowinski J, et al. G(olf) and Gs in rat basal ganglia: possible involvement of G(olf) in the coupling of dopamine D1 receptor with adenylyl cyclase. *J Neurosci* 1993;**13**:2237–48.
42. Schwindinger WF, Mihalcik LJ, Giger KE, Betz KS, Stauffer AM, Linden J, et al. Adenosine A2A receptor signaling and golf assembly show a specific requirement for the gamma7 subtype in the striatum. *J Biol Chem* 2010;**285**:29787–96.
43. Corvol JC, Studler JM, Schonn JS, Girault JA, Herve D. Galpha(olf) is necessary for coupling D1 and A2a receptors to adenylyl cyclase in the striatum. *J Neurochem* 2001;**76**:1585–8.
44. Herve D, Le Moine C, Corvol JC, Belluscio L, Ledent C, Fienberg AA, et al. Galpha(olf) levels are regulated by receptor usage and control dopamine and adenosine action in the striatum. *J Neurosci* 2001;**21**:4390–9.
45. Schwindinger WF, Betz KS, Giger KE, Sabol A, Bronson SK, Robishaw JD. Loss of G protein gamma 7 alters behavior and reduces striatal alpha(olf) level and cAMP production. *J Biol Chem* 2003;**278**:6575–9.
46. Zhuang X, Belluscio L, Hen R. GOLF alpha mediates dopamine D1 receptor signaling. *J Neurosci* 2000;**20**:RC91.
47. Kheirbek MA, Britt JP, Beeler JA, Ishikawa Y, McGehee DS, Zhuang X. Adenylyl cyclase type 5 contributes to corticostriatal plasticity and striatum-dependent learning. *J Neurosci* 2009;**29**:12115–24.
48. Fujishige K, Kotera J, Michibata H, Yuasa K, Takebayashi S, Okumura K, et al. Cloning and characterization of a novel human phosphodiesterase that hydrolyzes both cAMP and cGMP (PDE10A). *J Biol Chem* 1999;**274**:18438–45.
49. Polli JW, Kincaid RL. Expression of a calmodulin-dependent phosphodiesterase isoform (PDE1B1) correlates with brain regions having extensive dopaminergic innervation. *J Neurosci* 1994;**14**:1251–61.
50. Nishi A, Kuroiwa M, Miller DB, O'Callaghan JP, Bateup HS, Shuto T, et al. Distinct roles of PDE4 and PDE10A in the regulation of cAMP/PKA signaling in the striatum. *J Neurosci* 2008;**28**:10460–71.
51. Reed TM, Repaske DR, Snyder GL, Greengard P, Vorhees CV. Phosphodiesterase 1B knock-out mice exhibit exaggerated locomotor hyperactivity and DARPP-32 phosphorylation in response to dopamine agonists and display impaired spatial learning. *J Neurosci* 2002;**22**:5188–97.
52. Beaulieu JM, Sotnikova TD, Marion S, Lefkowitz RJ, Gainetdinov RR, Caron MG. An Akt/beta-arrestin 2/PP2A signaling complex mediates dopaminergic neurotransmission and behavior. *Cell* 2005;**122**:261–73.
53. Hasbi A, Fan T, Alijaniaram M, Nguyen T, Perreault ML, O'Dowd BF, et al. Calcium signaling cascade links dopamine D1-D2 receptor heteromer to striatal BDNF production and neuronal growth. *Proc Natl Acad Sci USA* 2009;**106**:21377–82.
54. Paoletti P. Molecular basis of NMDA receptor functional diversity. *Eur J Neurosci* 2011;**33**:1351–65.

55. Traynelis SF, Wollmuth LP, McBain CJ, Menniti FS, Vance KM, Ogden KK, et al. Glutamate receptor ion channels: structure, regulation, and function. *Pharmacol Rev* 2010;**62**:405–96.
56. Seeburg PH, Hartner J. Regulation of ion channel/neurotransmitter receptor function by RNA editing. *Curr Opin Neurobiol* 2003;**13**:279–83.
57. Sobolevsky AI, Rosconi MP, Gouaux E. X-ray structure, symmetry and mechanism of an AMPA-subtype glutamate receptor. *Nature* 2009;**462**:745–56.
58. Johnson JW, Ascher P. Glycine potentiates the NMDA response in cultured mouse brain neurons. *Nature* 1987;**325**:529–31.
59. Nowak L, Bregestovski P, Ascher P, Herbet A, Prochiantz A. Magnesium gates glutamate-activated channels in mouse central neurones. *Nature* 1984;**307**:462–5.
60. Hebb DO. *The organization of behavior.* New York: Wiley & Sons; 1949.
61. Kessels HW, Malinow R. Synaptic AMPA receptor plasticity and behavior. *Neuron* 2009;**61**:340–50.
62. Luscher C, Malenka RC. Drug-evoked synaptic plasticity in addiction: from molecular changes to circuit remodeling. *Neuron* 2011;**69**:650–63.
63. Freund TF, Katona I, Piomelli D. Role of endogenous cannabinoids in synaptic signaling. *Physiol Rev* 2003;**83**:1017–66.
64. Holtmaat A, Svoboda K. Experience-dependent structural synaptic plasticity in the mammalian brain. *Nat Rev Neurosci* 2009;**10**:647–58.
65. Kasai H, Fukuda M, Watanabe S, Hayashi-Takagi A, Noguchi J. Structural dynamics of dendritic spines in memory and cognition. *Trends Neurosci* 2010;**33**:121–9.
66. Taylor SS, Kim C, Cheng CY, Brown SH, Wu J, Kannan N. Signaling through cAMP and cAMP-dependent protein kinase: diverse strategies for drug design. *Biochim Biophys Acta* 2008;**1784**:16–26.
67. Logue JS, Scott JD. Organizing signal transduction through A-kinase anchoring proteins (AKAPs). *FEBS J* 2010;**277**:4370–5.
68. Gloerich M, Bos JL. Epac: defining a new mechanism for cAMP action. *Annu Rev Pharmacol Toxicol* 2010;**50**:355–75.
69. Walaas SI, Nairn AC, Greengard P. Regional distribution of calcium- and cyclic adenosine 3′:5′-monophosphate-regulated protein phosphorylation systems in mammalian brain. I. Particulate systems. *J Neurosci* 1983;**3**:291–301.
70. Walaas SI, Nairn AC, Greengard P. Regional distribution of calcium- and cyclic adenosine 3′:5′-monophosphate-regulated protein phosphorylation systems in mammalian brain. II. Soluble systems. *J Neurosci* 1983;**3**:302–11.
71. Walaas SI, Aswad DW, Greengard P. A dopamine- and cyclic AMP-regulated phosphoprotein enriched in dopamine-innervated brain regions. *Nature* 1983;**301**:69–71.
72. Girault JA, Horiuchi A, Gustafson EL, Rosen NL, Greengard P. Differential expression of ARPP-16 and ARPP-19, two highly related cAMP-regulated phosphoproteins, one of which is specifically associated with dopamine-innervated brain regions. *J Neurosci* 1990;**10**:1124–33.
73. Horiuchi A, Williams KR, Kurihara T, Nairn AC, Greengard P. Purification and cDNA cloning of ARPP-16, a cAMP-regulated phosphoprotein enriched in basal ganglia, and of a related phosphoprotein, ARPP-19. *J Biol Chem* 1990;**265**:9476–84.
74. Hemmings HC, Jr., Greengard P. ARPP-21, a cAMP-regulated phosphoprotein Mr=21,000 enriched in dopamine-innervated brain regions. I. Purification and characterization of the protein from bovine caudate nucleus. *J Neurosci* 1989;**9**:851–64.
75. Ouimet CC, Hemmings HC, Jr., Greengard P. ARPP-21, a cyclic AMP-regulated phosphoprotein enriched in dopamine-innervated brain regions. II. Immunocytochemical localization in rat brain. *J Neurosci* 1989;**9**:865–75.

76. Williams KR, Hemmings HC, Jr., LoPresti MB, Konigsberg WH, Greengard P. DARPP-32, a dopamine- and cyclic AMP-regulated neuronal phosphoprotein. Primary structure and homology with protein phosphatase inhibitor-1. *J Biol Chem* 1986;**261**:1890–903.
77. Yger M, Girault JA. DARPP-32, Jack of all trades... master of which? *Front Behav Neurosci* 2011;**5**:56.
78. Egloff MP, Johnson DF, Moorhead G, Cohen PTW, Cohen P, Barford D. Structural basis for the recognition of regulatory subunits by the catalytic subunit of protein phosphatase 1. *EMBO J* 1997;**16**:1876–87.
79. Bollen M, Peti W, Ragusa MJ, Beullens M. The extended PP1 toolkit: designed to create specificity. *Trends Biochem Sci* 2010;**35**:450–8.
80. Hemmings HC, Jr., Nairn AC, Greengard P. DARPP-32, a dopamine- and adenosine 3′:5′-monophosphate-regulated neuronal phosphoprotein. II. Comparison of the kinetics of phosphorylation of DARPP-32 and phosphatase inhibitor 1. *J Biol Chem* 1984;**259**:14491–7.
81. Hemmings HC, Jr., Greengard P, Tung HYL, Cohen P. DARPP-32, a dopamine-regulated neuronal phosphoprotein, is a potent inhibitor of protein phosphatase-1. *Nature* 1984;**310**:503–5.
82. Kwon YG, Huang HB, Desdouits F, Girault JA, Greengard P, Nairn AC. Characterization of the interaction between DARPP-32 and protein phosphatase 1 (PP-1): DARPP-32 peptides antagonize the interaction of PP-1 with binding proteins. *Proc Natl Acad Sci USA* 1997;**94**:3536–41.
83. King MM, Huang CY, Chock PB, Nairn AC, Hemmings HC, Jr., Chan KF, et al. Mammalian brain phosphoproteins as substrates for calcineurin. *J Biol Chem* 1984;**259**:8080–3.
84. Halpain S, Girault JA, Greengard P. Activation of NMDA receptors induces dephosphorylation of DARPP-32 in rat striatal slices. *Nature* 1990;**343**:369–72.
85. Nishi A, Snyder GL, Nairn AC, Greengard P. Role of calcineurin and protein phosphatase-2A in the regulation of DARPP-32 dephosphorylation in neostriatal neurons. *J Neurochem* 1999;**72**:2015–21.
86. Bibb JA, Snyder GL, Nishi A, Yan Z, Meijer L, Fienberg AA, et al. Phosphorylation of DARPP-32 by Cdk5 modulates dopamine signalling in neurons. *Nature* 1999;**402**:669–71.
87. Girault JA, Hemmings HC, Jr., Williams KR, Nairn AC, Greengard P. Phosphorylation of DARPP-32, a dopamine- and cAMP-regulated phosphoprotein, by casein kinase II. *J Biol Chem* 1989;**264**:21748–59.
88. Desdouits F, Cohen D, Nairn AC, Greengard P, Girault JA. Phosphorylation of DARPP-32, a dopamine- and cAMP-regulated phosphoprotein, by casein kinase I *in vitro* and *in vivo*. *J Biol Chem* 1995;**270**:8772–8.
89. Desdouits F, Siciliano JC, Greengard P, Girault JA. Dopamine- and cAMP-regulated phosphoprotein DARPP-32: phosphorylation of Ser-137 by casein kinase I inhibits dephosphorylation of Thr-34 by calcineurin. *Proc Natl Acad Sci USA* 1995;**92**:2682–5.
90. Stipanovich A, Valjent E, Matamales M, Nishi A, Ahn JH, Maroteaux M, et al. A phosphatase cascade by which rewarding stimuli control nucleosomal response. *Nature* 2008;**453**:879–84.
91. Janssens V, Longin S, Goris J. PP2A holoenzyme assembly: in cauda venenum (the sting is in the tail). *Trends Biochem Sci* 2008;**33**:113–21.
92. Ahn JH, McAvoy T, Rakhilin SV, Nishi A, Greengard P, Nairn AC. Protein kinase A activates protein phosphatase 2A by phosphorylation of the B56delta subunit. *Proc Natl Acad Sci USA* 2007;**104**:2979.
93. Ahn JH, Sung JY, McAvoy T, Nishi A, Janssens V, Goris J, et al. The B″/PR72 subunit mediates Ca2+-dependent dephosphorylation of DARPP-32 by protein phosphatase 2A. *Proc Natl Acad Sci USA* 2007;**104**:9876–81.

94. Desdouits F, Siciliano JC, Nairn AC, Greengard P, Girault JA. Dephosphorylation of Ser-137 in DARPP-32 by protein phosphatases 2A and 2C: different roles *in vitro* and in striatonigral neurons. *Biochem J* 1998;**330**:211.
95. Qi Z, Miller GW, Voit EO. The internal state of medium spiny neurons varies in response to different input signals. *BMC Syst Biol* 2010;**4**:26.
96. Nakano T, Doi T, Yoshimoto J, Doya K. A kinetic model of dopamine- and calcium-dependent striatal synaptic plasticity. *PLoS Comput Biol* 2010;**6**:e1000670.
97. Lindskog M, Kim M, Wikström MA, Blackwell KT, Kotaleski JH. Transient calcium and dopamine increase PKA activity and DARPP-32 phosphorylation. *PLoS Comput Biol* 2006;**2**: e119.
98. Le Novere N, Li L, Girault JA. DARPP-32: molecular integration of phosphorylation potential. *Cell Mol Life Sci* 2008;**65**:2125–7.
99. Fernandez E, Schiappa R, Girault JA, Novère NL. DARPP-32 is a robust integrator of dopamine and glutamate signals. *PLoS Comput Biol* 2006;**2**:e176.
100. Barbano PE, Spivak M, Flajolet M, Nairn AC, Greengard P, Greengard L. A mathematical tool for exploring the dynamics of biological networks. *Proc Natl Acad Sci USA* 2007;**104**:19169–74.
101. Svenningsson P, Nishi A, Fisone G, Girault J-A, Nairn AC, Greengard P. DARPP-32: an integrator of neurotransmission. *Annu Rev Pharmacol Toxicol* 2004;**44**:269–96.
102. Greengard P, Allen PB, Nairn AC. Beyond the dopamine receptor: the DARPP-32/protein phosphatase-1 cascade. *Neuron* 1999;**23**:435–47.
103. Girault JA, Greengard P. The neurobiology of dopamine signaling. *Arch Neurol* 2004;**61**:641–4.
104. Bateup HS, Svenningsson P, Kuroiwa M, Gong S, Nishi A, Heintz N, et al. Cell type-specific regulation of DARPP-32 phosphorylation by psychostimulant and antipsychotic drugs. *Nat Neurosci* 2008;**11**:932–9.
105. Schultz W. Getting formal with dopamine and reward. *Neuron* 2002;**36**:241–63.
106. Grace AA. Phasic versus tonic dopamine release and the modulation of dopamine system responsivity: a hypothesis for the etiology of schizophrenia. *Neuroscience* 1991;**41**:1–24.
107. Svenningsson P, Tzavara E, Carruthers R, Rachleff I, Wattler S, Nehls M, et al. Diverse psychotomimetics act through a common signaling pathway. *Science* 2003;**302**:1412–5.
108. Fienberg AA, Hiroi N, Mermelstein PG, Song WJ, Snyder GL, Nishi A, et al. DARPP-32: regulator of the efficacy of dopaminergic neurotransmission. *Science* 1998;**281**:838–9.
109. Schiffmann SN, Desdouits F, Menu R, Greengard P, Vincent JD, Vanderhaeghen JJ, et al. Modulation of the voltage-gated sodium current in rat striatal neurons by DARPP-32, an inhibitor of protein phosphatase. *Eur J Neurosci* 1998;**10**:1312–20.
110. Surmeier DJ, Bargas J, Hemmings HC, Jr., Nairn AC, Greengard P. Modulation of calcium currents by a $D_1$ dopaminergic protein kinase/phosphatase cascade in rat neostriatal neurons. *Neuron* 1995;**14**:385–97.
111. Flores-Hernandez J, Hernandez S, Snyder GL, Yan Z, Fienberg AA, Moss SJ, et al. D(1) dopamine receptor activation reduces GABA(A) receptor currents in neostriatal neurons through a PKA/DARPP-32/PP1 signaling cascade. *J Neurophysiol* 2000;**83**:2996–3004.
112. Snyder GL, Allen PB, Fienberg AA, Valle CG, Huganir RL, Nairn AC, et al. Regulation of phosphorylation of the GluR1 AMPA receptor in the neostriatum by dopamine and psychostimulants *in vivo*. *J Neurosci* 2000;**20**:4480–8.
113. Calabresi P, Gubellini P, Centonze D, Picconi B, Bernardi G, Chergui K, et al. Dopamine and cAMP-regulated phosphoprotein 32 kDa controls both striatal long-term depression and long-term potentiation, opposing forms of synaptic plasticity. *J Neurosci* 2000;**20**:8443–51.

114. Zhang Y, Svenningsson P, Picetti R, Schlussman SD, Nairn AC, Ho A, et al. Cocaine self-administration in mice is inversely related to phosphorylation at Thr34 (protein kinase A site) and Ser130 (kinase CK1 site) of DARPP-32. *J Neurosci* 2006;**26**:2645–51.
115. Zachariou V, Sgambato-Faure V, Sasaki T, Svenningsson P, Berton O, Fienberg AA, et al. Phosphorylation of DARPP-32 at Threonine-34 is required for cocaine action. *Neuropsychopharmacology* 2006;**31**:555–62.
116. Zachariou V, Benoit-Marand M, Allen PB, Ingrassia P, Fienberg AA, Gonon F, et al. Reduction of cocaine place preference in mice lacking the protein phosphatase 1 inhibitors DARPP 32 or Inhibitor 1. *Biol Psychiatry* 2002;**51**:612–20.
117. Valjent E, Pascoli V, Svenningsson P, Paul S, Enslen H, Corvol JC, et al. Regulation of a protein phosphatase cascade allows convergent dopamine and glutamate signals to activate ERK in the striatum. *Proc Natl Acad Sci USA* 2005;**102**:491–6.
118. Bateup HS, Santini E, Shen W, Birnbaum S, Valjent E, Surmeier DJ, et al. Distinct subclasses of medium spiny neurons differentially regulate striatal motor behaviors. *Proc Natl Acad Sci USA* 2010;**107**:14845–50.
119. Mani SK, Fienberg AA, O'Callaghan JP, Snyder GL, Allen PB, Dash PK, et al. Requirement for DARPP-32 in progesterone-facilitated sexual receptivity in female rats and mice. *Science* 2000;**287**:1053–6.
120. Kolata S, Light K, Wass CD, Colas-Zelin D, Roy D, Matzel LD. A dopaminergic gene cluster in the prefrontal cortex predicts performance indicative of general intelligence in genetically heterogeneous mice. *PLoS One* 2010;**5**:e14036.
121. Meyer-Lindenberg A, Straub RE, Lipska BK, Verchinski BA, Goldberg T, Callicott JH, et al. Genetic evidence implicating DARPP-32 in human frontostriatal structure, function, and cognition. *J Clin Invest* 2007;**117**:672.
122. El-Rifai W, Smith MF, Jr., Li G, Beckler A, Carl VS, Montgomery E, et al. Gastric cancers overexpress DARPP-32 and a novel isoform, t-DARPP. *Cancer Res* 2002;**62**:4061–4.
123. Girault JA, Shalaby IA, Rosen NL, Greengard P. Regulation by cAMP and vasoactive intestinal peptide of phosphorylation of specific proteins in striatal cells in culture. *Proc Natl Acad Sci USA* 1988;**85**:7790–4.
124. Tsou K, Girault JA, Greengard P. Dopamine $D_1$ agonist SKF 38393 increases the state of phosphorylation of ARPP-21 in substantia nigra. *J Neurochem* 1993;**60**:1043–6.
125. Dulubova I, Horiuchi A, Snyder GL, Girault JA, Czernik AJ, Shao L, et al. ARPP-16/ARPP-19: a highly conserved family of cAMP-regulated phosphoproteins. *J Neurochem* 2001;**77**:229–38.
126. Walaas SI, Hemmings Jr. HC, Greengard P, Nairn AC. Beyond the dopamine receptor: regulation and roles of serine/threonine protein phosphatases. *Front Neuroanat.* 2011;**5**:50.
127. Rakhilin SV, Olson PA, Nishi A, Starkova NN, Fienberg AA, Nairn AC, et al. A network of control mediated by regulator of calcium/calmodulin-dependent signaling. *Science* 2004;**306**:698–701.
128. Mochida S, Maslen SL, Skehel M, Hunt T. Greatwall phosphorylates an inhibitor of protein phosphatase 2A that is essential for mitosis. *Science* 2010;**330**:1670–3.
129. Gharbi-Ayachi A, Labbe JC, Burgess A, Vigneron S, Strub JM, Brioudes E, et al. The substrate of Greatwall kinase, Arpp 19, controls mitosis by inhibiting protein phosphatase 2A. *Science* 2010;**330**:1673–7.
130. Girault JA, Valjent E, Caboche J, Herve D. ERK2: a logical AND gate critical for drug-induced plasticity? *Curr Opin Pharmacol* 2007;**7**:77–85.
131. Thomas GM, Huganir RL. MAPK cascade signalling and synaptic plasticity. *Nat Rev Neurosci* 2004;**5**:173–83.
132. Valjent E, Corvol JC, Pages C, Besson MJ, Maldonado R, Caboche J. Involvement of the extracellular signal-regulated kinase cascade for cocaine-rewarding properties. *J Neurosci* 2000;**20**:8701–9.

133. Valjent E, Pages C, Rogard M, Besson MJ, Maldonado R, Caboche J. Delta 9-tetrahydrocannabinol-induced MAPK/ERK and Elk-1 activation *in vivo* depends on dopaminergic transmission. *Eur J Neurosci* 2001;**14**:342–52.
134. Valjent E, Pages C, Herve D, Girault JA, Caboche J. Addictive and non-addictive drugs induce distinct and specific patterns of ERK activation in mouse brain. *Eur J Neurosci* 2004;**19**:1826–36.
135. Ibba F, Vinci S, Spiga S, Peana AT, Assaretti AR, Spina L, et al. Ethanol-induced extracellular signal regulated kinase: role of dopamine D1 receptors. *Alcohol Clin Exp Res* 2009;**33**:858–67.
136. Salzmann J, Marie-Claire C, Le Guen S, Roques BP, Noble F. Importance of ERK activation in behavioral and biochemical effects induced by MDMA in mice. *Br J Pharmacol* 2003;**140**:831–8.
137. Valjent E, Corvol JC, Trzaskos JM, Girault JA, Herve D. Role of the ERK pathway in psychostimulant-induced locomotor sensitization. *BMC Neurosci* 2006;**7**:20.
138. Valjent E, Corbille AG, Bertran-Gonzalez J, Herve D, Girault JA. Inhibition of ERK pathway or protein synthesis during reexposure to drugs of abuse erases previously learned place preference. *Proc Natl Acad Sci USA* 2006;**103**:2932–7.
139. Miller CA, Marshall JF. Molecular substrates for retrieval and reconsolidation of cocaine-associated contextual memory. *Neuron* 2005;**47**:873–84.
140. Pascoli V, Besnard A, Herve D, Pages C, Heck N, Girault JA, et al. Cyclic adenosine monophosphate-independent tyrosine phosphorylation of NR2B mediates cocaine-induced extracellular signal-regulated kinase activation. *Biol Psychiatry* 2011;**69**:218–27.
141. Sgambato V, Pagès C, Rogard M, Besson MJ, Caboche J. Extracellular signal-regulated kinase (ERK) controls immediate early gene induction on corticostriatal stimulation. *J Neurosci* 1998;**18**:8814–25.
142. Scott L, Zelenin S, Malmersjo S, Kowalewski JM, Markus EZ, Nairn AC, et al. Allosteric changes of the NMDA receptor trap diffusible dopamine 1 receptors in spines. *Proc Natl Acad Sci USA* 2006;**103**:762–7.
143. Lee FJS, Xue S, Pei L, Vukusic B, Chery N, Wang Y, et al. Dual regulation of NMDA receptor functions by direct protein-protein interactions with the dopamine D1 receptor. *Cell* 2002;**111**:219–30.
144. Hallett PJ, Spoelgen R, Hyman BT, Standaert DG, Dunah AW. Dopamine D1 activation potentiates striatal NMDA receptors by tyrosine phosphorylation-dependent subunit trafficking. *J Neurosci* 2006;**26**:4690–700.
145. Dunah AW, Sirianni AC, Fienberg AA, Bastia E, Schwarzschild MA, Standaert DG. Dopamine D1-dependent trafficking of striatal N-methyl-D-aspartate glutamate receptors requires Fyn protein tyrosine kinase but not DARPP-32. *Mol Pharmacol* 2004;**65**:121–9.
146. Fiorentini C, Mattanza C, Collo G, Savoia P, Spano P, Missale C. The tyrosine phosphatase Shp-2 interacts with the dopamine D(1) receptor and triggers D(1) -mediated Erk signaling in striatal neurons. *J Neurochem* 2011;**117**:253–63.
147. Ali DW, Salter MW. NMDA receptor regulation by Src kinase signalling in excitatory synaptic transmission and plasticity. *Curr Opin Neurobiol* 2001;**11**:336–42.
148. Snyder GL, Fienberg AA, Huganir RL, Greengard P. A dopamine D1 receptor protein kinase A dopamine- and cAMP-regulated phosphoprotein ($M_r$ 32 kDa) protein phosphatase-1 pathway regulates dephosphorylation of the NMDA receptor. *J Neurosci* 1998;**18**:10297–303.
149. Krapivinsky G, Krapivinsky L, Manasian Y, Ivanov A, Tyzio R, Pellegrino C, et al. The NMDA receptor is coupled to the ERK pathway by a direct interaction between NR2B and RasGRF1. *Neuron* 2003;**40**:775–84.

150. Fasano S, D'Antoni A, Orban PC, Valjent E, Putignano E, Vara H, et al. Ras-guanine nucleotide-releasing factor 1 (Ras-GRF1) controls activation of extracellular signal-regulated kinase (ERK) signaling in the striatum and long-term behavioral responses to cocaine. *Biol Psychiatry* 2009;**66**:758–68.
151. Toki S, Kawasaki H, Tashiro N, Housman DE, Graybiel AM. Guanine nucleotide exchange factors CalDAG-GEFI and CalDAG-GEFII are colocalized in striatal projection neurons. *J Comp Neurol.* 2001;**437**:398–407.
152. Mazzucchelli C, Vantaggiato C, Ciamei A, Fasano S, Pakhotin P, Krezel W, et al. Knockout of ERK1 MAP kinase enhances synaptic plasticity in the striatum and facilitates striatal-mediated learning and memory. *Neuron* 2002;**34**:807–20.
153. Sweatt JD. Mitogen-activated protein kinases in synaptic plasticity and memory. *Curr Opin Neurobiol* 2004;**14**:311–7.
154. Vanhoutte P, Barnier JV, Guibert B, Pagès C, Besson MJ, Hipskind RA, et al. Glutamate induces phosphorylation of Elk-1 and CREB, along with c-fos activation, via an extracellular signal-regulated kinase-dependent pathway in brain slices. *Mol Cell Biol* 1999;**19**:136–46.
155. Trifilieff P, Lavaur J, Pascoli V, Kappes V, Brami-Cherrier K, Pages C, et al. Endocytosis controls glutamate-induced nuclear accumulation of ERK. *Mol Cell Neurosci* 2009;**41**:325–36.
156. Lavaur J, Bernard F, Trifilieff P, Pascoli V, Kappes V, Pagès C, et al. A TAT-DEF-Elk-1 peptide regulates the cytonuclear trafficking of Elk-1 and controls cytoskeleton dynamics. *J Neurosci* 2007;**27**:14448–58.
157. Brami-Cherrier K, Valjent E, Herve D, Darragh J, Corvol JC, Pages C, et al. Parsing molecular and behavioral effects of cocaine in mitogen- and stress-activated protein kinase-1-deficient mice. *J Neurosci* 2005;**25**:11444–54.
158. Valjent E, Aubier B, Corbille AG, Brami-Cherrier K, Caboche J, Topilko P, et al. Plasticity-associated gene Krox24/Zif268 is required for long-lasting behavioral effects of cocaine. *J Neurosci* 2006;**26**:4956–60.
159. Bertran-Gonzalez J, Hakansson K, Borgkvist A, Irinopoulou T, Brami-Cherrier K, Usiello A, et al. Histone H3 phosphorylation is under the opposite tonic control of dopamine D2 and adenosine A2A receptors in striatopallidal neurons. *Neuropsychopharmacology* 2009;**34**:1710–20.
160. Flajolet M, Wang Z, Futter M, Shen W, Nuangchamnong N, Bendor J, et al. FGF acts as a co-transmitter through adenosine A(2A) receptor to regulate synaptic plasticity. *Nat Neurosci* 2008;**11**:1402–9.
161. Alexander GE, Crutcher MD. Functional architecture of basal ganglia circuits: neural substrates of parallel processing. *Trends Neurosci* 1990;**13**:266–71.
162. Cenci MA. Dopamine dysregulation of movement control in L-DOPA-induced dyskinesia. *Trends Neurosci* 2007;**30**:236–43.
163. Corvol JC, Muriel MP, Valjent E, Feger J, Hanoun N, Girault JA, et al. Persistent increase in olfactory type G-protein alpha subunit levels may underlie D1 receptor functional hypersensitivity in Parkinson disease. *J Neurosci* 2004;**24**:7007–14.
164. Santini E, Valjent E, Usiello A, Carta M, Borgkvist A, Girault JA, et al. Critical involvement of cAMP/DARPP-32 and extracellular signal-regulated protein kinase signaling in L-DOPA-induced dyskinesia. *J Neurosci* 2007;**27**:6995–7005.
165. Santini E, Alcacer C, Cacciatore S, Heiman M, Herve D, Greengard P, et al. L-DOPA activates ERK signaling and phosphorylates histone H3 in the striatonigral medium spiny neurons of hemiparkinsonian mice. *J Neurochem* 2009;**108**:621–33.
166. Gerfen CR, Paletzki R, Worley P. Differences between dorsal and ventral striatum in Drd1a dopamine receptor coupling of dopamine- and cAMP-regulated phosphoprotein-32 to activation of extracellular signal-regulated kinase. *J Neurosci* 2008;**28**:7113–20.

167. Darmopil S, Martin AB, De Diego IR, Ares S, Moratalla R. Genetic inactivation of dopamine D1 but not D2 receptors inhibits L-DOPA-induced dyskinesia and histone activation. *Biol Psychiatry* 2009;**66**:603–13.
168. Fasano S, Bezard E, D'Antoni A, Francardo V, Indrigo M, Qin L, et al. Inhibition of Ras-guanine nucleotide-releasing factor 1 (Ras-GRF1) signaling in the striatum reverts motor symptoms associated with L-dopa-induced dyskinesia. *Proc Natl Acad Sci USA* 2011;**107**:21824–9.
169. Santini E, Heiman M, Greengard P, Valjent E, Fisone G. Inhibition of mTOR signaling in Parkinson's disease prevents L-DOPA-induced dyskinesia. *Sci Signal* 2009;**2**:ra36.
170. Hoeffer CA, Klann E. mTOR signaling: at the crossroads of plasticity, memory and disease. *Trends Neurosci.* 2010;**33**:67–75.
171. Heiman M, Schaefer A, Gong S, Peterson JD, Day M, Ramsey KE, et al. A translational profiling approach for the molecular characterization of CNS cell types. *Cell.* 2008;**135**:738–48.
172. Nishi A, Bibb JA, Matsuyama S, Hamada M, Higashi H, Nairn AC, et al. Regulation of DARPP-32 dephosphorylation at PKA- and Cdk5-sites by NMDA and AMPA receptors: distinct roles of calcineurin and protein phosphatase-2A. *J Neurochem.* 2002;**81**:832–41.
173. Flores-Hernandez J, Cepeda C, Hernandez-Echeagaray E, Calvert CR, Jokel ES, Fienberg AA, et al. Dopamine enhancement of NMDA currents in dissociated medium-sized striatal neurons: role of D1 receptors and DARPP-32. *J Neurophysiol.* 2002;**88**:3010–20.
174. Price CJ, Kim P, Raymond LA. D1 dopamine receptor-induced cyclic AMP-dependent protein kinase phosphorylation and potentiation of striatal glutamate receptors. *J Neurochem.* 1999;**73**:2441–6.
175. Pascoli V, Turiault M. Luscher C. Reversal of cocaine-evoked synaptic potentiation resets drug-induced adaptive behavior. *Nature* 2011. [Epub ahead of print].

# Multiple Roles of STAT3 in Cardiovascular Inflammatory Responses

Cheh Peng Lim[*] and Xin-Yuan Fu[*,†,‡]

[*]Department of Biochemistry, National University of Singapore, Singapore

[†]Cancer Science Institute of Singapore, National University of Singapore, Singapore

[‡]Department of Microbiology and Immunology, Indiana University School of Medicine, Indiana, USA



Cardiovascular inflammation has been regarded as a major cause for heart failure. However, the mechanisms that are involved in the regulation of cardiovascular inflammation are not well understood. In this chapter, we review the recent progress in revealing the roles of STAT3 in cardiovascular inflammation. Particularly, we discuss cardiomyocyte-specific STAT3-deficient mice and their phenotypes. An emerging view is that STAT3 plays a key role not only in cardiac protection but also in the control of cardiac-related inflammation.

## I. Introduction

Proteins are the fundamental building blocks of an organism. Their expression and activation are regulated by a variety of mechanisms to render specificity by means of quantity, time, and space, including posttranslational modification such as ubiquitination or sumolyation, methylation or acetylation, and phosphorylation. Protein phosphorylation is a unique mode of posttranslational protein modification which enables a rapid activation or deactivation of a protein, via the addition of a phosphate group ($PO_4^{3-}$) to a protein. A series of protein phosphorylation events results in a signaling cascade that eventually culminates in other cellular events such as


DOI: 10.1016/B978-0-12-396456-4.00010-9


1877-1173/12 $35.00

irreversible transcription of target genes, proteolysis, and epigenetic events. A rapid, inducible signaling pathway regulated by protein phosphorylation is mediated by the signal transducer and activator of transcription (STAT) family of proteins.[1,2] In the 1980s, the Darnell laboratory in the Rockefeller University was investigating interferon-induced gene expression.[3–6] The STAT proteins were eventually biochemically identified as the key signaling molecules in interferon signaling in the early 1990s.[7,8] The striking feature of this signaling pathway is the SH2-domain-mediated inducible tyrosine phosphorylation of STAT proteins.[9–11] After protein tyrosine phosphorylation, these factors are able to transduce signals via translocation from the cell membrane to the nucleus and bind to DNA, thereby resulting in specific regulation of gene transcription. Seven mammalian STATs have been discovered, namely, STAT1, STAT2, STAT3, STAT4, STAT5a, STAT5b, and STAT6, which are encoded by different genes.[12]

STAT proteins are activated via phosphorylation by various kinases, including Janus kinase (JAK), growth factor receptors such as epidermal growth factor receptor (EGFR), nonreceptor tyrosine kinases such as v-src, and G-protein coupled receptors such as RANTES.[13–16] The classical JAK–STAT pathway, which actively mediates signals in cytokine-induced gene expression, involves a series of intricate steps. The binding of the ligand onto the cytokine receptor results in oligomerization of the receptor subunits, followed by phosphorylation and activation of receptor-associated JAKs. This results in transphosphorylation of the cytokine receptor at specific residues, which provides a unique recognition site for cytoplasmic STAT protein to dock. STATs bind to the phosphoreceptor via its SH2 domain, and are successively phosphorylated by JAKs at a critical tyrosine residue at the cytoplasmic tail. PhosphoSTATs homodimerize or heterodimerize via reciprocal phosphotyrosine and SH2 domains, and translocate into the nucleus to bind response elements in the promoters of target genes to stimulate transcription.

The signaling pathways of the STAT family of proteins are stimulated by distinct and overlapping ligands, including interferons, interleukins, growth factors, and hormones.[17,18] Hence the various members of STAT family exhibit certain discrete and yet some similar functions. STAT1 has been reported to play a role in antiviral, antibacterial, apoptotic, growth inhibitory, and tumor suppressive responses, whereas STAT3 generally has the opposite function, including antiapoptosis, cell proliferation, and oncogenesis. STAT4 and STAT6 are involved in the development of T helper 1 (Th1) and T helper 2 (Th2) cells in the immune system, whereas STAT5A and STAT5B are involved in hormone signaling.[16] STATs also play a role in inflammation, which is elaborated in this review, with a special emphasis in STAT3, particularly on STAT3 functions in cardiovascular inflammation.

## II. IL-6 Family of Cytokines and STAT3

IL-6 is a major cytokine generated during inflammation. The IL-6 receptor system consists two chains: the 80-kD IL-6 receptor for ligand binding and a 130-kD coreceptor (gp130) for signal transduction. It is clear that this gp130 signaling chain is used by IL-6, LIF, CT-1, oncostatin M (OSM), ciliary neurotrophic factor (CNTF), and IL-11.[19] These cytokines play roles in the development and normal functioning of some important organs and tissues such as heart and liver, as well as in lymphocyte differentiation.[20] Interruption of gp130 signaling in pan-knockout (all cell types affected) mice results in embryonic lethality with multiple organ defects, including thin-walled and dilated cardiac ventricles.[21] Although mice with cardiac myocyte-restricted deletion of gp130 have apparently normal hearts at 2 months of age, after proximal aortic banding, these animals fail to develop compensatory hypertrophy and rapidly progress to cardiac dilation and failure. Histological examination of these hearts shows massive cardiac myocyte apoptosis.[22] These findings support the idea that gp130 signaling is critical for both the hypertrophic stress response of cardiac myocytes to pressure overload and the prevention of heart failure secondary to apoptosis.

Activation of gp130 leads to downstream activation of at least three different signaling pathways, namely, the Ras–MAP kinase pathway, the PI-3–kinase-dependent pathway, and the JAK–STAT pathway.[23] In fact, STAT3 was originally discovered as a transcription factor activated by the IL-6 family of cytokines through gp130.[24,25]

## III. STAT3 in Inflammation and Heart Diseases

Ischemia–reperfusion-induced heart failure is a major cause of mortality. It has been found that the heart can adapt or gain tolerance to an ischemia event if a stressful stimulus has been given before the event. This stimuli-induced protection of heart against the subsequent injury is called "preconditioning."[26–29] Although significant progress has been made to demonstrate that preconditioning is well conserved among different species, can be observed in a variety of organs beyond heart (such as brain), and can be induced by a number of different stimuli, the molecular mechanism that governs preconditioning is still elusive. Previous studies have shown that innate immunity and inflammatory cytokines, such as TNF-$\alpha$ and IL-6, may have an important role in ischemia preconditioning.[30–32] Lipopolysaccharide (LPS), as a strong inducer of innate responses, also induces protection against ischemia–reperfusion injury.[33–35] Similar to cardiac inflammation as discussed above, nitric oxide (NO) plays a critical role in ischemia preconditioning.[36,37] Furthermore, as the major mediator during ischemia, oxygen species are suggested as activators of innate receptors: Toll-like receptors that may further trigger generation of proinflammatory cytokines inducing preconditioning protection against injury.[38–40] Another shared event by inflammatory responses and preconditioning is the activation of signaling protein NF-κB. Many studies have implicated the importance of NF-κB during preconditioning.[41–47] An important function of NF-κB during preconditioning is through the regulation of NO.[36,48,49]

There are many other pathways that may affect ischemia preconditioning. One well-studied case is the role of protein kinase C (PKC). The PKC family of proteins has been shown to have crucial functions in preconditioning.[50–54] It is noted that several different PKC isoforms exist and they may have distinct roles in ischemia. For example, PKC-δ was shown to play a pivotal role in the preconditioning, and phosphorylation of PKC-δ on serine 643 may be of particular relevance in transferring signals to mitochondrial K(ATP) channels.[55] Furthermore, another member of the PKC family, PKC-ε, forms multiprotein complexes in the heart and acts as a functional signaling module with the protein tyrosine kinase Lck. Interestingly, disruption of the interaction of PKC-ε and Lck abolished the cardioprotective function during ischemic injury.[56]

Other signaling proteins that play a role in both inflammation and ischemia belong to the family of MAP kinases (MAPK). Innate responses can activate a cascade of kinases, and p38 MAPK, p46 JNKs, and ERKs could all be involved in innate responses. However, p38 MAPK might be a major kinase that may have an essential function during ischemia preconditioning: its activity is increased during preconditioning, and the p38 inhibitor SB203580 can inhibit the protective function in preconditioning heart.[29,42,57,58] It may not just be an accident that p38 MAPK is also a critical mediator for innate immune responses from Toll-like receptors and inflammatory cytokines,[59] further suggesting that a shared mechanism between ischemia preconditioning and cardiac inflammation likely exists.

In a cardiac hypertrophy model using pressure overload, the abdominal aorta of rats was constricted, and Western Blot showed that JAK1, JAK2, Tyk2, STAT1, STAT2, and STAT3 were tyrosine-phosphorylated rapidly after exposure to the pressure overload, indicating that the JAK–STAT pathway was activated by acute pressure overload.[60] Several studies have further confirmed that STAT3 is essential for gp130-mediated hypertrophy and cytoprotection in the heart. STAT3 is activated under a variety of stress conditions, such as pressure overload, hypoxia, and acute myocardial infarction.[60–62] The expression of several genes involved in hypertrophy (e.g., c-fos, ANP), cytoprotection (Bcl-$_{xl}$, MnSOD), and angiogenesis (VEGF) is also regulated by STAT3.[63–66] The observation that transgenic mice with cardiac-specific overexpression of STAT3 show not only myocardial hypertrophy but are also protected against doxorubicin-induced apoptosis provides evidence that STAT3 can protect the heart from stress-induced injury *in vivo*.[67] Other investigators have shown, using cultured cardiomyocytes, that blockade of MAPK activation can prevent the antiapoptotic effect of gp130 activation in response to cardiotrophin-1 (CT-1).[68] Knockout of individual MAPKs does not produce a known cardiac defect in mice, likely because of redundancies in the MAPK signaling system. Therefore, assessing the relative importance of MAPKs using this strategy is difficult.

As mentioned above, the Gp130–STAT3 pathway is required for cardiac protection. The JAK–STAT pathway was also examined in acute myocardial infarction to determine its pathophysiological roles in ischemic heart disease. Roles of STAT3 in ischemic preconditioning were first observed by several groups.[69,70] In this acute myocardial infarction model, the proximal left anterior descending coronary artery of male rats was ligated and the rats were sacrificed at various time points. Results showed tyrosine phosphorylation of STAT3 in the myocardium obtained from the healthy border area adjacent to the infarcted myocardium,[62] indicating the activation of STAT signaling in acute myocardial infarction.

Earlier work by Takeda *et al.* showed that deletion of STAT3 resulted in embryonic lethality.[71] To overcome this technical hurdle, genetically engineered mice with organ-specific overexpression or mice with tissue-specific knockout of STAT3 were generated by several groups, including ours. In the case of heart, cardiac-specific transgenic mice with overexpression of phospho-STAT3 showed increased vascular formation compared with nontransgenic mice, suggesting that STAT3 is a modulator of vessel growth.[66] Further analyses revealed that these transgenic mice exhibited a smaller infarct myocardium compared to nontransgenic littermates after ischemia and reperfusion injury, suggesting that STAT3-mediated myocardial protection is sufficient to prevent the injury expansion induced by the inflammatory reaction following the ischemia and reperfusion injury.[72] It was also demonstrated that this transgene showed an upregulation of metallothionein 1 and 2, which are reactive oxygen species (ROS) scavengers, and correlated with a reduction in ROS generation upon ischemia and reperfusion injury. These studies indicate that activation of STAT3 protects the myocardium from ischemia and reperfusion injury *in vivo*, which may be partially through metallothionein.

The protective role of STAT3 in heart is in agreement with studies from cardiac-specific STAT3 knockout mice. Mice with cardiomyocyte-restricted deletion of STAT3 were generated by Jacoby *et al.*[73] and these mice showed increased susceptibility to doxorubicin-induced heart failure, indicating that STAT3 has a cardioprotective function against doxorubicin-induced injury. Furthermore, with advanced age at 9 months, cardiac-specific STAT3-null mice exhibited severe fibrosis whereas wild-type mice did not show any significant fibrosis. More interestingly, these studies indicate that cardiomyocyte-restricted knockout of STAT3 results in higher sensitivity to inflammation, and that STAT pathway is required for regulation of inflammatory responses of cardiomyocytes, which may contribute to the development of cardiac fibrosis. Overall, these results demonstrated that STAT3 is critical in protecting inflammation-induced heart damage.

While the above studies were animal based, Ng *et al.* examined STAT activation in ischemic heart disease and dilated cardiomyopathy samples collected from human patients, with normal donor hearts as controls.[74] Elevated tyrosine phosphorylation of STAT1, STAT3b, STAT5, and STAT6 was observed in dilated cardiomyopathy, whereas only STAT1 and STAT5 phosphorylation were significantly increased in ischemic heart disease, compared to normal donor hearts. It was speculated that, while STAT3a elicits cardioprotection in response to ischemia and reperfusion, STAT3b may be proapoptotic during the progression to heart failure. This work showed that STATs might be playing a role in regulating progression of human heart failure.

We would point out that activation of STAT3 is associated with oncogenesis, and decreasing STAT3 activity would be the target of cancer therapy. In contrast, activation of STAT3 is important for cardioprotection and favorable in therapy for heart failure, and therefore targeting STAT3 in heart diseases must be carefully controlled for a balanced STAT3 activation.[75,76]

## IV. STAT3 in Endothelia-Related Inflammation

It was postulated that intracellular signaling in endothelial cells could affect the immune responses via cell surface or secreted factors.[77] To investigate the role of STAT3 in such a system, an endothelial conditional knockout of STAT3 was generated and the innate immune function was examined by treatment with LPS, which is an endotoxin that elicits strong immune responses via the toll-like receptor 4 (TLR4).[78] After 40 h of LPS challenge, 60% homozygous-deleted mice ($STAT3^{E-/-}$) failed to survive, whereas all the homozygous littermates survived.[79] Cardiomyocyte-specific STAT3 knockout and hepatocyte-specific STAT3 knockout mice did not show such susceptibility, indicating that this endotoxic shock-specific lethality is rather restricted to endothelial cells upon LPS challenge. These $STAT3^{E-/-}$ mice had severe infiltration of leukocytes into multiple organs at 17h of LPS challenge, suggesting severe systemic inflammation. These findings correlated with the elevated levels of serum AST, which is an indicator of multiple organ damage. Moreover, the serum levels of inflammatory cytokines such as TNF-$\alpha$, IL-6, and IFN-$\gamma$ increased upon LPS treatment in $STAT3^{E-/-}$ mice. Taken together, this shows that protection against endotoxin-induced inflammation is dependent on STAT3 in endothelial cells.

The role of endothelial STAT3 was further analyzed on myocardial response to ischemia and reperfusion injury. Endothelial-restricted STAT3 knockout mice showed poorer postischemic myocardial functional recovery.[80] This was evidenced by increased lactate dehydrogenase release in coronary effluent in these STAT3 knockout compared to wild-type mice upon ischemia and reperfusion injury, indicating impaired myocardial function. Ischemia and reperfusion injury also caused higher expression of caspase 8 in myocardium around the capillary upon ischemia and reperfusion injury in the STAT3 knockout mice, which may contribute to the compromised integrity of endothelium and blood vessels. Elevated myocardial IL-6 was secreted in the endothelial-restricted STAT3 knockout mice compared to wild-type mice after ischemia and reperfusion injury. These results suggest that endothelial STAT3 plays a critical role in myocardial proinflammatory signaling and in myocardial recovery after ischemia and reperfusion injury by protecting the myocardium from apoptotic response and maintaining capillary integrity.

In summary, here a brief review of the roles of STAT3 in cardiovascular inflammation has been presented. This is an emerging research field involving complicated mechanisms regulating both cardiac inflammation and ischemia. In the future, it is essential to provide further information on the molecular mechanisms of STAT3-regulated innate immune responses and induced alterations in cardiac gene expression, as well as epigenetic remodeling that may cause pathogenesis associated with cardiovascular inflammation and ischemia.

## References


1. Darnell JE, Jr., Kerr IM, Stark GR. Jak-STAT pathways and transcriptional activation in response to IFNs and other extracellular signaling proteins. *Science* 1994;**264**:1415–21.
2. Stark GR, Kerr IM, Williams BR, Silverman RH, Schreiber RD. How cells respond to interferons. *Annu Rev Biochem* 1998;**67**:227–64.
3. Larner AC, et al. Transcriptional induction of two genes in human cells by beta interferon. *Proc Natl Acad Sci USA* 1984;**81**:6733–7.
4. Reich N, et al. Interferon-induced transcription of a gene encoding a 15-kDa protein depends on an upstream enhancer element. *Proc Natl Acad Sci USA* 1987;**84**:6394–8.
5. Levy D, Reich N, Kessler D, Pine R, Darnell JE. Jr., Transcriptional regulation of interferon-stimulated genes: a DNA response element and induced proteins that recognize it. *Cold Spring Harb Symp Quant Biol* 1988;**53**(Pt 2):799–802.
6. Decker T, Lew DJ, Cheng YS, Levy DE, Darnell JE. Jr., Interactions of alpha- and gamma-interferon in the transcriptional regulation of the gene encoding a guanylate-binding protein. *EMBO J* 1989;**8**:2009–14.
7. Fu XY, Schindler C, Improta T, Aebersold R, Darnell Jr. JE. The proteins of ISGF-3, the interferon alpha-induced transcriptional activator, define a gene family involved in signal transduction. *Proc Natl Acad Sci USA* 1992;**89**:7840–3.

8. Schindler C, Fu XY, Improta T, Aebersold R, Darnell JE. Jr., Proteins of transcription factor ISGF-3: one gene encodes the 91- and 84-kDa ISGF-3 proteins that are activated by interferon alpha. *Proc Natl Acad Sci USA* 1992;**89**:7836–9.
9. Fu XY. A transcription factor with SH2 and SH3 domains is directly activated by an interferon alpha-induced cytoplasmic protein tyrosine kinase(s). *Cell* 1992;**70**:323–35.
10. Schindler C, Shuai K, Prezioso VR, Darnell JE. Jr., Interferon-dependent tyrosine phosphorylation of a latent cytoplasmic transcription factor. *Science* 1992;**257**:809–13.
11. Velazquez L, Fellous M, Stark GR, Pellegrini S. A protein tyrosine kinase in the interferon a/b signaling pathway. *Cell* 1992;**70**:313–22.
12. Darnell JE. Jr., STATs and gene regulation. *Science* 1997;**277**:1630–5.
13. O'Shea JJ. Jaks, STATs, cytokine signal transduction, and immunoregulation: are we there yet? *Immunity* 1997;**7**:1–11.
14. Akira S. Functional roles of STAT family proteins: lessons from knockout mice. *Stem Cells* 1999;**17**:138–46.
15. Levy DE, Darnell JE. Jr., Stats: transcriptional control and biological impact. *Nat Rev Mol Cell Biol* 2002;**3**:651–62.
16. Lim CP, Cao X. Structure, function, and regulation of STAT proteins. *Mol Biosyst* 2006;**2**:536–50. doi:10.1039/b606246f.
17. Leonard WJ, O'Shea JJ. Jaks and STATs: biological implications. *Annu Rev Immunol* 1998;**16**:293–322. doi:10.1146/annurev.immunol.16.1.293.
18. Bromberg JF. Activation of STAT proteins and growth control. *Bioessays* 2001;**23**:161–9. doi:10.1002/1521-1878(200102)23:2<161::AID-BIES1023>3.0.CO;2-0 [pii]10.1002/1521-1878 (200102)23:2<161::AID-BIES1023>3.0.CO;2-0.
19. Akira S. IL-6-regulated transcription factors. *Int J Biochem Cell Biol* 1997;**29**:1401–18.
20. Taga T, Kishimoto T. Gp130 and the interleukin-6 family of cytokines. *Annu Rev Immunol* 1997;**15**:797–819.
21. Yoshida K, et al. Targeted disruption of gp130, a common signal transducer for the interleukin 6 family of cytokines, leads to myocardial and hematological disorders. *Proc Natl Acad Sci USA* 1996;**93**:407–11.
22. Hirota H, et al. Loss of a gp130 cardiac muscle cell survival pathway is a critical event in the onset of heart failure during biomechanical stress. *Cell* 1999;**97**:189–98.
23. Hirano T, Nakajima K, Hibi M. Signaling mechanisms through gp130: a model of the cytokine system. *Cytokine Growth Factor Rev* 1997;**8**:241–52.
24. Zhong Z, Wen Z, Darnell JE. Stat3: a stat family member activated by tyrosine phosphorylation in response to epidermal growth factor and interleukin-6. *Science* 1994;**264**:95–8.
25. Akira S, et al. Molecular cloning of APRF, a novel IFN-stimulated gene factor 3 p91-related transcription factor involved in the gp130-mediated signaling pathway. *Cell* 1994;**77**:63–71.
26. Murry CE, Jennings RB, Reimer KA. Preconditioning with ischemia: a delay of lethal cell injury in ischemic myocardium. *Circulation* 1986;**74**:1124–36.
27. Yellon DM, Alkhulaifi AM, Pugsley WB. Preconditioning the human myocardium. *Lancet* 1993;**342**:276–7.
28. Yellon DM, Downey JM. Preconditioning the myocardium: from cellular physiology to clinical cardiology. *Physiol Rev* 2003;**83**:1113–51.
29. Tsai BM, et al. Preconditioning: evolution of basic mechanisms to potential therapeutic strategies. *Shock* 2004;**21**:195–209.
30. Kurrelmeyer KM, et al. Endogenous tumor necrosis factor protects the adult cardiac myocyte against ischemic-induced apoptosis in a murine model of acute myocardial infarction. *Proc Natl Acad Sci USA* 2000;**97**:5456–61.

31. Craig R, et al. p38 MAPK and NF-kappa B collaborate to induce interleukin-6 gene expression and release. Evidence for a cytoprotective autocrine signaling pathway in a cardiac myocyte model system. *J Biol Chem* 2000;**275**:23814–24.
32. Smith RM, Lecour S, Sack MN. Innate immunity and cardiac preconditioning: a putative intrinsic cardioprotective program. *Cardiovasc Res* 2002;**55**:474–82.
33. Eising GP, Mao L, Schmid-Schonbein GW, Engler RL, Ross J. Effects of induced tolerance to bacterial lipopolysaccharide on myocardial infarct size in rats. *Cardiovasc Res* 1996;**31**:73–81.
34. Rowland RT, et al. LPS-induced delayed myocardial adaptation enhances acute preconditioning to optimize postischemic cardiac function. *Am J Physiol* 1997;**272**:H2708–15.
35. Meng X, et al. Myocardial gene reprogramming associated with a cardiac cross-resistant state induced by LPS preconditioning. *Am J Physiol* 1998;**275**:C475–83.
36. Bolli R, et al. The nitric oxide hypothesis of late preconditioning. *Basic Res Cardiol* 1998;**93**:325–38.
37. Bolli R. Cardioprotective function of inducible nitric oxide synthase and role of nitric oxide in myocardial ischemia and preconditioning: an overview of a decade of research. *J Mol Cell Cardiol* 2001;**33**:1897–918.
38. Frantz S, et al. Toll4 (TLR4) expression in cardiac myocytes in normal and failing myocardium. *J Clin Invest* 1999;**104**:271–80.
39. Forbes RA, Steenbergen C, Murphy E. Diazoxide-induced cardioprotection requires signaling through a redox-sensitive mechanism. *Circ Res* 2001;**88**:802–9.
40. Li C, et al. Modulating Toll-like receptor mediated signaling by (1–>3)-beta-D-glucan rapidly induces cardioprotection. *Cardiovasc Res* 2004;**61**:538–47.
41. Meldrum DR, et al. Alpha-adrenergic activation of myocardial NF kappa B during hemorrhage. *J Surg Res* 1997;**69**:268–76.
42. Maulik N, Sato M, Price BD, Das DK. An essential role of NFkappaB in tyrosine kinase signaling of p38 MAP kinase regulation of myocardial adaptation to ischemia. *FEBS Lett* 1998;**429**:365–9.
43. Morgan EN, et al. An essential role for NF-kappaB in the cardioadaptive response to ischemia. *Ann Thorac Surg* 1999;**68**:377–82.
44. Das DK, Maulik N, Sato M, Ray PS. Reactive oxygen species function as second messenger during ischemic preconditioning of heart. *Mol Cell Biochem* 1999;**196**:59–67.
45. Zhao TC, Taher MM, Valerie KC, Kukreja RC. p38 Triggers late preconditioning elicited by anisomycin in heart: involvement of NF-kappaB and iNOS. *Circ Res* 2001;**89**:915–22.
46. Sasaki H, et al. Hypoxia/reoxygenation promotes myocardial angiogenesis via an NF kappa B-dependent mechanism in a rat model of chronic myocardial infarction. *J Mol Cell Cardiol* 2001;**33**:283–94.
47. Jones WK, Brown M, Ren X, He S, McGuinness M. NF-kappaB as an integrator of diverse signaling pathways: the heart of myocardial signaling? *Cardiovasc Toxicol* 2003;**3**:229–54.
48. Xuan YT, et al. Nuclear factor-kappaB plays an essential role in the late phase of ischemic preconditioning in conscious rabbits. *Circ Res* 1999;**84**:1095–109.
49. Li G, et al. Myocardial protection by remote preconditioning: the role of nuclear factor kappa-B p105 and inducible nitric oxide synthase. *Eur J Cardiothorac Surg* 2004;**26**:968–73.
50. Speechly-Dick ME, Mocanu MM, Yellon DM. Protein kinase C. Its role in ischemic preconditioning in the rat. *Circ Res* 1994;**75**:586–90.
51. Liu Y, Ytrehus K, Downey JM. Evidence that translocation of protein kinase C is a key event during ischemic preconditioning of rabbit myocardium. *J Mol Cell Cardiol* 1994;**26**:661–8.
52. Mitchell MB, et al. Preconditioning of isolated rat heart is mediated by protein kinase C. *Circ Res* 1995;**76**:73–81.
53. Meldrum DR, et al. Protein kinase C isoform diversity in preconditioning. *J Surg Res* 1997;**69**:183–7.

54. Ping P, et al. PKC-dependent activation of p44/p42 MAPKs during myocardial ischemia-reperfusion in conscious rabbits. *Am J Physiol* 1999;**276**:H1468–81.
55. Uecker M, et al. Translocation of protein kinase C isoforms to subcellular targets in ischemic and anesthetic preconditioning. *Anesthesiology* 2003;**99**:138–47.
56. Ping P, et al. Formation of protein kinase C(epsilon)-Lck signaling modules confers cardioprotection. *J Clin Invest* 2002;**109**:499–507.
57. Weinbrenner C, Liu GS, Cohen MV, Downey JM. Phosphorylation of tyrosine 182 of p38 mitogen-activated protein kinase correlates with the protection of preconditioning in the rabbit heart. *J Mol Cell Cardiol* 1997;**29**:2383–91.
58. Mocanu MM, Baxter GF, Yue Y, Critz SD, Yellon DM. The p38 MAPK inhibitor, SB203580, abrogates ischaemic preconditioning in rat heart but timing of administration is critical. *Basic Res Cardiol* 2000;**95**:472–8.
59. Kyriakis JM, Avruch J. Mammalian mitogen-activated protein kinase signal transduction pathways activated by stress and inflammation. *Physiol Rev* 2001;**81**:807–69.
60. Pan J, et al. Role of angiotensin II in activation of the JAK/STAT pathway induced by acute pressure overload in the rat heart. *Circ Res* 1997;**81**:611–7.
61. Hishinuma S, Funamoto M, Fujio Y, Kunisada K, Yamauchi-Takihara K. Hypoxic stress induces cardiotrophin-1 expression in cardiac myocytes. *Biochem Biophys Res Commun* 1999;**264**:436–40.
62. Negoro S, et al. Activation of JAK/STAT pathway transduces cytoprotective signal in rat acute myocardial infarction. *Cardiovasc Res* 2000;**47**:797–805. doi:10.1016/S0008-6363(00)00138-3 [pii].
63. Kunisada K, et al. Activation of gp130 transduces hypertrophic signals via STAT3 in cardiac myocytes. *Circulation* 1998;**98**:346–52.
64. Negoro S, et al. Activation of signal transducer and activator of transcription 3 protects cardiomyocytes from hypoxia/reoxygenation-induced oxidative stress through the upregulation of manganese superoxide dismutase. *Circulation* 2001;**104**:979–81.
65. Funamoto M, et al. Signal transducer and activator of transcription 3 is required for glycoprotein 130-mediated induction of vascular endothelial growth factor in cardiac myocytes. *J Biol Chem* 2000;**275**:10561–6.
66. Osugi T, et al. Cardiac-specific activation of signal transducer and activator of transcription 3 promotes vascular formation in the heart. *J Biol Chem* 2002;**277**:6676–81. doi:10.1074/jbc.M108246200 [pii].
67. Kunisada K, et al. A novel role for STAT3 in cardiac remodeling. *Cytokine* 2000;**12**:1512–8.
68. Sheng Z, et al. Cardiotrophin 1 (CT-1) inhibition of cardiac myocyte apoptosis via a mitogen-activated protein kinase-dependent pathway. *J Biol Chem* 1997;**272**:5783–91.
69. Hoshijima M, et al. Role of STAT3 in ischemic preconditioning. *J Clin Invest* 2001;**108**:1459–67.
70. Hattori R, et al. Role of STAT3 in ischemic preconditioning. *J Mol Cell Cardiol* 2001;**33**:1929–36.
71. Takeda K, et al. Targeted disruption of the mouse Stat3 gene leads to early embryonic lethality. *Proc Natl Acad Sci USA* 1997;**94**:3801–4.
72. Oshima Y, et al. STAT3 mediates cardioprotection against ischemia/reperfusion injury through metallothionein induction in the heart. *Cardiovasc Res* 2005;**65**:428–35. doi:10.1016/j.cardiores.2004.10.021 [pii] S0008-6363(04)00468-7.
73. Jacoby JJ, et al. Cardiomyocyte-restricted knockout of STAT3 results in higher sensitivity to inflammation, cardiac fibrosis, and heart failure with advanced age. *Proc Natl Acad Sci USA* 2003;**100**:12929–34. doi:10.1073/pnas.2134694100 [pii].
74. Ng DC, Court NW, dos Remedios CG, Bogoyevitch MA. Activation of signal transducer and activator of transcription (STAT) pathways in failing human hearts. *Cardiovasc Res* 2003;**57**:333–46. doi:10.1016/S0008-6363(02)00664-8 [pii].

75. Boengler K, Hilfiker-Kleiner D, Drexler H, Heusch G, Schulz R. The myocardial JAK/STAT pathway: from protection to failure. *Pharmacol Ther* 2008;**120**:172–85. doi:10.1016/j.pharmthera.2008.08.002 [pii] S0163-7258(08)00140-X.
76. Lavecchia A, Di Giovanni C, Novellino E. STAT-3 inhibitors: state of the art and new horizons for cancer treatment. *Curr Med Chem* 2011;**18**:2359–75. doi:BSP/CMC/E-Pub/2011/ 168 [pii].
77. Cotran RS, Pober JS. Cytokine-endothelial interactions in inflammation, immunity, and vascular injury. *J Am Soc Nephrol* 1990;**1**:225–35.
78. Poltorak A, et al. Defective LPS signaling in C3H/HeJ and C57BL/10ScCr mice: mutations in Tlr4 gene. *Science* 1998;**282**:2085–8.
79. Kano A, et al. Endothelial cells require STAT3 for protection against endotoxin-induced inflammation. *J Exp Med* 2003;**198**:1517–25. doi:10.1084/jem.20030077 jem.20030077 [pii].
80. Wang M, et al. Endothelial STAT3 plays a critical role in generalized myocardial proinflammatory and proapoptotic signaling. *Am J Physiol Heart Circ Physiol* 2007;**293**:H2101–8. doi:10.1152/ajpheart.00125.2007 [pii] 00125.2007.

# Targeting Phosphorylation of Eukaryotic Initiation Factor-2α to Treat Human Disease

Melissa J. Fullwood,[*] Wei Zhou,[*,†] and Shirish Shenolikar[*,†]

[*]*Program in Neuroscience and Behavioral Disorders, Duke-NUS Graduate Medical School Singapore, Singapore*
[†]*Program in Cardiovascular and Metabolic Disorders, Duke-NUS Graduate Medical School Singapore, Singapore*



The unfolded protein response, also known as endoplasmic reticulum (ER) stress, has been implicated in numerous human diseases, including atherosclerosis, cancer, diabetes, and neurodegenerative disorders. Protein misfolding activates one or more of the three ER transmembrane sensors to initiate a complex network of signaling that transiently suppresses protein translation while also enhancing protein folding and proteasomal degradation of misfolded proteins to ensure full recovery from ER stress. Gene disruption studies in mice have provided critical insights into the role of specific signaling components and pathways in the differing responses of animal tissues to ER stress. These studies have emphasized an important contribution of translational repression to sustained insulin synthesis and β-cell viability in experimental models of type-2 diabetes. This has focused attention on the recently discovered small-molecule inhibitors of eIF2α phosphatases that prolong eIF2α phosphorylation

Progress in Molecular Biology
and Translational Science, Vol. 106
DOI: 10.1016/B978-0-12-396456-4.00005-5




1877-1173/12 $35.00

to reduce cell death in several animal models of human disease. These compounds show significant cytoprotection in cellular and animal models of neurodegenerative disorders, highlighting a potential strategy for future development of drugs to treat human protein misfolding disorders.

## I. Introduction

Cells face frequent fluctuations in the availability of nutrients and other environmental factors. Thus, mammalian cells have developed complex signaling mechanisms to detect such environmental disturbances and adjust their physiology appropriately to survive these stresses. Common to all eukaryotes is their ability to attenuate protein translation in response to stress. As protein synthesis places a high demand on their energy supply, this allows cells to temporarily focus their attention on the task of overcoming the stress. Analyzing the cell's response to misfolded proteins, a common everyday occurrence, can elucidate the signaling pathways that oversee protein synthesis and quality control. These stress response pathways are greatly exacerbated in diseases associated with protein overload, expression of folding-defective proteins, or errors in protein processing. In this review, we discuss these signaling mechanisms and, in particular, highlight the phosphorylation of the eukaryotic translation initiation factor eIF2α, which results in translational attenuation, an early event triggered by many different forms of cell stress. Genetic evidence in mice and humans suggest that the inability to catalyze eIF2α phosphorylation may contribute to diseases such as diabetes. Finally, we discuss the recent discovery of small molecules that enhance eIF2α phosphorylation and protect cells from the damaging effects of protein misfolding. These discoveries raise new hopes for the development of novel treatments for human diseases, including diabetes, cancer, and neurodegenerative disorders.

## II. Control of Protein Translation

Timely synthesis of proteins and their folding into active three-dimensional structures are absolutely critical for the normal functioning and growth of all eukaryotic cells. Mammalian cells utilize a complex array of regulatory mechanisms, including the reversible phosphorylation of several initiation and elongation factors, as well as many ribosomal proteins, to control protein synthesis.[1]

Iron deficiency is one the most prevalent nutritional deficiencies globally and is a major cause of life-threatening anemias. Nearly 40 years ago, Irving London and colleagues studying the synthesis of the globin polypeptides *in vitro* in rabbit reticulocyte lysates established the presence of a heme-regulated inhibitor (HRI) that is activated by iron deficiency or the absence of heme, a member of the porphyrin family and a cofactor for hemoglobin.[2] Subsequent work showed that HRI was a protein kinase that phosphorylated the translational initiation factor eIF2$\alpha$ and prevented the assembly of an initiation complex comprised of mRNA, ribosomal subunits, and methionyl-tRNA, thereby inhibiting the synthesis of $\alpha$- and $\beta$-globins. When the rabbit HRI cDNA was cloned,[3] it was noted that the primary sequence encoded by this cDNA bore remarkable homology to the yeast GCN2 (general control nonderepressible) protein. The *Saccharomyces cerevisiae* GCN2 gene also encoded a protein kinase, which was activated by amino acid starvation. The buildup of uncharged tRNAs that occurs as a result of amino acid deprivation activates Gcn2, which phosphorylates the yeast eIF2$\alpha$ (also known as Sui2) to inhibit general protein synthesis and elicit cell cycle arrest. At the same time, the increased eIF2$\alpha$ phosphorylation enhances the translation of the transcription factor Gcn4 and the transcription of yeast enzymes that promote amino acid biosynthesis.[4,5] In this manner, eIF2$\alpha$ phosphorylation allows yeast to overcome the deficiency in amino acids in the external media. Later studies have established that GCN2 is also present in mammalian cells and is activated by nutrient deprivation and other stresses.[6,7]

The primary structure of the HRI protein also shares sequence homology with an inhibitory protein factor termed dsI, which is found in cells infected with viruses. This factor also displays the properties of a protein kinase and is activated by double-stranded RNAs present in the virus-infected cells. The dsI protein is also induced by interferons, cytokines released by host cells to communicate the presence of a viral (or bacterial) infection to recruit immune cells, and aids the eradication of these pathogens. The primary role of dsI, also known as double-stranded RNA-dependent eIF2$\alpha$ kinase (or PKR), is to phosphorylate eIF2$\alpha$ and inhibit protein synthesis in the virus-infected cells.[8] Thus, PKR functions as a component of the host surveillance system that inhibits viral replication and limits the spread of infection to other cells and tissues. Interferons and PKR also play a role in fighting cancers.[7] Facing this challenge, viruses have evolved multiple strategies to circumvent PKR, including the expression of pseudosubstrate inhibitors, degradation of PKR protein, expression of proteins that sequestrate dsRNA, as well as the expression of a component of eIF2$\alpha$ phosphatase (discussed below) that overrides PKR function.[8] Interestingly, PKR can substitute for HRI in inhibiting globin synthesis in reticulocyte lysates, and both mammalian HRI and PKR can functionally replace GCN2 in budding yeast,[9] highlighting the common role of these protein kinases in translational repression.

Nearly 30 years after the discovery of HRI, the fourth eIF2α kinase was identified and termed PEK or pancreatic eIF2α kinase because of its high expression in pancreas and, more specifically, islets. Like HRI and PKR, PEK could also inhibit globin synthesis in reticulocyte lysates and mediate GCN2 signaling in yeast.[10] More recently renamed double-stranded RNA-activated protein kinase-like endoplasmic reticulum kinase or PERK, this kinase responds to the presence of unfolded proteins in the endoplasmic reticulum (ER)[11] and is a major focus of discussion in this review.

In summary, while more than a dozen protein components undergo reversible phosphorylation to regulate protein translation, either initiation or elongation, in mammalian cells, only one, namely the phosphorylation of eIF2α at serine-51, is evolutionarily conserved from yeast to humans. Moreover, in contrast to budding yeast, which responds to nutritional deprivation by activating a single eIF2α kinase, namely Gcn2, higher eukaryotes utilize four distinct eIF2α kinases (also discussed in Chapter 8) to repress protein translation in response to a wide variety of physiological and pathological stresses. Moreover, increase in eIF2α phosphorylation plays a critical role in the response of eukaryotic cells to environmental stress.

## III. Cellular Mechanisms for Protein Folding

Newly synthesized proteins must be folded into their native macromolecular structure to acquire their full biological function. In this regard, the stochastic folding of a linear polypeptide chain is dictated partly by the sequence of amino acids. The primary sequence of all polypeptides confers some propensity for folding, but, in general, the energetics of protein folding is too high to permit most proteins to spontaneously adopt their native structures. The inherent folding capability of the polypeptide chain is further enhanced by molecular crowding, achieved by the high concentration of proteins in cytoplasm of all eukaryotic cells, which demands that all proteins occupy minimal space to function. The process of protein folding is, however, intrinsically inefficient and must be aided by the activity of protein chaperones present in all eukaryotic cells. Many of these are heat shock proteins whose levels increase in response to elevated temperature and other physiological stresses. Cells express a multitude of heat shock proteins, collectively described as Hsp60, Hsp70, Hsp90, and Hsp100 based on their apparent molecular size. These chaperones do not carry any specific information to fold the diverse array of cellular proteins into their native state. Rather, most chaperones associate with the hydrophobic surfaces of newly synthesized proteins to prevent their premature aggregation. Following the complete folding of their substrates, the chaperones hydrolyze adenosine triphosphate (ATP) to release the folded proteins.[12] Thus, both protein translation and folding make heavy demands on the cell's energy. In addition, cells express lectin chaperones, protein disulfide isomerases, and protein prolyl isomersases, which also participate in the correct folding of cellular proteins.[13] The energy used for protein translation and folding, as well as the oxidative stress created by the insertion of disulfide bridges, can accumulate mitochondrial damage or dysfunction and trigger programmed cell death.[14,15]

When faced with the task of synthesizing mutant proteins that are inherently defective in protein folding, under conditions of protein overload (e. g., viral infection), or if the expression of cellular chaperones fails to keep pace with ongoing protein synthesis, intracellular protein aggregates and oxidative stress or mitochondrial dysfunction are frequently observed. These protein aggregates are often polyubiquitinated and can potentially overwhelm the protein degradation machinery, comprising the 26S proteasome, to further enhance protein aggregation.[16] The accumulation of intracellular protein aggregates are readily observed in histological samples and are a hallmark of many human diseases, including diabetes and neurodegenerative disorders.[17] When faced with elevated levels of unfolded proteins, cells activate a complex cellular response, known as the "unfolded protein response" or UPR, to promote protein folding, reduce oxidative stress, and facilitate cell survival. However, if this process is unsuccessful and excessive cell damage occurs, UPR also activates programmed cell death or apoptosis to eliminate the dysfunctional or damaged cell.[18]

## IV. The Unfolded Protein Response

Proteins are the workhorses of all cellular processes and are synthesized by both cytosolic and membrane-bound ribosomes. The membrane-bound ribosomes stud the outer surface of the ER, described as the "rough" ER, which is readily visible by electron microscopy. These ribosomes are tethered to the ER via signal peptides present at the N-termini of secreted or transmembrane proteins, which are folded and further processed in the ER. The signal peptides bind signal recognition particles that transport the associated ribosome to ER. Here, the signal peptide crosses the ER membrane and is proteolytically cleaved in the lumen. Following complete synthesis of the secreted or transmembrane protein, the ribosome is released from the ER membrane and returns to the cytosol.[19]

The presence of ER and distribution of membrane-bound and cytosolic ribosomes vary widely from cell to cell. For example, erythrocytes synthesize hemoglobin, a cytoplasmic protein, and possess an abundance of ribosomes but very little ER. By contrast, plasma cells synthesize and secrete antibodies and contain an extensive, densely packed rough ER. The traditional view that cytosolic (and nuclear) proteins are translated on cytosolic ribosomes while membrane (and secreted) proteins are translated on membrane-bound ribosomes has been challenged with the recent findings that, in addition to synthesizing secreted and transmembrane membrane proteins, the ER-bound ribosomes also translate mRNAs encoding cytoplasmic and nuclear proteins.[20] Even more remarkable is the observation that ER binding significantly enhances the efficiency of mRNA translation by ribosomes.[21] Together, this suggests that ER-bound ribosomes are a major site of protein synthesis in many mammalian cells. The smooth ER lacks ribosomes and plays a key role in the metabolism of lipids, steroids, and carbohydrates, as well as in controlling calcium homeostasis and drug metabolism. Although the molecular basis is still poorly understood, increasing evidence points to the modulation of these unrelated biochemical processes also contributing to altered protein synthesis and quality control in the ER.

The ER is a highly dynamic organelle, so that drugs that alter the cellular levels of ER-bound proteins also modify the content and morphology of the ER. For example, statins inhibit the ER-bound enzyme HMG-CoA reductase to inhibit cholesterol biosynthesis and activate the ER-associated sterol response element-binding protein (SREBP) transcription factors to induce further expression of HMG-CoA reductase. This is turn induces UPR and expansion of a crystalloid ER in many tissues.[22,23] Similarly, problems in protein misfolding lead to dramatic changes in ER content and morphology. For example, pancreatic β-cells in both diabetic mice and humans with type-2 diabetes display an extensively dilated ER with the presence of amorphous protein aggregates in the lumen.[24,25] Thus, both protein misfolding and changes in ER-localized metabolic processes induce UPR. For these reasons, the unfolded protein response is also referred to as "ER stress."

Recent studies have highlighted a membranous connection between ER and mitochondria. This mitochondrial-associated ER-membrane compartment, known as MAMs or mitochondrial-associated membranes, may transduce signals from the ER to mitochondria to dictate cell fate or the decision to initiate cell death or ensure cell survival.[26] This is a very active area of research and future studies will likely provide a better understanding of the contribution of MAMs to UPR.

## V. UPR Signaling Pathways

BiP or binding immunoglobulin protein (also known as GRP78, a glucose-regulated protein) is a major ER luminal protein and among the first sensors to recognize newly synthesized or unfolded proteins in the ER lumen.[27] BiP is a member of the Hsp70 family of chaperone proteins. Although an abundant ER protein, BiP levels are further increased by UPR and other forms of stress. This increases the overall protein folding capacity in the cell. It is noteworthy that BiP serves many other functions in the ER besides binding to unfolded proteins. For example, BiP aids the entry or translocation of proteins newly synthesized by the ER-bound ribosomes into the ER lumen. BiP recruits PDIs (protein disulfide reductases) to catalyze the formation of disulfide bridges. Elevated BiP levels facilitate the retrograde transport of the misfolded proteins back across the ER membrane to facilitate their degradation by the cytoplasmic 26S proteasome.[28] Finally, Bip functions as a regulator of three ER transmembrane proteins that serve as cellular sensors of UPR. These proteins, PERK, ATF6, and IRE1, all share homology in their luminal domains that bind BiP (Fig. 1). When there is a buildup of misfolded proteins in the ER, the disordered proteins compete for BiP binding. Thus, BiP is sequestered away from the three signal transducers,[29] activating UPR signaling to initiate the complex translational and transcriptional program, sometimes referred to as the "integrated stress response".[30,31]

## A. Transient Repression of Protein Translation

One of the earliest events that follow UPR activation is the dimerization and autophosphorylation of PERK. This activates PERK, which then phosphorylates eIF2α at serine-51 and attenuates protein translation. In all eukaryotes, eIF2α associates with eIF2β and eIF2γ to form an eIF2 complex. In the presence of GTP, the eIF2 complex assembles Met-tRNA, the large and small ribosomal subunits at the ATG start site on mRNA, to form the translation initiation complex.[32] Subsequently, GTP is hydrolyzed to GDP, and the recycling of GDP to GTP catalyzed by eIF2B is critical for new rounds of translation initiation. The phosphorylation of serine-51 on eIF2α increases its affinity for eIF2B and inhibits GDP–GTP exchange.[33] Thus, the covalent modification of eIF2α attenuates general protein translation. However, translation of a subset of mRNAs remains unhindered. Previous studies have shown that the mRNA encoding the transcription factor ATF4 (activating transcription factor 4) contains upstream open reading frames (uORFs) and is more efficiently translated following eIF2α phosphorylation.[34] Under conditions of low eIF2α phosphorylation, the presence of uORFs slows the ribosome scanning, and ATF4 mRNA is either poorly translated or not at all. In the presence of elevated eIF2α phosphorylation, however, ribosomes are preferentially recruited to the downstream ATG and mRNAs containing uORFs, which include ATF4 and its downstream targets, CHOP (CAAT/enhancer binding protein homologous transcription factor, also known as GADD153 or growth arrest and DNA damage-inducible gene 153)[35] and GADD34,[36] and are more efficiently translated.

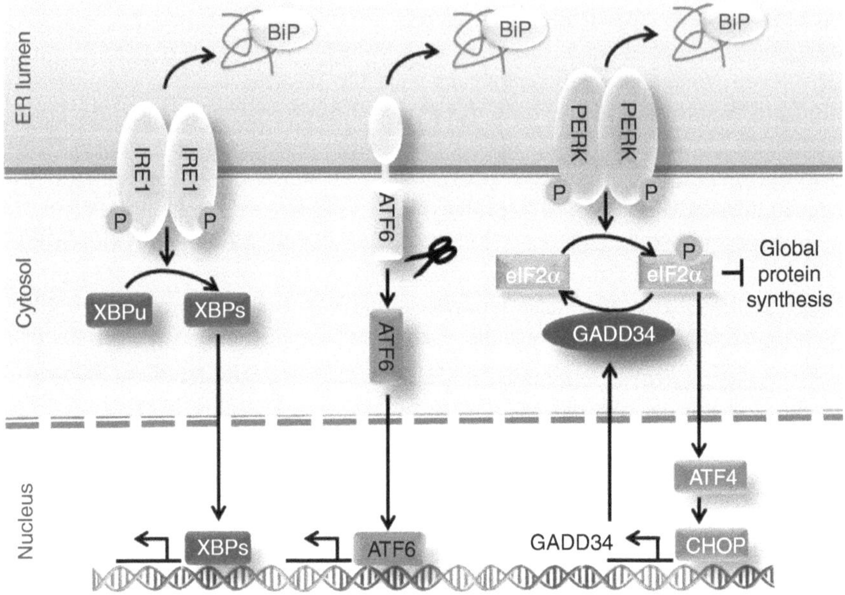


FIG. 1. The unfolded protein response (UPR). The schematic shows the activation of the three ER transmembrane sensors to transduce UPR signaling, following the sequestration of BiP by an increase in misfolded proteins in the ER. The transcriptional pathways are activated by IRE1 dimerization and autophosphorylation by the IRE1 kinase. This activates the nuclease activity associated with IRE1 to promote the splicing of the Xbp1 mRNA and association of Xbp1s transcription factor to UPRE sequences in the promoters of ERAD genes to enhance their transcription. ATF6 is transported from ER to golgi (not shown), where it is proteolysed (depicted by scissors) to release the mature ATF6 transcription factor. ATF6 translocates to the nucleus to activate chaperone genes containing ERSE sequences. The translational repression pathway is activated by PERK dimerization and autophosphorylation. PERK phosphorylates eIF2α to inhibit global protein translation. The phosphorylated eIF2α enhances the translation of ATF4, a transcription factor that promotes the expression of CHOP and GADD34, which feeds back to recruit PP1α (not shown) and enhance the dephosphorylation of eIF2α. (See Color Insert.)

The primary goal of translation repression is to reduce the protein load in the ER and buy cells time to deal with the acquired burden of misfolded proteins. However, in cells experiencing severe or chronic ER stress, CHOP/GADD153, a transcription factor, activates the transcription of proapoptotic genes, such as Bim (and caspase-12), to trigger programmed cell death and thereby eliminate the irreparably damaged cells.[37] CHOP also activates the transcription of ERO1 (ER oxidoreductin I), which catalyzes the thiol–disulfide exchange with protein disufide isomerase to promote protein disufide bridges.[38] This increases oxidative stress, which can reach lethal levels in the face of excessive protein misfolding. However, under conditions of acute or reversible ER stress, ATF4, either acting alone[39] or in conjunction with CHOP,[38] activates the transcription of GADD34. GADD34 is an ER-associated phosphatase regulatory subunit, which binds the alpha isoform of protein phosphatase-1 (PP1) to generate an active eIF2α phosphatase. Thus, GADD34 functions in a feedback loop to restore general protein synthesis,[39,40] which is essential for the continued cell growth and viability.

The eIF2α phosphatase places a strong brake on the phosphorylation of eIF2α by PERK (and other eIF2α kinases), impairing their ability to repress translation in response to a subsequent stress. Moreover, the persistent or prolonged expression of GADD34 by enhancing general protein translation has the potential to further exacerbate protein misfolding. Mathematical modeling of the PERK pathway activated by acute, reversible, or chronic terminal ER stress suggested that the instability or rapid turnover of the GADD34 protein plays a critical role in the cell's decision to die or survive.[41] This was supported by studies that highlighted the hypersensitivity of GADD34-overexpressing cells to cell death induced by oxidative stress[42] and irradiation.[43] Thus, mammalian cells have developed elaborate mechanisms, including the trafficking of GADD34 protein from cytosol to ER[44] and polyubiquitination of the α-amino group at the N-terminus of GADD34,[45] to actively degrade the protein using the 26S proteasome. The GADD34 protein is also highly phosphorylated, but the physiological role of these covalent modifications has not been defined. We speculate that GADD34 phosphorylation may also control its turnover and/or function to determine cell fate. It is worth noting that stress-activated BiP expression was markedly reduced in the GADD34-null mouse embryo fibroblasts,[46] suggesting that GADD34 may be required to complete the later transcriptional phase of UPR signaling that results in the increased expression of BiP and protection of cells from apoptosis in response to protein misfolding.

Emerging studies have indicated a critical role for PERK in the activation of autophagy,[47] which may also promote cell survival following ER stress. The precise mechanistic link between eIF2α phosphorylation and the expression of ATF4, CHOP, and GADD34 in the activation of autophagy proteins remains to be investigated.

## B. Enhancing Protein Folding Capacity

The ER transmembrane protein ATF6 (activating transcription factor 6) is another UPR sensor, which when dissociated from BiP is trafficked from ER to golgi, where it is successively cleaved by site-1 and site-2 proteases, which are better known for their role in processing SREBPs,[48] transcription factors that control fatty acid and cholesterol biosynthesis.[49] The cleaved ATF6 migrates to the nucleus and activates the transcription of genes encoding BiP and other chaperones.[50] The multiple steps involved in trafficking of ATF6 from ER to golgi and, following its proteolytic processing, entry into the nucleus mean that the ATF6 pathway is slower to activate than the PERK pathway. Moreover, BiP is a longer lived protein than ATF4, CHOP, or GADD34. Thus, the activation of the ATF6 pathway and expression of BiP precondition cells[41] and allow them to adapt better to subsequent bouts of protein misfolding, thereby ensuring their survival. High levels of BiP are often seen in diseased tissues and may reflect the presence of a chronic ER stress.

Two isoforms, ATF6$\alpha$ and ATF6$\beta$, are both induced by stress, with ATF6$\alpha$ being the predominant isoform in most cells. While targeted gene disruption in mice has suggested that ATF6$\alpha$ and ATF6$\beta$ display redundant function,[51] other studies have shown that ATF6$\beta$ expression is slightly delayed compared to ATF6$\alpha$ and that ATF6$\beta$ attenuates ATF6$\alpha$ function.[52] This hints at yet another feedback loop that regulates the duration of ER stress. Studies of the ATF6$\alpha$-null mice have suggested that GADD34 (and CHOP) levels are higher in tissues of the mutant mice following ER stress compared to wild-type mice, hinting that ATF6$\alpha$ may repress GADD34 gene transcription.[53] This also suggests that the ATF6$\alpha$ may sustain eIF2$\alpha$ phosphorylation and translational repression while increasing the cellular protein folding capacity through expression of BiP and other chaperones to ensure cell survival. Under circumstances in which the elevated levels of BiP and other chaperones fail to clear the buildup of misfolded proteins, their futile efforts at refolding proteins will likely generate an oxidative stress to damage mitochondria and trigger apoptosis.

## C. Increased Degradation of Misfolded Proteins

The third ER stress sensor, IRE1, also contains a luminal domain that binds BiP. In the presence of unfolded proteins, as described above for PERK and ATF6, BiP is competed away from IRE1. IRE1 is a dual-function enzyme, containing both kinase and endonuclease activity. Upon the release of BiP, IRE1 dimerizes and autophosphorylates.[27] This results in its increased endonuclease activity. Two isoforms, IRE1$\alpha$ and IRE1$\beta$, are expressed in many cells, with IRE1$\alpha$ being the predominant isoform. IRE1$\alpha$ degrades ER-associated mRNAs and, thus, may also contribute to translational repression.[54] However, the major function of IRE1 is to catalyze the splicing of the mRNA encoding the transcription factor XBP1, which is the product of an ATF6-induced gene.[31] This allows the mature XBP1 protein to be translated and subsequently to activate the transcription of genes encoding components of the ER-associated protein degradation (ERAD) machinery. ERAD is associated with the translocation of misfolded proteins from the ER lumen to cytosol for degradation by the 26S proteasome.[55]

One gene activated by XBP1 is p58IPK, a member of the Dnaj family of proteins.[56] p58IPK was first identified as a cellular factor utilized by influenza virus to antagonize PKR and promote viral replication.[57,58] Subsequent studies showed that p58IPK targets both PKR and PERK to inhibit eIF2$\alpha$ phosphorylation.[59,60] In addition, p58IPK contains a signal peptide sequence that targets it to the ER lumen, where it binds BiP and functions as a cochaperone to enhance protein folding.[61] Thus, in addition to the activation of ERAD, the IRE1 signaling pathway may also terminate eIF2$\alpha$ phosphorylation and activate BiP function, highlighting the crosstalk and coordination of the three UPR signaling pathways.

In summary, mammalian cells use a three-pronged approach to reduce protein load, increase protein folding, and degrade misfolded proteins to ensure their survival. Emerging data suggest that the crosstalk between the three UPR signaling pathways is critical for cells to effectively manage the stress. In particular, distinct components of the three pathways work to dictate the extent and duration of eIF2$\alpha$ phosphorylation and translational repression required for cell recovery.[31] It is also clear that the three pathways allow different cell types to respond differently to stress. For example, the ATF6 pathway is preferentially activated by ischemia.[62] This diversity also plays a role in the susceptibility of tissues to different protein misfolding disorders.

## VI. UPR and Human Disease

Numerous human diseases arise from problems in protein folding. While many mechanisms contribute to protein misfolding, these disorders can be clustered into three major groups (Table I). These include diseases caused by cells experiencing a protein overload which overwhelms the ER protein quality control machinery that normally manages the day-to-day protein folding. Mutations in proteins that compromise their ability to be folded and errors in the cellular protein processing machinery also create a chronic UPR that results in disease. Such mutations when introduced into mice can be used to model the human disease, and these mouse models highlight the dramatic activation of ER stress in diseased tissues. However, these experimental models have not provided clear information about the stress signaling pathways most relevant to the disease process, nor any hints on whether pharmacological targeting of any one pathway can achieve the desired therapeutic benefit without toxicity. For example, mice expressing the various mutant human superoxidase dismutases (SOD1) associated with human amylotrophic lateral sclerosis (also known as Lou Gehrig's disease) acquire motor neuron disease similar to that seen in humans.[63] Likewise, the transgenic expression of mutant human Huntingtin[64] or $\alpha$-synuclein[65] have yielded useful mouse models of Huntington's and Parkinson's disease, respectively. All these models highlight

TABLE I
Human Diseases Associated with Protein Misfolding

| | |
|---|---|
| Protein overload | Alzheimer's disease |
| | Parkinson's disease |
| | Hereditary hyperthyroidism |
| | Creutzfeld–Jacob disease (prion disease) |
| | Type-2 diabetes (hyperinsulinemia) |
| | Familial amyelotrophic lateral sclerosis |
| Mutant proteins | Cystic fibrosis (mutant CFTR) |
| | Nephrogenic diabetes insipidus (mutant aquaporin) |
| | Hereditary hypercholesterolemia (mutant LDLR) |
| | Wallcott–Rallison syndrome (mutant PERK) |
| | Osteogenic imperfecta (mutant collagen) |
| | Retinitis pigementosa (mutant rhodopsin) |
| Protein processing | Albinism (tyrosine deficiency) |
| | Obesity (prohormone convertase-1 deficiency) |
| | α1-Antitrypsin deficiency |
| | Thyroglobulin deficiency |
| | Lipoprotein lipase deficiency |

Three major contributors to protein misfolding include protein overload, mutations resulting in folding-defective proteins, and errors in protein processing. The diseases listed exemplify human disorders associated with protein misfolding. While diseases such as Parkinson's are linked to gene mutations, these account for only a small fraction of the disease population, and the mechanism underlying in accumulation of Lewy bodies in most patients may be indirect, such mitochondrial damage and high levels of reactive oxygen species. Yet others, such as retinitis pigmentosa, result from many mutations, and for simplicity, only one example is shown. Finally, errors in UPR, such as mutations in PERK, can also contribute to protein misfolding.

the remarkable increase in ER stress that accompanies the expression of mutant proteins in the target tissues.[66–68] However, it has still been difficult to decipher the precise role of ER stress in the disease process, and, most importantly, these models have provided only limited insights into the risks and benefits of pharmacologically targeting individual UPR signaling pathways. With this in mind, our discussion will focus on gene disruption or knockout studies in mice that may help us to evaluate the contribution of individual UPR signaling components in the physiology and pathophysiology of different cells and organs (Table II). It is anticipated that, when combined with the aforementioned mouse models of human disease, these animal models will yield invaluable information on the most appropriate therapeutic strategy for the treatment of diseases associated with chronic or severe ER stress.

## A. Genetic Targeting of Transcriptional Pathways Activated by UPR

Following the detection of misfolded proteins in the ER, ATF6 and IRE1 activate parallel transcriptional pathways that are crucial for the long-term adaptation of cells to ER stress. As both pathways enhance the expression of

TABLE II
DISRUPTION OF GENES ENCODING MOUSE UPR SIGNALING COMPONENTS

| Component | Function | Outcome of mouse mutations |
|---|---|---|
| eIF2α | Eukaryotic translation intitation factor | Homozygous eIF2α (S51A) mice died 1 day after birth.[69] |
| GCN2 | eIF2α kinase | GCN2-null mice were viable but showed sensitivity to lack of dietary amino acids.[5,6] |
| HRI | eIF2α kinase | HRI-null mice were viable but had low red blood cells and erythroid progenitors.[70] |
| PKR | eIF2α kinase | PKR-null mice were viable and showed increased susceptibility to virus infections.[71,72] |
| PERK | ER-transmembrane eIF2α kinase | PERK-null mice were viable but displayed early onset diabetes and mild skeletal defects.[11,73] |
| GADD34 | Stress-induced ER-associated subunit of eIF2α phosphatase | GADD34-null mice were viable and had no overt phenotype. One strain showed aberrant red cell count and morphology.[74] |
| CReP | ER-associated subunit of a constitutive eIF2α phosphatase | CReP-null mice were pale (reduced red blood count), stunted and died within 1 day of birth.[75] |
| ATF4 | Transcription factor | ATF4-null mice were viable but blind due to lens defects.[76,77] |
| CHOP | Transcription factor | CHOP-null mice were viable and with no overt phenotype.[38] |
| ERO1 | ER luminal oxidoreductin | ERO1-α-null mice had mild perturbations in cardiac calcium homeostasis.[78] ERO1-β mutant mice had mild diabetes.[79] |
| Caspase-12 | ER-associated caspase | Caspase-12 mutant mice were viable and displayed no overt phenotype.[80] |
| ATF6 | ER-transmembrae protein and transcription factor | ATF6α and ATF6β mutant mice were both normal, but the double knockouts were embryonic lethal.[52,53] |
| BiP | ATF6-induced ER luminal chaperone | BiP-null mice were embryonic lethal.[81] |
| IRE1 | ER-transmembrane kinase/nuclease | IRE1α-null mice were embryonic lethal with defects in B cell lymphopoesis while IRE1β-null mice were viable with no obvious phenotype.[82,83] |
| Xbp1 | ATF6-induced transcription factor processed by IRE1 | Xbp1-null mice were embryonic lethal with defects in liver and exocrine gland development.[84–86] |

A number of UPR components have been mutated in mice and provided insights into the tissues whose physiology is most reliant on these components. Yet others highlight the risk associated with impairment of their function. These mutant mice provide an excellent experimental platform to test the physiological impact of specific disease genes and alterations in diet and other factors implicated in human protein misfolding disorders.

chaperones and components of the ERAD machinery, together they enhance the cell's tolerance for misfolded proteins and protect cells from death induced by chronic stress.

ATF6α is the major isoform expressed in most mammalian cells but the disruption of the mouse ATF6α gene has no discernable effect on mouse development.[51,53] When presented with a transient pharmacological stress that induces liver dysfunction and steatosis (fatty liver) in mice, however, the ATF6α-null mice showed significant difficulty in recovering from this insult compared to wild-type mice.[87] In contrast to ATF6α, which is a short-lived protein and a strong activator of ERSE (ER stress response element) containing genes, ATF6β is a more stable protein and a weaker transcriptional activator.[52,88] The loss of function of the mouse ATFβ gene was also without any notable phenotype.[51] The disruption of both ATF6 genes, however, resulted in embryonic lethality.[51] This suggested a critical role for the combined actions of both ATF6 isoforms in mammalian development but provided no hint of their role in human disease. Analysis of UPR signaling in the ATF6α-null mouse embryo fibroblasts (MEFs) highlighted a severe deficit in BiP induction in response to various pharmacological stresses, which was consistent with ATF6α's known role in activating BiP transcription.[51,53] By contrast, ATF6β MEFs were indistinguishable from wild-type MEFs in their response to ER stress.[51] These data questioned the proposed role for ATF6β as a repressor of ATF6α function[52] but, instead, suggested a nonredundant role for the two ATF6 isoforms in transducing UPR signals. The function of ATF6β in UPR signaling remains unknown.

Interestingly, ATF6α-null MEFs also showed an attenuated induction of CHOP and p58IPK by ER stress.[51,53] These data were consistent with the notion that ATF6α cooperates with ATF4 to induce CHOP expression and raised the possibility that ATF6α may also collaborate with Xbp1, a known transcriptional activator of p58IPK.[56] Indeed, ATF6α formed a heterodimer with Xbp1 to bind UPRE elements in promoters of ERAD genes. Moreover, the induction of ERAD genes by ER stress was either strongly attenuated or completely lost in the ATF6α-null MEFs.[51] Another surprise was that GADD34 induction was greatly enhanced in the ATF6α-null MEFs,[53] which was inconsistent with GADD34 simply being an ATF4-[40] or CHOP-induced gene[38]; the transcriptional regulation of the GADD34 gene is still poorly understood. On the other hand, GADD34 mRNA levels were unchanged in yet other ATF6α-null MEFS, suggesting that the mode of gene disruption, deletion of exon 4[53] or exons 8 and 9,[51] or the particular mouse strain influenced these results.

As with the embryonic lethality seen in the ATF6α/β double knockout, disruption of the mouse BiP gene, a major target of ATF6α, was also embryonic lethal.[81] Loss-of-function mutations in the mouse IRE1α[82,83] and Xbp1[84] genes also resulted in early embryonic lethality. These studies suggested a critical role for the UPR-activated transcriptional pathways, specifically ATF6/BiP and IRE1α/Xbp1, in mouse development. Most notably, Xbp1 was required for the differentiation of hepatocytes[85] and the lethality of the Xbp1-null mice was rescued by the liver-specific reexpression of Xbp1.[86] However, the role of UPR in tissue differentiation and development is still poorly understood. Remarkably, gene expression analyses in the knockout MEFs suggested that ATF6α and Xbp1 served overlapping or redundant functions, activating a subset of UPRE (UPR element) containing genes.[56,89] Similarly, there was significant overlap in the genes whose expression required functional ATF6α and PERK pathways.[89] Genome-wide analysis of Xbp1 binding sites highlighted genes associated in protein synthesis, folding, and trafficking as expected, but also identified DNA damage and repair genes and genes associated with neurodegenerative and myodegenerative disorders.[89] Most importantly, these studies pointed to genes that were expressed in a tissue- and condition-specific manner and suggested that Xbp1 regulated the differentiation of muscle, pancreas, and salivary glands.

In contrast to BiP knockout mice, homozygous "knockin" mice expressing a mutant BiP and lacking the KDEL ER-targeting sequence, survived several hours after birth. These mutant animals died of respiratory failure due to chronic ER stress induced by the impaired synthesis of pulmonary surfactants.[90] By comparison, the heterozygous mutant BiP (KDEL) mice developed age-dependent renal lesions that were exacerbated by protein overload.[91] These data suggested that BiP played a critical role in protecting the kidney from chronic disease or injury resulting from an excessively high protein diet.

## B. Gene Targeting of Translational Pathway Activated by UPR

Phosphorylation of eIF2α is an early response to misfolded proteins and is activated by osmotic stress,[92] glucose deficiency,[93] and a wide range of other stresses. To define the physiological importance of eIF2α phosphorylation at serine-51, Scheuner et al.[69] generated a "knockin" mouse in which serine-51 was substituted by alanine to abolish eIF2α phosphorylation. The homozygous eIF2α (S51A) mutant mice died within the first day of life, as a result of a severe loss in pancreatic β-cells and hyperglycemia. While the heterozygous eIF2α (S51A) mutant mice appeared normal on a chow diet, these animals also developed severe β-cell dysfunction when placed on a high-fat diet.[94] These data hinted at a critical role for eIF2α phosphorylation in the development and functioning of pancreatic β-cells, in which the elevated basal translation or proinsulin induced by eIF2α (S51A) likely elicited a chronic ER stress[94] and subsequent apoptosis.

As discussed earlier, four distinct kinases catalyze eIF2α phosphorylation at serine-51 in mammalian cells (Chapter 8). GCN2 is the primary sensor of nutrient availability.[95] It promotes eIF2α phosphorylation in liver of mice fed a diet lacking leucine.[96] While homozygous GCN2-null mice were viable and fertile, when presented with a diet deficient in amino acids, many stillborn pups were delivered.[6,96] Injection of threoninol, a drug that blocks threonyl-tRNA charging in wild-type mice, altered their preference for a threonine-rich diet over a threonine-depleted diet. By contrast, threoninol had no impact on the dietary preference of the GCN2-null mice.[97] Finally, amino acid deprivation failed to increase eIF2α phosphorylation in tissues of the GCN2 knockout mice,[97,98] providing strong support for a pivotal role of GCN2 in nutrient sensing.[95] Indeed, GCN2 promoted the survival of cancer cells derived from solid tumors that were poorly perfused and therefore deprived of nutrients.[99]

HRI is the major eIF2α kinase in erythroid cells and inhibits the synthesis of α and β globins in response to iron deficiency.[70] Consistent with this observation, HRI-null mice have an abnormal blood count and, when stressed, erythrocytes from the mutant mice overproduced globins and experienced hemolysis.[70] Thus, HRI appears primarily essential for the viability of red cell precursors.

PKR has broader substrate specificity than other eIF2α kinases, and its substrates include many transcription factors and other proteins. Thus, PKR not only controls cell growth and differentiation but also activates apoptosis.[8] Deletion of the double-stranded RNA-binding domain in PKR mutant mice resulted in no apparent phenotype other than a modest increase in their susceptibility to viral infection.[71,72] Deletion of the PKR kinase domain also yields mice with no overt phenotype,[100] but analysis of these PKR-null MEFs identified a pathway by which fatty acids activated PKR, which then phosphorylated IRS1 (insulin receptor substrate-1) to inhibit insulin signaling. Consistent with these findings, the PKR-null mice are significantly protected from the development of insulin resistance when placed on a high-fat diet.[93] These data suggest an intriguing link between obesity and viral infections, both of which activate PKR, and the development of diabetes.[93] These studies also highlighted the challenges of generating genetically modified mice in which different functional domains are selected for gene disruption, yielding different phenotypes.

PERK is the major eIF2α kinase in most tissues, particularly secretory tissues, and the only UPR sensor known to be mutated in humans. The Wolcott–Rallison syndrome (WRS) is a rare genetic disorder associated with missense mutation in the human PERK gene that prevents the expression of a functional kinase.[101] WRS is characterized by early onset diabetes with a severe loss of β-cell mass, aberrant skeletal development, mental retardation, cardiovascular abnormalities, and other defects.[101] Interestingly, SNPs (single nucleotide polymorphisms) in the human PERK gene[102] are also linked to an increased risk for type-2 diabetes. Homozygous PERK-null mice were born with a normal pancreas but soon lost β-cells to develop early diabetes.[11,103] These animals also developed skeletal dysplasia associated with defective bone mineralization similar to WRS patients. Yet other studies of PERK-null MEFs highlighted the severely attenuated eIF2α phosphorylation in response to a variety of stresses, although compensatory increases in IRE1α activation and reduced cell survival when exposed to ER stress were also observed.[11] Interestingly, cells and tissues from the PERK-null mice displayed altered ER morphology and defective calcium mobilization.[104] These data suggest that PERK activity is required for calcium signaling and insulin secretion and that the loss of PERK function impacts multiple cellular processes to elicit diabetes. Finally, PERK was required for the induction of autophagy, a pathway that contributes to cell survival under conditions of severe nutrient starvation.[47] More recent studies suggest that PERK (and ATF4) is necessary for ATF6 transcription and translation[105] and, therefore, for downstream events such as expression of BiP and other chaperones. These data argue for a protective role for PERK in chronic ER stress and the development of human disease, particularly type-2 diabetes.

Additional evidence supporting a role for UPR signaling in β-cell viability and human diabetes comes from the analysis of Wolfram syndrome, an autosomal recessive disorder associated with juvenile-onset diabetes. This disease results from loss-of-function mutations in the WFS1 gene encoding Wolframin, a β-cell ER membrane protein.[106,107] ER-stress-induced Wolframin expression and the levels of Wolframin were greatly reduced in both PERK-null and IRE1α-null MEFs following ER stress. This suggested that both these UPR signaling pathways contributed to WFS1 gene activation. The WFS1 mutations increased ER stress in β-cells and suggested a chaperone role for Wolframin in proinsulin folding.[108] Polymorphisms in WFS1 gene were also linked to type-2 diabetes.[109,110] Homozygous WFS1-null mice are small and male sterile, making it difficult to accurately model Wolfram syndrome in mice.[111] The conditional deletion of WFS1 in pancreatic β-cells yielded mice with overt diabetes resulting from increased ER stress and apoptosis in β-cells.[112]

The Xbp1-induced gene p58IPK encodes an ER-associated Hsp40 family protein induced by ER stress and inhibits eIF2α phosphorylation by PKR and PERK.[59,60] Deletion of the mouse p58IPK gene had no impact on the pancreatic development but the mutant animals displayed a progressive loss of β-cells, with concomitant reduction in insulin production and diabetes.[113] As p58IPK serves many functions in the ER (e.g., cochaperone for BiP), it is difficult to attribute these findings solely to changes in PERK activity. Nevertheless, the above data support a key role for PERK and eIF2α phosphorylation in insulin synthesis and β-cell survival.

Phosphorylation of eIF2α in mammalian cells is opposed by a signaling complex composed of protein phosphatase-1α bound to either GADD34 (protein encoded by the PPP1R15A gene[114]) or CReP (the product of the PPP1R15B gene[115]) (Fig. 2). Initially discovered as an mRNA transcript induced in Chinese hamster ovary (CHO) cells by ultraviolet irradiation,[116] GADD34 was postulated to be a proapoptotic gene and, consistent with this, GADD34 overexpression sensitized cells to apoptosis induced by ionizing radiation.[43,117] GADD34 was also identified as Myd116, a gene required for myeloid differentiation.[118,119] Later studies established GADD34 as a component of a stress-activated eIF2α phosphatase.[12,46] Three distinct GADD34 knockout mice have been generated,[46,74,120] and in one of these (deleting

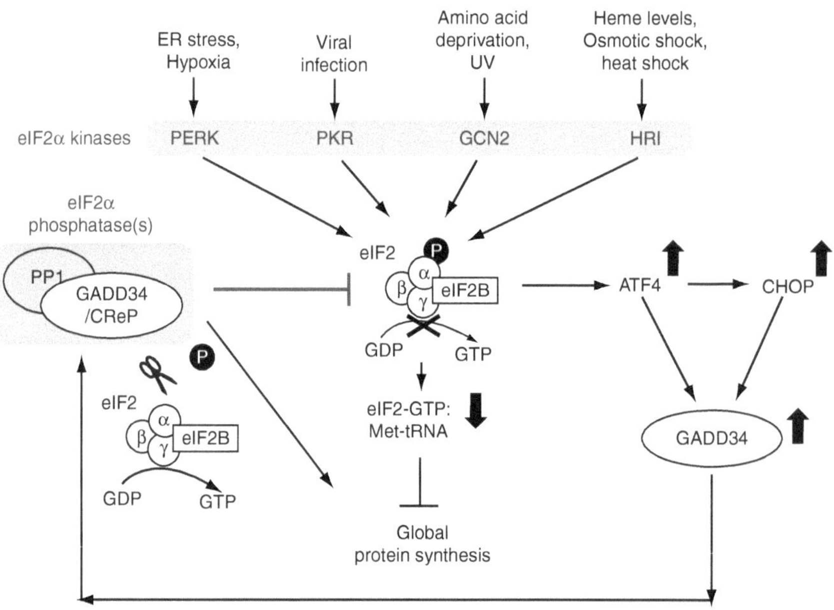


FIG. 2. Cellular control of eIF2α phosphorylation. Four eIF2α kinases, PERK, PKR, GCN2, and HRI, are activated by distinct stresses to phosphorylate eIF2α at serine 51 and inhibit the recycling or exchange of GDP for GTP from the heterotrimeric eIF2 complex. This prevents the assembly of the eIF2 complex as well as initiation of translation at other mRNA start sites and attenuates general protein synthesis. The phosphorylated eIF2α, however, permits the translation of ATF4 a well as its downstream targets CHOP and GADD34. GADD34 gene expression may utilize both ATF4 and CHOP. Once expressed, GADD34 binds the PP1α catalytic subunit. The PP1 complexes containing GADD34 and its homolog CreP catalyze the dephosphorylation of eIF2α to promote GDP/GTP exchange on the eIF2 complex and reinitiate protein synthesis. (See Color Insert.)

exons 2 and 3) a thalassemia-like phenotype with hypersplenism and decreased hemoglobin content in erythrocytes were observed, suggesting a key role for GADD34 in hemoglobin synthesis.[74] This phenotype was remarkably similar to that of the HRI-null mouse,[70] suggesting a requirement for both kinase and phosphatase to control eIF2α phosphorylation and protein translation in red blood cells. In two other GADD34 mutant mice, one deleting exon 1[120] and the other exon 3,[46] the blood count was normal. ER-stress-induced eIF2α phosphorylation was prolonged in all three GADD34-null MEFs, supporting GADD34's role as a key component of an eIF2α phosphatase. Intraperitoneal injection of tunicamycin in wild-type mice-induced ER stress and renal lesions resembling human acute tubular necrosis, while the kidneys of GADD34-null mice were remarkably spared under similar circumstances.[46] GADD34-null mice also showed a remarkably enhanced capacity for wound healing,[121] suggesting that GADD34 attenuates cell migration by inhibiting myosin IIA expression; but the underlying mechanism remains unclear. GADD34 overexpression also activated autophagy and protected cells expressing mutant Huntingtin from ER-stress-induced programmed cell death.[122] These studies hinted at a cytoprotective role for GADD34 in ER stress, linking changes in eIF2α phosphorylation to mTOR signaling.[123] Transgenic mice expressing an activated form of GADD34 in the liver maintained insulin sensitivity and diminished hepatic steatosis induced by a high-fat diet,[124] demonstrating an additional protective role for GADD34 in the liver. On the other hand, these mutant mice showed reduced insulin sensitivity in skeletal muscle and fat tissue, possibly resulting from elevated circulating IGFBP-3 (insulin-like growth factor binding protein 3).[125] These data suggest that changes in eIF2α phosphorylation in liver are communicated via plasma IGFBP-3 to other tissues to regulate their insulin sensitivity. Thus, elevated GADD34 levels either directly or indirectly alter the physiology of other tissues to modulate insulin resistance, hyperinsulinemia, and eventually type-2 diabetes. In this regard, it is noteworthy that new studies point to the ability of stressed cells to transmit ER stress to other as yet unstressed neighboring cells.[126]

The cellular content of GADD34 protein is regulated by multiple mechanisms, including the stress-induced transcription of the GADD34 gene possibly via the combined actions of ATF4 and CHOP.[38,39] The GADD34 mRNA is highly labile,[36,41] though the molecular basis for its instability is still unclear. ER-stress-induced phosphorylation of eIF2α promotes the translation of GADD34 mRNA, which contains two uORFs[36] as noted for CHOP[35] and ATF4.[34] Finally, the GADD34 protein is actively degraded by a mechanism involving its translocation to ER,[44] polyubiquitination of its N-terminal α-amino group,[45] and degradation by the proteasome. Mutations that stabilized GADD34 resulted in the accumulation of intracellular protein aggregates[45] and cell death (Brush and Shenolikar, unpublished data).

CReP, a less well-studied homolog of GADD34, assembles a constitutive eIF2α phosphatase that is unresponsive to ER stress.[115] Homozygous CReP-null mice showed a reduced red cell count and were pale and stunted at birth. These mutant animals failed to nurse and died within 1 day of birth.[75] Interestingly, the double knockout of GADD34 and CReP genes in mice resulted in a more severe phenotype, with embryonic lethality occurring prior to the implantation stage.[75] This latter phenotype was largely rescued by crossing the GADD34/CReP double knockout mice with the homozygous eIF2α S51A mice. These data demonstrated that the function of GADD34 and CReP as eIF2α phosphatases was not shared by other cellular phosphatases. The role of eIF2α phosphatases in mammalian development was transmitted primarily by the dephosphorylation at a single phosphoserine residue, serine-51, in a single phosphoprotein substrate, namely eIF2α.[75]

Examining some of the downstream targets of eIF2α phosphorylation provides further insights into the physiological role of the translation pathway activated by UPR. For example, eIF2α phosphorylation results in the enhanced translation of ATF4 (activating transcription factor 4), which in turn activates the transcription of CHOP[127] and GADD34.[40] Homozygous ATF4 knockout mice are born at less than expected Mendelian ratios and have defects in eye lens development.[76,77] Moreover, ATF4-null MEFs failed to induce GADD34 in response to stress, thereby showing a prolonged eIF2α phosphorylation and delayed translation recovery.[40] Interestingly, ATF4−/− MEFs, like the GADD34−/− MEFs, showed significantly reduced BiP expression following ER stress, emphasizing the role for GADD34 in the translation (or transcription) of BiP. CHOP has been implicated in the expression of GADD34[38] as well as proapoptotic genes, such as Bim and caspase-12.[37] Consistent with these observations, the disruption of mouse CHOP gene protected both cells and animals against protein aggregation and cell death induced by ER stress.[38]

Remarkably, CHOP activates the transcription of genes implicated in both cell survival, such as the ER-associated oxidoreductin 1 (ERO1), and cell death, such as the ER-localized caspase-12. The mice lacking ERO1-α, which is expressed in most tissues, were phenotypically normal, although calcium homeostasis in cardiomyocytes of the mutant animals was perturbed.[78] By contrast, mice lacking ERO1-β, selectively expressed in pancreatic β-cells, showed a mild diabetic phenotype with modest fasting hyperglycemia.[79] Surprisingly, ERO1-α was also present in β-cells but did not compensate for the loss of ERO1-β, suggesting nonredundant roles for the two ERO1 isoforms. Cellular studies showed that the loss of function of ERO1-β enhanced the ability of β-cells to cope with ER stress induced by the expression of a folding-defective mutant insulin *Akita Ins2*(C96Y). However, this benefit was not realized in mice, where the loss of a single ERO1-β allele exacerbated the diabetic phenotype associated with the *Akita Ins2*(C96Y) mutation.[79] These results highlight the necessity to utilize animal models rather than cell lines to reliably assess the role of UPR signaling in disease.

The caspase-12-null mice were resistant to ER-stress-induced apoptosis, namely renal toxicity induced by tunicamycin injection,[80] and the cortical neurons from these mutant mice were also protected from neurotoxicity and death induced by amyloid-β. However, caspase-12-null mice were not protected from infectious prion disease,[128] indicating that caspase-12 is not essential for the pathogenesis of all neurological protein misfolding disorders.

Together, the above studies revealed a complex interplay between the translational and transcriptional pathways that mediate UPR and control the function and viability of different tissues. The data also highlighted that cellular studies, using either overexpression or knockdown, suggested roles for UPR signaling components that frequently did not replicate in animals, Nevertheless, the broad conclusions of the gene disruption studies in mice were that suppression of the transcriptional pathways that impaired chaperone expression or ERAD might incur severe toxicity or side effects. By comparison, modulation of the translational pathway, specifically eIF2α phosphorylation, provided a safer outcome from the pharmacological targeting of UPR. Even here, targeting specific isoforms, for example, GADD34 versus CReP, may be important to ensure greater safety.

## VII. Modulation of eIF2α Phosphorylation: Potential for Treating Human Disease

Although ER stress or protein misfolding has been implicated in a wide range of disorders, the strategy for targeting UPR signaling may differ depending on whether the therapeutic aim is to protect cells, such as β-cells, neurons, and cardiomyocytes, from cell death in the face of chronic ER stress, or to evoke terminal ER stress and apoptosis in cancers that are difficult to treat by current therapies. For simplicity, we will limit this discussion to potential cytoprotective strategies, but even here the strategy may differ depending on whether ER stress is incurred by genetic mutations that result in folding-defective proteins or one wishes to target diseases resulting from protein overload or deficits in protein processing (Table II). We speculate that chemical chaperones, drugs that elevate cellular expression of heat shock proteins or antioxidants that reduce oxidative stress caused by futile protein folding, may be better suited for the treatment of diseases arising from expression of folding-defective proteins. By comparison, protein overload associated with viral infections or hyperinsulinemia associated with insulin resistance that contributes to β-cell exhaustion and type-2 diabetes might be better managed by attenuating protein translation. In the latter situation, the modulation of eIF2α phosphorylation may offer an attractive opportunity for the development of novel antidiabetic (or antiviral) drugs.

Salubrinal was discovered while screening for small molecules that protected rat PC12 cells from death caused by a pharmacological ER stress.[129] Further analyses established that salubrinal, which elevated eIF2α phosphorylation, inhibited one or more cellular eIF2α phosphatase(s). Indeed, salubrinal inhibited eIF2α phosphatases containing GADD34, CReP (Fig. 3), or ICP34.5, the HSV-1 protein that overcomes PKR activation.[129,130] These data suggested that salubrinal targeted the conserved PP1-binding domain in these three proteins and disrupted their ability to scaffold PP1α, and possibly eIF2α,[131] and assemble a functional eIF2α phosphatase.

Salubrinal showed remarkable protective effects in a number of experimental models of neurodegenerative disease. For example, amyloid β-protein (Aβ) aggregates, seen as plaques in postmortem brain of individuals with Alzheimer's disease, induce ER stress in cells.[132] In this regard, salubrinal increased eIF2α phosphorylation and attenuated apoptosis induced by Aβ in human neuroblastoma cells.[133] In cells expressing mutant Huntingtin, the primary cause of protein aggregates and neuronal death in Huntington's disease,[134] salubrinal abrogated apoptotic signaling and increased cell survival.[135] Salubrinal also protected PC12 cells from cell death induced by α-synuclein,[135] a component of Lewy bodies seen in Parkinson's disease. In animals, salubrinal protected against neuronal death caused by ischemia and excitotoxicity in the rat brain.[136] In the transgenic mouse models of familial amyotrophic lateral sclerosis (fALS), salubrinal reduced the buildup of polyubiquitinated proteins in motor neurons, delayed the progression of motoneuron disease, improved muscle strength, and extended the lifespan of the mutant mice.[137]

Interestingly, salubrinal inhibited the replication of HSV1,[129,130] dengue,[138] and other viruses in cultured cells, indicating the potential of eIF2α phosphatase inhibitors as pan antiviral agents. Salubrinal also overcame the apparent resistance of multiple myeloma cells to bortezomib (also known as velcade), a proteasomal inhibitor clinically approved for these cancers. Bortezomib elevates GADD34 in target cells, antagonizing eIF2α phosphorylation and leading to cell cycle arrest.[139] Salubrinal, by inhibiting GADD34 function, synergized with bortezomib to enhance the death of multiple myeloma[139] and leukemic cells.[140]

One concern associated with the use of salubrinal and possibly other eIF2a phosphatase inhibitors is that, by increasing eIF2a phosphorylation, these drugs will likely increase cellular levels of their target GADD34 and thereby limit the duration of their action. While this could make it difficult to develop

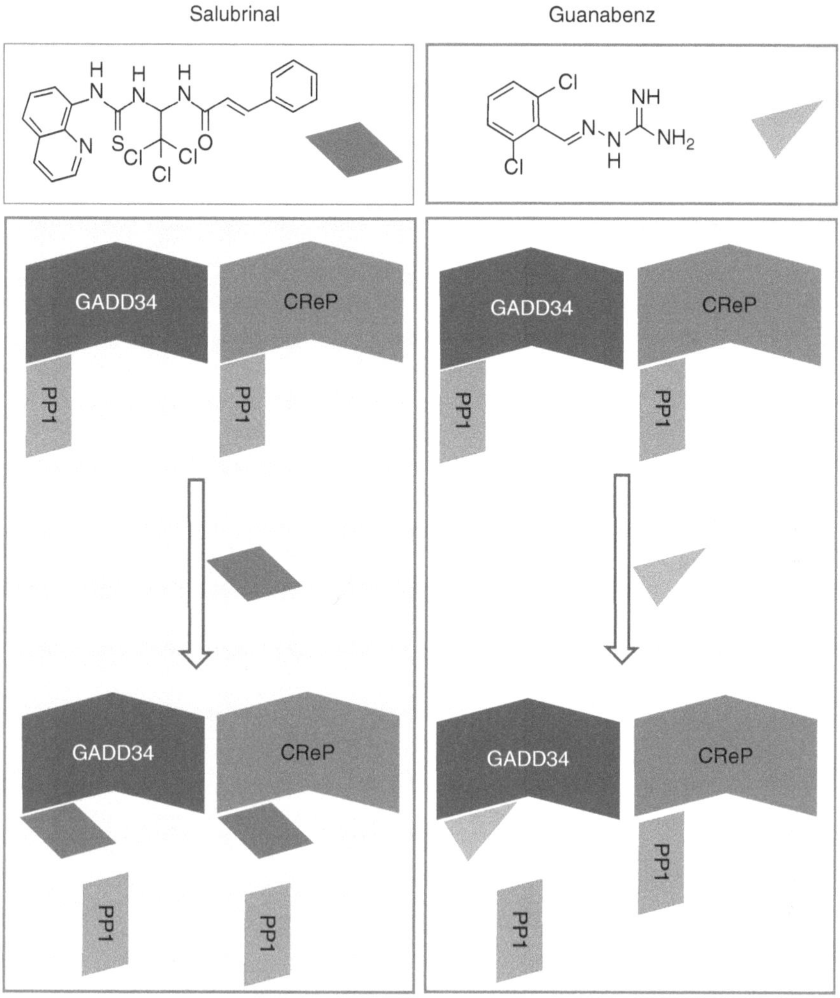


FIG. 3. Mode of action of Salubrinal and Guanabenz as inhibitors of cellular eIF2α phosphatases. Salubrinal (green diamond) and Guanabenz (orange triangle) are structurally unrelated compounds. Recent studies show that they both target cellular eIF2α phosphatases comprised of PP1 bound to either GADD34 or CreP. The prevailing evidence suggests that salubrinal disrupts the PP1 complex containing both GADD34 and CReP (left panel), while guanabenz selectively binds to GADD34 and disrupts its association with PP1 catalytic subunit. In this manner, both drugs elevate the phosphorylation of eIF2α, particularly in response to ER stress. (See Color Insert.)

an effective dosing regimen, these same properties would reduce the possibility of overdosing, making these drugs safer. Salubrinal also has low solubility and bioavailability[137] and needs to be modified to improve its pharmacokinetic properties and druggability. Initial attempts at synthesizing more efficacious salubrinal analogs were unsuccessful.[141]

Guanabenz (also known as Wytensin), a clinically approved α2-adrenergic agonist used to treat hypertension,[142] was shown to possess an antiprion activity[143] that was independent of the α2-adrenergic receptor. Bertolotti and colleagues recently established that Guanabenz bound GADD34, but not CReP, and disrupted PP1α binding.[144] Guanabenz protected β-cells from death induced by expression of the mutant insulin, *Akita Ins2*(C96Y).[144] This novel property of Guanabenz raises exciting possibilities, as the drug has been in use in patients for several decades without notable side effects, suggesting a significant margin of safety. By comparison, gene disruption studies hinted that inhibiting both GADD34 and CReP may have much greater liability. Nevertheless, salubrinal and guanabenz represent an exciting opportunity for drug development aimed at enhancing eIF2α phosphorylation to treat human protein misfolding disorders. Indeed, this has encouraged new small-molecule screens to identify novel modulators of ER stress.[145,146]

In conclusion, UPR signaling pathways are critical for cells to manage protein folding associated with everyday protein synthesis. Extensive work over the last two decades have identified components of this complex translational and transcriptional regulatory network and analyzed their role in cells and animals. While transient translational repression allows cells time to more effectively cope with protein folding, the expression of numerous protein chaperones and activation of ERAD further aid this process. The latter have the added advantage of preconditioning cells to tolerate repeated bouts of protein misfolding. Our understanding of how transient, reversible ER stress differs from chronic terminal ER stress that contributes to cell death and disease is still incomplete, but the use of genetically modified mice have provided new insights into the potential beneficial effects of modulating UPR components as well as highlighting the hazards of targeting others. The discovery of small molecules that disrupt the PP1/GADD34 complex to enhance eIF2α phosphorylation may, for the first time, allow us to mimic physiological perturbations, such as hibernation and ischemia,[147,148,149] as well as endogenous mechanisms such as the activation of GLP1 receptors,[73] which disrupt this eIF2α phosphatase complex. Thus, these experimental tools offer the opportunity to analyze the physiological impact of acute, reversible modulation of UPR signaling. Finally, Salubrinal and Guanabenz provide two distinct chemical templates from which to design novel drugs to treat a variety of human diseases associated with ER stress.

## Acknowledgments

Our research is supported by an Individual Research Grant from the National Medical Research Council (NMRC/GMS/1252/2010) and start-up funds from Duke-NUS Graduate Medical School, Singapore (to S.S.). M.J.F. is supported by a Lee Kuan Yew Postdoctoral Fellowship.

## References


1. Holcik M, Sonenberg N. Translational control in stress and apoptosis. *Nat Rev Mol Cell Biol* 2005;**6**:318–27.
2. Hunt T, Vanderhoff G, London IM. Control of globin synthesis: role of heme. *J Mol Biol* 1971;**66**:471–81.
3. Chen JJ, Throop MS, Gehrke L, Kuo I, Pal JK, Brodsky M, et al. Cloning of the cDNA of the heme-regulated eukaryotic initiation factor 2 alpha (eIF-2 alpha) kinase of rabbit reticulocytes: homology to yeast GCN2 protein kinase and human double-stranded-RNA-dependent eIF-2 alpha kinase. *Proc Natl Acad Sci USA* 1991;**88**:7729–33.
4. Hinnebusch AG. Evidence for translational regulation of the activator of general amino acid control in yeast. *Proc Natl Acad Sci USA* 1984;**81**:6442–6.
5. Dever TE, Feng L, Wek RC, Cigan AM, Donahue TF, Hinnebusch AG. Phosphorylation of initiation factor 2 alpha by protein kinase GCN2 mediates gene-specific translational control of GCN4 in yeast. *Cell* 1992;**68**:585–96.
6. Zhang P, McGrath BC, Reinert J, Olsen DS, Lei L, Gill S, et al. The GCN2 eIF2alpha kinase is required for adaptation to amino acid deprivation in mice. *Mol Cell Biol* 2002;**22**:6681–8.
7. Narasimhan J, Staschke KA, Wek RC. Dimerization is required for activation of eIF2 kinase Gcn2 in response to diverse environmental stress conditions. *J Biol Chem* 2004;**279**:22820–32.
8. Garcia MA, Gil J, Ventoso I, Guerra S, Domingo E, Rivas C, et al. Impact of protein kinase PKR in cell biology: from antiviral to antiproliferative action. *Microbiol Mol Biol Rev* 2006;**70**:1032–60.
9. Dever TE, Chen JJ, Barber GN, Cigan AM, Feng L, Donahue TF, et al. Mammalian eukaryotic initiation factor 2 alpha kinases functionally substitute for GCN2 protein kinase in the GCN4 translational control mechanism of yeast. *Proc Natl Acad Sci USA* 1993;**90**:4616–20.
10. Shi Y, Vattem KM, Sood R, An J, Liang J, Stramm L, et al. Identification and characterization of pancreatic eukaryotic initiation factor 2 alpha-subunit kinase, PEK, involved in translational control. *Mol Cell Biol* 1998;**18**:7499–509.
11. Harding HP, Zhang Y, Bertolotti A, Zeng H, Ron D. Perk is essential for translational regulation and cell survival during the unfolded protein response. *Mol Cell* 2000;**5**:897–904.
12. Hartl FU, Bracher A, Hayer-Hartl M. Molecular chaperones in protein folding and proteostasis. *Nature* 2011;**475**:324–32.
13. Hebert DN, Molinari M. In and out of the ER: protein folding, quality control, degradation, and related human diseases. *Physiol Rev* 2007;**87**:1377–408.
14. Hotamisligil GS. Endoplasmic reticulum stress and the inflammatory basis of metabolic disease. *Cell* 2010;**140**:900–17.
15. Otsu M, Sitia R. Diseases originating from altered protein quality control in the endoplasmic reticulum. *Curr Med Chem* 2007;**14**:1639–52.
16. Kopito RR. Aggresomes, inclusion bodies and protein aggregation. *Trends Cell Biol* 2000;**10**:524–30.
17. Aguzzi A, O'Connor T. Protein aggregation diseases: pathogenicity and therapeutic perspectives. *Nat Rev Drug Discov* 2010;**9**:237–48.
18. Scheuner D, Kaufman RJ. The unfolded protein response: a pathway that links insulin demand with beta-cell failure and diabetes. *Endocr Rev* 2008;**29**:317–33.
19. Rapoport TA. Protein translocation across the eukaryotic endoplasmic reticulum and bacterial plasma membranes. *Nature* 2007;**450**:663–9.
20. Lerner RS, Seiser RM, Zheng T, Lager PJ, Reedy MC, Keene JD, et al. Partitioning and translation of mRNAs encoding soluble proteins on membrane-bound ribosomes. *RNA* 2003;**9**:1123–37.

21. Stephens SB, Nicchitta CV. Divergent regulation of protein synthesis in the cytosol and endoplasmic reticulum compartments of mammalian cells. *Mol Biol Cell* 2008;**19**:623–32.
22. Chen JC, Wu ML, Huang KC, Lin WW. HMG-CoA reductase inhibitors activate the unfolded protein response and induce cytoprotective GRP78 expression. *Cardiovasc Res* 2008;**80**:138–50.
23. Chin DJ, Luskey KL, Anderson RG, Faust JR, Goldstein JL, Brown MS. Appearance of crystalloid endoplasmic reticulum in compactin-resistant Chinese hamster cells with a 500-fold increase in 3-hydroxy-3-methylglutaryl-coenzyme A reductase. *Proc Natl Acad Sci USA* 1982;**79**:1185–9.
24. Momose K, Nunomiya S, Nakata M, Yada T, Kikuchi M, Yashiro T. Immunohistochemical and electron-microscopic observation of beta-cells in pancreatic islets of spontaneously diabetic Goto-Kakizaki rats. *Med Mol Morphol* 2006;**39**:146–53.
25. Zuber C, Fan JY, Guhl B, Roth J. Misfolded proinsulin accumulates in expanded pre-Golgi intermediates and endoplasmic reticulum subdomains in pancreatic beta cells of Akita mice. *FASEB J* 2004;**18**:917–9.
26. Malhotra JD, Kaufman RJ. ER stress and its functional link to mitochondria: role in cell survival and death. *Cold Spring Harb Perspect Biol* 2011;**3**:a004424.
27. Bertolotti A, Zhang Y, Hendershot LM, Harding HP, Ron D. Dynamic interaction of BiP and ER stress transducers in the unfolded-protein response. *Nat Cell Biol* 2000;**2**:326–32.
28. Glickman MH, Ciechanover A. The ubiquitin-proteasome proteolytic pathway: destruction for the sake of construction. *Physiol Rev* 2002;**82**:373–428.
29. Liu CY, Schroder M, Kaufman RJ. Ligand-independent dimerization activates the stress response kinases IRE1 and PERK in the lumen of the endoplasmic reticulum. *J Biol Chem* 2000;**275**:24881–5.
30. Harding HP, Calfon M, Urano F, Novoa I, Ron D. Transcriptional and translational control in the Mammalian unfolded protein response. *Annu Rev Cell Dev Biol* 2002;**18**:575–99.
31. Ron D, Walter P. Signal integration in the endoplasmic reticulum unfolded protein response. *Nat Rev Mol Cell Biol* 2007;**8**:519–29.
32. Roy AL, Chakrabarti D, Gupta NK. Protein synthesis in rabbit reticulocytes: Mg2+-inhibition of ternary complex (Met-tRNA(f).eIF-2.GTP) formation by reticulocyte eIF-2. *Biochem Biophys Res Commun* 1987;**146**:114–20.
33. Krishnamoorthy T, Pavitt GD, Zhang F, Dever TE, Hinnebusch AG. Tight binding of the phosphorylated alpha subunit of initiation factor 2 (eIF2alpha) to the regulatory subunits of guanine nucleotide exchange factor eIF2B is required for inhibition of translation initiation. *Mol Cell Biol* 2001;**21**:5018–30.
34. Vattem KM, Wek RC. Reinitiation involving upstream ORFs regulates ATF4 mRNA translation in mammalian cells. *Proc Natl Acad Sci USA* 2004;**101**:11269–74.
35. Palam LR, Baird TD, Wek RC. Phosphorylation of eIF2 facilitates ribosomal bypass of an inhibitory upstream ORF to enhance *CHOP* translation. *J Biol Chem* 2011;**286**:10939–49.
36. Lee YY, Cevallos RC, Jan E. An upstream open reading frame regulates translation of GADD34 during cellular stresses that induce eIF2alpha phosphorylation. *J Biol Chem* 2009;**284**:6661–73.
37. Puthalakath H, O'Reilly LA, Gunn P, Lee L, Kelly PN, Huntington ND, et al. ER stress triggers apoptosis by activating BH3-only protein Bim. *Cell* 2007;**129**:1337–49.
38. Marciniak SJ, Yun CY, Oyadomari S, Novoa I, Zhang Y, Jungreis R, et al. CHOP induces death by promoting protein synthesis and oxidation in the stressed endoplasmic reticulum. *Genes Dev* 2004;**18**:3066–77.
39. Novoa I, Zeng H, Harding HP, Ron D. Feedback inhibition of the unfolded protein response by GADD34-mediated dephosphorylation of eIF2alpha. *J Cell Biol* 2001;**153**:1011–22.
40. Ma Y, Hendershot LM. Delineation of a negative feedback regulatory loop that controls protein translation during endoplasmic reticulum stress. *J Biol Chem* 2003;**278**:34864–73.

41. Rutkowski DT, Arnold SM, Miller CN, Wu J, Li J, Gunnison KM, et al. Adaptation to ER stress is mediated by differential stabilities of pro-survival and pro-apoptotic mRNAs and proteins. *PLoS Biol* 2006;**4**(11):e374.
42. Lu PD, Jousse C, Marciniak SJ, Zhang Y, Novoa I, Scheuner D, et al. Cytoprotection by pre-emptive conditional phosphorylation of translation initiation factor 2. *EMBO J* 2004;**23**:169–79.
43. Adler HT, Chinery R, Wu DY, Kussick SJ, Payne JM, Fornace AJ, Jr., et al. Leukemic HRX fusion proteins inhibit GADD34-induced apoptosis and associate with the GADD34 and hSNF5/INI1 proteins. *Mol Cell Biol* 1999;**19**:7050–60.
44. Zhou W, Brush MH, Choy MS, Shenolikar S. Association with endoplasmic reticulum promotes proteasomal degradation of GADD34 protein. *J Biol Chem* 2011;**286**:21687–96.
45. Brush MH, Shenolikar S. Control of cellular GADD34 levels by the 26S proteasome. *Mol Cell Biol* 2008;**28**:6989–7000.
46. Novoa I, Zhang Y, Zeng H, Jungreis R, Harding HP, Ron D. Stress-induced gene expression requires programmed recovery from translational repression. *EMBO J* 2003;**22**:1180–7.
47. Avivar-Valderas A, Eduardo Salas E, Bobrovnikova-Marjon E, Diehl JA, Nagi C, Debnath IJ, et al. PERK integrates autophagy and oxidative stress responses to promote survival during extracellular matrix detachment. *Mol Cell Biol* 2011;**31**:3616–29.
48. Ye J, Rawson RB, Komuro R, Chen X, Dave UP, Prywes R, et al. ER stress induces cleavage of membrane-bound ATF6 by the same proteases that process SREBPs. *Mol Cell* 2000;**6**:1355–64.
49. Eberlé D, Hegarty B, Bossard P, Ferré P, Foufelle F. SREBP transcription factors: master regulators of lipid homeostasis. *Biochimie* 2004;**86**:839–48.
50. Shen J, Chen X, Hendershot L, Prywes R. ER stress regulation of ATF6 localization by dissociation of BiP/GRP78 binding and unmasking of Golgi localization signals. *Dev Cell* 2002;**3**:99–111.
51. Yamamoto K, Sato T, Matsui T, Sato M, Okada T, Yoshida H, et al. Transcriptional induction of mammalian ER quality control proteins is mediated by single or combined action of ATF6alpha and XBP1. *Dev Cell* 2007;**13**:365–76.
52. Thuerauf DJ, Morrison L, Glembotski CC. Opposing roles for ATF6alpha and ATF6beta in endoplasmic reticulum stress response gene induction. *J Biol Chem* 2004;**279**:21078–84.
53. Wu J, Rutkowski DT, Dubois M, Swathirajan J, Saunders T, Wang J, et al. ATF6alpha optimizes long-term endoplasmic reticulum function to protect cells from chronic stress. *Dev Cell* 2007;**13**:351–64.
54. Hollien J, Weissman JS. Decay of endoplasmic reticulum-localized mRNAs during the unfolded protein response. *Science* 2006;**313**:104–7.
55. Vembar SS, Brodsky JL. One step at a time: endoplasmic reticulum-associated degradation. *Nat Rev Mol Cell Biol* 2008;**9**:944–57.
56. Lee AH, Iwakoshi NN, Glimcher LH. XBP-1 regulates a subset of endoplasmic reticulum resident chaperone genes in the unfolded protein response. *Mol Cell Biol* 2003;**23**:7448–59.
57. Lee TG, Tang N, Thompson S, Miller J, Katze MG. The 58,000-dalton cellular inhibitor of the interferon-induced double-stranded RNA-activated protein kinase (PKR) is a member of the tetratricopeptide repeat family of proteins. *Mol Cell Biol* 1994;**14**:2331–42.
58. Lee TG, Tomita J, Hovanessian AG, Katze MG. Purification and partial characterization of a cellular inhibitor of the interferon-induced protein kinase of Mr 68,000 from influenza virus-infected cells. *Proc Natl Acad Sci USA* 1990;**87**:6208–12.
59. Van Huizen R, Martindale JL, Gorospe M, Holbrook NJ. P58IPK, a novel endoplasmic reticulum stress-inducible protein and potential negative regulator of eIF2alpha signaling. *J Biol Chem* 2003;**278**:15558–64.

60. Yan W, Frank CL, Korth MJ, Sopher BL, Novoa I, Ron D, et al. Control of PERK eIF2alpha kinase activity by the endoplasmic reticulum stress-induced molecular chaperone P58IPK. *Proc Natl Acad Sci USA* 2002;**99**:15920–5.
61. Rutkowski DT, Kang SW, Goodman AG, Garrison JL, Taunton J, Katze MG, et al. The role of p58IPK in protecting the stressed endoplasmic reticulum. *Mol Biol Cell* 2007;**18**:3681–91.
62. Doroudgar S, Thuerauf DJ, Marcinko MC, Belmont PJ, Glembotski CC. Ischemia activates the ATF6 branch of the endoplasmic reticulum stress response. *J Biol Chem* 2009;**284**:29735–45.
63. Jonsson PA, Graffmo KS, Brannstrom T, Nilsson P, Andersen PM, Marklund SL. Motor neuron disease in mice expressing the wild type-like D90A mutant superoxide dismutase-1. *J Neuropathol Exp Neurol* 2006;**65**:1126–36.
64. Hodgson JG, Agopyan N, Gutekunst CA, Leavitt BR, LePiane F, Singaraja R, et al. A YAC mouse model for Huntington's disease with full-length mutant huntingtin, cytoplasmic toxicity, and selective striatal neurodegeneration. *Neuron* 1999;**23**:181–92.
65. Giasson BI, Duda JE, Quinn SM, Zhang B, Trojanowski JQ, Lee VM. Neuronal alpha-synucleinopathy with severe movement disorder in mice expressing A53T human alpha-synuclein. *Neuron* 2002;**34**:521–33.
66. Atkin JD, Farg MA, Turner BJ, Tomas D, Lysaght JA, Nunan J, et al. Induction of the unfolded protein response in familial amyotrophic lateral sclerosis and association of protein-disulfide isomerase with superoxide dismutase 1. *J Biol Chem* 2006;**281**:30152–65.
67. Carnemolla A, Fossale E, Agostoni E, Michelazzi S, Calligaris R, De Maso L, et al. Rrs1 is involved in endoplasmic reticulum stress response in Huntington disease. *J Biol Chem* 2009;**284**:18167–73.
68. Jiang P, Gan M, Ebrahim AS, Lin WL, Melrose HL, Yen SH. ER stress response plays an important role in aggregation of alpha-synuclein. *Mol Neurodegener* 2010;**5**:56–65.
69. Scheuner D, Song B, McEwen E, Liu C, Laybutt R, Gillespie P, et al. Translational control is required for the unfolded protein response and in vivo glucose homeostasis. *Mol Cell* 2001;**7**:1165–76.
70. Han AP, Yu C, Lu L, Fujiwara Y, Browne C, Chin G, et al. Heme-regulated eIF2alpha kinase (HRI) is required for translational regulation and survival of erythroid precursors in iron deficiency. *EMBO J* 2001;**20**:6909–18.
71. Yang YL, Reis LF, Pavlovic J, Aguzzi A, Schafer R, Kumar A, et al. Deficient signaling in mice devoid of double-stranded RNA-dependent protein kinase. *EMBO J* 1995;**14**:6095–106.
72. Abraham N, Stojdl DF, Duncan PI, Methot N, Ishii T, Dube M, et al. Characterization of transgenic mice with targeted disruption of the catalytic domain of the double-stranded RNA-dependent protein kinase, PKR. *J Biol Chem* 1999;**274**:5953–62.
73. Zhang W, Feng D, Li Y, Iida K, McGrath B, Cavener DR. PERK EIF2AK3 control of pancreatic beta cell differentiation and proliferation is required for postnatal glucose homeostasis. *Cell Metab* 2006;**4**:491–7.
74. Patterson AD, Hollander MC, Miller GF, Fornace AJ. Gadd34 requirement for normal hemoglobin synthesis. *Mol Cell Biol* 2006;**26**:1644–53.
75. Harding HP, Zhang Y, Scheuner D, Chen JJ, Kaufman RJ, Ron D. Ppp 1r15 gene knockout reveals an essential role for translation initiation factor 2 alpha (eIF2alpha) dephosphorylation in mammalian development. *Proc Natl Acad Sci USA* 2009;**106**:1832–7.
76. Tanaka T, Tsujimura T, Takeda K, Sugihara A, Maekawa A, Terada N, et al. Targeted disruption of ATF4 discloses its essential role in the formation of eye lens fibres. *Genes Cells* 1998;**3**:801–10.
77. Hettmann T, Barton K, Leiden JM. Microphthalmia due to p53-mediated apoptosis of anterior lens epithelial cells in mice lacking the CREB-2 transcription factor. *Dev Biol* 2000;**222**:110–23.

78. Chin KT, Kang G, Qu J, Gardner LB, Coetzee WA, Zito E, et al. The sarcoplasmic reticulum luminal thiol oxidase ERO1 regulates cardiomyocyte excitation-coupled calcium release and response to hemodynamic load. *FASEB J* 2011;**25**:2583–91.
79. Zito E, Chin KT, Blais J, Harding HP, Ron D. ERO1-beta, a pancreas-specific disulfide oxidase, promotes insulin biogenesis and glucose homeostasis. *J Cell Biol* 2010;**188**:821–32.
80. Nakagawa T, Zhu H, Morishima N, Li E, Xu J, Yankner BA, et al. Caspase-12 mediates endoplasmic-reticulum-specific apoptosis and cytotoxicity by amyloid-beta. *Nature* 2000;**403**:98–103.
81. Luo S, Mao C, Lee B, Lee AS. GRP78/BiP is required for cell proliferation and protecting the inner cell mass from apoptosis during early mouse embryonic development. *Mol Cell Biol* 2006;**26**:5688–97.
82. Zhang K, Wong HN, Song B, Miller CN, Scheuner D, Kaufman RJ. The unfolded protein response sensor IRE1alpha is required at 2 distinct steps in B cell lymphopoiesis. *J Clin Invest* 2005;**115**:268–81.
83. Iwawaki T, Akai R, Yamanaka S, Kohno K. Function of IRE1 alpha in the placenta is essential for placental development and embryonic viability. *Proc Natl Acad Sci USA* 2009;**106**:16657–62.
84. Reimold AM, Etkin A, Clauss I, Perkins A, Friend DS, Zhang J, et al. An essential role in liver development for transcription factor XBP-1. *Genes Dev* 2000;**14**:152–7.
85. Lee AH, Scapa EF, Cohen DE, Glimcher LH. Regulation of hepatic lipogenesis by the transcription factor XBP1. *Science* 2008;**320**:1492–6.
86. Lee AH, Chu GC, Iwakoshi NN, Glimcher LH. XBP-1 is required for biogenesis of cellular secretory machinery of exocrine glands. *EMBO J* 2005;**24**:4368–80.
87. Yamamoto K, Takahara K, Oyadomari S, Okada T, Sato T, Harada A, et al. Induction of liver steatosis and lipid droplet formation in ATF6alpha-knockout mice burdened with pharmacological endoplasmic reticulum stress. *Mol Biol Cell* 2010;**21**:2975–86.
88. Thuerauf DJ, Marcinko M, Belmont PJ, Glembotski CC. Effects of the isoform-specific characteristics of ATF6 alpha and ATF6 beta on endoplasmic reticulum stress response gene expression and cell viability. *J Biol Chem* 2007;**282**:22865–78.
89. Acosta-Alvear D, Zhou Y, Blais A, Tsikitis M, Lents NH, Arias C, et al. XBP1 controls diverse cell type- and condition-specific transcriptional regulatory networks. *Mol Cell* 2007;**27**:53–66.
90. Mimura N, Hamada H, Kashio M, Jin H, Toyama Y, Kimura K, et al. Aberrant quality control in the endoplasmic reticulum impairs the biosynthesis of pulmonary surfactant in mice expressing mutant BiP. *Cell Death Differ* 2007;**14**:1475–85.
91. Kimura K, Jin H, Ogawa M, Aoe T. Dysfunction of the ER chaperone BiP accelerates the renal tubular injury. *Biochem Biophys Res Commun* 2008;**366**:1048–53.
92. Bevilacqua E, Wang X, Majumder M, Gaccioli F, Yuan CL, Wang C, et al. eIF2alpha phosphorylation tips the balance to apoptosis during osmotic stress. *J Biol Chem* 2010;**285**:17098–111.
93. Nakamura T, Furuhashi M, Li P, Cao H, Tuncman G, Sonenberg N, et al. Double-stranded RNA-dependent protein kinase links pathogen sensing with stress and metabolic homeostasis. *Cell* 2010;**140**:338–48.
94. Scheuner D, Vander Mierde D, Song B, Flamez D, Creemers JW, Tsukamoto K, et al. Control of mRNA translation preserves endoplasmic reticulum function in beta cells and maintains glucose homeostasis. *Nat Med* 2005;**11**:757–64.
95. Dever TE, Hinnebusch AG. GCN2 whets the appetite for amino acids. *Mol Cell* 2005;**18**:141–2.
96. Anthony TG, McDaniel BJ, Byerley RL, McGrath BC, Cavener DR, McNurlan MA, et al. Preservation of liver protein synthesis during dietary leucine deprivation occurs at the expense of skeletal muscle mass in mice deleted for eIF2 kinase GCN2. *J Biol Chem* 2004;**279**:36553–61.

97. Hao S, Sharp JW, Ross-Inta CM, McDaniel BJ, Anthony TG, Wek RC, et al. Uncharged tRNA and sensing of amino acid deficiency in mammalian piriform cortex. *Science* 2005;**307**:1776–8.
98. Maurin AC, Jousse C, Averous J, Parry L, Bruhat A, Cherasse Y, et al. The GCN2 kinase biases feeding behavior to maintain amino acid homeostasis in omnivores. *Cell Metab* 2005;**1**:273–7.
99. Ye J, Kumanova M, Hart LS, Sloane K, Zhang H, De Panis DN, et al. The GCN2-ATF4 pathway is critical for tumour cell survival and proliferation in response to nutrient deprivation. *EMBO J* 2010;**29**:2082–96.
100. Baltzis D, Li S, Koromilas AE. Functional characterization of pkr gene products expressed in cells from mice with a targeted deletion of the N terminus or C terminus domain of PKR. *J Biol Chem* 2002;**277**:38364–72.
101. Delepine M, Nicolino M, Barrett T, Golamaully M, Lathrop GM, Julier C. EIF2AK3, encoding translation initiation factor 2-alpha kinase 3, is mutated in patients with Wolcott-Rallison syndrome. *Nat Genet* 2000;**25**:406–9.
102. Allotey RA, Mohan V, McDermott MF, Deepa R, Premalatha G, Hassan Z, et al. The EIF2AK3 gene region and type I diabetes in subjects from South India. *Genes Immun* 2004;**5**:648–52.
103. Harding HP, Zeng H, Zhang Y, Jungries R, Chung P, Plesken H, et al. Diabetes mellitus and exocrine pancreatic dysfunction in perk−/− mice reveals a role for translational control in secretory cell survival. *Mol Cell* 2001;**7**(6):1153–63.
104. Guojin Huang, Jian Yao, Weizhong Zeng, Yusuke Mizuno, Kristine E. Kamm, James T. Stull, et al. ER stress disrupts $Ca^{2+}$-signaling complexes and $Ca^{2+}$ regulation in secretory and muscle cells from PERK-knockout mice. *J Cell Sci* 2006;**119**:153–61.
105. Teske BF, Wek SA, Bunpo P, Cundiff JK, McClintick JN, Anthony TG, et al. The eIF2 kinase PERK and the integrated stress response facilitate activation of ATF6 during endoplasmic reticulum stress. *Mol Biol Cell* 2011; Inpress −10.1091/mbc.E11-06-0510.
106. Inoue H, Tanizawa Y, Wasson J, Behn P, Kalidas K, Bernal-Mizrachi E, et al. A gene encoding a transmembrane protein is mutated in patients with diabetes mellitus and optic atrophy (Wolfram syndrome). *Nat Genet* 1998;**20**:143–8.
107. Strom TM, Hortnagel K, Hofmann S, Gekeler F, Scharfe C, Rabl W, et al. Diabetes insipidus, diabetes mellitus, optic atrophy and deafness (DIDMOAD) caused by mutations in a novel gene (wolframin) coding for a predicted transmembrane protein. *Hum Mol Genet* 1998;**7**:2021–8.
108. Fonseca SG, Fukuma M, Lipson KL, Nguyen LX, Allen JR, Oka Y, et al. WFS1 is a novel component of the unfolded protein response and maintains homeostasis of the endoplasmic reticulum in pancreatic beta-cells. *J Biol Chem* 2005;**280**:39609–15.
109. Khanim F, Kirk J, Latif F, Barrett TG. WFS1/wolframin mutations, Wolfram syndrome, and associated diseases. *Hum Mutat* 2001;**17**:357–67.
110. Sandhu MS, Weedon MN, Fawcett KA, Wasson J, Debenham SL, Daly A, et al. Common variants in WFS1 confer risk of type 2 diabetes. *Nat Genet* 2007;**39**:951–3.
111. Noormets K, Koks S, Kavak A, Arend A, Aunapuu M, Keldrimaa A, et al. Male mice with deleted Wolframin (Wfs1) gene have reduced fertility. *Reprod Biol Endocrinol* 2009;**7**:82–8.
112. Riggs AC, Bernal-Mizrachi E, Ohsugi M, Wasson J, Fatrai S, Welling C, et al. Mice conditionally lacking the Wolfram gene in pancreatic islet beta cells exhibit diabetes as a result of enhanced endoplasmic reticulum stress and apoptosis. *Diabetologia* 2005;**48**:2313–21.
113. Ladiges WC, Knoblaugh SE, Morton JF, Korth MJ, Sopher BL, Baskin CR, et al. Pancreatic beta-cell failure and diabetes in mice with a deletion mutation of the endoplasmic reticulum molecular chaperone gene P58IPK. *Diabetes* 2005;**54**:1074–81.
114. Brush MH, Weiser DC, Shenolikar S. Growth arrest and DNA damage-inducible protein GADD34 targets protein phosphatase 1 alpha to the endoplasmic reticulum and promotes

dephosphorylation of the alpha subunit of eukaryotic translation initiation factor 2. *Mol Cell Biol* 2003;**23**:1292–303.

115. Jousse C, Oyadomari S, Novoa I, Lu P, Zhang Y, Harding HP, et al. Inhibition of a constitutive translation initiation factor 2alpha phosphatase, CReP, promotes survival of stressed cells. *J Cell Biol* 2003;**163**:767–75.
116. Fornace Jr. AJ, Nebert DW, Hollander MC, Luethy JD, Papathanasiou M, Fargnoli J, et al. Mammalian genes coordinately regulated by growth arrest signals and DNA-damaging agents. *Mol Cell Biol* 1989;**9**:4196–203.
117. Hollander MC, Zhan Q, Bae I, Fornace AJ. Mammalian GADD34, an apoptosis- and DNA damage-inducible gene. *J Biol Chem* 1997;**272**:13731–7.
118. Zhan Q, Lord KA, Alamo Jr. I, Hollander MC, Carrier F, Ron D, et al. The gadd and MyD genes define a novel set of mammalian genes encoding acidic proteins that synergistically suppress cell growth. *Mol Cell Biol* 1994;**14**:2361–71.
119. Abdollahi A, Lord KA, Hoffman-Liebermann B, Liebermann DA. Sequence and expression of a cDNA encoding MyD118: a novel myeloid differentiation primary response gene induced by multiple cytokines. *Oncogene* 1991;**6**:165–7.
120. Kojima E, Takeuchi A, Haneda M, Yagi A, Hasegawa T, Yamaki K, et al. The function of GADD34 is a recovery from a shutoff of protein synthesis induced by ER stress: elucidation by GADD34-deficient mice. *FASEB J* 2003;**17**:1573–5.
121. Tanaka C, Ito S, Nishio N, Kodera Y, Sakurai H, Suzuki H, et al. GADD34 suppresses wound healing by upregulating expression of myosin IIA. *Transgenic Res* 2010;**19**:637–45.
122. Hyrskyluoto A, Reijonen S, Kivinen J, Lindholm D, Korhonen L. GADD34 mediates cytoprotective autophagy in mutant huntingtin expressing cells via the mTOR pathway. *Exp Cell Res* 2011; In press.
123. Uddin MN, Ito S, Nishio N, Suganya T, Isobe K. Gadd34 induces autophagy through the suppression of the mTOR pathway during starvation. *Biochem Biophys Res Commun* 2011;**407**:692–8.
124. Oyadomari S, Harding HP, Zhang Y, Oyadomari M, Ron D. Dephosphorylation of translation initiation factor 2alpha enhances glucose tolerance and attenuates hepatosteatosis in mice. *Cell Metab* 2008;**7**:520–32.
125. Birkenfeld AL, Lee HY, Majumdar S, Juzcak MJ, Camporez J, Jornayvaz FR, et al. Influence of the hepatic eIF2{alpha} ER stress response pathway on insulin mediated ER stress, Hepatic and Peripheral Glucose Metabolism. *J Biol Chem* 2011;**286**:36163–70.
126. Mahadevan NR, Rodvold J, Sepulveda H, Rossi S, Drew AF, Zanetti M. Transmission of endoplasmic reticulum stress and pro-inflammation from tumor cells to myeloid cells. *Proc Natl Acad Sci USA* 2011;**108**:6561–6.
127. Harding HP, Novoa I, Zhang Y, Zeng H, Wek R, Schapira M, et al. Regulated translation initiation controls stress-induced gene expression in mammalian cells. *Mol Cell* 2000;**6**:1099–108.
128. Steele AD, Hetz C, Yi CH, Jackson WS, Borkowski AW, Yuan J, et al. Prion pathogenesis is independent of caspase-12. *Prion* 2007;**1**:243–7.
129. Boyce M, Bryant KF, Jousse C, Long K, Harding HP, Scheuner D, et al. A selective inhibitor of eIF2alpha dephosphorylation protects cells from ER stress. *Science* 2005;**307**:935–9.
130. Bryant KF, Macari ER, Malik N, Boyce M, Yuan J, Coen DM. ICP34.5-dependent and -independent activities of salubrinal in herpes simplex virus-1 infected cells. *Virology* 2008;**379**:197–204.
131. Li Y, Zhang C, Chen X, Yu J, Wang Y, Yang Y, et al. ICP34.5 protein of herpes simplex virus facilitates the initiation of protein translation by bridging eukaryotic initiation factor 2α (eIF2α) and protein phosphatase 1. *J Biol Chem* 2011;**286**:24785–92.

132. Hoozemans JJ, van Haastert ES, Nijholt DA, Rozemuller AJ, Eikelenboom P, Scheper W. The unfolded protein response is activated in pretangle neurons in Alzheimer's disease hippocampus. *Am J Pathol* 2009;**174**:1241–51.
133. Lee do Y, Lee KS, Lee HJ, Kim do H, Noh YH, Yu K, et al. Activation of PERK signaling attenuates Abeta-mediated ER stress. *PLoS One* 2010;**5**:e10489.
134. Zoghbi HY, Orr HT. Glutamine repeats and neurodegeneration. *Annu Rev Neurosci* 2000;**23**:217–47.
135. (a) Reijonen S, Putkonen N, Norremolle A, Lindholm D, Korhonen L. Inhibition of endoplasmic reticulum stress counteracts neuronal cell death and protein aggregation caused by N-terminal mutant huntingtin proteins. *Exp Cell Res* 2008;**314**:950–60; (b) Smith WW, Jiang H, Pei Z, Tanaka Y, Morita H, Sawa A, et al. Endoplasmic reticulum stress and mitochondrial cell death pathways mediate A53T mutant alpha-synuclein-induced toxicity. *Hum Mol Genet* 2005;**14**:3801–11.
136. Sokka AL, Putkonen N, Mudo G, Pryazhnikov E, Reijonen S, Khiroug L, et al. Endoplasmic reticulum stress inhibition protects against excitotoxic neuronal injury in the rat brain. *J Neurosci* 2007;**27**:901–8.
137. Saxena S, Cabuy E, Caroni P. A role for motoneuron subtype-selective ER stress in disease manifestations of FALS mice. *Nat Neurosci* 2009;**12**:627–36.
138. Umareddy I, Pluquet O, Wang QY, Vasudevan SG, Chevet E, Gu F. Dengue virus serotype infection specifies the activation of the unfolded protein response. *Virol J* 2007;**4**:91–6.
139. Schewe DM, Aguirre-Ghiso JA. Inhibition of eIF2alpha dephosphorylation maximizes bortezomib efficiency and eliminates quiescent multiple myeloma cells surviving proteasome inhibitor therapy. *Cancer Res* 2009;**69**:1545–52.
140. Drexler HC. Synergistic apoptosis induction in leukemic cells by the phosphatase inhibitor salubrinal and proteasome inhibitors. *PLoS One* 2009;**4**:e4161.
141. Long K, Boyce M, Lin H, Yuan J, Ma D. Structure-activity relationship studies of salubrinal lead to its active biotinylated derivative. *Bioorg Med Chem Lett* 2005;**15**:3849–52.
142. Holmes B, Brogden RN, Heel RC, Speight TM, Avery GS. Guanabenz. A review of its pharmacodynamic properties and therapeutic efficacy in hypertension. *Drugs* 1983;**26**:212–29.
143. Tribouillard-Tanvier D, Beringue V, Desban N, Gug F, Bach S, Voisset C, et al. Antihypertensive drug guanabenz is active in vivo against both yeast and mammalian prions. *PLoS One* 2008;**3**:e1981.
144. Tsaytler P, Harding HP, Ron D, Bertolotti A. Selective inhibition of a regulatory subunit of protein phosphatase 1 restores proteostasis. *Science* 2011;**332**:91–4.
145. Harding HP, Zhang Y, Khersonsky S, Marciniak S, Scheuner D, Kaufman RJ, et al. Bioactive small molecules reveal antagonism between the integrated stress response and sterol-regulated gene expression. *Cell Metab* 2005;**2**:361–71.
146. Fribley AM, Cruz PG, Miller JR, Callaghan MU, Cai P, Narula N, et al. Complementary cell-based high-throughput screens identify novel modulators of the unfolded protein response. *J Biomol Screen* 2011;**16**:825–35.
147. Connor JH, Weiser DC, Li S, Hallenbeck JM, Shenolikar S. Growth arrest and DNA damage-inducible protein GADD34 assembles a novel signaling complex containing protein phosphatase 1 and inhibitor 1. *Mol Cell Biol* 2001;**21**:6841–50.
148. Cunha DA, Ladriere L, Ortis F, Igoillo-Esteve M, Gurzov EN, Lupi R, et al. Glucagon-like peptide-1 agonists protect pancreatic beta-cells from lipotoxic endoplasmic reticulum stress through upregulation of BiP and JunB. *Diabetes* 2009;**58**:2851–62.
149. Nakka VP, Gusain A, Raghubir R. Endoplasmic reticulum stress plays critical role in brain damage after cerebral ischemia/reperfusion in rats. *Neurotox Res* 2010;**17**:189–202.

# The Role of Protein Phosphorylation in Therapy Resistance and Disease Progression in Chronic Myelogenous Leukemia

WEN CHUN JUAN* AND
S. TIONG ONG*,†

*Program in Cancer and Stem Cell Biology, Duke-NUS Graduate Medical School, Singapore, Singapore

†Division of Medical Oncology, Department of Medicine, Duke University Medical Center, Durham, North Carolina, USA



Progress in Molecular Biology
and Translational Science, Vol. 106
DOI: 10.1016/B978-0-12-396456-4.00007-9




1877-1173/12 $35.00

This review focuses on the central role that protein phosphorylation plays in the pathogenesis of chronic myelogenous leukemia (CML). It will cover the signaling pathways that are dysregulated by the oncogenic tyrosine kinase, BCR-ABL1, which both defines and drives the disease, and the barriers to disease control. These will include the mechanisms that underlie drug resistance, as well as the features of CML that prevent its cure by tyrosine kinase inhibitors. In the second section, we will cover the proteins and pathways that lead to the transformation of early chronic-phase CML to the more advanced blast phase of the disease. Here, we will outline the key pathophysiologic differences between the chronic and the blast phase, the mechanisms that contribute to these differences, and how these might be therapeutically targeted in patients. In the final section, we will summarize the major lessons learnt from the CML clinic. We will focus on how these observations have impacted our understanding of the therapeutic potential of modulating protein phosphorylation in human diseases and areas in which future research in CML pathophysiology may be important.

# I. Introduction

Chronic myelogenous leukemia (CML) is a myeloproliferative disorder that is defined by the presence of the fusion gene *BCR-ABL1* in primitive hematopoietic progenitors. In 90% of cases, this is appreciable at the cytogenetic level as the Philadelphia chromosome (Ph), which juxtaposes the *BCR* (breakpoint cluster region) gene next to the *ABL1* (Abelson murine leukemia viral oncogene homolog 1) gene. In the remaining 10% of cases, the same translocation occurs at a subkaryotypic level and requires detection by either Southern blot or polymerase chain reaction (PCR) based techniques. In contrast to the native c-ABL protein, the fusion gene encodes for a constitutively active tyrosine kinase that is both necessary and sufficient to transform normal cells.[1] Together, these critical advances in understanding have led to the dramatic success story of the tyrosine kinase inhibitor, Gleevec™ (imatinib), in the control of this disease, which in effect turned a previously deadly disorder into a chronic illness.[2] Indeed, the dramatic success of imatinib has been often used to highlight the importance of studying protein phosphorylation in human disease. However, the use of tyrosine kinase inhibitors in CML remains a work in progress, and several clinically driven questions remain at the front and center of CML research. These include the issue of whether patients in early stage or chronic-phase CML can be cured with tyrosine kinase inhibitor therapy alone, and relatedly, how to overcome clinical tyrosine kinase inhibitor resistance. However, probably, the most pressing issue is the fact that late-stage or blast crisis CML remains a highly aggressive and fatal disease, despite the advent of tyrosine kinase inhibitors.[3–5] Accordingly, this review will focus on the role of protein phosphorylation in cellular transformation (Section II), tyrosine kinase inhibitor resistance (Section III), and transformation to blast crisis (Section IV). In Section V, we will summarize the major lessons learnt in the CML clinic and how these impact our understanding of the therapeutic potential of modulating protein phosphorylation in human diseases. In doing so, we hope to demonstrate that an increased understanding of dysregulated protein phosphorylation, as well as the cellular processes they mediate, will continue to play a critical role in improving the management of patients with CML.

## II. Pathways and Proteins Implicated in CML Pathophysiology

During the decades that led up to pivotal trial demonstrating the superiority of imatinib over standard CML therapy,[6] much work focused on understanding both the protein targets and the signaling pathways downstream of BCR-ABL1. An increased understanding of the pathways activated by BCR-ABL1 also improved our appreciation of the cellular processes that phosphorylation controlled in the context of hematopoietic cells and how such alterations contributed to transformation.[7] In addition to promoting cell proliferation and survival, these studies demonstrated that BCR-ABL1 also disrupted normal cell trafficking and promoted genetic instability, and that it influenced these processes by modulating gene expression at several levels, including transcriptionally, translationally, and posttranslationally. Thus, it came to be appreciated that BCR-ABL1 was harnessing multiple proteins, pathways, and processes to promote transformation, and that targeting BCR-ABL1 function itself (vs. each pathway in a piecemeal fashion) would be necessary to achieve an optimal therapeutic outcome.

In the following sections, we will also highlight several important concepts in protein phosphorylation in CML and, when appropriate, relate how these concepts impact either the clinical management of patients or clinical outcomes. The first is that, in CML, many of the pathways used in physiologic growth factor signaling are co-opted to transmit inappropriate proliferation and survival messages to the cellular machinery.[7] This is most likely the means by which BCR-ABL1-containing hematopoietic cells exert a proliferative and survival advantage over their normal counterparts. This concept has become critically important to our understanding of disease persistence and the inability of tyrosine kinase inhibitor therapy to eradicate CML stem cells. Thus, under certain circumstances, physiologic growth factor signaling has been shown to result in redundancy of BCR-ABL1 signaling for maintaining CML stem cell survival, thereby allowing such cells to survive in the absence of BCR-ABL1 kinase activity.[8] Here, it is important to mention that it is probably a combination of growth factors and/or cytokines that are important in the persistence of CML cells, as the use of therapeutic doses of G-CSF (granulocyte colony-stimulating factor) to prevent neutropenia in CML patients on imatinib appears to be safe and, in fact, enhances imatinib responses.[9]

Another important concept is that the malignant compartment in CML mirrors the normal hematopoietic system, in that there are BCR-ABL1-containing stem cells (defined as the ability to regenerate the disease in immunodeficient mice),[10] as well as BCR-ABL1-containing differentiated progeny within the different lineages. Very likely, CML stem cells, as well as their differentiated progeny, retain many, if not most, of the differentiation stage- and lineage-specific features of their normal counterparts. Thus, the activation status of a specific pathway in differentiated CML cell may not be mirrored in the CML stem cell, with important functional consequences.[11] Further, it may be the similarity *per se* between CML stem cells and normal hematopoietic stem cells (in terms of being naturally long-lived, quiescent, possessing an enhanced ability to pump cytotoxic compounds out of the cell,[12] and endowed with the ability to self-renew) that renders CML stem cells highly resistant to drug-based therapy.[13]

In the following section, we will describe the key differences in BCR-ABL1 versus c-ABL that lead to its constitutive activity, highlight the downstream targets that are phosphorylated by BCR-ABL1, together with the cellular processes they regulate, and how these, in turn, contribute to BCR-ABL1-driven transformation.

## A. Transforming Properties of the BCR-ABL1 Protein

The human *c-ABL* gene, which encodes for a nonreceptor tyrosine kinase, is the homolog of the oncogenic *v-abl* gene carried by the Abelson murine leukemia virus. The c-ABL protein contains several structural domains including the SRC homology domains (SH1, SH2, and SH3). Intramolecular interactions between these domains create an autoinhibitory conformation that limits the tyrosine kinase activity of the protein.[14] In addition, c-ABL also contains nuclear localization and export signals, and hence the protein can be found both in the nucleus and in the cytoplasm.[15] In contrast, BCR-ABL1 has constitutively active tyrosine kinase activity because it does not adopt a conformation that suppresses its kinase activity.[14] Further, BCR-ABL1 is found primarily in the cytoplasm where it can activate cell survival pathways, even though it has retained a nuclear localization signal.[16] Previous work has shown that the ability to be retained in the cytoplasm is critical for BCR-ABL1's transforming ability, as tethering BCR-ABL1 in the nucleus abolished its ability to mediate cell survival.[16] Definitive evidence for the role of BCR-ABL1 in the development of a myeloproliferative disorder has been shown using animal models. Here, mice transplanted with murine bone marrow cells transduced with *BCR-ABL1* retroviruses led to the development of a CML-like myeloproliferative disease.[17] These and other similar studies demonstrating the sufficiency of BCR-ABL1 for cellular transformation suggested that a logical approach to treat CML is to develop inhibitors that can suppress the tyrosine kinase activity of BCR-ABL1. However, such an approach was initially thought to be impracticable because tyrosine kinase inhibitors were considered to be nonspecific and, therefore, potentially too toxic for use in humans.[2] Another related concern came from the analysis of *c-Abl* knockout mice. Here, studies revealed that c-Abl was required for normal eye development, as well as development of the hematopoietic and reproductive system, suggesting that the inhibition of endogenous c-Abl might itself be toxic.[18] However, it was eventually demonstrated that relative specific tyrosine kinase inhibitors could be chemically synthesized,[19] a result which suggested that targeting of specific kinases in human disease states was indeed possible. In this respect, it is important to highlight that, when the phase 1 dose-escalating study of imatinib was eventually conducted, serious side effects were infrequent and a maximum tolerated dose could not be defined, which was a very unusual result for a phase 1 study at the time.[20] The unexpected lack of toxicity may be due to several reasons, which include the possibility that the essential functions of c-ABL do not require it to be constitutively active, the fact that it may require higher concentrations to inhibit c-ABL versus BCR-ABL1,[21] and trough plasma tyrosine kinase inhibitor levels are such that continuous inhibition of either BCR-ABL1 or c-ABL signaling does not actually occur.[22]

## B. Downstream Pathways Activated by BCR-ABL1

### 1. MAP Kinase Pathway

Receptor tyrosine kinases belong to a class of cell-surface receptors that respond to extracellular cues and mediate cell growth, survival, proliferation, and differentiation. These receptors possess intrinsic tyrosine kinase activity, which, in turn, triggers a phosphorylation cascade that mediates changes in gene expression in the nucleus. The core module of the phosphorylation cascade is regulated by three protein kinases. The last of these is known as mitogen-activated protein (MAP) kinase. There are three well-characterized families of MAP kinases, which include (1) extracellular signal-regulated kinases 1 and 2 (ERK1/2), (2) c-Jun N-terminal kinase (JNK), and (3) p38 families.

In CML, the MAP kinase signaling pathway mediated by ERK1/2 is constitutively activated. The mechanism of constitutive ERK1/2 activation involves phosphorylation of tyrosine 177 of BCR, which functions as a binding site for the SH2 domain of the growth factor receptor-bound protein 2 (GRB2).[23] GRB2 is a small adapter protein that recruits other proteins to itself, resulting in the activation of a GTPase known as the RAt sarcoma (RAS) protein.

Activated RAS triggers a signaling cascade that leads to the activation of ERK1/2. Phosphorylation of tyrosine 177 plays a critical role in malignant transformation mediated by BCR-ABL1 because mutation of tyrosine 177 to phenylalanine prevents GRB-2 binding and impairs the transformation of primary bone marrow cultures.[23] Importantly, another group has also confirmed the importance of tyrosine 177 phosphorylation in the efficient induction of a CML-like disease *in vivo*,[24] while others have sought to take advantage of these findings by therapeutically interfering with Grb2 function.[25]

The Rac subfamily of Rho guanosine triphosphatases (GTPases) also plays a critical role in BCR-ABL1-mediated leukemogenesis[26] and the maintenance of BCR-ABL1-positive leukemic stem cells.[27] Rac GTPases are molecular targets of BCR-ABL1 and the combined loss of Rac1 and Rac2 suppresses the development of BCR-ABL1-induced myeloproliferative disease in mice. In Rac-deficient mice, activation of the various MAP kinases such as ERK1/2, JNK, and p38 were suppressed, which indicates that activation of the MAP kinase signaling pathways are dependent on Rac1 and Rac2.[26]

Apart from the activation of RAS and RAC, inactivation of negative regulators of these signaling cascades have also been described to promote the development of myeloproliferative disorders. *JUNB*, a negative regulator of the JNK downstream target *JUN*, was found to be inactivated by promoter methylation in primary samples derived from CML patients.[28] This observation argued for the notion that *JUNB* could function as a tumor suppressor in myeloid cells. Indeed, transgenic mice that lacked *junB* expression in the myeloid lineage were observed to develop a myeloproliferative disease that was transplantable and that could also progress to blast crisis.[29] Further, only *junB* deficiency in long-term hematopoietic stem cells could transplant CML-like disease in recipient mice.[30]

Distally, the activation of MAP kinases has been shown to lead to changes in gene expression due to the phosphorylation and activation of transcription factors such as activator protein 1 (AP-1)[31] as well as the increased translation of *MYC* mRNA via heterogeneous nuclear ribonucleoprotein K (HNRPK).[32] In turn, these transcriptional factors induce the expression of genes required for cell cycle progression, such as cyclins,[33] cyclin-dependent kinases,[34] and growth factors.[35]

### 2. JAK–STAT Pathway

Cytokine receptors lack tyrosine kinase activity, and hence depend on associated cytoplasmic tyrosine kinases known as Janus kinases (JAKs) to mediate changes in gene expression. Upon ligand binding, the cytokine receptors oligomerize and allow the associated JAKs to cross-phosphorylate and become activated. The activated JAKs phosphorylate the cytokine receptors, which create a docking site for the SH2 domain present among members of the signal transducers and activators of transcription (STAT) family of transcription factors. These STAT proteins dimerize after they are phosphorylated by the JAKs, after which they enter the nucleus to regulate gene expression. Constitutive activation of STATs, especially STAT5, has been described in several $Ph^+$ cell lines, and also in peripheral blood mononuclear cells derived from patients with CML.[36] Importantly, the activation of STAT5 is known to play an important role in leukemogenesis, as expression of a dominant negative mutant of STAT5 inhibited the growth and survival of BCR-ABL1-transformed hematopoietic cells.[37] Further, cells derived from mice with the entire Stat5a/b locus removed were observed to be resistant to BCR-ABL1-induced transformation.[38] In CML, the activation of STAT5 does not appear to require the JAKs.[39] Instead, BCR-ABL1 phosphorylates HCK, a member of the SRC family kinases, which then recruits and activates STAT5.[40] The mechanism by which STAT5 contributes to malignant transformation involves the upregulation of STAT5 target genes such as the serine–threonine kinase, *PIM-1*,[41] and B-cell lymphoma-extra large (*BCL-xL*) protein, an antiapoptotic member of the Bcl-2 family of proteins.[42]

### 3. Phosphatidylinositol 3-Kinase Pathway

In addition to activating the MAP kinase pathway, activated receptor tyrosine kinases can also trigger a signaling cascade that is mediated by phosphatidylinositol 3-kinase (PI3K). PI3K is a heterodimer that is composed of p85 and p110 subunits and is activated upon association with activated receptor tyrosine kinases. Subsequently, PI3K phosphorylates phosphatidylinositols to generate phosphatidylinositol-3-triphosphates that serve as second messengers for numerous targets including the serine–threonine kinase, AKT.[43] In CML, the formation of a complex between BCR-ABL1 and GRB2-associated-binding protein 2 (GAB2) recruits and activates PI3K in the absence of growth signals.[44] Further, the constitutive activation of PI3K is required for BCR-ABL1-mediated leukemogenesis and the growth of $Ph^+$ cells.[45,46] Interestingly, studies using mice that lack individual PI3K genes indicate that BCR-ABL1-dependent transformation in B lineage CML depends on specific PI3K gene products, including those of the Pik3r1 gene.[47] These studies suggest that isoform-specific PI3K inhibitors may offer more target selectivity and, potentially, fewer side effects.[48]

The molecular mechanism leading to enhanced cell growth and proliferation by AKT is due to the activation of its downstream effectors, including the mammalian target of rapamycin (mTOR). mTOR is a serine–threonine kinase that stimulates protein synthesis and ribosome biogenesis by forming two distinct multimeric complexes termed mTORC1 and mTORC2.[49,50] In response to growth factors, mTORC1 regulates cap-dependent translation by phosphorylating the translation inhibitor, the eukaryotic translation initiation factor 4E-binding protein 1 (4E-BP1). Phosphorylated 4E-BP1 dissociates from eukaryotic initiation factor 4E (eIF4E), which enables eIF4E to form a complex with eIF4G and eIF4A. This multimeric complex is termed eIF4F, which allows translation initiation to take place by melting secondary structures within the mRNA for the association of the 43S pre-initiation complex.[51] Apart from 4E-BP1, S6 kinases (S6K1 and S6K2) are also downstream targets of mTOR. S6K promotes cell growth by enhancing translation of 5'TOP (terminal oligopyrimidine tract) mRNAs that contain a short stretch of polypyrimidines next to the 5' cap.[52] However, S6K is not thought to mediate mRNA translation directly, but does so indirectly via downstream effectors of S6K such as ribosomal protein S6 and eIF4B. It has been proposed that phosphorylation enhances the activity of these effectors, leading to the increased translation of 5'TOP mRNAs.[53,54] In contrast, the function of mTORC2 is not well studied. Studies have shown that mTORC2 plays a role in cell survival and actin cytoskeleton reorganization.[55,56] Further, mTORC2 phosphorylates AKT and protein kinase C and protects them from proteasomal degradation.[57] In CML, previous studies have shown that both mTORC1 and mTORC2 were activated and played critical roles in the growth and survival of BCR-ABL1-expressing leukemic cells.[58,59]

AKT is also involved in antiapoptosis signaling by phosphorylating the proapoptotic protein, Bcl-2-associated agonist of cell death (BAD). Phosphorylated BAD is inactive because it is sequestered by the 14-3-3 proteins, which prevent BAD from interacting with the antiapoptotic members of the Bcl-2 family such as B-cell lymphoma 2 (BCL-2).[60] Evasion of apoptosis can also be mediated by the inhibition of the transcription factor, FOXO3A, by AKT. In fact, inactivation of FOXO transcription factors are critical for the survival of BCR-ABL1-transformed B cells.[47] One of the known targets of FOXO3A is BIM, a potent inducer of apoptosis termed Bcl-2-interacting mediator of cell death, and in BCR-ABL1-expressing cells, a survival advantage is conferred on CML cells by FOXO3A-dependent BIM suppression.[61]

## C. The Clinical Success and Limitations of Imatinib Therapy in Chronic Phase CML

From the above, it is clear that BCR-ABL1 activates multiple signaling pathways, each of which contributes to transformation through enlisting the help of existing physiologic growth and survival pathways. From a therapeutic perspective, these data indicate that inhibition of BCR-ABL1 directly would be the most direct and efficient way to counter the transforming properties of the oncoprotein. The dramatic success of the early phase clinical trials using imatinib validated this thinking. In addition, the studies were notable for the speed of responses and also for the durability of the responses among the chronic-phase patients but not blast crisis patients. Following on from these initial studies, a randomized clinical trial was initiated in 2000 comparing single-agent imatinib to interferon-α and Ara-C (the International Randomized Study of Interferon and STI571 [IRIS] study). This study has continued to yield important long-term information regarding the safety and efficacy of imatinib, which, at the sixth year of reporting, has demonstrated an estimated rate of freedom from progression of 93% and an estimated overall survival of 88%.[62] This remarkable survival rate is double that of historical controls (43%) prior to the imatinib era.[63]

However, while imatinib has been singularly effective in exerting control in the majority of patients with chronic-phase CML, it does not appear to be able to cure the disease. Thus, only about 10% of chronic-phase patients achieve a complete molecular remission, and of these, only about half are able to discontinue imatinib without the disease recurring, although follow-up of these patients remains short.[64] This phenomenon has been described as disease persistence and is likely related to the continued survival of CML stem cells in patients. In fact, even among patients who sustain a complete cytogenetic response (the disappearance of all $Ph^+$ cells in the bone marrow), the persistence of a small but identifiable number of primitive CML progenitors appears to be a consistent finding.[65]

Several reviews have been written about the phenomenon of CML stem cell persistence,[66–68] which highlight a number of features of CML stem cells that may enable them to survive in the presence of tyrosine kinase inhibitors. Most striking among the recent findings is the ability of quiescent, nondividing primitive CML progenitors to survive *in vitro* in the presence of tyrosine kinase inhibitors, suggesting that such cells do not require BCR-ABL1 to remain in this state.[69–74] Whether this situation exists *in vivo* in patients or not is beyond the ability of current technologies to determine, but it is a critical question in CML therapeutics. Still, mounting evidence continues to support the notion that BCR-ABL1 may be redundant when physiologic growth factors are present, including in primitive CML progenitors which are insensitive to imatinib-induced cell death despite inhibition of BCR-ABL1 activity.[8] It is therefore plausible that stromal cell interactions, as well as other bone marrow microenvironmental factors, will also be found to contribute to the phenomenon of CML progenitor persistence.[75,76] Several strategies have therefore been proposed to overcome CML cell persistence, which include the use of growth factors to stimulate cell cycle entry, interrupting stromal-leukemia cell interactions and identifying signaling pathways that may yet differentiate normal versus CML stem cells. In this respect, the recent observation that decreased AKT phosphorylation and the resulting nuclear localization (and activation) of the transcription factor Foxo3a occur in populations enriched for CML stem cells is intriguing. Here, the authors showed that TGF-β inhibition led to increased AKT signaling, decreased nuclear Foxo3a, and depletion of CML cells *in vivo*.[11] The possibility of targeting kinase activity of leukemia stem cells (LSCs) while leaving normal stem cells untouched has also been highlighted by others. Here, using phosphatase and tensin homolog (PTEN) inactivation to generate a murine leukemia model, it was shown that mTORC1 inhibitors can extinguish LSCs and at the same time restore the growth of normal stem cells.[77,78] These data therefore hint at the possibility of targeting protein phosphorylation to eliminate CML stem cells and effect a potential cure.

## III. Pathways, Kinases, and Proteins Mediating Drug Resistance

Although imatinib has been extremely successful in the treatment of most patients in chronic-phase CML, a proportion of patients will either fail to respond altogether or have a suboptimal response to tyrosine kinase inhibitors.[79] Such patients are classified as having primary tyrosine kinase inhibitor resistance, in contrast to those with secondary resistance, where the disease is seen to respond initially to tyrosine kinase inhibitors but then recurs. Importantly, patients with primary resistance, as well as those with suboptimal responses to tyrosine kinase inhibitors,[79] are at significantly increased risk of disease progression and poorer clinical outcomes.[62,80]

In general, the mechanisms of imatinib resistance can be classified into two main categories. First, factors intrinsic to the leukemic cell can lead to imatinib resistance. These intrinsic factors can be BCR-ABL1 dependent (e.g., a kinase domain mutation that decreases the binding of tyrosine kinase inhibitors) or BCR-ABL1 independent (activation of SRC kinases, differential activity of drug influx/efflux pumps,[81] and the apparent intrinsic resistance of CML stem cells to tyrosine kinase inhibitors[13]). Interestingly, BCR-ABL1 kinase domain mutations are relatively common among patients with secondary resistance, but less common among patients with primary resistance, suggesting the existence of different mechanisms of resistance between the two groups of patients.[82–84] Second, mechanisms of resistance extrinsic to the cell have also been proposed. These include factors in the bone marrow microenvironment, including growth factor- or cytokine-mediated activation of survival pathways that render the CML cell less dependent on BCR-ABL1.[8] Here, we will be reviewing some of the proteins, kinases, and cellular processes that have been shown to mediate tyrosine kinase inhibitor resistance in CML.

### A. BCR-ABL1

The development of mutations in the kinase domain of *BCR-ABL1* is the most frequent mechanism of resistance to imatinib treatment in CML and is often associated with secondary (acquired) resistance.[85] Because mutations in the *BCR-ABL1* kinase domain have been found in patients prior to the initiation of tyrosine kinase inhibitor therapy, it is thought that the selective pressure of tyrosine kinase inhibitor usage allows such subclones to expand and gradually outgrow cells that are sensitive to the drug.[86] Mutations conferring resistance are often mapped to the P-loop region of the kinase domain, which functions as a binding site for the phosphate moieties of ATP. Mutations also occur within the activation loop, preventing the kinase from adopting the inactive conformation to which imatinib binds, as well as the catalytic domain.[87] Although numerous mutations have been discovered using *in vitro* screens, it is important to note that not all mutations are clinically relevant because they may not be associated with clinical resistance. Therefore, the optimal evidence to assess resistance against BCR-ABL1 inhibitors should come from clinical studies rather than *in vitro* assessment.[88]

Intriguingly, previous work has provided evidence that not all *BCR-ABL1* kinase domain mutations are biologically equivalent, and that they may have other functions than mediating drug resistance. For example, it has been demonstrated that different *BCR-ABL1* mutants have varied oncogenic potentials with respect to their transformation potency, and that this is unrelated to the degree of tyrosine kinase inhibitor resistance conferred by the specific kinase domain mutation in question. By using mass spectrometry, these investigators found that different kinase domain mutants exhibited differential phosphorylation of the phosphotyrosine proteome, suggesting the intriguing possibility of altered substrate specificity between the different kinase domain mutants.[89]

Finally, overexpression of an unmutated BCR-ABL1 protein is another mechanism that contributes to imatinib resistance and occurs in approximately 18% of CML patients.[85] Here, the increased expression of BCR-ABL1 protein is thought to overcome, by mass effect, any tyrosine kinase inhibitor that has been able to enter the cells. Mechanistically, BCR-ABL1 overexpression is likely to occur at the transcriptional and posttranscriptional levels, as increased BCR-ABL1 protein expression has also been described in the absence of *BCR-ABL1* gene amplification. Importantly, others have suggested that differential levels of BCR-ABL1 kinase activity may exist in different subsets of CML progenitors, with higher levels of activity among the more primitive $Lin^{-}CD34^{+}CD38^{-}$ population, which corresponds to a population that is enriched for CML stem cells.[90]

## B. Drug Influx and Efflux Pumps

Resistance against tyrosine kinase inhibitors can also occur as a result of low intracellular availability of the drug. Intracellular availability of imatinib, as well as other tyrosine kinase inhibitors, is a balance between influx and efflux.[91–93] Studies have shown that inhibition of drug influx through the human organic cation transporter (hOCT1) is a critical factor that regulates availability of imatinib in the cell.[94] Importantly, in advanced-phase CML, the GG allele at *HOCT1* (rs683369) has been shown to be significantly associated with poor response to imatinib.[81] Apart from the inhibition of imatinib influx, genetic aberrations that promote drug efflux could also be linked to imatinib resistance. Consistent with this hypothesis, increased expression of the multidrug resistance gene (*MDR-1*) has been observed in a CML cell line that is resistant to imatinib.[95]

## C. *SRC* Family Kinases

The *SRC* family kinases are a group of genes that encode for cytoplasmic nonreceptor protein kinases (*FYN*, *SRC*, *YES*, *BLK*, *FGR*, *LCK*, *YRK*, *HCK*, *LYN*) that regulate cell growth. These genes are classical proto-oncogenes that are frequently mutated or overexpressed in cancers, leading to uncontrolled cell proliferation.[96] Studies have shown that the BCR-ABL1 oncoprotein activates HCK through direct interaction via a mechanism that requires several domains of BCR-ABL1 and the SH2 and SH3 domains of HCK.[97] Activation of HCK promotes cell proliferation by activating STAT5.[98] Notably, overexpression of LYN has also been associated with disease progression and imatinib resistance in CML.[99] Moreover, persistent activation of LYN has been observed in some CML patients who do not carry *BCR-ABL1* mutations.[100] These results suggest that an approach that inhibits both BCR-ABL1 and LYN is required to overcome imatinib resistance associated with persistent LYN activation.

## D. Inactivation of Genes That Activate Apoptosis

The *p53* gene encodes a transcription factor that is involved in preventing malignant transformation. Following DNA damage, the p53 protein can induce cell cycle arrest at the G1/S phase so as to allow the damage to be repaired.[101] However, if the DNA damage cannot be repaired, p53 induces apoptosis by upregulating proapoptotic members of the Bcl-2 family such as p53 upregulated modulator of apoptosis (*PUMA*) and *NOXA*.[102] In CML, progression from the chronic phase to the blast phase is often associated with increased resistance to imatinib. Approximately 25–30% of patients with myeloid blast phase have mutations in the *p53* gene, which suggests that *p53* may play an important role in response to imatinib therapy.[103] Indeed, when *p53* was inactivated in murine cells that express BCR-ABL1, these cells were observed to be more resistant to imatinib *in vitro* and *in vivo*.[104]

The *BIM* gene encodes a potent proapoptotic member of the Bcl-2 family of proteins. The protein contains a Bcl-2 homology 3 (BH3) domain that is important for the induction of apoptosis, as it is the BH3 domain that directly engages antiapoptotic members of the Bcl-2 family such as BCL-2.[105] In CML, constitutive activation of the MAP kinase signaling pathway can suppress the expression of the major BIM protein isoform, $BIM_{EL}$, due to increased proteasomal degradation as a result of phosphorylation of $BIM_{EL}$ by ERK1/2 on serine 69.[106] Numerous studies have reported that imatinib-induced apoptosis in CML requires the upregulation of BIM, as BIM downregulation can prevent imatinib-induced apoptosis in CML cells.[107–109] Consistent with the notion that BIM expression mediates response to imatinib, epigenetic downregulation of *BIM* expression has been observed in some CML patients and is also associated with suboptimal responses to imatinib therapy.[110]

## E. Autophagy

Autophagy is a catabolic process that involves sequestration and degradation of damaged proteins and organelles through the lysosomal machinery. Degradation of these intracellular components can serve as an energy source during periods of starvation and hypoxia to maintain cellular homeostasis for survival.[111] Loss of *BECLIN1*, a gene that is required for the induction of autophagy, has been shown to promote the formation of tumors, which strongly suggests that autophagy is required to suppress tumor formation.[112] However, others have demonstrated that autophagy can also play a protective role in established tumors because the inhibition of autophagy can enhance the cell-killing effects of cytotoxic drugs.[113] Therefore, autophagy has a dual role in tumorigenesis and cancer progression by promoting both cell survival and cell death. In CML, the induction of autophagy serves to protect CML cells against the cell-killing effects of imatinib.[114] Here, imatinib was first shown to induce autophagy in K562 cells, while inhibiting autophagy with chloroquine enhanced the cell-killing effects of imatinib. The cellular protection of autophagy may also extend to primitive CML progenitors, as the addition of autophagy inhibitors to tyrosine kinase inhibitors led to the increased induction of cell death in primary CML stem cells.[115]

## F. Role of Phosphatases in Tyrosine Kinase Inhibitor Resistance

Normal cell homeostasis is regulated by a complex network of signal transduction pathways that are activated by protein phosphorylation initiated by protein kinases. These activated pathways can be reversibly inactivated by protein dephosphorylation catalyzed by protein phosphatases.[116] Aberrant phosphatase activity has been implicated in a number of human diseases. These include the constitutive activation of cell survival pathways and aberrant phosphorylation of microtubule-associated proteins found in the central nervous system, leading to the development of cancer and Alzheimer's disease, respectively.[117,118] Here, we describe some of the key phosphatases that have been shown to play important roles in the development and progression of CML.

## 1. SHP2

The SH2 domain-containing phosphatases (Shps) belong to a class of non-transmembrane protein tyrosine phosphatases that contain two N-terminal SH2 domains.[119] There are two vertebrate Shps: SHP1 and SHP2. SHP2 acts upstream of RAS and is required for the full activation of the RAS/ERK signaling pathway.[120] Several models describing the role of SHP2 in the activation of RAS/ERK signaling have been described. One of these models proposes that SHP2 limits the recruitment of RasGAP to receptor tyrosine kinases, and thereby promotes RAS activation.[121] Another model suggests that SHP2 enhances SRC activity by dephosphorylating the negative regulatory tyrosyl phosphorylation sites of SRC at the C-terminus.[119,122] A third possibility is that SHP2 activates RAS/ERK signaling by inactivating inhibitors of the RAS/ERK pathway. Indeed, recent work has shown that Sprouty-related Ena/vasodilator-stimulated phosphoprotein homology 1-domain-containing protein (SPRED1), a negative regulator of RAS/ERK signaling, is a substrate of SHP2. Further, tyrosine dephosphorylation of SPRED1 is needed to limit the inhibitory function of SPRED1 in the RAS/ERK signaling pathway.[123]

Activating mutations of the *SHP2* gene are observed in approximately 50% of individuals with Noonan syndrome (NS), a genetic disease that is characterized by facial defects, short stature, and cardiac abnormalities.[124] Interestingly, NS patients are also at a greater risk for developing sporadic juvenile myelomonocytic leukemia.[125] In CML, association of SHP2 and the BCR-ABL1–GAB2 complex plays an important role for the constitutive activation of the RAS/ERK signaling pathway.[44] Further, overexpression of SHP2 has also been observed in CML patients.[126] Taken together, these findings suggest that SHP2 could play a role in the development of CML.

## 2. PTEN

PTEN is a phosphatase that dephosphorylates phosphatidylinositol-3-triphosphate and, hence, functions as a negative regulator of the PI3K signaling pathway. PTEN is often deleted or inactivated in human cancers, including brain, breast, and prostate cancer.[127] In a murine model of BCR-ABL1-induced leukemias, *Pten* transcripts were downregulated by BCR-ABL1 in the LSC population and *Pten* deletion accelerated the development of CML.[128] However, while functional PTEN inactivation contributes to the aggressiveness of CML, PTEN mutations at the genetic level may be relatively uncommon in CML patients.[129] Taken together, these results strongly argued for the hypothesis that PTEN functions as a tumor suppressor in BCR-ABL1-induced leukemias by regulating the PI3K signaling pathway.

### 3. PP2A

PP2A is a major serine–threonine phosphatase that is involved in the dephosphorylation of specific substrates. PP2A is made up of a heterotrimeric complex, which includes (1) a structural subunit ($PP2A_A$), (2) a regulatory subunit ($PP2A_B$), and (3) a catalytic subunit ($PP2A_c$).[117] The regulatory subunits (also known as B subunits) determine substrate specificity. Because there are more than 20 interchangeable B subunits, the number of substrates that could be targeted by PP2A is large.[117] PP2A downregulates numerous signal transduction pathways initiated and/or sustained by several protein kinases, including the RAS/ERK,[130] PI3K,[131] and WNT signaling[132] pathways. The regulation of PP2A activity is dependent on posttranslational PP2A modifications. For instance, PP2A is inactivated by the phosphorylation of tyrosine 307 on the catalytic subunit.[117] Other posttranslational mechanisms include inhibitory proteins that suppress PP2A's phosphatase activity such as SET[133] and the cancerous inhibitor of PP2A (CIP2A).[134]

PP2A's ability to suppress mitogenic signaling suggests that it could function as a tumor suppressor. Evidence supporting this hypothesis came from an early observation that the DNA tumor virus, Simian virus 40 (SV40), expresses the small-t antigen that suppresses PP2A's phosphatase activity by displacing B subunits from the heterotrimeric complex.[135] Second, somatic mutations in PP2A subunits have also been observed in numerous solid tumors.[136]

In CML, PP2A activity decreases slightly in chronic-phase $CD34^+$ progenitors but decreases significantly in blast-phase $CD34^+$ progenitors. Increased PP2A activity has also been shown to inhibit BCR-ABL1 oncogenesis *in vitro* and *in vivo*, which suggests that PP2A functions as a tumor suppressor in CML.[133] The molecular mechanisms leading to decreased PP2A activity in CML involve the upregulation of SET[133] and CIP2A[137] as well as signaling mediated by sphingosine kinase-1 and sphingosine 1-phosphate receptor 2.[138]

## G. Overcoming Therapy Resistance Through Modulating Protein Phosphorylation

The development of imatinib resistance has become a major problem in the treatment of CML. Therefore, there is a critical need to develop strategies to overcome resistance and to eradicate the leukemic stem cells. One approach to overcome resistance is to design more potent BCR-ABL1 inhibitors that can inhibit the activity of BCR-ABL1 mutants. Another approach is to use inhibitors to target proteins found in signaling pathways downstream of BCR-ABL1. Additionally, compounds that enhance cell death, such as drugs that inhibit the cytoprotective effects of autophagy or the use of BH3 mimetics (ABT-737) to promote apoptosis, can also be employed to treat CML patients. In this review, we will be focusing on overcoming therapy resistance through modulating protein phosphorylation. In doing so, we should also highlight that it is becoming increasingly appreciated that phosphorylation patterns between leukemic stem cells and the more committed BCR-ABL1-positive progenitor cells are different, with important therapeutic implications.[11,90]

## H. Second- and Third-Generation BCR-ABL1 Inhibitors

Mutations in the BCR-ABL1 kinase domain that prevent effective binding of imatinib enable the oncoprotein to sustain the activation of cell survival signal transduction pathways during drug therapy. Therefore, there is a need to develop novel BCR-ABL1 inhibitors to overcome resistance associated with these mutations. Nilotinib is an example of a second-generation BCR-ABL1 inhibitor. It is approximately 10–50 times more potent than imatinib and is able to inhibit the tyrosine kinase activity and proliferation of most cell lines expressing BCR-ABL1 mutants, except the T315I mutant.[139] However, nilotinib is unable to eradicate the primitive, quiescent population of CML cells.[70] In a clinical study, 60% of patients with BCR-ABL1 mutations had a hematologic response, and 41% of these patients showed cytogenetic response after they were treated with nilotinib.[140]

Overexpression of SRC family kinases such as LYN has also been associated with disease progression and imatinib resistance in CML.[99] Therefore, effective eradication of all CML cells could require the inhibition of both BCR-ABL1 and SRC kinases. Dasatinib is a compound that has been developed as a dual-specific SRC- and ABL kinase inhibitor. It is at least 100-fold more potent than imatinib in the inhibition of BCR-ABL1, and it is also effective against the active conformation of the kinase domain, which is resistant to imatinib.[141] Similar to nilotinib, dasatinib is also able to inhibit the proliferation of most cell lines expressing BCR-ABL1 mutants except the T315I mutant.[141] However, dasatinib is also unable to eliminate the primitive, quiescent population of CML cells.[71]

In general, when used as first-line therapy, the second-generation tyrosine kinase inhibitors (dasatinib and nilotinib) have resulted in faster and deeper responses compared to imatinib.[142,143] When used for patients who have failed in, or are intolerant to, imatinib therapy, these drugs have also shown significant activity and elicited improved responses.[3,4] However, whether they will lead to improvements in long-term clinical outcomes, such as improved survival, remains to be seen.

Because second-generation tyrosine kinase inhibitors are unable to overcome all clinically important mutations, a third generation of "pan-BCR-ABL1" inhibitor compounds has begun testing in the clinic. Importantly, these pan-BCR-ABL1 inhibitors are able to overcome resistance associated with the cross-resistant T315I mutation. The two leading compounds are ponatinib (AP24534) and DCC-2036, which are currently in phase 1 and phase 2 clinical trials, respectively.

## I. Farnesyl Transferase Inhibitors

The development of second-generation BCR-ABL1 inhibitors enabled the effective targeting of most CML cells harboring BCR-ABL1 mutations. However, there are patients who are resistant to BCR-ABL1 inhibitors in the absence of BCR-ABL1 mutations. The application of BCR-ABL1 inhibitors in combination with drugs that target signaling pathways downstream of BCR-ABL1 may help to overcome resistance. The RAS protein is activated by BCR-ABL1, leading to the activation of signaling transduction pathways that are critical for malignant transformation.[23] Therefore, targeting oncogenic RAS could be a useful approach to enhance the cell-killing effects of BCR-ABL1 inhibitors. Another reason inhibitors of this pathway may be useful in CML, particularly in combating CML stem cell persistence, is to overcome the phenomenon of MAP kinase activation that has been found in CML progenitors following imatinib exposure.[144] Here, the authors showed that combined treatment with a MEK-1/2 inhibitor and imatinib significantly decreased CML progenitor growth.

The proper function of RAS requires prenylation, which is a posttranslational modification that involves the addition of a 15-carbon farnesyl isoprenoid moiety to a conserved cysteine residue in a carboxy-terminal CAAX motif.[145] Prenylation enables the RAS protein to be recruited and to be associated with proteins found on the cell membrane. The process of prenylation is mediated by farnesyltransferase (FT). Drugs targeting FT (tipifarnib, lonafarnib, BMS-214662) have been developed to inhibit RAS signaling, as well as other clinically approved drugs, including the bisphosphonates, which were subsequently been found to have anti-RAS activity.[146]

When used as a single agent, lonafarnib and tipifarnib were demonstrated to inhibit the growth of tumor cells *in vitro* and *in vivo*.[147,148] However, the effect of either tipifarnib or lonafarnib was modest when they were administered as a single agent to CML patients who were not responsive to imatinib or interferon-alpha therapy.[149,150] BMS-214662 holds a greater promise for the treatment of CML. When used as a single agent, BMS-214662 was able to kill the quiescent CML stem cell population by apoptosis. Further, BMS-214662 was extremely effective against cells from blast crisis and cell lines expressing BCR-ABL1 mutants when used in combination with BCR-ABL1 inhibitors.[151]

However, it is important to note that the antiproliferative and cell-killing effects of FT inhibitors may not be based only on RAS inhibition because these compounds can inhibit the farnesylation of other proteins as well.

## J. mTOR Inhibitors

Activation of the PI3K signaling pathway is critical for the survival and proliferation of BCR-ABL1-transformed cells. Therefore, the development of compounds that inhibit PI3K or proteins downstream of PI3K, such as mTOR, could enhance responses to imatinib as well as to overcome resistance. Rapamycin is a macrolide antibiotic that forms a complex with the immunophilin molecule FKBP12. The resulting complex inhibits mTORC1. Recent work has shown that the combination of rapamycin and imatinib synergistically decreases the proliferation of BCR-ABL1-transformed Ba/F3 cells and the K562 cell line.[152] Further, in a murine model of CML, mice treated with both rapamycin and imatinib showed improved survival compared to those treated with either rapamycin alone or imatinib alone.[58] The molecular mechanism for the synergistic effects of imatinib and rapamycin combination likely involves a greater inhibition of translation than either agent alone, and may be due to an increased activity of the translation inhibitor 4E-BP1 when both BCR-ABL1 and mTORC1 are inhibited.[153]

Recent work has also suggested that the rapamycin-insensitive mTORC2 complex also plays a critical role in the growth and survival of BCR-ABL1-expressing cells. Here, the use of OSI-027, an inhibitor that targets both mTORC2 and mTORC1, resulted in an effective antileukemic response against primitive progenitors from CML patients. In addition, OSI-027 could also induce apoptosis in cells expressing the BCR-ABL1 T315I mutation that are refractory to imatinib. Taken together, these findings suggest that dual inhibition of mTORC2 and mTORC1 could overcome imatinib resistance associated with *BCR-ABL1* mutations.[59]

## K. Identification of Synthetic Lethal Proteins and Novel Pathways to Overcome Tyrosine Kinase Inhibitor Resistance

As applied to cancer cells, the concept of synthetic lethality is that oncogene-addicted cancer cells may rely on additional pathways for their survival when the original oncogene is inhibited, and that the simultaneous inhibition of the additional pathway will result in cancer cell, but not normal cell, death.[154,155] Based on this understanding, and by employing RNAi-based screening, one group has uncovered genes from the noncanonical WNT/$Ca^{2+}$/NFAT pathway as being synthetic lethal with BCR-ABL1 inhibition. Importantly, the authors found that NFAT inhibition, using cyclosporine A, sensitized CML cells to imatinib-induced cell death.[156] Because the second-generation tyrosine kinase inhibitors have a different spectrum of kinase inhibitory activity compared to imatinib, it will be interesting to determine if such screens might yield different "hits."

## IV. Protein Phosphorylation in Disease Progression

Disease progression in CML is variously defined by different clinical classification systems,[152,157,158] but all share in common a criterion requiring an accumulation in the percentage of undifferentiated cells in either blood or bone marrow. In two-thirds of cases, blasts are myeloid in phenotype, while the remaining cases are lymphoid. Importantly, myeloid and lymphoid blast phases are thought to be genetically distinct, with the latter type associated with shorter survival time.[159] Depending on the actual number of blasts, progression is classed as either accelerated or blast phase, although recent gene expression studies suggest that only two stages can be clearly distinguished, chronic phase versus accelerated and blast phase.[160] It is generally assumed that the accumulation of both genetic and/or epigenetic "hits" leads to disease progression, and indeed some classification systems also require the documentation of additional karyotypic abnormalities, a feature known as "clonal evolution." Clinically, disease progression is usually heralded by the onset of constitutional symptoms and peripheral blood counts that are increasingly difficult to control. With the onset of blast crisis, patient survival is short, usually measured in months, despite the use of combined tyrosine kinase inhibitor-based multiagent chemotherapy. Indeed, by this stage of the disease, even high-dose chemotherapy with stem cell transplantation is usually not curative, and almost all patients will eventually succumb from CML itself or complications of therapy.

At the biologic level, the clinical features of blast crisis are accompanied by a differentiation block, resistance to tyrosine kinase inhibitors, and activation of self-renewal pathways. Importantly, the last feature is thought to reside in a blast crisis progenitor cell population that immunophenotypically resembles a committed myeloid progenitor termed the granulocyte–macrophage progenitor (GMP).[161,162] Currently, it is not clear if the blast crisis "GMP" population arises from the original CML stem cell or from a more differentiated progenitor. Be that as it may, significant advances have been made in the understanding of the proteins and pathways that underline several of the key biologic changes, including the cell type they occur in. In the following sections, we will review the processes that have been found to be consistent with changes in blast crisis CML. With the identification of these processes, it is hoped that the imatinib equivalent(s) for advanced-stage CML might be found. However, whether one drug "targeting" one protein or process will hold the key to reversing all three processes of differentiation block, resistance to tyrosine kinase inhibitors, and activation of β-catenin-driven self-renewal pathways remains to be seen. Thankfully though, in most patients who present in chronic phase and are treated with imatinib, fewer appear destined to transform to blast crisis compared to historical controls,[62] demonstrating that tyrosine kinase inhibitors have changed the natural history of the disease, while also supporting the idea that BCR-ABL1 does indeed promote genetic instability and the mutator phenotype in CML.[158,163–165]

## A. BCR-ABL1 Protein Levels Parallel Disease Progression

Recent work has shown that BCR-ABL1 protein levels in $CD34^+$ CML cells progressively increase with the chronic phase to blast crisis transition, and thus track disease stage.[166] Importantly, the increase in BCR-ABL1 protein levels was also found to be closely associated with the activation of downstream mediators of BCR-ABL1-dependent signaling, including CrkL, STAT5, and MAP kinase,[32] suggesting that the majority of the known signaling pathways described above would be similarly affected. Functionally, the increase in BCR-ABL1 protein also encouraged the development of kinase domain mutations *in vitro*, including mutations that are known to be associated with clinical tyrosine kinase inhibitor resistance. Thus, increases in BCR-ABL1 protein may be critical to activating the pathways responsible of the mutator phenotype that others have found to be dependent on BCR-ABL1.

Mouse models of CML have also suggested that both BCR-ABL1 protein levels and the cell lineage they are expressed in are important in determining the tempo of the disease, blast crisis lineage phenotype, and resistance to tyrosine kinase inhibitors. The clearest example that levels of BCR-ABL1 protein itself are functionally important comes from the use of bone marrow retroviral transduction models of CML. Here, using the same bicistronic vector, when higher levels of BCR-ABL1 protein are expressed under control of regular cap-dependent translation versus less efficient IRES-mediated translation, animals developed a CML-like disease earlier and were more resistant to therapy with imatinib. Others have in the meantime gone on to determine the mechanisms by which elevated levels of BCR-ABL1 protein affect cell proliferation, survival, and differentiation, which are detailed below.

What gives rise to the increase in BCR-ABL1 protein levels? Several mechanisms have been proposed and likely act independently from the instances of BCR-ABL1 gene amplification described above. These include increased activity by the *BCR* promoter,[167] increased BCR-ABL1 mRNA translation, decreased PP2A activity (leading to BCR-ABL1 protein stabilization),[133] and possibly inhibition of SHP-1 phosphatase. However, this is clearly an underinvestigated area of study, given the importance of increased BCR-ABL1 protein levels in blast crisis and the relatively few articles on *BCR-ABL1* expression regulation.

## B. Proteins and Pathways Implicated in Acquired Self-Renewal in Blast Crisis CML

A defining feature of cancer or LSCs is the ability to self-renew and propagate the tumor in immunocompromised animals.[10] These characteristics are fulfilled by chronic-phase CML stem cells, which, in contrast to their differentiated progeny, remain difficult to eradicate and are thought to give rise to recrudescent disease when tyrosine kinase inhibitor therapy is stopped. Since then, the discovery of a distinct LSC in blast crisis, which fulfilled the criteria of a LSC and were enriched for in a GMP-like population (blast crisis GMP), suggested that this population might be responsible for the almost universal tyrosine kinase inhibitor resistance seen in blast crisis patients. Importantly, it was also shown that β-catenin activation is required for the acquired LSC function of blast crisis GMPs, as genetic inhibition of β-catenin (through expression of axin 2) prevented LSC function. These data begged the questions of whether BCR-ABL1 and/or other factors were required to activate self-renewal in blast crisis cells and whether such factors might be druggable. In this respect, the studies in murine models of CML are informative in that they demonstrate that *BCR-ABL1 per se*, unlike other leukemia fusion genes, is unable to confer LSC function on committed progenitors.[168] This conclusion is supported by the consistent clinical observation that patients with blast crisis CML almost never have enduring response to second-generation BCR-ABL1 tyrosine kinase inhibitors. Thus, the available data suggest that factors beyond BCR-ABL1 are responsible for the acquired self-renewal found in blast crisis cells.

Since the initial description of blast crisis GMP LSCs, at least three different mechanisms have been proposed to contribute toward their LSC function and center on the reactivation of β-catenin signaling. The first report described the direct tyrosine phosphorylation at Y86 and Y654 on β-catenin itself by BCR-ABL1, which, in turn, led to the nuclear localization of β-catenin, a necessary step for its transcriptional activity.[169] Specifically, the authors showed that BCR-ABL1-dependent phosphorylation of β-catenin allowed it to bind to TCF and activate transcription. Inhibition of BCR-ABL1 led to the cytoplasmic accumulation of β-catenin, increased binding to the Axin/GSK3β destruction complex, and its subsequent proteasomal degradation. Whether this mechanism alone can account for the full clinical picture of blast crisis is unclear, particularly since the considerably more potent second-generation tyrosine kinase inhibitors are unable to exert long-term control over blast crisis.

These clinical observations suggest that other factors may also contribute to blast crisis pathophysiology in a BCR-ABL1-independent manner. Evidence for this has emerged from a second study, which also implicated proteins involved in canonical Wnt signaling. Here, using blast crisis patient material, missplicing of GSK3β cDNA was found in five of eight samples examined, which was associated with the production of a misspliced isoform of GSK3β missing exons 8 and 9.[162] Importantly, these exons encode a critical axin-binding domain, and while expression of regular full-length GSK3β led to the expected decrease in β-catenin activation in chronic-phase CML progenitors, expression of the misspliced isoform led to β-catenin activation.

A third study, published in abstract form, provides important evidence that the inactivation of PP2A described above is also an important factor in β-catenin activation in blast crisis GMP.[170] Importantly, this group has shown that the PP2A activator FTY720 (fingolimod) is capable of extinguishing β-catenin activation and the LSC function of blast crisis GMPs. Because FTY720 is an FDA-approved drug for use in multiple sclerosis, these studies may lead to the use of this, or related drugs, in patients with blast crisis CML.

In addition, our own group has demonstrated that the mRNA cap-binding protein eIF4E is overexpressed and phosphorylated at S209 in blast crisis GMPs, and that eIF4E phosphorylation at S209 was necessary to activate β-catenin signaling in blast crisis GMPs. Because the MAP kinase interacting kinases 1 and 2 (MNK1/2) are required for *in vivo* eIF4E phosphorylation,[171] we also found that small-molecule inhibition of MNK1/2 could prevent β-catenin activation, an intervention that also abrogated the LSC function of blast crisis GMPs.[172] Interestingly, we also found that PP2A activation, using FTY720, resulted in the inhibition of eIF4E phosphorylation in blast crisis cells, a finding that was consistent with other reports identifying eIF4E as a target of PP2A in cancer cells. Thus, eIF4E may also be an important protein target of FTY720-mediated PP2A reactivation.

As well as activation of canonical Wnt signaling, other pathways utilized by LSCs have been delineated in murine models of CML. These include the involvement of the Hedgehog pathway, which is a highly conserved pathway that regulates the cell proliferation, differentiation, and migration during development.[173,174] Two independent studies have contributed to the understanding that Smoothened (Smo), a transmembrane receptor in the Hh signaling pathway, is important in the CML LSC function.[175,176] Both studies present data indicating that loss of Smo decreased the incidence of a CML-like disease using a murine model with BCR-ABL1. This decrease in disease incidence was also associated with a reduction in LSK ($Lin^{-}Sca\text{-}1^{+}$ $c\text{-}Kit^{+}$) cells that contain the LSC population. Pharmacological inhibition of the Hh pathway in this model with cyclopamine had similar effects to loss of Smo and was also found to inhibit colony formation by primary blast crisis patient cells. The mechanism of action of cyclopamine occurs by direct binding to Smo, resulting in its inactivation. These data have led to a number of clinical trials using inhibitors to Hh signaling in cancer. Because prior work had shown that PI3K/AKT signaling could enhance signaling in this pathway, particularly when levels of Sonic hedgehog were low,[177] it could also be the case that inhibitors in the PI3K/AKT pathway might modulate Hh signaling.

## C. Proteins and Pathways Implicated in Differentiation Block in Blast Crisis CML

The differentiation block that accompanies the onset of blast crisis CML is a defining feature of the disease. Several mechanisms have been described to contribute to the block, including both genetic and epigenetic changes. In this review, only those that are primarily mediated by changes in protein phosphorylation will be covered.

As described above, a central feature of blast crisis cells is the increase in BCR-ABL1 protein that accompanies the transition from chronic phase. Data from several lines of evidence suggest that this increase in BCR-ABL1 protein is critically important in mediating differentiation arrest in blast crisis cells. Concomitant with the increase in BCR-ABL1 protein is an increase in MAP kinase signaling downstream of BCR-ABL1.[32] While MAP kinase signaling in chronic-phase CML $CD34^+$ cells can be accounted for by physiologic growth factor-mediated signaling, MAP kinase signaling in blast crisis is not only constitutively increased but also results in the increased expression of the RNA-binding protein, hnRNP-E2. hnRNP-E2 is directly phosphorylated by ERK1/2 at S173, S189, S272, and T213, and thereby stabilized. This leads to increased hnRNP-E2 levels, as well as increased binding of hnRNP-E2 to the 5'UTR of the CCAAT/enhancer binding protein α (C/EBPα), which results in decreased translation and expression of C/EBPα. The transcriptional function of C/EBPα is essential for granulocytic differentiation of granulocyte/monocyte precursors, and as expected, the differentiation of blast crisis cells was unable to proceed because of the decrease in C/EBPα levels. Importantly, pharmacologic MAP kinase inhibition using clinically relevant inhibitors was able to overcome the differentiation arrest of primary blast crisis blasts.[32] As further evidence for the role of increased BCR-ABL1 protein in differentiation arrest, it has also been shown that PP2A inactivation in blast crisis leads to increased MAP kinase signaling, and, when this is relieved by PP2A activation, leads to the differentiation of blast crisis precursors.[133] Thus, the activity of the MAP kinase signaling pathway, through regulating a key transcription factor, can directly mediate blast crisis cell differentiation arrest and be therapeutically targeted to promote differentiation.

More recently, this pathway was again highlighted to be involved in a novel function of the miR-328.[177a] Here, it was found that miR-328 was repressed in a BCR-ABL1-dependent manner, and, when the levels were restored, resulted in differentiation of blast crisis cells. This was found to occur by miR-328 binding to hnRNP-E2, causing the release of C/EBPα, which could then exert its transcriptional effects to promote differentiation.

## V. What Clinical Experience Has Taught CML Biologists

The FDA approval of imatinib in 2001 followed a clear demonstration of its efficacy in CML, which at the same time underlined the central role of protein phosphorylation in a well-defined human cancer. Since then, ongoing studies have yielded a rich trove of additional data that has shed further light on the complexities in the relationship between protein phosphorylation and human disease states. Some of these lessons are relevant to CML alone, while others have broader implications regarding the therapeutic targeting of other cancerous and noncancerous states.

In the broadest terms, we can consider the data in terms of the major lesson learnt: that tyrosine kinase inhibitors are sufficient to produce long-lasting responses in chronic-phase CML, and that they can elicit such responses with minimal toxicity. The fact that single-agent imatinib has exerted considerable disease control is of course remarkable, and a number of important practical lessons can be drawn from its success.[2] These include the existence of a robust biomarker that predicts for sensitivity. In diseases associated with aberrant protein phosphorylation, these will usually consist of overexpression and/or mutation of a kinase. Ideally, the kinase should be both the therapeutic target as well as the "driver" of the disease. The importance of identifying a patient population that would be sensitive to the drug being tested was further appreciated in the development of EGFR inhibitors in non-small-cell lung cancers.[178] A second important point is that the therapeutic intervention should be applied early in the course of the disease, when additional transforming events are less likely to have occurred, and response can be more easily attributed to target inhibition.

The availability of second- and now third-generation tyrosine kinase inhibitors has also provided some interesting observations. Most important is the finding that the more potent inhibitors of BCR-ABL1, namely, nilotinib and dasatinib (nanomolar potency), are able to exert quicker and deeper clinical responses compared to imatinib (micromolar potency). However, while very encouraging, it remains to be seen if these "Next-Gen" tyrosine kinase inhibitors are more effective in delivering complete molecular responses, and possibly cure. If the rates of "cure" are similar to that of imatinib,[64] it may indicate that we have reached a biologic ceiling for the effect of BCR-ABL1 inhibitors in CML. This would suggest that alternative approaches to improve treatment, this time directed against non-BCR-ABL1 kinase targets, might be required. Indeed, such an approach was recently taken when investigators found that the combination of interferon-$\alpha$ (which may stimulate hematopoietic stem cells into the cell cycle)[179,180] with imatinib yielded improved molecular responses.[181] Another interesting clinical observation has been the nonoverlapping toxicity that occurs with the use of the different agents, presumably a result of individual off-target effects. Importantly, such off-target effects will need to be taken into consideration in the development of combination studies.

We have also learnt a tremendous amount from the limitations of tyrosine kinase inhibitor therapy. First, for most patients, it has become clear that single-agent tyrosine kinase inhibitor cannot cure CML and will likely require them to take tyrosine kinase inhibitors indefinitely.[64] The persistence of CML progenitors in patients who have achieved deep, and even complete molecular responses suggests that the CML stem cell is resistant to tyrosine kinase inhibitor therapy. The identification of factors that differentiate normal versus CML stem cells therefore remains a very active area of ongoing research, and current efforts are focusing on microenvironmental factors in the bone marrow. Here, it is important to mention that one microenvironmental factor that is often overlooked in leukemia studies is the fact that the bone marrow microenvironment is hypoxic. Since hypoxia helps to maintain certain hematopoietic stem cell functions,[182] it may also be the case that critical LSC functions are similarly dependent on physiologic hypoxia.[8] Second, the failure of tyrosine kinase inhibitor-based therapy to control blast crisis is a significant reminder of the limitations of targeting BCR-ABL1 alone in CML. The dismal situation here is most likely because other cancer-promoting pathways have been activated and will require multipronged approaches tackling the different features of blast crisis.[158] Analogous to the situation in the 1960s and 1970s when multiagent chemotherapy was being developed, we may have to resort to multiagent targeted therapy to effectively control the "multiheaded beast" of blast crisis CML. Importantly, such approaches, using a minimum of three agents, may also help in preventing the emergence of drug-resistant clones.[183,184]

One surprising but interesting observation is that transient, but potent, BCR-ABL1 inhibition appears to be sufficient to induce cancer cell death and control disease *in vivo*.[22] This conclusion was based on the ability of second-generation and the more potent tyrosine kinase inhibitor, dasatinib, to kill CML cells *in vitro*, despite intermittent BCR-ABL1 kinase suppression. This finding provides an explanation for the clinical observation that a single daily dose dasatinib (serum half-life of 3–5h), which results in transient BCR-ABL1 inhibition in CML mononuclear cells, is sufficient to control the disease.[185]

These data, in turn, have led to the concept that targeted anticancer efficacy is related to both the duration of target inhibition as well as the depth of inhibition. However, it remains to be seen whether the use of the significantly more potent second-generation tyrosine kinase inhibitors will lead to improved responses, increased killing of more primitive CML progenitors, and ultimately the cure of CML.

An important constant in the results of CML clinical trials, even when patients are molecularly defined and targeted, is the fact that clinical responses remain extremely heterogeneous. Thus, while most CML patients will have a complete cytogenetic response in the bone marrow to imatinib, with the clearance of all cytogenetically detectable $Ph^+$ cells, almost 30% will not.[62] The use of second-generation tyrosine kinase inhibitors, such as nilotinib and dasatinib, has improved overall responses. Nevertheless, a significant amount of patient-to-patient differences remain with respect to the kinetics of response, as well as the depth of response, and both these factors may influence long-term outcomes.[79] It is important to highlight that while much of the therapeutic efforts in CML has focused appropriately on the tumor genome (i.e., BCR-ABL1 alone), relatively little effort has focused on the germline genome (~25,000 genes). The field of pharmacogenomics, the study of how inherited and acquired genetic differences in drug response, has gone part way to addressing some of the heterogeneity in patient drug tolerance[186] but has not yet explained differences in intrinsic tumor responses. Because polymorphisms affecting the function of many proteins are known to exist, we anticipate that normal germline variations, or polymorphisms, will be discovered which directly affect protein phosphorylation and function, and thereby the intrinsic response to targeted therapies.

Given the above clinical and preclinical observations, we believe a case may be made that the next stage in the battle against CML will be waged against BCR-ABL1-independent factors. As mentioned in this review, this may include unexpected targets revealed by synthetic lethal screens, factors in the CML stem cell niche in the bone marrow, additional pathways to self-renewal co-opted by blast crisis GMPs, and possibly polymorphisms in the normal human genome.[187] While it is difficult to predict the exact details of this next lap of research, we believe, given the advances over the past 10 years, that it is sure to be interesting, exciting, and rewarding.

## References

1. Lugo TG, Witte ON. The BCR-ABL oncogene transforms Rat-1 cells and cooperates with v-myc. *Mol Cell Biol* 1989;**9**:1263–70.
2. Druker BJ. Translation of the Philadelphia chromosome into therapy for CML. *Blood* 2008;**112**:4808–17.

3. Kantarjian H, Giles F, Wunderle L, Bhalla K, O'Brien S, Wassmann B, et al. Nilotinib in imatinib-resistant CML and Philadelphia chromosome-positive ALL. *N Engl J Med* 2006;**354**:2542–51.
4. Talpaz M, Shah NP, Kantarjian H, Donato N, Nicoll J, Paquette R, et al. Dasatinib in imatinib-resistant Philadelphia chromosome-positive leukemias. *N Engl J Med* 2006;**354**:2531–41.
5. Saglio G, Hochhaus A, Goh YT, Masszi T, Pasquini R, Maloisel F, et al. Dasatinib in imatinib-resistant or imatinib-intolerant chronic myeloid leukemia in blast phase after 2 years of follow-up in a phase 3 study: efficacy and tolerability of 140 milligrams once daily and 70 milligrams twice daily. *Cancer* 2010;**116**:3852–61.
6. O'Brien SG, Guilhot F, Larson RA, Gathmann I, Baccarani M, Cervantes F, et al. Imatinib compared with interferon and low-dose cytarabine for newly diagnosed chronic-phase chronic myeloid leukemia. *N Engl J Med* 2003;**348**:994–1004.
7. Deininger MW, Goldman JM, Melo JV. The molecular biology of chronic myeloid leukemia. *Blood* 2000;**96**:3343–56.
8. Corbin AS, Agarwal A, Loriaux M, Cortes J, Deininger MW, Druker BJ. Human chronic myeloid leukemia stem cells are insensitive to imatinib despite inhibition of BCR-ABL activity. *J Clin Invest* 2011;**121**:396–409.
9. Quintas-Cardama A, Kantarjian H, O'Brien S, Garcia-Manero G, Rios MB, Talpaz M, et al. Granulocyte-colony-stimulating factor (filgrastim) may overcome imatinib-induced neutropenia in patients with chronic-phase chronic myelogenous leukemia. *Cancer* 2004;**100**:2592–7.
10. Clarke MF, Dick JE, Dirks PB, Eaves CJ, Jamieson CH, Jones DL, et al. Cancer stem cells—perspectives on current status and future directions: AACR Workshop on cancer stem cells. *Cancer Res* 2006;**66**:9339–44.
11. Naka K, Hoshii T, Muraguchi T, Tadokoro Y, Ooshio T, Kondo Y, et al. TGF-beta-FOXO signalling maintains leukaemia-initiating cells in chronic myeloid leukaemia. *Nature* 2010;**463**:676–80.
12. Challen GA, Little MH. A side order of stem cells: the SP phenotype. *Stem Cells* 2006;**24**:3–12.
13. Nicholson E, Holyoake T. The chronic myeloid leukemia stem cell. *Clin Lymphoma Myeloma* 2009;**9**(Suppl. 4):S376–81.
14. Hantschel O, Superti-Furga G. Regulation of the c-Abl and Bcr-Abl tyrosine kinases. *Nat Rev Mol Cell Biol* 2004;**5**:33–44.
15. Taagepera S, McDonald D, Loeb JE, Whitaker LL, McElroy AK, Wang JY, et al. Nuclear-cytoplasmic shuttling of C-ABL tyrosine kinase. *Proc Natl Acad Sci USA* 1998;**95**:7457–62.
16. Vigneri P, Wang JY. Induction of apoptosis in chronic myelogenous leukemia cells through nuclear entrapment of BCR-ABL tyrosine kinase. *Nat Med* 2001;**7**:228–34.
17. Kelliher MA, McLaughlin J, Witte ON, Rosenberg N. Induction of a chronic myelogenous leukemia-like syndrome in mice with v-abl and BCR/ABL. *Proc Natl Acad Sci USA* 1990;**87**:6649–53.
18. Tybulewicz VL, Crawford CE, Jackson PK, Bronson RT, Mulligan RC. Neonatal lethality and lymphopenia in mice with a homozygous disruption of the c-abl proto-oncogene. *Cell* 1991;**65**:1153–63.
19. Yaish P, Gazit A, Gilon C, Levitzki A. Blocking of EGF-dependent cell proliferation by EGF receptor kinase inhibitors. *Science* 1988;**242**:933–5.
20. Druker BJ, Talpaz M, Resta DJ, Peng B, Buchdunger E, Ford JM, et al. Efficacy and safety of a specific inhibitor of the BCR-ABL tyrosine kinase in chronic myeloid leukemia. *N Engl J Med* 2001;**344**:1031–7.
21. Finn AJ, Feng G, Pendergast AM. Postsynaptic requirement for Abl kinases in assembly of the neuromuscular junction. *Nat Neurosci* 2003;**6**:717–23.

22. Shah NP, Kasap C, Weier C, Balbas M, Nicoll JM, Bleickardt E, et al. Transient potent BCR-ABL inhibition is sufficient to commit chronic myeloid leukemia cells irreversibly to apoptosis. *Cancer Cell* 2008;**14**:485–93.
23. Pendergast AM, Quilliam LA, Cripe LD, Bassing CH, Dai Z, Li N, et al. BCR-ABL-induced oncogenesis is mediated by direct interaction with the SH2 domain of the GRB-2 adaptor protein. *Cell* 1993;**75**:175–85.
24. Million RP, Van Etten RA. The Grb2 binding site is required for the induction of chronic myeloid leukemia-like disease in mice by the Bcr/Abl tyrosine kinase. *Blood* 2000;**96**:664–70.
25. Kardinal C, Konkol B, Lin H, Eulitz M, Schmidt EK, Estrov Z, et al. Chronic myelogenous leukemia blast cell proliferation is inhibited by peptides that disrupt Grb2-SoS complexes. *Blood* 2001;**98**:1773–81.
26. Thomas EK, Cancelas JA, Chae HD, Cox AD, Keller PJ, Perrotti D, et al. Rac guanosine triphosphatases represent integrating molecular therapeutic targets for BCR-ABL-induced myeloproliferative disease. *Cancer Cell* 2007;**12**:467–78.
27. Sengupta A, Arnett J, Dunn S, Williams DA, Cancelas JA. Rac2 GTPase deficiency depletes BCR-ABL+ leukemic stem cells and progenitors *in vivo*. *Blood* 2010;**116**:81–4.
28. Yang MY, Liu TC, Chang JG, Lin PM, Lin SF. JunB gene expression is inactivated by methylation in chronic myeloid leukemia. *Blood* 2003;**101**:3205–11.
29. Passegue E, Jochum W, Schorpp-Kistner M, Mohle-Steinlein U, Wagner EF. Chronic myeloid leukemia with increased granulocyte progenitors in mice lacking junB expression in the myeloid lineage. *Cell* 2001;**104**:21–32.
30. Passegue E, Wagner EF, Weissman IL. JunB deficiency leads to a myeloproliferative disorder arising from hematopoietic stem cells. *Cell* 2004;**119**:431–43.
31. Chang F, Steelman LS, Lee JT, Shelton JG, Navolanic PM, Blalock WL, et al. Signal transduction mediated by the Ras/Raf/MEK/ERK pathway from cytokine receptors to transcription factors: potential targeting for therapeutic intervention. *Leukemia* 2003;**17**:1263–93.
32. Notari M, Neviani P, Santhanam R, Blaser BW, Chang JS, Galietta A, et al. A MAPK/HNRPK pathway controls BCR/ABL oncogenic potential by regulating MYC mRNA translation. *Blood* 2006;**107**:2507–16.
33. Bakiri L, Lallemand D, Bossy-Wetzel E, Yaniv M. Cell cycle-dependent variations in c-Jun and JunB phosphorylation: a role in the control of cyclin D1 expression. *EMBO J* 2000;**19**:2056–68.
34. Hermeking H, Rago C, Schuhmacher M, Li Q, Barrett JF, Obaya AJ, et al. Identification of CDK4 as a target of c-MYC. *Proc Natl Acad Sci USA* 2000;**97**:2229–34.
35. Rooney JW, Hoey T, Glimcher LH. Coordinate and cooperative roles for NF-AT and AP-1 in the regulation of the murine IL-4 gene. *Immunity* 1995;**2**:473–83.
36. Chai SK, Nichols GL, Rothman P. Constitutive activation of JAKs and STATs in BCR-Abl-expressing cell lines and peripheral blood cells derived from leukemic patients. *J Immunol* 1997;**159**:4720–8.
37. Sillaber C, Gesbert F, Frank DA, Sattler M, Griffin JD. STAT5 activation contributes to growth and viability in Bcr/Abl-transformed cells. *Blood* 2000;**95**:2118–25.
38. Hoelbl A, Kovacic B, Kerenyi MA, Simma O, Warsch W, Cui Y, et al. Clarifying the role of Stat5 in lymphoid development and Abelson-induced transformation. *Blood* 2006;**107**:4898–906.
39. Ilaria RL, Jr., Van Etten RA. P210 and P190(BCR/ABL) induce the tyrosine phosphorylation and DNA binding activity of multiple specific STAT family members. *J Biol Chem* 1996;**271**:31704–10.
40. Klejman A, Schreiner SJ, Nieborowska-Skorska M, Slupianek A, Wilson M, Smithgall TE, et al. The Src family kinase Hck couples BCR/ABL to STAT5 activation in myeloid leukemia cells. *EMBO J* 2002;**21**:5766–74.

41. Nosaka T, Kawashima T, Misawa K, Ikuta K, Mui AL, Kitamura T. STAT5 as a molecular regulator of proliferation, differentiation and apoptosis in hematopoietic cells. *EMBO J* 1999;**18**:4754–65.
42. Horita M, Andreu EJ, Benito A, Arbona C, Sanz C, Benet I, et al. Blockade of the Bcr-Abl kinase activity induces apoptosis of chronic myelogenous leukemia cells by suppressing signal transducer and activator of transcription 5-dependent expression of Bcl-xL. *J Exp Med* 2000;**191**:977–84.
43. Vivanco I, Sawyers CL. The phosphatidylinositol 3-kinase AKT pathway in human cancer. *Nat Rev Cancer* 2002;**2**:489–501.
44. Sattler M, Mohi MG, Pride YB, Quinnan LR, Malouf NA, Podar K, et al. Critical role for Gab2 in transformation by BCR/ABL. *Cancer Cell* 2002;**1**:479–92.
45. Skorski T, Bellacosa A, Nieborowska-Skorska M, Majewski M, Martinez R, Choi JK, et al. Transformation of hematopoietic cells by BCR/ABL requires activation of a PI-3k/Akt-dependent pathway. *EMBO J* 1997;**16**:6151–61.
46. Skorski T, Kanakaraj P, Nieborowska-Skorska M, Ratajczak MZ, Wen SC, Zon G, et al. Phosphatidylinositol-3 kinase activity is regulated by BCR/ABL and is required for the growth of Philadelphia chromosome-positive cells. *Blood* 1995;**86**:726–36.
47. Kharas MG, Deane JA, Wong S, O'Bosky KR, Rosenberg N, Witte ON, et al. Phosphoinositide 3-kinase signaling is essential for ABL oncogene-mediated transformation of B-lineage cells. *Blood* 2004;**103**:4268–75.
48. Ihle NT, Powis G. Take your PIK: phosphatidylinositol 3-kinase inhibitors race through the clinic and toward cancer therapy. *Mol Cancer Ther* 2009;**8**:1–9.
49. Gingras AC, Raught B, Sonenberg N. Regulation of translation initiation by FRAP/mTOR. *Genes Dev* 2001;**15**:807–26.
50. Wullschleger S, Loewith R, Hall MN. TOR signaling in growth and metabolism. *Cell* 2006;**124**:471–84.
51. Haghighat A, Mader S, Pause A, Sonenberg N. Repression of cap-dependent translation by 4E-binding protein 1: competition with p220 for binding to eukaryotic initiation factor-4E. *EMBO J* 1995;**14**:5701–9.
52. Hay N, Sonenberg N. Upstream and downstream of mTOR. *Genes Dev* 2004;**18**:1926–45.
53. Thomas G. An encore for ribosome biogenesis in the control of cell proliferation. *Nat Cell Biol* 2000;**2**:E71–2.
54. Raught B, Peiretti F, Gingras AC, Livingstone M, Shahbazian D, Mayeur GL, et al. Phosphorylation of eucaryotic translation initiation factor 4B Ser422 is modulated by S6 kinases. *EMBO J* 2004;**23**:1761–9.
55. Sarbassov DD, Ali SM, Kim DH, Guertin DA, Latek RR, Erdjument-Bromage H, et al. Rictor, a novel binding partner of mTOR, defines a rapamycin-insensitive and raptor-independent pathway that regulates the cytoskeleton. *Curr Biol* 2004;**14**:1296–302.
56. Loewith R, Jacinto E, Wullschleger S, Lorberg A, Crespo JL, Bonenfant D, et al. Two TOR complexes, only one of which is rapamycin sensitive, have distinct roles in cell growth control. *Mol Cell* 2002;**10**:457–68.
57. Facchinetti V, Ouyang W, Wei H, Soto N, Lazorchak A, Gould C, et al. The mammalian target of rapamycin complex 2 controls folding and stability of Akt and protein kinase C. *EMBO J* 2008;**27**:1932–43.
58. Mohi MG, Boulton C, Gu TL, Sternberg DW, Neuberg D, Griffin JD, et al. Combination of rapamycin and protein tyrosine kinase (PTK) inhibitors for the treatment of leukemias caused by oncogenic PTKs. *Proc Natl Acad Sci USA* 2004;**101**:3130–5.
59. Carayol N, Vakana E, Sassano A, Kaur S, Goussetis DJ, Glaser H, et al. Critical roles for mTORC2- and rapamycin-insensitive mTORC1-complexes in growth and survival of BCR-ABL-expressing leukemic cells. *Proc Natl Acad Sci USA* 2010;**107**:12469–74.

60. del Peso L, Gonzalez-Garcia M, Page C, Herrera R, Nunez G. Interleukin-3-induced phosphorylation of BAD through the protein kinase Akt. *Science* 1997;**278**:687–9.
61. Essafi A, Fernandez de Mattos S, Hassen YA, Soeiro I, Mufti GJ, Thomas NS, et al. Direct transcriptional regulation of Bim by FoxO3a mediates STI571-induced apoptosis in Bcr-Abl-expressing cells. *Oncogene* 2005;**24**:2317–29.
62. Hochhaus A, O'Brien SG, Guilhot F, Druker BJ, Branford S, Foroni L, et al. Six-year follow-up of patients receiving imatinib for the first-line treatment of chronic myeloid leukemia. *Leukemia* 2009;**23**:1054–61.
63. Kantarjian HM, Cortes JE, O'Brien S, Luthra R, Giles F, Verstovsek S, et al. Long-term survival benefit and improved complete cytogenetic and molecular response rates with imatinib mesylate in Philadelphia chromosome-positive chronic-phase chronic myeloid leukemia after failure of interferon-alpha. *Blood* 2004;**104**:1979–88.
64. Mahon FX, Rea D, Guilhot J, Guilhot F, Huguet F, Nicolini F, et al. Discontinuation of imatinib in patients with chronic myeloid leukaemia who have maintained complete molecular remission for at least 2 years: the prospective, multicentre Stop Imatinib (STIM) trial. *Lancet Oncol* 2010;**11**:1029–35.
65. Bhatia R, Holtz M, Niu N, Gray R, Snyder DS, Sawyers CL, et al. Persistence of malignant hematopoietic progenitors in chronic myelogenous leukemia patients in complete cytogenetic remission following imatinib mesylate treatment. *Blood* 2003;**101**:4701–7.
66. Helgason GV, Young GA, Holyoake TL. Targeting chronic myeloid leukemia stem cells. *Curr Hematol Malig Rep* 2010;**5**:81–7.
67. Deininger MW, Holyoake TL. Can we afford to let sleeping dogs lie? *Blood* 2005;**105**:1840–1.
68. Deininger M, Buchdunger E, Druker BJ. The development of imatinib as a therapeutic agent for chronic myeloid leukemia. *Blood* 2005;**105**:2640–53.
69. Konig H, Holyoake TL, Bhatia R. Effective and selective inhibition of chronic myeloid leukemia primitive hematopoietic progenitors by the dual Src/Abl kinase inhibitor SKI-606. *Blood* 2008;**111**:2329–38.
70. Jorgensen HG, Allan EK, Jordanides NE, Mountford JC, Holyoake TL. Nilotinib exerts equipotent antiproliferative effects to imatinib and does not induce apoptosis in CD34+ CML cells. *Blood* 2007;**109**:4016–9.
71. Copland M, Hamilton A, Elrick LJ, Baird JW, Allan EK, Jordanides N, et al. Dasatinib (BMS-354825) targets an earlier progenitor population than imatinib in primary CML but does not eliminate the quiescent fraction. *Blood* 2006;**107**:4532–9.
72. Graham SM, Jorgensen HG, Allan E, Pearson C, Alcorn MJ, Richmond L, et al. Primitive, quiescent, Philadelphia-positive stem cells from patients with chronic myeloid leukemia are insensitive to STI571 *in vitro*. *Blood* 2002;**99**:319–25.
73. Holyoake TL, Jiang X, Jorgensen HG, Graham S, Alcorn MJ, Laird C, et al. Primitive quiescent leukemic cells from patients with chronic myeloid leukemia spontaneously initiate factor-independent growth *in vitro* in association with up-regulation of expression of interleukin-3. *Blood* 2001;**97**:720–8.
74. Holyoake T, Jiang X, Eaves C, Eaves A. Isolation of a highly quiescent subpopulation of primitive leukemic cells in chronic myeloid leukemia. *Blood* 1999;**94**:2056–64.
75. Lane SW, Scadden DT, Gilliland DG. The leukemic stem cell niche: current concepts and therapeutic opportunities. *Blood* 2009;**114**:1150–7.
76. Konopleva M, Tabe Y, Zeng Z, Andreeff M. Therapeutic targeting of microenvironmental interactions in leukemia: mechanisms and approaches. *Drug Resist Updat* 2009;**12**:103–13.
77. Yilmaz OH, Valdez R, Theisen BK, Guo W, Ferguson DO, Wu H, et al. Pten dependence distinguishes haematopoietic stem cells from leukaemia-initiating cells. *Nature* 2006;**441**:475–82.
78. Lee JY, Nakada D, Yilmaz OH, Tothova Z, Joseph NM, Lim MS, et al. mTOR activation induces tumor suppressors that inhibit leukemogenesis and deplete hematopoietic stem cells after Pten deletion. *Cell Stem Cell* 2010;**7**:593–605.

79. Baccarani M, Cortes J, Pane F, Niederwieser D, Saglio G, Apperley J, et al. Chronic myeloid leukemia: an update of concepts and management recommendations of European LeukemiaNet. *J Clin Oncol* 2009;**27**:6041–51.
80. Hughes TP, Hochhaus A, Branford S, Muller MC, Kaeda JS, Foroni L, et al. Long-term prognostic significance of early molecular response to imatinib in newly diagnosed chronic myeloid leukemia: an analysis from the International Randomized Study of Interferon and STI571 (IRIS). *Blood* 2010;**116**:3758–65.
81. Crossman LC, Druker BJ, Deininger MW, Pirmohamed M, Wang L, Clark RE. hOCT 1 and resistance to imatinib. *Blood* 2005;**106**:1133–4. (author reply 1134).
82. Jabbour E, Kantarjian H, Jones D, Talpaz M, Bekele N, O'Brien S, et al. Frequency and clinical significance of BCR-ABL mutations in patients with chronic myeloid leukemia treated with imatinib mesylate. *Leukemia* 2006;**20**:1767–73.
83. Soverini S, Colarossi S, Gnani A, Rosti G, Castagnetti F, Poerio A, et al. Contribution of ABL kinase domain mutations to imatinib resistance in different subsets of Philadelphia-positive patients: by the GIMEMA Working Party on Chronic Myeloid Leukemia. *Clin Cancer Res* 2006;**12**:7374–9.
84. Khorashad JS, de Lavallade H, Apperley JF, Milojkovic D, Reid AG, Bua M, et al. Finding of kinase domain mutations in patients with chronic phase chronic myeloid leukemia responding to imatinib may identify those at high risk of disease progression. *J Clin Oncol* 2008;**26**:4806–13.
85. Quintas-Cardama A, Kantarjian HM, Cortes JE. Mechanisms of primary and secondary resistance to imatinib in chronic myeloid leukemia. *Cancer Control* 2009;**16**:122–31.
86. Luzzatto L, Melo JV. Acquired resistance to imatinib mesylate: selection for pre-existing mutant cells. *Blood* 2002;**100**:1105.
87. O'Hare T, Walters DK, Stoffregen EP, Jia T, Manley PW, Mestan J, et al. *In vitro* activity of Bcr-Abl inhibitors AMN107 and BMS-354825 against clinically relevant imatinib-resistant Abl kinase domain mutants. *Cancer Res* 2005;**65**:4500–5.
88. Branford S, Melo JV, Hughes TP. Selecting optimal second-line tyrosine kinase inhibitor therapy for chronic myeloid leukemia patients after imatinib failure: does the BCR-ABL mutation status really matter? *Blood* 2009;**114**:5426–35.
89. Griswold IJ, MacPartlin M, Bumm T, Goss VL, O'Hare T, Lee KA, et al. Kinase domain mutants of Bcr-Abl exhibit altered transformation potency, kinase activity, and substrate utilization, irrespective of sensitivity to imatinib. *Mol Cell Biol* 2006;**26**:6082–93.
90. Jiang X, Zhao Y, Smith C, Gasparetto M, Turhan A, Eaves A, et al. Chronic myeloid leukemia stem cells possess multiple unique features of resistance to BCR-ABL targeted therapies. *Leukemia* 2007;**21**:926–35.
91. White DL, Saunders VA, Dang P, Engler J, Zannettino AC, Cambareri AC, et al. OCT-1-mediated influx is a key determinant of the intracellular uptake of imatinib but not nilotinib (AMN107): reduced OCT-1 activity is the cause of low *in vitro* sensitivity to imatinib. *Blood* 2006;**108**:697–704.
92. Hiwase DK, Saunders V, Hewett D, Frede A, Zrim S, Dang P, et al. Dasatinib cellular uptake and efflux in chronic myeloid leukemia cells: therapeutic implications. *Clin Cancer Res* 2008;**14**:3881–8.
93. Eechoute K, Sparreboom A, Burger H, Franke RM, Schiavon G, Verweij J, et al. Drug transporters and imatinib treatment: implications for clinical practice. *Clin Cancer Res* 2011;**17**:406–15.
94. Thomas J, Wang L, Clark RE, Pirmohamed M. Active transport of imatinib into and out of cells: implications for drug resistance. *Blood* 2004;**104**:3739–45.
95. Mahon FX, Deininger MW, Schultheis B, Chabrol J, Reiffers J, Goldman JM, et al. Selection and characterization of BCR-ABL positive cell lines with differential sensitivity to the tyrosine kinase inhibitor STI571: diverse mechanisms of resistance. *Blood* 2000;**96**:1070–9.

96. Abram CL, Courtneidge SA. Src family tyrosine kinases and growth factor signaling. *Exp Cell Res* 2000;**254**:1–13.
97. Stanglmaier M, Warmuth M, Kleinlein I, Reis S, Hallek M. The interaction of the Bcr-Abl tyrosine kinase with the Src kinase Hck is mediated by multiple binding domains. *Leukemia* 2003;**17**:283–9.
98. Hu Y, Liu Y, Pelletier S, Buchdunger E, Warmuth M, Fabbro D, et al. Requirement of Src kinases Lyn, Hck and Fgr for BCR-ABL1-induced B-lymphoblastic leukemia but not chronic myeloid leukemia. *Nat Genet* 2004;**36**:453–61.
99. Donato NJ, Wu JY, Stapley J, Gallick G, Lin H, Arlinghaus R, et al. BCR-ABL independence and LYN kinase overexpression in chronic myelogenous leukemia cells selected for resistance to STI571. *Blood* 2003;**101**:690–8.
100. Wu J, Meng F, Lu H, Kong L, Bornmann W, Peng Z, et al. Lyn regulates BCR-ABL and Gab2 tyrosine phosphorylation and c-Cbl protein stability in imatinib-resistant chronic myelogenous leukemia cells. *Blood* 2008;**111**:3821–9.
101. Kuerbitz SJ, Plunkett BS, Walsh WV, Kastan MB. Wild-type p53 is a cell cycle checkpoint determinant following irradiation. *Proc Natl Acad Sci USA* 1992;**89**:7491–5.
102. Villunger A, Michalak EM, Coultas L, Mullauer F, Bock G, Ausserlechner MJ, et al. p53- and drug-induced apoptotic responses mediated by BH3-only proteins puma and noxa. *Science* 2003;**302**:1036–8.
103. Calabretta B, Perrotti D. The biology of CML blast crisis. *Blood* 2004;**103**:4010–22.
104. Wendel HG, de Stanchina E, Cepero E, Ray S, Emig M, Fridman JS, et al. Loss of p53 impedes the antileukemic response to BCR-ABL inhibition. *Proc Natl Acad Sci USA* 2006;**103**:7444–9.
105. O'Connor L, Strasser A, O'Reilly LA, Hausmann G, Adams JM, Cory S, et al. Bim: a novel member of the Bcl-2 family that promotes apoptosis. *EMBO J* 1998;**17**:384–95.
106. Luciano F, Jacquel A, Colosetti P, Herrant M, Cagnol S, Pages G, et al. Phosphorylation of Bim-EL by Erk1/2 on serine 69 promotes its degradation via the proteasome pathway and regulates its proapoptotic function. *Oncogene* 2003;**22**:6785–93.
107. Aichberger KJ, Mayerhofer M, Krauth MT, Vales A, Kondo R, Derdak S, et al. Low-level expression of proapoptotic Bcl-2-interacting mediator in leukemic cells in patients with chronic myeloid leukemia: role of BCR/ABL, characterization of underlying signaling pathways, and reexpression by novel pharmacologic compounds. *Cancer Res* 2005;**65**:9436–44.
108. Kuroda J, Puthalakath H, Cragg MS, Kelly PN, Bouillet P, Huang DC, et al. Bim and Bad mediate imatinib-induced killing of Bcr/Abl+ leukemic cells, and resistance due to their loss is overcome by a BH3 mimetic. *Proc Natl Acad Sci USA* 2006;**103**:14907–12.
109. Kuribara R, Honda H, Matsui H, Shinjyo T, Inukai T, Sugita K, et al. Roles of Bim in apoptosis of normal and Bcr-Abl-expressing hematopoietic progenitors. *Mol Cell Biol* 2004;**24**:6172–83.
110. San Jose-Eneriz E, Agirre X, Jimenez-Velasco A, Cordeu L, Martin V, Arqueros V, et al. Epigenetic down-regulation of BIM expression is associated with reduced optimal responses to imatinib treatment in chronic myeloid leukaemia. *Eur J Cancer* 2009;**45**:1877–89.
111. Mizushima N. Autophagy: process and function. *Genes Dev* 2007;**21**:2861–73.
112. Liang XH, Jackson S, Seaman M, Brown K, Kempkes B, Hibshoosh H, et al. Induction of autophagy and inhibition of tumorigenesis by beclin 1. *Nature* 1999;**402**:672–6.
113. Rubinsztein DC, Gestwicki JE, Murphy LO, Klionsky DJ. Potential therapeutic applications of autophagy. *Nat Rev Drug Discov* 2007;**6**:304–12.
114. Mishima Y, Terui Y, Taniyama A, Kuniyoshi R, Takizawa T, Kimura S, et al. Autophagy and autophagic cell death are next targets for elimination of the resistance to tyrosine kinase inhibitors. *Cancer Sci* 2008;**99**:2200–8.
115. Bellodi C, Lidonnici MR, Hamilton A, Helgason GV, Soliera AR, Ronchetti M, et al. Targeting autophagy potentiates tyrosine kinase inhibitor-induced cell death in Philadelphia chromosome-positive cells, including primary CML stem cells. *J Clin Invest* 2009;**119**:1109–23.

116. Virshup DM, Shenolikar S. From promiscuity to precision: protein phosphatases get a makeover. *Mol Cell* 2009;**33**:537–45.
117. Perrotti D, Neviani P. Protein phosphatase 2A (PP2A), a drugable tumor suppressor in Ph1 (+) leukemias. *Cancer Metastasis Rev* 2008;**27**:159–68.
118. Tian Q, Wang J. Role of serine/threonine protein phosphatase in Alzheimer's disease. *Neurosignals* 2002;**11**:262–9.
119. Neel BG, Gu H, Pao L. The 'Shp'ing news: SH2 domain-containing tyrosine phosphatases in cell signaling. *Trends Biochem Sci* 2003;**28**:284–93.
120. Shi ZQ, Yu DH, Park M, Marshall M, Feng GS. Molecular mechanism for the Shp-2 tyrosine phosphatase function in promoting growth factor stimulation of Erk activity. *Mol Cell Biol* 2000;**20**:1526–36.
121. Cleghon V, Feldmann P, Ghiglione C, Copeland TD, Perrimon N, Hughes DA, et al. Opposing actions of CSW and RasGAP modulate the strength of Torso RTK signaling in the Drosophila terminal pathway. *Mol Cell* 1998;**2**:719–27.
122. Cunnick JM, Meng S, Ren Y, Desponts C, Wang HG, Djeu JY, et al. Regulation of the mitogen-activated protein kinase signaling pathway by SHP2. *J Biol Chem* 2002;**277**:9498–504.
123. Quintanar-Audelo M, Yusoff P, Sinniah S, Chandramouli S, Guy GR. Sprouty-related Ena/vasodilator-stimulated phosphoprotein homology 1-domain-containing protein (SPRED1), a tyrosine-protein phosphatase non-receptor type 11 (SHP2) substrate in the Ras/extracellular signal-regulated kinase (ERK) pathway. *J Biol Chem* 2011;**286**:23102–12.
124. Tartaglia M, Mehler EL, Goldberg R, Zampino G, Brunner HG, Kremer H, et al. Mutations in PTPN11, encoding the protein tyrosine phosphatase SHP-2, cause Noonan syndrome. *Nat Genet* 2001;**29**:465–8.
125. Choong K, Freedman MH, Chitayat D, Kelly EN, Taylor G, Zipursky A. Juvenile myelomonocytic leukemia and Noonan syndrome. *J Pediatr Hematol Oncol* 1999;**21**:523–7.
126. Xu R, Yu Y, Zheng S, Zhao X, Dong Q, He Z, et al. Overexpression of Shp2 tyrosine phosphatase is implicated in leukemogenesis in adult human leukemia. *Blood* 2005;**106**:3142–9.
127. Li J, Yen C, Liaw D, Podsypanina K, Bose S, Wang SI, et al. PTEN, a putative protein tyrosine phosphatase gene mutated in human brain, breast, and prostate cancer. *Science* 1997;**275**:1943–7.
128. Peng C, Chen Y, Yang Z, Zhang H, Osterby L, Rosmarin AG, et al. PTEN is a tumor suppressor in CML stem cells and BCR-ABL-induced leukemias in mice. *Blood* 2010;**115**:626–35.
129. Aggerholm A, Gronbaek K, Guldberg P, Hokland P. Mutational analysis of the tumour suppressor gene MMAC1/PTEN in malignant myeloid disorders. *Eur J Haematol* 2000;**65**:109–13.
130. Sontag E. Protein phosphatase 2A: the Trojan Horse of cellular signaling. *Cell Signal* 2001;**13**:7–16.
131. Ivaska J, Nissinen L, Immonen N, Eriksson JE, Kahari VM, Heino J. Integrin alpha 2 beta 1 promotes activation of protein phosphatase 2A and dephosphorylation of Akt and glycogen synthase kinase 3 beta. *Mol Cell Biol* 2002;**22**:1352–9.
132. Li X, Yost HJ, Virshup DM, Seeling JM. Protein phosphatase 2A and its B56 regulatory subunit inhibit Wnt signaling in Xenopus. *EMBO J* 2001;**20**:4122–31.
133. Neviani P, Santhanam R, Trotta R, Notari M, Blaser BW, Liu S, et al. The tumor suppressor PP2A is functionally inactivated in blast crisis CML through the inhibitory activity of the BCR/ABL-regulated SET protein. *Cancer Cell* 2005;**8**:355–68.
134. Junttila MR, Puustinen P, Niemela M, Ahola R, Arnold H, Bottzauw T, et al. CIP2A inhibits PP2A in human malignancies. *Cell* 2007;**130**:51–62.
135. Arroyo JD, Hahn WC. Involvement of PP2A in viral and cellular transformation. *Oncogene* 2005;**24**:7746–55.

136. Ruediger R, Pham HT, Walter G. Alterations in protein phosphatase 2A subunit interaction in human carcinomas of the lung and colon with mutations in the A beta subunit gene. *Oncogene* 2001;**20**:1892–9.
137. Lucas CM, Harris RJ, Giannoudis A, Copland M, Slupsky JR, Clark RE. Cancerous inhibitor of PP2A (CIP2A) at diagnosis of chronic myeloid leukemia is a critical determinant of disease progression. *Blood* 2011;**117**:6660–8.
138. Salas A, Ponnusamy S, Senkal CE, Meyers-Needham M, Selvam SP, Saddoughi SA, et al. Sphingosine kinase-1 and sphingosine 1-phosphate receptor 2 mediate Bcr-Abl1 stability and drug resistance by modulation of protein phosphatase 2A. *Blood* 2011;**117**:5941–52.
139. Weisberg E, Manley PW, Breitenstein W, Bruggen J, Cowan-Jacob SW, Ray A, et al. Characterization of AMN107, a selective inhibitor of native and mutant Bcr-Abl. *Cancer Cell* 2005;**7**:129–41.
140. Quintas-Cardama A, Cortes JE. Chronic myeloid leukemia: diagnosis and treatment. *Mayo Clin Proc* 2006;**81**:973–88.
141. Shah NP, Tran C, Lee FY, Chen P, Norris D, Sawyers CL. Overriding imatinib resistance with a novel ABL kinase inhibitor. *Science* 2004;**305**:399–401.
142. Saglio G, Kim DW, Issaragrisil S, le Coutre P, Etienne G, Lobo C, et al. Nilotinib versus imatinib for newly diagnosed chronic myeloid leukemia. *N Engl J Med* 2010;**362**:2251–9.
143. Kantarjian H, Shah NP, Hochhaus A, Cortes J, Shah S, Ayala M, et al. Dasatinib versus imatinib in newly diagnosed chronic-phase chronic myeloid leukemia. *N Engl J Med* 2010;**362**:2260–70.
144. Chu S, Holtz M, Gupta M, Bhatia R. BCR/ABL kinase inhibition by imatinib mesylate enhances MAP kinase activity in chronic myelogenous leukemia CD34+ cells. *Blood* 2004;**103**:3167–74.
145. Melo JV, Chuah C. Novel agents in CML therapy: tyrosine kinase inhibitors and beyond. *Hematology Am Soc Hematol Educ Program* 2008;**2008**:427–35.
146. Chuah C, Barnes DJ, Kwok M, Corbin A, Deininger MW, Druker BJ, et al. Zoledronate inhibits proliferation and induces apoptosis of imatinib-resistant chronic myeloid leukaemia cells. *Leukemia* 2005;**19**:1896–904.
147. Peters DG, Hoover RR, Gerlach MJ, Koh EY, Zhang H, Choe K, et al. Activity of the farnesyl protein transferase inhibitor SCH66336 against BCR/ABL-induced murine leukemia and primary cells from patients with chronic myeloid leukemia. *Blood* 2001;**97**:1404–12.
148. End DW, Smets G, Todd AV, Applegate TL, Fuery CJ, Angibaud P, et al. Characterization of the antitumor effects of the selective farnesyl protein transferase inhibitor R115777 *in vivo* and *in vitro*. *Cancer Res* 2001;**61**:131–7.
149. Cortes J, Albitar M, Thomas D, Giles F, Kurzrock R, Thibault A, et al. Efficacy of the farnesyl transferase inhibitor R115777 in chronic myeloid leukemia and other hematologic malignancies. *Blood* 2003;**101**:1692–7.
150. Borthakur G, Kantarjian H, Daley G, Talpaz M, O'Brien S, Garcia-Manero G, et al. Pilot study of lonafarnib, a farnesyl transferase inhibitor, in patients with chronic myeloid leukemia in the chronic or accelerated phase that is resistant or refractory to imatinib therapy. *Cancer* 2006;**106**:346–52.
151. Copland M, Pellicano F, Richmond L, Allan EK, Hamilton A, Lee FY, et al. BMS-214662 potently induces apoptosis of chronic myeloid leukemia stem and progenitor cells and synergizes with tyrosine kinase inhibitors. *Blood* 2008;**111**:2843–53.
152. Ly C, Arechiga AF, Melo JV, Walsh CM, Ong ST. Bcr-Abl kinase modulates the translation regulators ribosomal protein S6 and 4E-BP1 in chronic myelogenous leukemia cells via the mammalian target of rapamycin. *Cancer Res* 2003;**63**:5716–22.
153. Prabhu S, Saadat D, Zhang M, Halbur L, Fruehauf JP, Ong ST. A novel mechanism for Bcr-Abl action: Bcr-Abl-mediated induction of the eIF4F translation initiation complex and mRNA translation. *Oncogene* 2007;**26**:1188–200.

154. Kaelin Jr. WG. The concept of synthetic lethality in the context of anticancer therapy. *Nat Rev Cancer* 2005;**5**:689–98.
155. Chan DA, Giaccia AJ. Harnessing synthetic lethal interactions in anticancer drug discovery. *Nat Rev Drug Discov* 2011;**10**:351–64.
156. Gregory MA, Phang TL, Neviani P, Alvarez-Calderon F, Eide CA, O'Hare T, et al. Wnt/Ca2+/NFAT signaling maintains survival of Ph+ leukemia cells upon inhibition of Bcr-Abl. *Cancer Cell* 2010;**18**:74–87.
157. Silver RT. The blast phase of chronic myeloid leukaemia. *Best Pract Res Clin Haematol* 2009;**22**:387–94.
158. Perrotti D, Jamieson C, Goldman J, Skorski T. Chronic myeloid leukemia: mechanisms of blastic transformation. *J Clin Invest* 2010;**120**:2254–64.
159. Mullighan CG, Miller CB, Radtke I, Phillips LA, Dalton J, Ma J, et al. BCR-ABL1 lymphoblastic leukaemia is characterized by the deletion of Ikaros. *Nature* 2008;**453**:110–4.
160. Radich JP, Dai H, Mao M, Oehler V, Schelter J, Druker B, et al. Gene expression changes associated with progression and response in chronic myeloid leukemia. *Proc Natl Acad Sci USA* 2006;**103**:2794–9.
161. Jamieson CH, Ailles LE, Dylla SJ, Muijtjens M, Jones C, Zehnder JL, et al. Granulocyte-macrophage progenitors as candidate leukemic stem cells in blast-crisis CML. *N Engl J Med* 2004;**351**:657–67.
162. Abrahamsson AE, Geron I, Gotlib J, Dao KH, Barroga CF, Newton IG, et al. Glycogen synthase kinase 3beta missplicing contributes to leukemia stem cell generation. *Proc Natl Acad Sci USA* 2009;**106**:3925–9.
163. Cramer K, Nieborowska-Skorska M, Koptyra M, Slupianek A, Penserga ET, Eaves CJ, et al. BCR/ABL and other kinases from chronic myeloproliferative disorders stimulate single-strand annealing, an unfaithful DNA double-strand break repair. *Cancer Res* 2008;**68**:6884–8.
164. Skorski T. BCR/ABL, DNA damage and DNA repair: implications for new treatment concepts. *Leuk Lymphoma* 2008;**49**:610–4.
165. Skorski T. Chronic myeloid leukemia cells refractory/resistant to tyrosine kinase inhibitors are genetically unstable and may cause relapse and malignant progression to the terminal disease state. *Leuk Lymphoma* 2011;**52**(Suppl. 1):23–9.
166. Barnes DJ, Palaiologou D, Panousopoulou E, Schultheis B, Yong AS, Wong A, et al. Bcr-Abl expression levels determine the rate of development of resistance to imatinib mesylate in chronic myeloid leukemia. *Cancer Res* 2005;**65**:8912–9.
167. Marega M, Piazza RG, Pirola A, Redaelli S, Mogavero A, Iacobucci I, et al. BCR and BCR-ABL regulation during myeloid differentiation in healthy donors and in chronic phase/blast crisis CML patients. *Leukemia* 2010;**24**:1445–9.
168. Huntly BJ, Shigematsu H, Deguchi K, Lee BH, Mizuno S, Duclos N, et al. MOZ-TIF2, but not BCR-ABL, confers properties of leukemic stem cells to committed murine hematopoietic progenitors. *Cancer Cell* 2004;**6**:587–96.
169. Coluccia AM, Vacca A, Dunach M, Mologni L, Redaelli S, Bustos VH, et al. Bcr-Abl stabilizes beta-catenin in chronic myeloid leukemia through its tyrosine phosphorylation. *EMBO J* 2007;**26**:1456–66.
170. Neviani P, Santhanam R, Ma Y, Marcucci G, Byrd JC, Chen CS, et al. Activation of PP2A by FTY720 inhibits survival and self-renewal of the Ph(+) chronic myelogenous leukemia (CML) CD34+/CD38− stem cell through the simultaneous suppression of BCR/ABL and BCR/ABL-independent signals. *Blood* 2008;**112**:77 (Abstract 189).
171. Ueda T, Watanabe-Fukunaga R, Fukuyama H, Nagata S, Fukunaga R. Mnk2 and Mnk1 are essential for constitutive and inducible phosphorylation of eukaryotic initiation factor 4E but not for cell growth or development. *Mol Cell Biol* 2004;**24**:6539–49.

172. Lim S, Saw TY, Chang S, Zhang M, Janes M, Fruman D, Rizzieri D, Tan SY, Chuah C, Ong ST. Targeting of a novel MNK-eIF4E-β-catenin axis in blast crisis chronic myelogenous leukemia inhibits leukemia stem cell function. *Blood* 2011;**118**:21 (Abstract 963).
173. D'Angelo RC, Wicha MS. Stem cells in normal development and cancer. *Prog Mol Biol Transl Sci* 2010;**95**:113–58.
174. Merchant AA, Matsui W. Targeting Hedgehog—a cancer stem cell pathway. *Clin Cancer Res* 2010;**16**:3130–40.
175. Zhao C, Chen A, Jamieson CH, Fereshteh M, Abrahamsson A, Blum J, et al. Hedgehog signalling is essential for maintenance of cancer stem cells in myeloid leukaemia. *Nature* 2009;**458**:776–9.
176. Dierks C, Beigi R, Guo GR, Zirlik K, Stegert MR, Manley P, et al. Expansion of Bcr-Abl-positive leukemic stem cells is dependent on Hedgehog pathway activation. *Cancer Cell* 2008;**14**:238–49.
177. Riobo NA, Lu K, Ai X, Haines GM, Emerson Jr. CP. Phosphoinositide 3-kinase and Akt are essential for Sonic Hedgehog signaling. *Proc Natl Acad Sci USA* 2006;**103**:4505–10.
177a. Eiring AM, Harb JG, Neviani P, Garton C, Oaks JJ, Spizzo R, et al. miR-328 functions as an RNA decoy to modulate hnRNP E2 regulation of mRNA translation in leukemic blasts. *Cell* 2010;**140**:652–65.
178. Pao W, Miller VA. Epidermal growth factor receptor mutations, small-molecule kinase inhibitors, and non-small-cell lung cancer: current knowledge and future directions. *J Clin Oncol* 2005;**23**:2556–68.
179. Essers MA, Offner S, Blanco-Bose WE, Waibler Z, Kalinke U, Duchosal MA, et al. IFNalpha activates dormant haematopoietic stem cells *in vivo*. *Nature* 2009;**458**:904–8.
180. Sato T, Onai N, Yoshihara H, Arai F, Suda T, Ohteki T. Interferon regulatory factor-2 protects quiescent hematopoietic stem cells from type I interferon-dependent exhaustion. *Nat Med* 2009;**15**:696–700.
181. Preudhomme C, Guilhot J, Nicolini FE, Guerci-Bresler A, Rigal-Huguet F, Maloisel F, et al. Imatinib plus peginterferon alfa-2a in chronic myeloid leukemia. *N Engl J Med* 2010;**363**:2511–21.
182. Eliasson P, Jonsson JI. The hematopoietic stem cell niche: low in oxygen but a nice place to be. *J Cell Physiol* 2010;**222**:17–22.
183. Komarova NL, Wodarz D. Combination therapies against chronic myeloid leukemia: short-term versus long-term strategies. *Cancer Res* 2009;**69**:4904–10.
184. Wodarz D, Komarova NL. Emergence and prevention of resistance against small molecule inhibitors. *Semin Cancer Biol* 2005;**15**:506–14.
185. Shah NP, Kantarjian HM, Kim DW, Rea D, Dorlhiac-Llacer PE, Milone JH, et al. Intermittent target inhibition with dasatinib 100 mg once daily preserves efficacy and improves tolerability in imatinib-resistant and -intolerant chronic-phase chronic myeloid leukemia. *J Clin Oncol* 2008;**26**:3204–12.
186. Wang L, McLeod HL, Weinshilboum RM. Genomics and drug response. *N Engl J Med* 2011;**364**:1144–53.
187. Housman D. Human DNA polymorphism. *N Engl J Med* 1995;**332**:318–20.

# Section II

## Protein Kinases

# Role of the JNK Pathway in Human Diseases

Kanaga Sabapathy[*,†,‡]

[*]*Division of Cellular & Molecular Research, Humphrey Oei Institute of Cancer Research, National Cancer Centre, Singapore*

[†]*Cancer and Stem Cell Biology Program, Duke-NUS Graduate Medical School, Singapore*

[‡]*Department of Biochemistry, National University of Singapore, Singapore*



The c-Jun-$NH_2$-terminal kinase (JNK) signaling pathway plays a critical role in regulating cell fate, being implicated in a multitude of diseases ranging from cancer to neurological and immunological/inflammatory conditions. Not surprisingly, therefore, it has been sought after for therapeutic intervention, and its inhibition has been shown to ameliorate many pathological conditions in experimental systems, paving the way for initial clinical trials. However, the fundamental problem in fully harnessing the potential provided by the JNK pathway has been the lack of specificity, due to the multiple JNK forms that are involved in multiple cellular processes in various cell types. Moreover, lack of sufficient knowledge of all JNK-interacting proteins and substrates has also hindered progress. This review will therefore focus on the role of the JNKs in human diseases and appraise the efforts to inhibit JNK signaling to ameliorate disease conditions, assessing potential challenges and providing insights into possible future directions to efficiently target this pathway for therapeutic use.

Progress in Molecular Biology
and Translational Science, Vol. 106
DOI: 10.1016/B978-0-12-396456-4.00013-4

# I. The c-Jun Amino Terminal Kinases—Background

c-Jun-$NH_2$ terminal kinases (JNKs) belong to the mitogen-activated protein kinase (MAPK) family of proteins.[1–3] The JNK MAPK pathway is predominantly activated by stress stimuli and plays important roles in development, apoptosis, cell growth, inflammatory, and immune responses.[4–7] Extracellular signals received by JNK via its upstream kinases through the appropriate MAPK modules are generally transmitted to the nucleus by phosphorylation of its nuclear substrates.[1] Notwithstanding the transmission of signals to the nucleus, JNKs have also been implicated in associating with proteins in the plasma membrane, thereby regulating cellular movement and other hitherto unidentified functions independent of regulation of gene expression.[8] Nonetheless, the central theme that has emerged in the regulation of JNK activity and its functionality is the critical importance of the regulated interplay of several other proteins that are part of the assembled signaling modules that interact with JNK. These include the upstream JNK2Ks and JNK3Ks, as well as the scaffold proteins that serve as platforms for the assembly of the modules.[4,5] In addition, the presence of multiple forms of JNKs adds to the complexity of JNK signaling.[6,9] The molecular determinants underlying the assembly of JNK signaling molecules and their interactions with other proteins in a cell type and signal-specific manner, and the specific roles of each JNK isozyme in the cell, are at present not fully understood. Despite the complexity, much information has been gathered over the past two decades on the transduction of signals through appropriate substrates to regulate cellular functions.

At the cellular level, JNK activation broadly results in apoptosis induction or promotion of cellular survival, and the choice of cell fate decision appears to depend on the cell type and on the stimulating signal. There are three highly homologous *JNK* genes (*JNK1*, *JNK2*, and *JNK3*), with a total of 10 splice variants, and several substrates of the JNKs have been identified.[8,9] While JNK1 and JNK2 are expressed ubiquitously, JNK3 is expressed mainly in brain and heart.[6,10] JNKs were first identified as enzymes that phosphorylate c-Jun, and hence, much of the earlier work was focused on understanding the kinase–substrate interaction using c-Jun as the prototype substrate.[11,12] All three JNKs were found to phosphorylate substrates such as c-Jun and ATF-2 to varying extents in *in vitro* kinase assays.[9] Upon signal stimulation, the JNKs have been shown to bind to the docking "delta" domain of c-Jun, which is essential for subsequent phosphorylation on the serines 63 and 73, leading to its activation.[1,5,9,13,14] Genetic analyses using individual and compound JNK knockout mice that we and others have generated have revealed specific and common roles for JNKs in many developmental processes and disease states (reviewed in Refs. 6,15). Combinatorial use of the various JNKs and their upstream kinases is thought to lead to differential regulation of substrate proteins in response to multiple stimuli, thereby establishing signal specificity.[2,5] Nevertheless, the specific roles of each of the JNK isozymes and their respective isoforms are not clearly understood.

## II. JNKs—How They Function

JNK's ability to regulate substrate function has primarily come from the study of c-Jun. JNK-mediated phosphorylation on the dual serine residues leads to enhancement of transactivation potential and stabilization of c-Jun[16,17] as a result of the inhibition of the latter's ubiquitination.[16–18] Loss of phosphorylation on these residues resulted in the decrease of transactivation and a concomitant reduction in apoptosis and proliferation, highlighting that phosphorylation by the JNKs is essential for the efficient functionality of c-Jun.[19] c-Jun itself is necessary for cellular proliferation and apoptosis,[20] and studies using genetically modified mice and cells have indicated that some of the effects of JNK activation can be recapitulated by c-Jun activation,[16,20,21] establishing the relevance of this pathway *in vivo*.

The positive effect of JNK on substrates has been the theme for many substrates including c-Jun, ATF-2, Foxo4, Stat3, etc. (reviewed in Ref. 8). However, JNK phosphorylation has also been shown to lead to inhibition of substrate function, such as in the case of NFATc3, Nur77, and Tau.[22–24] Thus, the current paradigm is that phosphorylation by JNKs can lead to both activation and inhibition of substrate function.[6,25]

After the initial identification of transcription factors such as c-Jun and ATF-2 as substrates, a large amount of work by many research groups have identified a host of JNK phosphorylation targets and interacting partners, numbering over 50 (reviewed in Ref. 8). These include transcription factors such as ATF-2, Elk-1, p53, and NFAT; apoptotic and survival molecules such as Bim, Bad, Bcl-2, and Mcl-1; migratory proteins such as paxillin and microtubule-associated protein; kinases such as Akt and $p90^{RSK}$; and E3 ligase Itch, to name a few.[8] Besides these, other JNK-interacting proteins such as JIP-1 and 14-3-3 have also been identified, which act as scaffolds/adaptors that bring various JNK–substrate complexes together.[8]

Other than regulating substrate function through phosphorylation, JNKs have also been associated with the degradation of substrates in the absence of cellular stimulation.[18] The degradation-inducing role of the JNKs is dependent on binding of the inactive enzyme to the substrates (e.g., c-Jun, ATF2, and p53) and occurs in the absence of substrate phosphorylation.[18] Degradation is inhibited upon JNK activation, which results in substrate phosphorylation.[17] How exactly JNK binding leads to target destabilization is poorly understood. However, genetic studies by us and others have demonstrated that lack of JNK2 leads to enhanced stability of c-Jun and that its stable reexpression leads to ubiquitination and subsequent destabilization of c-Jun, confirming JNK's ability to degrade targets, though probably in an indirect manner.[16] In addition, JNK1 has been shown to phosphorylate and promote the function of ubiquitin E3-ligase Itch.[26] Thus, cumulative evidence points to JNK's role in destabilizing targets, though much work is required to understand the molecular mechanisms behind this process. Nonetheless, the dual ability of the JNKs in targeting c-Jun and other substrates to degrade or protect them upon activation suggests an intricate role of the JNKs in regulating substrate activity.

In addition to regulating the activity and stability of targets, JNKs have also been reported to regulate the localization of substrates such as Net and NFATc3. Phosphorylation by the JNKs has been shown to lead to the nuclear exclusion of both Net and NFATc3, leading to inhibition of their effects,[22,27] though the mechanistic details remain to be explored.

Altogether, the accumulated data highlight the multifunctional aspect of JNKs, which not only act as kinases leading to phosphorylation of substrates but also lead to other modifications and can also function in scaffolding complexes. These varied roles of the JNKs have not surprisingly endowed them with a host of properties in regulating cellular physiology, which are probably perturbed in a range of pathological conditions leading to disease manifestation. Nonetheless, the bulk of our knowledge of JNK comes from its activity as a kinase, and most disease associations of JNK have been currently attributed to its kinase activity.

## III. Human Diseases Where JNK Is Implicated

Given the vast number of instances in which JNK has been shown to play a crucial role in regulating cellular functions in cellular and animal models, JNK activity has also been analyzed in human disease states and other conditions. For example, even eccentric exercise and marathon running have been shown to induce a transient but robust burst of JNK activity in human skeletal muscles.[28,29] However, bulk of the reported literature is on determining JNK activity in disease states, which are summarized in Table I. The disease groups can be divided into the following main areas: neurological, coronary, hepatobiliary, and respiratory diseases; and autoimmune, inflammatory, and cancer conditions (Fig. 1). Remarkably, many of these conditions have mirrored the data from the JNK knockout animals (Ref. 6 and see next section), indicating the high level of consistency between the human and mouse systems.

Among the major disease groups in which JNK activity is affected, at least four major tissues have been repeatedly implicated. They include neuronal, cardiac, hepatobiliary, and respiratory tissues. Among the neuronal diseases, JNK has been shown to be activated in neurons of Alzheimer's disease (AD),

TABLE I

HUMAN DISEASE/CONDITIONS WHERE JNK ACTIVATION IS IMPLICATED

| Disease/condition | Tissue analyzed | Effect on JNK activity | Reference |
|---|---|---|---|
| Eccentric exercise | Skeletal muscle | Transient and robust increase | 28 |
| Marathon running | Skeletal muscle | Transient and robust increase | 29 |
| Parkinson's disease | Postmortem brain tissue | Upregulated | 30 |
| Parkinson's disease | Leukocyte of PD with G2019S mutation of LRRK2 | Downregulated | 31 |
| Alzheimer's disease | Neurons (neurofibrillary tangles) | Up | 32 |
| Alzheimer's disease | Cerebral neurons | Up | 33 |
| Alzheimer's disease | Hippocampus (neurofibrillary tangles) | Up | 34 |
| Stroke | Postmortem brain tissue | Up | 35 |
| Coronary artery disease | Serum and adipose tissues | Up | 36 |
| Heart failure | Heart tissue of failing heart | Up | 37 |
| Abdominal aortic aneurysm | Aortic samples | Up | 38 |
| Abdominal aortic aneurysm | Aortic wall samples | Up | 39 |
| Noonan syndrome | | Up | 40 |
| Chronic HCV infection | Liver samples | Up | 41 |
| Acute liver injury | Liver | Up | 42 |
| Non-alcoholic fatty liver disease | Liver, muscle | Up | 43 |
| Asthma | Endobronchial biopsy | Down | 44 |
| Chronic obstructive pulmonary disease of smokers | Leukocytes | Up | 45 |
| Inflammatory bowel disease | Colonic tissue | Up | 46 |
| Inflammatory bowel disease | Colonic mucosa | No change | 47 |
| Crohn's disease | Colonic biopsies | Up | 48 |

(*Continues*)

TABLE I (*Continued*)

| Disease/condition | Tissue analyzed | Effect on JNK activity | Reference |
|---|---|---|---|
| Rheumatoid arthritis | Synovial tissue | Up | 49 |
| Rheumatoid arthritis | Fibroblast-like synoviocytes and synovial fluid | Up | 50 |
| Systemic lupus eryththematosus (SLE) | Peripheral blood | Up | 51 |
| SLE | Bone marrow mononuclear cells | Up | 52 |
| SLE | PBMC | Up | 53 |
| SLE | PBMC | Up | 54 |
| Celiac disease | PBMC | Up | 55 |
| Behcet's disease | Endothelial cells by the serum (autoantibodies) | Up | |
| Colorectal cancer | Cancer mucosa | Down (2–4×) | 56 |
| Retinoblastoma | Retinoblastoma tissue | Up | 57 |
| Melanoma | Nevi | Up | 58 |
| Breast carcinoma | Triple negative breast cancer | Up | 59 |
| Ovarian cancer | Invasive ovarian carcinoma | Up | 60 |
| Obesity/insulin resistant | Skeletal muscle | Up | 61 |
| Type 2 diabetes | PBMC | Up | 62 |
| Insulin resistance | Adipose tissue | Up | 63 |

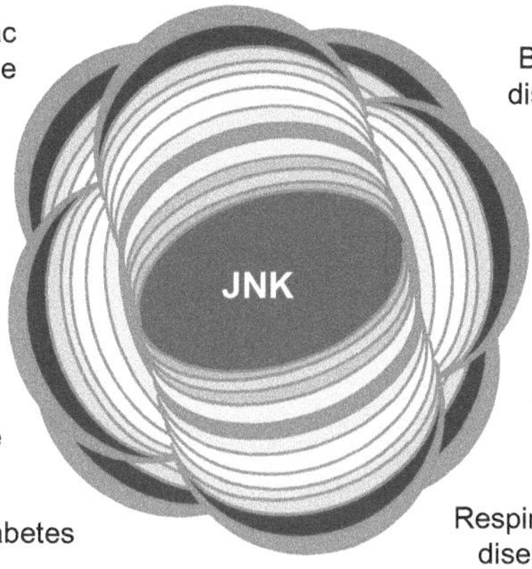


FIG. 1. JNKs in disease—the plethora of diseases in which JNK activity is affected is shown. (See Color Insert.)

Parkinson's disease (PD), stroke, polyglutamine disease, Pick's disease, corticobasal degeneration, dementia, argyrophilic grain disease, and amyotrophic lateral sclerosis, to name a few (Table I).[30–35] In cardiac conditions such as coronary heart disease, heart failure, and abdominal aortic aneurysm, aortic wall samples, adipose tissues around the heart, and serum from patients displayed elevated JNK activity.[36–40] Further, liver samples from chronic HCV-infected patients and after acute liver injury, as well as samples from nonalcoholic fatty liver disease, were found to have elevated JNK activity.[41–43] Moreover, endobronchial tissues from asthma patients and leukocytes from pulmonary disease patients were found to have altered JNK activity.[44,45] Colonic biopsies from inflammatory bowel disease and Crohn's disease, synovial tissues from rheumatoid arthritis (RA) patients, and peripheral blood mononuclear cells from autoimmune systemic lupus erythematosus and Behcet's disease patients displayed higher JNK activity.[46–55,64] Many cancer samples, including retinoblastoma, melanoma nevi, colonic mucosa from colorectal cancer patients, and breast and ovarian cancers, exhibit elevated JNK activity.[56–60] Besides, increased JNK activity has been strongly associated with obesity and Type 2 diabetes.[61–63] What is striking is that in almost all human disease cases, JNK activity has been shown to be upregulated (rather than downregulated), indicating that activated JNK pathway may be at least be in part causal to the disease state.

The critical question that arises from these observations is whether the elevated JNK activity in human diseases is the sole causal factor and determinant of the disease state, or just a reflection of the activated state of the cellular stress pathway. Studies from the knockout models have indicated that loss of JNK can indeed ameliorate some of the diseases noted in humans, suggesting a causal role for JNK in disease pathology (described in the following section). Nonetheless, direct demonstration of JNK activity leading to disease manifestation is still lacking.

## IV. Functions Uncovered by the Use of Knockout Mice—JNKs' Role in Development and Disease

It has been over a decade since mice lacking the individual *JNK* genes were initially reported, and the field has learned enormously from their analyses. Although individual knockouts are born and do not show any overt defects, compound JNK1/JNK2 deficiency leads to embryonic lethality, highlighting an essential role for JNK activity during development.[65] Moreover, varying degrees of loss of *JNK* alleles have shown defects in neuronal development, eyelid closure, and epidermal morphogenesis during embryogenesis,[65,66] further suggesting that a critical level of JNK activity is required to maintain normal homeostatic balance and thus ensure proper development of tissues. Consistent with a crucial role for the JNKs to safeguard embryonic development, it is noteworthy that there are hardly any human developmental defects that have been directly associated with JNK deficiency or mutations, indicating that this pathway is probably not amenable to disturbance without affecting development and life.

Genetic analyses using individual and compound JNK knockout mice that we and others have generated have revealed specific and common roles for JNKs in both developmental processes and upon exposure to extracellular stress stimuli. These include neuronal apoptosis, neuronal microtubule maintenance, T-cell activation and apoptosis, acute liver inflammation and failure, regulation of insulin resistance and obesity, bile acid production, osteoclastogenesis, fibroblasts apoptosis, proliferation, and senescence, as well as transformation and tumourigenesis (reviewed in Refs. 6 and 16,25,65–78). These findings have highlighted that the individual JNK proteins can have both overlapping and redundant functions, as well as specific functions that are not substitutable by the other JNKs in a cell type- and signal-dependent context (reviewed in Ref. 6). Moreover, we have also shown that despite their biochemical and structural similarities and earlier expectations, JNK1 and JNK2 exert different and opposite effects on c-Jun turnover and activity which correlates with their differential effects on fibroblast proliferation.[16,25] JNK1 appears to be a positive regulator of cellular proliferation, being required for phosphorylation and activation of c-Jun. On the contrary, JNK2 appears to primarily function as a negative regulator of c-Jun stability, correlating with its negative effect on cell proliferation. Similarly, JNK1 and JNK2 have been shown to have opposite effects on p53 levels.[79] Altogether, these studies have revealed that the JNK proteins can indeed be very different and specific in their functions and, therefore, cannot be classified in the same category when trying to modulate their activity for therapeutic purposes.

Besides regulating developmental processes, the JNK signaling pathway is frequently found perturbed in many disease conditions compared to other MAPKs in animal models.[1,5,6] Immunologically, these include T- and B-cell lymphoma, X-linked agammaglobulinemia (XLA) and X-linked immunodeficiency, and other autoimmune diseases.[80–87] Several neurological diseases have been shown to be regulated by JNK signaling. For example, in a PD model, absence of JNK2 and JNK3 led to greater protection against dopaminergic cell death.[30] In an AD model, *Jnk3*$^{-/-}$ mice were found resistant to β-amyloid protein (Aβ)-induced apoptosis.[88] Further, JNK mutant mice have also been used to demonstrate JNK's role in chronic inflammatory diseases. For instance, *Jnk1*$^{-/-}$ mice were protected from joint damage in a RA model.[83] In addition, atheleresclerosis appears to depend on JNK signaling. Using the *ApoE*$^{-/-}$ mice model for atheleroscleros*is*, absence of JNK was shown to decrease the disease manifestation.[89,90] This was specific to JNK2, as JNK1 deficiency did not affect the disease.

Besides the effects on neurological and inflammatory models, JNK mutant mice have been utilized to demonstrate the relevance of JNKs in diabetes, hepatic disease, and cardiac pathologies. Consistent with the fact that JNKs phosphorylate IRS1,[91] and are activated by high-fat diet, absence of JNK1 improved insulin sensitivity and diabetic condition.[74] Moreover, *Jnk1*$^{-/-}$ mice were found to have reduced steatosis and hepatitis[92] and also reduced hepatic fibrosis in a nonsteatotic model of fibrosis,[93] suggesting a causal role for the JNKs in liver pathologies. Similarly, using *Jnk2*$^{-/-}$*Jnk1*$^{+/-}$ mice, it was demonstrated that cardiac hypertrophy was exaggerated in response to transverse aortic constriction (TAC).[94] Further, *Jnk1*$^{-/-}$ mice displayed abnormal response to TAC as exemplified by reduced left ventricular (LV) systolic functions,[95] suggesting together that JNKs play protective roles in maintaining LV systolic functions. Finally, abundant evidence exists for the role of JNKs in carcinogenesis, and many mouse models have confirmed the important role of JNK signaling in the development and progression of cancer (reviewed in Ref. 96). For a comprehensive summary of the phenotypes of the JNK knockouts over the years, the reader is referred to some of the earlier reviews.[6,15]

Although these knockout studies have demonstrated a crucial and contributory role of the JNK signaling pathway in many diseases, one has to be cognizant of the fact that absence of JNK from birth may lead to adaptation and, hence, may not be truly reflective of a clinical scenario. Thus, such JNK inhibition may not necessarily lead to beneficial efforts therapeutically when applied at the pathological state. Conditional, time-based depletion of JNK proteins in disease states have yet to be modeled elaborately in mouse systems, which may be useful to shed further light on the relevance of the JNK pathway in initiating and maintaining the disease phenotype. Further, transgenic, time-based overexpression of activated JNK needs to be employed to study or demonstrate causal role of JNK activation in various diseases.

## V. The Relevance of JNK-Mediated Phosphorylation in Regulating Biological Outcome *In Vivo*

Although enormous amount of data have been derived from the complete JNK knockout mice, the question that has remained elusive is whether the kinase activity of JNK is absolutely required for its proper functioning, and whether phosphorylation of its substrates is indeed causal to disease states. Given that most of the work on JNK alludes to its activity in phosphorylating substrates, it has been taken to mean that most of the JNK knockout-dependent phenotypes are indeed due to loss of kinase activity. Nonetheless, mice deficient in JNK kinase activity have not yet been generated to directly prove this possibility. However, mice that lack the JNK phosphorylation sites on c-Jun have been made (due to alanine substitutions of the serines 63 and 73; $c\text{-}Jun^{AA}$ mice), which, to some extent, mimic lack of JNK activity on c-Jun.[19,97,98] These mice are viable in contrast to the $c\text{-}Jun^{-/-}$ case, indicating that c-Jun phosphorylation on the serines 63 and 73 are not required for embryonic development. Nonetheless, these $c\text{-}Jun^{AA}$ mice do exhibit defects in neuronal apoptosis and lead to reduction of tumor formation in the $APC^{min}$ background and in other models, to cite a few examples, indicating that c-Jun phosphorylation is indeed required for efficient functioning, and, hence, indirectly supports the relevance of JNK-mediated phosphorylation of c-Jun in regulating biological functions and disease conditions.[97,98] These data also highlight that phosphorylation by the JNKs may be crucial in fine-tuning the functions of their target substrates and, hence, leading to optimal functions of the proteins, without which cellular outcome could be perturbed and lead to promotion of disease conditions.

## VI. JNK Inhibition in Cells, Mice and Man

As the JNK pathway is deregulated in many disease states, tremendous efforts have gone into trying to inhibit the pathway, and, not surprisingly therefore, a large collection of various types of JNK inhibitors has been generated that affect JNK enzymatic activity or are inhibitory peptides.[99] Among them are two major classes of compounds: ATP-competitive and ATP-noncompetitive inhibitors as well as peptide inhibitors.[99] The first type of inhibitors, identified and tested extensively, are the reversible ATP-competitive inhibitors, such as SP600125 and CEP1347, of which SP600125 has been utilized in many *in vitro* and *in vivo* studies extensively[99] and has shown success both in cell culture and in mice models (Tables II and III). For example, several studies have shown that inhibiting JNK activity can ameliorate JNK-dependent apoptosis, prevent cytokine production from synoviocytes, endothelial cells, and macrophages, and enhance recovery from ischemia/reperfusion damage (Table II).[50,100–110] As these studies were conducted *in vitro* only with the specialized cells in culture, attempts have been made to reproduce them in animal models. JNK inhibition has consistently shown to decrease cytokine production in colonic tissues in an inflammatory bowel disease model and also in the case of RA and asthma (Table III).[111–129] Similar studies have also been successfully carried out in neuronal disease models including PD models and others, highlighting not only the efficacy of JNK inhibitors in controlling disease progression but also tolerability of these inhibitors in an organismal setting. However, many of these compounds have varying degrees of toxicity and lack the required specificity,[99] as such ATP-competitive inhibitors would indiscriminately inhibit phosphorylation of all substrates. Consistently, there are many reports of the failure of such competitive JNK inhibitors in animal models. For example, JNK inhibition has not been found to be successful in some cases of hepatic injury, especially after concanavalin-A or carbon tetrachloride treatment.[130] Moreover, SP600125 treatment exacerbated liver injury after ischemia,[131] increased cardiac myocyte apoptosis,[132] was not beneficial following CVB3 infection,[133] affected physiological JNK functions such as long-term potentiation,[114] etc. Part of the problem could be that the SP600125 compound is not very specific, as was found recently to be capable of inhibiting 13 other kinases.[134,135]

Therefore, second-generation ATP-competitive inhibitors such as CC-401 have been developed and have shown success in renal injury models and have entered clinical trials for AML (Ref. 99; http://clinicaltrials.gov/ct2/show/NCT00126893). Other CC-401-related compounds have also been generated that show promise. In addition, AS601245, another ATP-binding site inhibitor, has shown promise in protecting against neuronal death in ischemic models.[120,121] Moreover, it was able to decrease cardiomyocyte apoptosis and infarct size after myocardial ischemia,[119] as well as experimental RA.[83] Additionally, another ATP-competitive JNK inhibitor, compound A, was reported to decrease body weight, blood glucose, and triglyceride levels.[118] These data suggest that some of the initial toxicity can be alleviated by the subsequent generation of drugs, though these ATP-competitive inhibitors may not be all that valuable without much specificity.

Hence, initiatives have been made to generate other ATP-noncompetitive JNK inhibitors based on peptides of interacting partners such as JIP that are able to target the substrate-docking site rather than binding to the ATP site.[99]

TABLE II

EFFECT OF JNK INHIBITION ON CELLULAR MODELS

| Cellular model | Method of inhibition | Substrate/target affected | Outcome | Reference |
|---|---|---|---|---|
| Human brain endothelial cells | SP600125 | JNK | Prevention of amyloid beta-induced inflammatory cytokine production | 100 |
| HIV viral protein-induced apoptosis in human monocytic cells | SP600125 | JNK-mediated Bcl2/ c-IAP1 downregulation | Inhibition of apoptosis | 101 |
| Macrophages of HIV-associated dementia patients | SP600125 | JNK | Decreased CXCL8 production associated with disease | 102 |
| Kidney endothelial cells | SP600125 | JNK | Reduced autoantibody mediated cytokine production in Wegener's granulomatosis | 103 |
| Asthma model in rat | SP600125 | JNK | Reduced inflammatory cell egress into airway lumen | 104 |
| Fibroblast-like synoviocytes and synovial fluid | SP600125 | JNK/c-Jun | Prevention of increase in IL-32 | 50 |
| Cortical neurons | D-JNKI1 | JNK | Decreased APP processing | 105 |
| Intracerebral reovirus infection-mediated encephalitis | Cell permeating peptide | JNK | Inhibits JNK activity and prolongs survival | 106 |
| Neuropathic pain induced by spinal nerve ligation | D-JNK I1 | JNK | Prevented and reversed SNL-induced mechanical allodynia | 107 |
| LPS induced uveitis in ocular chamber (retinal and glial cells) | XG-102 (D-JNK I1) | JNK | Reduced intraocular inflammation | 108 |
| Cellular hyper-osmolarity | PYC71N (inhibitory peptide) | JNK | Decreased c-Jun phosphorylation during hyperosmotic stress | 109 |
| Human pancreatic islet cells | Dominant inhibitor JNK-binding domain of JIP-1 (JNKi) | JNK up upon leptin and glucose treatment | Reduced JNK and apoptosis | 110 |

TABLE III

Effect of JNK Inhibition on Disease Models

| Disease model | Method of inhibition | Substrate/target affected | Outcome | Reference |
|---|---|---|---|---|
| Brain damage | CEP1347 | MLK | Reduced loss of dopaminergic cell bodies | 111 |
| NeuroAIDs | CEP1347 | MLK | Neuroprotection | 112 |
| Colonic tissue of inflammatory bowel disease | SP600125 | JNK | Decreased cytokine production | 46 |
| Rheumatoid arthritis | SP600125 | JNK | Protection against joint damage | 83 |
| Asthma | SP600125 | JNK | Inhibition of lung inflammation | 113 |
| Long-term potentiation in adult cortex | SP600125 | JNK | Suppression of induction of cingulate LTP | 114 |
| Brain ischemia and reperfusion | SP600125 | JNK | Neuroprotection | 115 |
| HCMV viral Infection | SP600125 | JNK | Decreased virus-induced death of megakaryocytes | 116 |
| PD model | SP600125 | JNK | Protection of dopaminergic neurons | 117 |
| Obesity | Compound A | JNK | Decrease body weight and blood glucose and TG levels and improve insulin sensitivity | 118 |
| Cardiac ischemia and reperfusion | AS601245 | JNK | Reduced cardiomyocyte apoptosis and infarct size | 119 |
| Global cerebral ischemia and reperfusion | AS601245 | JNK | Reduced axon/dendrite damage and prevention of ischemia-induced impairment of memory | 120 |
| Global cerebral ischemia and reperfusion | AS601245 | JNK | Prevention of loss of hippocampal CA1 neurons | 121 |
| Hepatic ischemia and reperfusion | CC0209766, CC0223105, and CC-401 | JNK | Decreased necrosis and apoptosis of hepatocytes and sinusoidal endothelial cells | 122 |
| Cerebral artery occlusion | D-JNK I1 | JNK | Decreased neuronal death | 123 |
| Cerebral ischemia | D-JNK I1 | JNK | Decreased neuronal death | 124 |
| Experimental STZ-induced diabetes | TAT-JIP | JNK | Improved insulin resistance but worsen albuminuria | 125 |
| Diabetes | D-JNK I1 | JNK | Improved insulin sensitivity and glucose tolerance | 126 |
| Myocardial ischemia and reperfusion | D-JNK I1 | JNK | Reduced infarct size | 127 |
| Acoustic trauma and aminoglycoside-induced hearing loss | D-JNK I1 | JNK | Otoprotective benefit in organ cultures of cochlea | 128 |
| Sound trauma-induced hearing loss | D-JNK I1 | JNK | Prevention of cell death | 129 |

The D-JNK1-1 peptide inhibitor was effective in a variety of model systems, including diminishing the production of Aβ-precursor protein and Aβ fragments[105] and being neuroprotective against ischemic injury[124] and against myocardial ischemic.[127] Moreover, similar peptide inhibitors were shown to improve insulin sensitivity and glucose tolerance,[126] inhibit cell death, and restore auditory functions in sound-exposed hearing loss,[128,129] among others. These peptides therefore appear to be better than the ATP-competitive inhibitors, as they are more specific.[99] Other JNK inhibitory peptides have also been used to show efficacy in decreasing hyperosmotic stress and reducing pancreatic cell death.[109] In addition, the use of peptide inhibitors has also highlighted these specificity issues at the molecular level. For example, concurrent analysis of the effects of SP600125 and JNK inhibitory peptides revealed that only 4 of the total 20 interleukin-induced genes were regulated by both the compounds, whereas 10 genes were regulated only by SP600125.[136]

Collectively, the encouraging cellular and preclinical data with more specific inhibitors have motivated companies to attempt clinical trials with some of the compounds, to determine toxicity, as well as to determine therapeutic outcome in Phase 2/3 trials.[99] Up till now, several trials have been carried out with AM-111, CEP-1347, CNI-1493, XG-102 (D-JNK1-I), and CC401, to name a few (Table IV). The Phase I study using CC401 for myeloid leukemia has been completed, but discontinued (Ref. 99; http://clinicaltrials.gov/ct2/show/NCT00126893). Some initial studies have shown positive improvements with CNI-1493 (endoscopic improvement in Crohn's diseases)[48] and AM-111 (hearing improvement in acute acoustic trauma patients).[137] Other trials using the XG-102 peptide inhibitor for stroke and Alzheimer's (http://www.xigenpharma.com/) and the SP600125 derivatives CC359 (ischemia/reperfusion damage) and CC930 (fibrotic disease) are under way (http://www.celgene.com/research/kinase-inhibitors.aspx). However, CEP1347 was not successful in delaying disability in patients with PD,[138] though it was effective in animal PD models,[111] highlighting the need for further validation prior to successful embarkation onto JNK inhibition therapy. These initial clinical trials data suggest that, though promising, much more work is required to optimize JNK inhibitors in clinical settings.

## VII. Challenges in Modulating the JNK Pathway to Treat Human Diseases

### A. Specificity

Although several JNK inhibitors have entered clinical trials, the initial pace of advancement and potential success seems to be limited for a variety of reasons. The fundamental problem has been that these inhibitors cannot

TABLE IV

JNK Inhibitors and Mode of Action on Human Diseases—Clinical Data

| JNK inhibitor | Mode of action | Disease state | Clinical trials | Reference |
|---|---|---|---|---|
| CC-401 | JNK | AML | Phase 1 | http://clinicaltrials.gov/ct2/show/NCT00126893 |
| CNI-1493 | JNK and p38 | Crohn's disease | Endoscopic improvement | 48 |
| AM-111 (D-JNK1l) | JNK | Acute acoustic trauma | Phase 1/2 positive therapeutic effect on hearing recovery | 137 |
| XG-102 (D-JNK1l) | JNK | Stroke or a transient ischemic attack (TIA) | Phase 1 | http://www.xigenpharma.com/ |
| XG-102 (D-JNK1l) | JNK | Alzheimer's | Phase 2/3 | http://www.xigenpharma.com/ |
| CC359 | JNK | Ischemia/reperfusion damage | | http://www.celgene.com/research/kinase-inhibitors.aspx |
| CC930 | JNK | Pulmonary fibrosis | Phase 2 | http://www.celgene.com/research/kinase-inhibitors.aspx |
| CEP1347 | Inhibition of MLK | Parkinson disease | Failed to delay disability in PD | 138 |

directly serve the purpose of modulating JNK activity for treatment of all diseases without affecting other cell types, as the JNKs mediate several signaling pathways in multiple cell types, and hence, use of these compounds for treatment can lead to other side effects and toxicity. Further, the current JNK inhibitors indiscriminately inhibit all JNKs, which may lead to further complications, as underscored by the knockout studies. Although some JNK inhibitors have shown some level of specificity—more active against JNK3 than JNK1—many have not.[99] Hence, generation of JNK1, JNK2, and JNK3 specific inhibitors may be required to tackle the problem of specificity. In addition, very little information is available on the specific role of the various splice variants, which may work in a cell type- and signal-specific manner. What is therefore needed are specific JNK inhibitors that provide both cell type and signal specificity, which can be generated only if the specific interacting partners affecting specific cellular processes in diseases are clearly defined. New efforts to understand the JNK signaling module specific for each particular disease in which the pathway has been implicated and found responsible for should be initiated, such that specific targeting of the particular module, be it JNK/scaffold or JNK/substrate interaction, can be inhibited to correct the JNK overactivation in the particular tissue/cell type, which will also reduce the effect on other nontarget tissues where such modules may not be functional. A classical example of this kind of adverse effect was demonstrated when JNK inhibition was used to improve insulin sensitivity but which lead to albuminuria.[125] Thus, specificity is a key issue that needs proper evaluation for successful JNK inhibition therapies.

## B. Relevance of JNK Pathway to Disease and Selectivity

It is also important to note that most diseases are often not monogenic in origin and are a culmination of multiple signaling pathways that are affected. It is more so in the case where JNK is activated, which often occurs under cellular stress, during which multiple signaling pathways would be activated. Hence, primary among the factors that need to be considered prior to JNK targeting would be to determine whether JNK activation is indeed causal and dominant and contributes significantly to the disease, rather than a simple correlatory factor. With that information, success with JNK inhibitors will be more likely. This may be one of the major reasons for the lack of success of JNK inhibition therapies and requires justification prior to embarking on JNK inhibition therapies even in cases where JNK activity is deregulated. One example where a clinical correlation was not achieved with JNK inhibitors is in PD trials, although JNK inhibition was able to affect neuronal apoptosis in culture, as well as in PD mouse models,[111,138] emphasizing the need for direct demonstration of JNK's causal role in disease prior to successful use in clinical trials. Such failures also underscore that JNK activation may not be the sole and primary contributor to disease development, thereby highlighting concerns in tackling the JNK pathway alone in monotherapeutic trials. Hence, potential treatment regimens should consider combination therapy, rather than with JNK inhibitors alone. To our knowledge, no combination therapy with JNK inhibitors has been conducted in clinical trials.

## C. Transient Versus Constitutive JNK Activation

*In vitro* studies have shown that JNK activation can be transient, often leading to cellular survival signals, compared to sustained activation, often leading to cell death.[139] One potential reason why the inhibition of the JNK pathway, though having a positive effect on the desired outcome in *in vitro* studies, has failed *in vivo* is that the magnitude of JNK activation has not been well determined in disease states. For example, chronic conditions would be expected to lead to sustained JNK activation, likely leading to activation of multiple substrates and potentially involving multiple JNK modules. Thus, prolonged JNK inhibition may be necessary to show appreciable improvements in some cases, which would require that JNK inhibition be applied locally without much detrimental effects on other tissues. Currently, knowledge of this information is missing in the disease models, where the only attribute that has been determined prior to JNK targeting is that JNK is indeed activated.

## D. Biomarkers

Another challenge that requires careful consideration even when specific and highly effective JNK inhibitors have been generated is the availability of biomarkers that will precisely reflect the JNK activation state in diseases and, at the same time, reflect successful response. Currently, most of the studies that have determined JNK activity from human samples have utilized specific tissue biopsies. However, analysis of biopsies may not be suitable and amenable in all cases, and hence, alternate methods are required, which is a great challenge in the field. A clear demonstration of JNK activation in disease states in humans is essential prior to embarking on clinical trials aimed at inhibiting the JNK pathway. An example of absence of such data prior to clinical trials resulting in failure is the case of PD.[138] Although abundant information is available on JNK's role in regulating neuronal death in culture and PD animal model, there are no direct observations, to our knowledge, of activated JNK in human PD neurons from patients. Not surprisingly, therefore, JNK inhibition clinical trials failed to have any positive effect on patients with PD, highlighting the importance of having prior knowledge of JNK activity in human disease states.

## VIII. Future Outlook

Review of the literature on JNK signaling over the past two decades reveals clearly that JNK activity is indeed induced in many disease states, and that genetic deletion or chemical inhibition of JNK can lead to amelioration of several disease conditions in animal models. Given these basic pretexts, much work is still required to specifically and efficiently target the JNK pathway in clinical conditions. The primary focus should therefore be to determine the causal role of activated JNK in disease manifestation, thereby classifying diseases and conditions that are dependent on activated JNK signaling. Upon determination of diseases that are dominantly JNK dependent, identifying specific substrates/JNK modules for specific disease states/cell types would be of paramount importance in overcoming the nonspecificity of JNK targeting. Once these two major objectives are achieved, clinical success of JNK inhibition would become a reality.

### References

1. Chang L, Karin M. Mammalian MAP kinase signalling cascades. *Nature* 2001;**410**:37–40.
2. Karin M, Gallagher E. From JNK to pay dirt: jun kinases, their biochemistry, physiology and clinical importance. *IUBMB Life* 2005;**57**:283–95.
3. Minden A, Karin M. Regulation and function of the JNK subgroup of MAP kinases. *Biochim Biophys Acta* 1997;**24**:F85–F104.
4. Karin M. Mitogen-activated protein kinase cascades as regulators of stress responses. *Ann NY Acad Sci* 1998;**851**:139–46.
5. Weston CR, Davis RJ. The JNK signal transduction pathway. *Curr Opin Cell Biol* 2007;**19**:142–9.
6. Sabapathy K. *JNK signalling*. Texas: Landes BioSciences; 2005. pp. 29–39.
7. Barr RK, Bogoyevitch MA. The c-Jun N-terminal protein kinase family of mitogen-activated protein kinases (JNK MAPKs). *Int J Biochem Cell Biol* 2001;**33**:1047–63.
8. Bogoyevitch MA, Kobe B. Uses for JNK: the many and varied substrates of the c-Jun N-terminal kinases. *Microbiol Mol Biol Rev* 2006;**70**:1061–95.
9. Gupta S, Barrett T, Whitmarsh AJ, Cavanagh J, Sluss HK, Dérijard B, et al. Selective interaction of JNK protein kinase isoforms with transcription factors. *EMBO J* 1996;**15**:2760–70.
10. Mohit AA, Martin JH, Miller CA. p493F12 kinase: a novel MAP kinase expressed in a subset of neurons in the human nervous system. *Neuron* 1995;**14**:67–78.
11. Dérijard B, Hibi M, Wu IH, Barrett T, Su B, Deng T, et al. JNK1: a protein kinase stimulated by UV light and Ha-Ras that binds and phosphorylates the c-Jun activation domain. *Cell* 1994;**76**:1025–37.
12. Hibi M, Lin A, Smeal T, Minden A, Karin M. Identification of an oncoprotein- and UV-responsive protein kinase that binds and potentiates the c-Jun activation domain. *Genes Dev* 1993;**11**:2135–48.
13. Kallunki T, Deng T, Hibi M, Karin M. c-Jun can recruit JNK to phosphorylate dimerization partners via specific docking interactions. *Cell* 1996;**87**:929–39.

14. Kallunki T, Su B, Tsigelny I, Sluss HK, Dérijard B, Moore G, et al. JNK2 contains a specificity-determining region responsible for efficient c-Jun binding and phosphorylation. *Genes Dev* 1994;**24**:2996–3007.
15. Bogoyevitch MA. The isoform-specific functions of the c-Jun N-terminal kinases (JNKs): differences revealed by gene targeting. *Bioessays* 2006;**9**:923–34.
16. Sabapathy K, Hochedlinger K, Nam SY, Bauer A, Karin M, Wagner EF. Distinct roles for JNK1 and JNK2 in regulating JNK activity and c-Jun-dependent cell proliferation. *Mol Cell* 2004;**15**:713–25.
17. Musti AM, Treier M, Bohmann D. Reduced ubiquitin-dependent degradation of c-Jun after phosphorylation by MAP kinases. *Science* 1997;**275**:400–2.
18. Fuchs SY, Fried VA, Ronai Z. Stress-activated kinases regulate protein stability. *Oncogene* 1998;**17**:1483–90.
19. Behrens A, Sibilia M, Wagner EF. Amino-terminal phosphorylation of c-Jun regulates stress-induced apoptosis and cellular proliferation. *Nat Genet* 1999;**21**:326–9.
20. Schreiber M, Kolbus A, Piu F, Szabowski A, Möhle-Steinlein U, Tian J, et al. Control of cell cycle progression by c-Jun is p53 dependent. *Genes Dev* 1999;**13**:607–19.
21. Behrens A, Sabapathy K, Graef I, Cleary M, Crabtree GR, Wagner EF. Jun N-terminal kinase 2 modulates thymocyte apoptosis and T cell activation through c-Jun and nuclear factor of activated T cell (NF-AT). *Proc Natl Acad Sci USA* 2001;**98**:1769–74.
22. Chow CW, Rincón M, Cavanagh J, Dickens M, Davis RJ. Nuclear accumulation of NFAT4 opposed by the JNK signal transduction pathway. *Science* 1997;**278**:1638–41.
23. Han YH, Cao X, Lin B, Lin F, Kolluri SK, Stebbins J, et al. Regulation of Nur77 nuclear export by c-Jun N-terminal kinase and Akt. *Oncogene* 2006;**25**:2974–86.
24. Yoshida H, Hastie CJ, McLauchlan H, Cohen P, Goedert M. Phosphorylation of microtubule-associated protein tau by isoforms of c-Jun N-terminal kinase (JNK). *J Neurochem* 2004;**90**:352–8.
25. Sabapathy K, Wagner EF. JNK2: a negative regulator of cellular proliferation. *Cell Cycle* 2004;**3**:1520–3.
26. Gao M, Labuda T, Xia Y, Gallagher E, Fang D, Liu YC, et al. Jun turnover is controlled through JNK-dependent phosphorylation of the E3 ligase Itch. *Science* 2004;**306**:271–5.
27. Ducret C, Maira SM, Lutz Y, Wasylyk B. The ternary complex factor Net contains two distinct elements that mediate different responses to MAP kinase signaling cascades. *Oncogene* 2000;**19**:5063–72.
28. Boppart MD, Aronson D, Gibson L, Roubenoff R, Abad LW, Bean J, et al. Eccentric exercise markedly increases c-Jun NH(2)-terminal kinase activity in human skeletal muscle. *J Appl Physiol* 1999;**87**:1668–73.
29. Boppart MD, Asp S, Wojtaszewski JF, Fielding RA, Mohr T, Goodyear LJ. Marathon running transiently increases c-Jun NH2-terminal kinase and p38 activities in human skeletal muscle. *J Physiol* 2000;**526**:663–9.
30. Hunot S, Vila M, Teismann P, Davis RJ, Hirsch EC, Przedborski S, et al. JNK-mediated induction of cyclooxygenase 2 is required for neurodegeneration in a mouse model of Parkinson's disease. *Proc Natl Acad Sci USA* 2004;**101**:665–70.
31. White LR, Toft M, Kvam SN, Farrer MJ, Aasly JO. MAPK-pathway activity, Lrrk2 G2019S, and Parkinson's disease. *J Neurosci Res* 2007;**85**:1288–94.
32. Zhu X, Raina AK, Rottkamp CA, Aliev G, Perry G, Boux H, et al. Activation and redistribution of c jun N-terminal kinase/stress activated protein kinase in degenerating neurons in Alzheimer's disease. *J Neurochem* 2001;**76**:435–41.
33. Shoji M, Iwakami N, Takeuchi S, Waragai M, Suzuki M, Kanazawa I, et al. JNK activation is associated with intracellular beta-amyloid accumulation. *Brain Res Mol Brain Res* 2000;**85**:221–33.

34. Thakur A, Wang X, Siedlak SL, Perry G, Smith MA, Zhu X. c-Jun phosphorylation in Alzheimer disease. *J Neurosci Res* 2007;**85**:1668–73.
35. Mitsios N, Gaffney J, Krupinski J, Mathias R, Wang Q, Hayward S, et al. Expression of signaling molecules associated with apoptosis in human ischemic stroke tissue. *Cell Biochem Biophys* 2007;**47**:73–86.
36. Baker AR, Harte AL, Howell N, Pritlove DC, Ranasinghe AM, da Silva NF, et al. Epicardial adipose tissue as a source of nuclear factor-kappaB and c-Jun N-terminal kinase mediated inflammation in patients with coronary artery disease. *J Clin Endocrinol Metab* 2009;**94**:261–7.
37. Cook SA, Sugden PH, Clerk A. Activation of c-Jun N-terminal kinases and p38-mitogen-activated protein kinases in human heart failure secondary to ischaemic heart disease. *J Mol Cell Cardiol* 1999;**31**:1429–34.
38. Yoshimura K, Aoki H, Ikeda Y, Furutani A, Hamano K, Matsuzaki M. Identification of c-Jun N-terminal kinase as a therapeutic target for abdominal aortic aneurysm. *Ann NY Acad Sci* 2006;**1085**:403–6.
39. Diehm N, Di Santo S, Schaffner T, Schmidli J, Völzmann J, Jüni P, et al. Severe structural damage of the seemingly non-diseased infrarenal aortic aneurysm neck. *J Vasc Surg* 2008;**48**:425–34.
40. Longoni M, Moncini S, Cisternino M, Morella IM, Ferraiuolo S, Russo S, et al. Noonan syndrome associated with both a new Jnk-activating familial SOS1 and a de novo RAF1 mutations. *Am J Med Genet* 2010;**152A**:2176–84.
41. Kluwe J, Pradere JP, Gwak GY, Mencin A, De Minicis S, Osterreicher CH, et al. Modulation of hepatic fibrosis by c-Jun-N-terminal kinase inhibition. *Gastroenterology* 2011;**138**:347–59.
42. Hasselblatt P, Rath M, Komnenovic V, Zatloukal K, Wagner EF. Hepatocyte survival in acute hepatitis is due to c-Jun/AP-1-dependent expression of inducible nitric oxide synthase. *Proc Natl Acad Sci USA* 2007;**104**:17105–10.
43. Ferreira DM, Castro RE, Machado MV, Evangelista T, Silvestre A, Costa A, et al. Apoptosis and insulin resistance in liver and peripheral tissues of morbidly obese patients is associated with different stages of non-alcoholic fatty liver disease. *Diabetologia* 2011;**54**:1788–98.
44. Liu W, Liang Q, Balzar S, Wenzel S, Gorska M, Alam R. Cell-specific activation profile of extracellular signal-regulated kinase 1/2, Jun N-terminal kinase, and p38 mitogen-activated protein kinases in asthmatic airways. *J Allergy Clin Immunol* 2008;**121**:893–902.
45. Rumora L, Milevoj L, Popović-Grle S, Barisić K, Cepelak I, Grubisić TZ. Levels changes of blood leukocytes and intracellular signalling pathways in COPD patients with respect to smoking attitude. *Clin Biochem* 2008;**41**:387–94.
46. Mitsuyama K, Suzuki A, Tomiyasu N, Tsuruta O, Kitazaki S, Takeda T, et al. Pro-inflammatory signaling by Jun-N-terminal kinase in inflammatory bowel disease. *Int J Mol Med* 2006;**17**:449–55.
47. Malamut G, Cabane C, Dubuquoy L, Malapel M, Dérijard B, Gay J, et al. No evidence for an involvement of the p38 and JNK mitogen-activated protein in inflammatory bowel diseases. *Dig Dis Sci* 2006;**51**:1443–53.
48. Hommes D, van den Blink B, Plasse T, Bartelsman J, Xu C, Macpherson B, et al. Inhibition of stress-activated MAP kinases induces clinical improvement in moderate to severe Crohn's disease. *Gastroenterology* 2002;**122**:7–14.
49. Schett G, Tohidast-Akrad M, Smolen JS, Schmid BJ, Steiner CW, Bitzan P, et al. Activation, differential localization, and regulation of the stress-activated protein kinases, extracellular signal-regulated kinase, c-JUN N-terminal kinase, and p38 mitogen-activated protein kinase, in synovial tissue and cells in rheumatoid arthritis. *Arthritis Rheum* 2000;**43**:2501–12.
50. Mun SH, Kim JW, Nah SS, Ko NY, Lee JH, Kim JD, et al. Tumor necrosis factor alpha-induced interleukin-32 is positively regulated via the Syk/protein kinase Cdelta/JNK pathway in rheumatoid synovial fibroblasts. *Arthritis Rheum* 2009;**60**:678–85.

51. Molad Y, Amit-Vasina M, Bloch O, Yona E, Rapoport MJ. Increased ERK and JNK activities correlate with disease activity in patients with systemic lupus erythematosus. *Ann Rheum Dis* 2010;**69**:175–80.
52. Nakou M, Bertsias G, Stagakis I, Centola M, Tassiulas I, Hatziapostolou M, et al. Gene network analysis of bone marrow mononuclear cells reveals activation of multiple kinase pathways in human systemic lupus erythematosus. *PLoS One* 2010;**5**:e13351.
53. Wong CK, Wong PT, Tam LS, Li EK, Chen DP, Lam CW. Activation profile of intracellular mitogen-activated protein kinases in peripheral lymphocytes of patients with systemic lupus erythematosus. *J Clin Immunol* 2009;**29**:738–46.
54. Oikonomidou O, Vlachoyiannopoulos PG, Kominakis A, Kalofoutis A, Moutsopoulos HM, Moutsatsou P. Glucocorticoid receptor, nuclear factor kappaB, activator protein-1 and C-jun N-terminal kinase in systemic lupus erythematosus patients. *Neuroimmunomodulation* 2006;**13**:194–204.
55. Broide E, Scapa E, Bloch O, Shapiro M, Kimchi NA, Ben-Yehudah G, et al. Evidence for aberrant regulation of MAP kinase signal transduction pathway in peripheral blood mononuclear cells in patients with active celiac disease. *Dig Dis Sci* 2009;**54**:1270–5.
56. Gulmann C, Sheehan KM, Conroy RM, Wulfkuhle JD, Espina V, Mullarkey MJ, et al. Quantitative cell signalling analysis reveals down-regulation of MAPK pathway activation in colorectal cancer. *J Pathol* 2009;**218**:514–9.
57. Chen Z, Yang A, Xu C, Xing Y, Gong W, Li J. c-Jun N-terminal kinase is involved in the regulation of proliferation and apoptosis by integrin-linked kinase in human retinoblastoma cells. *Graefes Arch Clin Exp Ophthalmol* 2011;**249**:1399–407.
58. Jørgensen K, Davidson B, Flørenes VA. Activation of c-jun N-terminal kinase is associated with cell proliferation and shorter relapse-free period in superficial spreading malignant melanoma. *Mod Pathol* 2006;**19**:1446–55.
59. Wang X, Chao L, Li X, Ma G, Chen L, Zang Y, et al. Elevated expression of phosphorylated c-Jun NH2-terminal kinase in basal-like and "triple-negative" breast cancers. *Hum Pathol* 2010;**41**:401–6.
60. Ødegaard E, Staff AC, Abeler VM, Kopolovic J, Onsrud M, Lazarovici P, et al. The activated nerve growth factor receptor p-TrkA is selectively expressed in advanced-stage ovarian carcinoma. *Hum Pathol* 2007;**38**:140–6.
61. Chung J, Nguyen AK, Henstridge DC, Holmes AG, Chan MH, Mesa JL, et al. HSP72 protects against obesity-induced insulin resistance. *Proc Natl Acad Sci USA* 2008;**105**:1739–44.
62. Takamura T, Honda M, Sakai Y, Ando H, Shimizu A, Ota T, et al. Gene expression profiles in peripheral blood mononuclear cells reflect the pathophysiology of type 2 diabetes. *Biochem Biophys Res Commun* 2007;**361**:379–84.
63. Boden G, Duan X, Homko C, Molina EJ, Song W, Perez O, et al. Increase in endoplasmic reticulum stress-related proteins and genes in adipose tissue of obese, insulin-resistant individuals. *Diabetes* 2008;**57**:2438–44.
64. Margutti P, Matarrese P, Conti F, Colasanti T, Delunardo F, Capozzi A, et al. Autoantibodies to the C-terminal subunit of RLIP76 induce oxidative stress and endothelial cell apoptosis in immune-mediated vascular diseases and atherosclerosis. *Blood* 2008;**111**:4559–70.
65. Sabapathy K, Jochum W, Hochedlinger K, Chang L, Karin M, Wagner EF. Defective neural tube morphogenesis and altered apoptosis in the absence of both JNK1 and JNK2. *Mech Dev* 1999;**89**:115–24.
66. Weston CR, Wong A, Hall JP, Goad ME, Flavell RA, Davis RJ. The c-Jun NH2-terminal kinase is essential for epidermal growth factor expression during epidermal morphogenesis. *Proc Natl Acad Sci USA* 2004;**28**:14114–9.
67. Chen JT, Lu DH, Chia CP, Ruan DY, Sabapathy K, Xiao ZC. Impaired long-term potentiation in c Jun N-terminal kinase 2-deficient mice. *J Neurochem* 2005;**93**:463–73.

68. Sabapathy K, Hu Y, Kallunki T, Schreiber M, David JP, Jochum W, et al. JNK2 is required for efficient T-cell activation and apoptosis but not for normal lymphocyte development. *Curr Biol* 1999;**9**:116–25.
69. Sabapathy K, Kallunki T, David JP, Graef I, Karin M, Wagner EF. c-Jun NH2-terminal kinase (JNK)1 and JNK2 have similar and stage-dependent roles in regulating T cell apoptosis and proliferation. *J Exp Med* 2001;**193**:317–28.
70. David JP, Sabapathy K, Hoffmann O, Idarraga MH, Wagner EF. JNK1 modulates osteoclastogenesis through both c-Jun phosphorylation-dependent and -independent mechanisms. *J Cell Sci* 2002;**115**:4317–25.
71. Yang DD, Conze D, Whitmarsh AJ, Barrett T, Davis RJ, Rincón M, et al. Differentiation of CD4+ T cells to Th1 cells requires MAP kinase JNK2. *Immunity* 1998;**9**:575–85.
72. Dong C, Yang DD, Wysk M, Whitmarsh AJ, Davis RJ, Flavell RA. Defective T cell differentiation in the absence of Jnk1. *Science* 1998;**282**:2092–5.
73. Yang DD, Kuan CY, Whitmarsh AJ, Rincón M, Zheng TS, Davis RJ, et al. Absence of excitotoxicity-induced apoptosis in the hippocampus of mice lacking the Jnk3 gene. *Nature* 1997;**389**:865–70.
74. Hirosumi J, Tuncman G, Chang L, Görgün CZ, Uysal KT, Maeda K, et al. A central role for JNK in obesity and insulin resistance. *Nature* 2002;**420**:333–6.
75. Li XM, Li CC, Yu SS, Chen JT, Sabapathy K, Ruan DY. JNK1 contributes to metabotropic glutamate receptor-dependent long-term depression and short-term synaptic plasticity in the mice area hippocampal CA1. *Eur J Neurosci* 2007;**25**:391–6.
76. Xu P, Davis RJ. c-Jun NH2-terminal kinase is required for lineage-specific differentiation but not stem cell self-renewal. *Mol Cell Biol* 2010;**30**:1329–40.
77. Ries V, Silva RM, Oo TF, Cheng HC, Rzhetskaya M, Kholodilov N, et al. JNK2 and JNK3 combined are essential for apoptosis in dopamine neurons of the substantia nigra, but are not required for axon degeneration. *J Neurochem* 2008;**107**:1578–88.
78. Dong C, Yang DD, Tournier C, Whitmarsh AJ, Xu J, Davis RJ, et al. JNK is required for effector T-cell function but not for T-cell activation. *Nature* 2000;**405**:91–4.
79. Tafolla E, Wang S, Wong B, Leong J, Kapila YL. JNK1 and JNK2 oppositely regulate p53 in signaling linked to apoptosis triggered by an altered fibronectin matrix: JNK links FAK and p53. *J Biol Chem* 2005;**280**:19992–9.
80. Gururajan M, Chui R, Karuppannan AK, Ke J, Jennings CD, Bondada S. c-Jun N-terminal kinase (JNK) is required for survival and proliferation of B-lymphoma cells. *Blood* 2005;**106**:1382–91.
81. Lizundia R, Chaussepied M, Huerre M, Werling D, Di Santo JP, Langsley G. c-Jun NH2-terminal kinase/c-Jun signaling promotes survival and metastasis of B lymphocytes transformed by Theileria. *Cancer Res* 2006;**66**:6105–10.
82. Galley Y, Hagens G, Glaser I, Davis W, Eichhorn M, Dobbelaere D. Jun NH2-terminal kinase is constitutively activated in T cells transformed by the intracellular parasite Theileria parva. *Proc Natl Acad Sci USA* 1997;**94**:5119–24.
83. Han Z, Boyle DL, Chang L, Bennett B, Karin M, Yang L, et al. c-Jun N-terminal kinase is required for metalloproteinase expression and joint destruction in inflammatory arthritis. *J Clin Invest* 2001;**108**:73–81.
84. Shin T, Ahn M, Jung K, Heo S, Kim D, Jee Y, et al. Activation of mitogen-activated protein kinases in experimental autoimmune encephalomyelitis. *J Neuroimmunol* 2003;**14**:118–25.
85. Stambe C, Atkins RC, Hill PA, Nikolic-Paterson DJ. Activation and cellular localization of the p38 and JNK MAPK pathways in rat crescentic glomerulonephritis. *Kidney Int* 2003;**64**:2121–32.
86. Kawakami Y, Kitaura J, Hata D, Yao L, Kawakami T. Functions of Bruton's tyrosine kinase in mast and B cells. *J Leukoc Biol* 1999;**65**:86–90.

87. Constant SL, Dong C, Yang DD, Wysk M, Davis RJ, Flavell RA. JNK1 is required for T cell-mediated immunity against Leishmania major infection. *J Immunol* 2000;**165**:2671–6.
88. Morishima Y, Gotoh Y, Zieg J, Barrett T, Takano H, Flavell R, et al. Beta-amyloid induces neuronal apoptosis via a mechanism that involves the c-Jun N-terminal kinase pathway and the induction of Fas ligand. *J Neurosci* 2001;**21**:7551–60.
89. Sumara G, Belwal M, Ricci R. "Jnking" atherosclerosis. *Cell Mol Life Sci* 2005;**62**:487–94.
90. Ricci R, Sumara G, Sumara I, Rozenberg I, Kurrer M, Akhmedov A, et al. Requirement of JNK2 for scavenger receptor A-mediated foam cell formation in atherogenesis. *Science* 2004;**306**:1558–61.
91. Aguirre V, Uchida T, Yenush L, Davis R, White MF. The c-Jun NH(2)-terminal kinase promotes insulin resistance during association with insulin receptor substrate-1 and phosphorylation of Ser(307). *J Biol Chem* 2000;**275**:9047–54.
92. Schattenberg JM, Singh R, Wang Y, Lefkowitch JH, Rigoli RM, Scherer PE, et al. JNK1 but not JNK2 promotes the development of steatohepatitis in mice. *Hepatology* 2006;**43**:163–72.
93. Kluwe J, Pradere JP, Gwak GY, Mencin A, De Minicis S, Osterreicher CH, et al. Modulation of hepatic fibrosis by c-Jun-N-terminal kinase inhibition. *Gastroenterology* 2010;**138**:347–59.
94. Liang Q, Bueno OF, Wilkins BJ, Kuan CY, Xia Y, Molkentin JD. c-Jun N-terminal kinases (JNK) antagonize cardiac growth through cross-talk with calcineurin-NFAT signaling. *EMBO J* 2003;**22**:5079–89.
95. Tachibana H, Perrino C, Takaoka H, Davis RJ, Naga Prasad SV, Rockman HA. JNK1 is required to preserve cardiac function in the early response to pressure overload. *Biochem Biophys Res Commun* 2006;**343**:1060–6.
96. Wagner EF, Nebreda AR. Signal integration by JNK and p38 MAPK pathways in cancer development. *Nat Rev Cancer* 2009;**9**:537–49.
97. Behrens A, Jochum W, Sibilia M, Wagner EF. Oncogenic transformation by ras and fos is mediated by c-Jun N-terminal phosphorylation. *Oncogene* 2000;**19**:2657–63.
98. Nateri AS, Spencer-Dene B, Behrens A. Interaction of phosphorylated c-Jun with TCF4 regulates intestinal cancer development. *Nature* 2005;**437**:281–5.
99. Bogoyevitch MA, Arthur PG. Inhibitors of c-Jun N-terminal kinases: JuNK no more? *Biochim Biophys Acta* 2008;**1784**:76–93.
100. Vukic V, Callaghan D, Walker D, Lue LF, Liu QY, Couraud PO, et al. Expression of inflammatory genes induced by beta-amyloid peptides in human brain endothelial cells and in Alzheimer's brain is mediated by the JNK-AP1 signaling pathway. *Neurobiol Dis* 2009;**34**:95–106.
101. Mishra S, Mishra JP, Kumar A. Activation of JNK-dependent pathway is required for HIV viral protein R-induced apoptosis in human monocytic cells: involvement of antiapoptotic BCL2 and c-IAP1 genes. *J Biol Chem* 2007;**282**:4288–300.
102. Zheng JC, Huang Y, Tang K, Cui M, Niemann D, Lopez A, et al. HIV-1-infected and/or immune-activated macrophages regulate astrocyte CXCL8 production through IL-1beta and TNF-alpha: involvement of mitogen-activated protein kinases and protein kinase R. *J Neuroimmunol* 2008;**200**:100–10.
103. Holmén C, Elsheikh E, Christensson M, Liu J, Johansson AS, Qureshi AR, et al. Anti endothelial cell autoantibodies electively activate SAPK/JNK signalling in Wegener's granulomatosis. *J Am Soc Nephrol* 2007;**18**:2497–508.
104. Eynott PR, Xu L, Bennett BL, Noble A, Leung SY, Nath P, et al. Effect of an inhibitor of Jun N-terminal protein kinase, SP600125, in single allergen challenge in sensitized rats. *Immunology* 2004;**112**:446–53.
105. Colombo A, Repici M, Pesaresi M, Santambrogio S, Forloni G, Borsello T. The TAT-JNK inhibitor peptide interferes with beta amyloid protein stability. *Cell Death Differ* 2007;**14**:1845–8.

106. Beckham JD, Goody RJ, Clarke P, Bonny C, Tyler KL. Novel strategy for treatment of viral central nervous system infection by using a cell-permeating inhibitor of c-Jun N-terminal kinase. *J Virol* 2007;**81**:6984–92.
107. Zhuang ZY, Wen YR, Zhang DR, Borsello T, Bonny C, Strichartz GR, et al. A peptide c-Jun N-terminal kinase (JNK) inhibitor blocks mechanical allodynia after spinal nerve ligation: respective roles of JNK activation in primary sensory neurons and spinal astrocytes for neuropathic pain development and maintenance. *J Neurosci* 2006;**26**:3551–60.
108. Touchard E, Omri S, Naud MC, Berdugo M, Deloche C, Abadie C, et al. A peptide inhibitor of c-Jun N-terminal kinase for the treatment of endotoxin-induced uveitis. *Invest Ophthalmol Vis Sci* 2010;**51**:4683–93.
109. Ngoei KR, Catimel B, Church N, Lio DS, Dogovski C, Perugini MA, et al. Characterization of a novel JNK (c-Jun N-terminal kinase) inhibitory peptide. *Biochem J* 2011;**24**:399–413.
110. Maedler K, Schulthess FT, Bielman C, Berney T, Bonny C, Prentki M, et al. Glucose and leptin induce apoptosis in human beta-cells and impair glucose-stimulated insulin secretion through activation of c-Jun N-terminal kinases. *FASEB J* 2008;**22**:1905–13.
111. Saporito MS, Brown EM, Miller MS, Carswell S. CEP-1347/KT-7515, an inhibitor of c-jun N-terminal kinase activation, attenuates the 1-methyl-4-phenyl tetrahydropyridine-mediated loss of nigrostriatal dopaminergic neurons *in vivo*. *J Pharmacol Exp Ther* 1999;**288**:421–7.
112. Eggert D, Dash PK, Gorantla S, Dou H, Schifitto G, Maggirwar SB, et al. Neuroprotective activities of CEP-1347 in models of neuroAIDS. *J Immunol* 2010;**184**:746–56.
113. Chialda L, Zhang M, Brune K, Pahl A. Inhibitors of mitogen-activated protein kinases differentially regulate costimulated T cell cytokine production and mouse airway eosinophilia. *Respir Res* 2005;**15**:6–36.
114. Toyoda H, Zhao MG, Xu H, Wu LJ, Ren M, Zhuo M. Requirement of extracellular signal-regulated kinase/mitogen-activated protein kinase for long-term potentiation in adult mouse anterior cingulate cortex. *Mol Pain* 2007;**3**:36.
115. Guan QH, Pei DS, Liu XM, Wang XT, Xu TL, Zhang GY. Neuroprotection against ischemic brain injury by SP600125 via suppressing the extrinsic and intrinsic pathways of apoptosis. *Brain Res* 2006;**1092**:36–46.
116. Dou J, Li X, Cai Y, Chen H, Zhu S, Wang Q, et al. Human cytomegalovirus induces caspase-dependent apoptosis of megakaryocytic CHRF-288-11 cells by activating the JNK pathway. *Int J Hematol* 2010;**91**:620–9.
117. Wang Y, Zhang Y, Wei Z, Li H, Zhou H, Zhang Z, et al. JNK inhibitor protects dopaminergic neurons by reducing COX-2 expression in the MPTP mouse model of subacute Parkinson's disease. *J Neurol Sci* 2009;**285**:172–7.
118. Cho H, Black SC, Looper D, Shi M, Kelly-Sullivan D, Timofeevski S, et al. Pharmacological characterization of a small molecule inhibitor of c-Jun kinase. *Am J Physiol Endocrinol Metab* 2008;**295**:1142–51.
119. Ferrandi C, Ballerio R, Gaillard P, Giachetti C, Carboni S, Vitte PA, et al. Inhibition of c-Jun N-terminal kinase decreases cardiomyocyte apoptosis and infarct size after myocardial ischemia and reperfusion in anaesthetized rats. *Br J Pharmacol* 2004;**142**:953–60.
120. Carboni S, Boschert U, Gaillard P, Gotteland JP, Gillon JY, Vitte PA. AS601245, a c-Jun NH2-terminal kinase (JNK) inhibitor, reduces axon/dendrite damage and cognitive deficits after global cerebral ischaemia in gerbils. *Br J Pharmacol* 2008;**153**:157–63.
121. Carboni S, Hiver A, Szyndralewiez C, Gaillard P, Gotteland JP, Vitte PA. AS601245 (1,3-benzothiazol-2-yl (2-[[2-(3-pyridinyl) ethyl] amino]-4 pyrimidinyl) acetonitrile): a c-Jun NH2-terminal protein kinase inhibitor with neuroprotective properties. *J Pharmacol Exp Ther* 2004;**310**:25–32.
122. Uehara T, Bennett B, Sakata ST, Satoh Y, Bilter GK, Westwick JK, et al. JNK mediates hepatic ischemia reperfusion injury. *J Hepatol* 2005;**42**:850–9.

123. Borsello T, Clarke PG, Hirt L, Vercelli A, Repici M, Schorderet DF, et al. A peptide inhibitor of c-Jun N-terminal kinase protects against excitotoxicity and cerebral ischemia. *Nat Med* 2003;**9**:1180–6.
124. Hirt L, Badaut J, Thevenet J, Granziera C, Regli L, Maurer F, et al. D-JNKI1, a cell-penetrating c-Jun-N-terminal kinase inhibitor, protects against cell death in severe cerebral ischemia. *Stroke* 2004;**35**:1738–43.
125. Ijaz A, Tejada T, Catanuto P, Xia X, Elliot SJ, Lenz O, et al. Inhibition of C-jun N-terminal kinase improves insulin sensitivity but worsens albuminuria in experimental diabetes. *Kidney Int* 2009;**75**:381–8.
126. Kaneto H, Nakatani Y, Miyatsuka T, Kawamori D, Matsuoka TA, Matsuhisa M, et al. Possible novel therapy for diabetes with cell-permeable JNK-inhibitory peptide. *Nat Med* 2004;**10**:1128–32.
127. Milano G, Morel S, Bonny C, Samaja M, von Segesser LK, Nicod P, et al. A peptide inhibitor of c-Jun NH2-terminal kinase reduces myocardial ischemia-reperfusion injury and infarct size *in vivo*. *Am J Physiol Heart Circ Physiol* 2007;**292**:H1828–35.
128. Wang J, Van De Water TR, Bonny C, de Ribaupierre F, Puel JL, Zine A. A peptide inhibitor of c-Jun N-terminal kinase protects against both aminoglycoside and acoustic trauma-induced auditory hair cell death and hearing loss. *J Neurosci* 2003;**23**:8596–607.
129. Wang J, Ruel J, Ladrech S, Bonny C, van de Water TR, Puel JL. Inhibition of the c-Jun N-terminal kinase-mediated mitochondrial cell death pathway restores auditory function in sound-exposed animals. *Mol Pharmacol* 2007;**71**:654–66.
130. Gunawan BK, Liu ZX, Han D, Hanawa N, Gaarde WA, Kaplowitz N. c-Jun N-terminal kinase plays a major role in murine acetaminophen hepatotoxicity. *Gastroenterology* 2006;**131**:165–78.
131. Lee KH, Kim SE, Lee YS. SP600125, a selective JNK inhibitor, aggravate hepatic ischemia-reperfusion injury. *Exp Mol Med* 2006;**31**:408–16.
132. Kyoi S, Otani H, Matsuhisa S, Akita Y, Tatsumi K, Enoki C, et al. Opposing effect of p38 MAP kinase and JNK inhibitors on the development of heart failure in the cardiomyopathic hamster. *Cardiovasc Res* 2006;**69**:888–98.
133. Si X, Luo H, Morgan A, Zhang J, Wong J, Yuan J, et al. Stress-activated protein kinases are involved in coxsackievirus B3 viral progeny release. *J Virol* 2005;**79**:13875–81.
134. Tanemura S, Momose H, Shimizu N, Kitagawa D, Seo J, Yamasaki T, et al. Blockage by SP600125 of Fcepsilon receptor-induced degranulation and cytokine gene expression in mast cells is mediated through inhibition of phosphatidylinositol 3-kinase signalling pathway. *J Biochem* 2009;**145**:345–54.
135. Bain J, McLauchlan H, Elliott M, Cohen P. The specificities of protein kinase inhibitors: an update. *Biochem J* 2003;**371**:199–204.
136. Holzberg D, Knight CG, Dittrich-Breiholz O, Schneider H, Dörrie A, Hoffmann E, et al. Disruption of the c-JUN-JNK complex by a cell-permeable peptide containing the c-JUN delta domain induces apoptosis and affects a distinct set of interleukin-1-induced inflammatory genes. *J Biol Chem* 2003;**278**:40213–23.
137. Suckfuell M, Canis M, Strieth S, Scherer H, Haisch A. Intratympanic treatment of acute acoustic trauma with a cell-permeable JNK ligand: a prospective randomized phase I/II study. *Acta Otolaryngolc* 2007;**127**:938–42.
138. Parkinson Study Group PRECEPT Investigators. Mixed lineage kinase inhibitor CEP-1347 fails to delay disability in early Parkinson disease. *Neurology* 2007;**9**:69480–90.
139. Xie M, Sabapathy K. Tyrosine 170 is dispensable for c-Jun turnover. *Cell Signal* 2010;**22**:330–7.

# PAKs in Human Disease

Perry M. Chan and Ed Manser

*sGSK Group, A-Star Neuroscience Research Partnership, Singapore, Singapore*



The p21activated kinases (PAKs) are one of the first direct kinase targets of Ras-related small GTPases to be discovered and have emerged as central players in growth factor signaling networks that regulate morphogenetic processes. In some situations, PAKs control cell proliferation, but their wider role involves establishing cell polarity and promoting cellular plasticity via changes in the actin cytoskeleton. PAKs have been shown to impact on three important areas of human health, namely, cancer, brain function, and virus infection. We review the mechanisms and targets of PAKs in these contexts and provide an overview of the ways in which inhibitors might act to arrest tumor growth, combat virus infection, and promote cell apoptosis. Although PAKs are most abundant in the brain, there are few details of how they might be operating in this context. The advent of new and more selective PAK inhibitors promises new avenues of treatment and allows us to probe in greater detail the importance of PAK biology.

## I. Introduction

The Rho-family GTPases signal by direct binding to a variety of well-studied "effector" protein kinases including ACK1, MRCK, ROCK, MLKs, and PAKs, as reviewed in Ref. 1. Of these, the p21-activated kinases (PAKs) represent the prototype, as they were first to be identified,[2] and the direct mechanism of activation of PAK1 by Cdc42- or Rac1-related proteins is well understood.[3] In budding yeast, PAK-like kinases Ste20p and Cla4p play multiple roles in cell cycle and polarity downstream of Cdc42.[4,5] Mammalian PAKs also contribute to

Progress in Molecular Biology
and Translational Science, Vol. 106
DOI: 10.1016/B978-0-12-396456-4.00011-0

cell cycle progression and the plasticity of the actin cytoskeleton, with potential to contribute to both cancer initiation and progression.[6] The brain is the site of constant cytoskeletal reorganization, and therefore, alterations in PAK signaling in pathological states are not unexpected. PAK1 and PAK3 are highly expressed in the brain and promote synaptic plasticity.[7] It is notable that PAKs form a complex with two other proteins, namely, PAK-interacting exchange factor (PIX) and GIT1,[8] which allow the kinase to associate with cell adhesions,[9,10] the centrosome,[11,12] and specialized compartments such as the postsynaptic density membrane[13] and immune synapse.[14] The targeting of group I PAKs to specific cellular sites is shown schematically in Fig. 1. Both the PIX, through its C-terminal PDZ-binding motif, and the PIX partner GIT1, through its paxillin-binding domain, are able to take the kinase to cell–cell junctions and focal adhesions, respectively. PAK1 has also been found enriched at the microtubule organizing center or centrosome through GIT1.

The potential role of PAKs in cancer progression has been extensively reviewed in Ref. 15; in this review, we focus on the role of known substrates of PAKs in this process and highlight the controversies associated with some of these studies. The mammalian PAKs fall into two distinct subgroups based on distinct catalytic and regulatory domain homology. Mammalian PAK1-3 comprises the better understood group I PAKs that are directly activated by interaction with the GTP-bound forms of Rac1 and Cdc42 (and related family members such as Chp, Rac3, TL10, and TCL). Binding of these proteins releases the autoinhibitory domain (AID) from the kinase domain and permits autophosphorylation. PAKs 1–3 are coupled to an SH3-containing adaptor termed PIX, which is a Rac1 activator.[16]

The PAK4--6 (group II) kinases are differentiated because they neither bind to Rac1 nor undergo Cdc42-stimulated autophosphorylation and thus kinase activation.[17] These group II kinases are not known to be modulated by protein adaptors but do enter the nucleus due to an N-terminally located basic cluster. The PAKs are basophilic and phosphorylate consensus sites defined from peptide arrays as [R/K][K/R]×[S/T]×[Y/F] for the group I and [R/K][K/R]×[S/T][Y/F] for the group II kinases.[18] They are also unusual in tolerating (or preferring) acidic residues C-terminal to the phosphate acceptor site (i.e., at position +1 and +2) unlike other basic directed kinases such as PKC and CamKII. Aside from siRNA-mediated knockdown of one or multiple PAK isoforms, most credible information regarding PAK function has been gleaned through the use of the PAK1 AID which exhibits a $K_i$ of ~100nM toward PAK1 activation *in vitro*.[6] The GST-AID derived from PAK1 does exert effects independent of PAK1 inhibition[19] because it avidly binds fragile-X proteins.[20] Fortunately, the PAK2 AID does not have this unwanted side effect and is most potent when expressed as a GST fusion protein (P.M. Chan and E. Manser, unpublished observations).

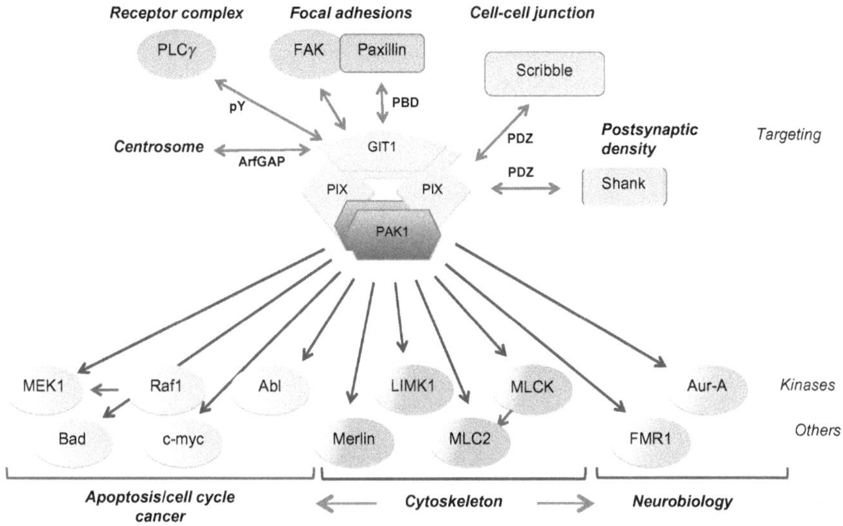


FIG. 1. The targeting and targets of group I PAKs. PAKs activate a variety of signaling networks, and as very little is known regarding the coupling of PAK1 to such complexes, these pathways are omitted for clarity. The group 1 PAKs (as dimers) form a complex with the SH3 domain of PIX: this Rac1 activation protein (RacGEF), in turn, binds the ArfGAP protein GIT1.[8] This allows the kinase the complex to associate via the protein–protein interaction motifs as indicated by cell adhesions or junctions,[9,10] the centrosome,[11,12] and specialized compartments such as the postsynaptic density membrane.[13] Group I PAKs (here shown as PAK1) can activate the mitogen-activated protein kinase (MAPK) pathway by phosphorylating Raf1 and MEK1. These signals are common in all cell types to allow cell proliferation. In addition, both groups of PAKs regulate survival pathways through inactivation of the proapoptotic protein mitochondrial protein BAD, which is acted on also by Raf1. PAK1 also phosphorylates a number of regulators of the cytoskeleton such as myosin light-chain kinase (MLCK) and its target the myosin II regulatory light chain (MLC2), LIM domain kinase (LIMK1), and the tumor suppressor merlin (NF2). PAKs also regulate centrosome duplication through activation of Aurora A (Aur-A), which is an important kinase in the brain to regulate local protein translation. The fragile-X mental retardation protein (FMR1) binds selectively to activated PAK1 as part of a complex that couples protein translation to synaptic activity. (See Color Insert.)

There has been significant interest in the development of drugs against PAKs for reasons that are outlined later in this chapter. The only group I-specific PAK inhibitor that has been broadly tested against other protein kinases is IPA-3. This small-molecule PAK inhibitor inhibits Cdc42-mediated PAK1 activation, rather than acting at the ATP binding site.[21] The mechanism of action of IPA-3 involves binding and covalent attachment to the PAK1 regulatory domain,[22] though this reaction is reversible. As the compound contains a disulfide bond, it is inactivated inside cells, rendering it unsuitable for therapeutic use. A recent paper[23] characterizes an orally administrable compound developed by Pfizer, termed PF-3758309. This exhibits potent binding and inhibition ($K_i$~5–15nM) of PAKs in an ATP-competitive manner. PF-3758309 acts on both group I and II PAKs as well as AMPK and RSK kinases at nanomolar concentrations.[23]

In this chapter, we outline the role of PAKs in various pathological conditions and, where known, point to the specific protein substrates that play roles in these processes. Because of ambiguity with respect to amino acid numbering of such substrates, where possible details of the phosphorylation site sequences are provided.

# II. PAKs and Cancer

## A. Is PAK an Oncogenic Driver?

Neurofibromatosis type 2 (NF2) involves Schwann cell hyperproliferation due to the lack of merlin, which is a negative regulator of PAK1: this disease might therefore benefit from PAK inhibition. Merlin Ser-518 (MKRLp**S**MEIE) is also a PAK1 target. Benign schwannomas that arise in patients with NF2 have elevated levels of phosphorylated (active) PAK1[24] and such phosphorylated PAK1 is found enriched in the nuclei of malignant cells.[25] A number of mutant cell lines developed for NF2 preclinical studies, including rat schwannoma cells, are sensitive to inhibition of PAKs.[24] These experiments are held back by the current lack of "specific" PAK inhibitor(s), which also restricts experimental scope. Nonetheless, merlin-deficient human mesothelioma cell lines are hypersensitive to treatments aimed at PAK1.[26]

In breast cancer, increased cytoplasmic levels of PAK1 have been correlated with recurrence rate and mortality.[27,28] Changes in the levels of PAKs can be due to amplification of the corresponding gene locus, or transcriptionally elevated levels of the mRNA and thus protein, as recently reviewed in Ref. 29. A more recent study in which PAK1 genomic copy number and expression were determined for a large panel of breast, lung, and head and neck tumors showed that PAK1 genomic amplification at 11q13 was prevalent in luminal breast cancer.[30] This study also found strong PAK1 expression in some squamous non-small-cell lung carcinomas. Similarly, in patients with gastric cancer, higher levels of PAK1 associate with advanced tumor stages and metastasis,[31] and overexpression of PAK1 is observed in 70% of colon cancer samples.[32] In summary, PAK1 levels are credibly increased in certain cancers. As PAK2 is the ubiquitous group 1 kinase, more attention needs to be paid to what is happening to this kinase isoform in the situation when PAK1 levels are altered.

How might group I PAKs be operating in cancer cells? It was long thought that PAK1 mediated activated Ras signaling to drive Akt activation.[33] Activated versions of Akt itself can relieve rodent fibroblasts from contact inhibition, and this process requires the activity of PAK1.[34] Reduction of PAK1 activity in these cells apparently dissociates the activation of Akt from that of the ERKs.[34] It is thus suggested that PAK1 links PI3K/Akt to the Raf/MEK/ERK pathways. In some cells, PAK1 could keep the MAP kinase cascade active by its ability to target Raf1 at Ser338[35,36] and MEK at Ser298.[37] It is notable that the Myc proto-oncogene has been invoked as downstream of PAK2.[38] The Myc Ser373 (ELKRp**S**FFAL) and Ser400 (LKKAp**T**AYIL) sites conform to canonical PAK1 target sequences, and interestingly, Myc can upregulate PAK1 and Cdc42 suggesting a feedback loop.[39] Similarly, Abl S619 (PKRSp**S**SFRE) has been identified as a PAK2 site[40] which could play a role in PAK-dependent proliferation. Some of the signals that emanate from PAK1 are illustrated in Fig. 1: clearly, changes to the cytoskeleton impinge directly on the growth and metastatic potential of cancer cells.

If PAKs can cooperate with other signals to drive oncogenic signaling pathways, their direct activation might recapitulate some features of oncogenic transformation. A number of studies have tested overexpressing an "activated" kinase PAK1(T423E), which is a phospho-mimetic replacing the sole phosphorylated residue in the kinase domain. PAK1(T423E) drives anchorage-independent growth in breast cancer cell lines,[41] hyperplasia of mammary epithelium in mice,[42] and breast carcinomas[43] in transgenic mice. Curiously, we have shown that this mutant PAK1(T423E) is actually not active in cells or *in vitro*.[44] The effects of this mutant kinase seen in transgenic animal may therefore arise as a result of compensation in the tissue, or rather be due to a blockage in the PAK1 signaling pathway.

There is good evidence for misregulated PAK4 in human cancer lines,[45] and in a large-scale study, PAK5 was identified as containing somatic mutations which could act as a cancer "driver".[46] Similar to PAK1 (as discussed above), the PAK4 locus is present on an amplicon associated with colorectal and pancreatic cancers.[47,48] In humans, PAK4 is expressed at low levels in normal tissue, but its expression is upregulated in Ras-dependent human tumors.[47–49]

## B. Is PAK Antiapoptotic?

The involvement of group I PAKs in protection from programmed cell death and anoikis is well documented.[50–53] The specific mechanism of anti-apoptotic protection widely reported involved direct inactivation of the proapoptotic Bcl-2-family member Bad[54] via phosphorylation of its Serine-112 and Ser-136. It was later suggested that these are actually poor targets for PAK1 *in vivo* and the kinase might be acting indirectly through cellular Raf-1.[55]

In normal mammary epithelium cells, the protective function of $\alpha6\beta4$ integrin (to block cell death) is mediated through Rac1 and PAK1, which, in turn, activate NFκB.[56]

PAK4 has also been invoked to protect cells against apoptotic cell death by phosphorylating the human Bad at the same sites (Ser 112/136) and by inhibiting caspase activation.[57,58] However, cell line studies indicate that the anti-apoptotic effect is kinase independent.[57,59] Pathway profiling of the Pfizer PAK inhibitor PF-3758309 suggested that kinase links to both the apoptosis and the cytoskeleton,[23] in keeping with a role for group I and group II PAKs in preventing apoptosis of cancer and perhaps normal cells.

## C. PAK Signaling and Cell Cycle Progression

Phosphorylation plays a pivotal role in the cell cycle progression and checkpoints, with kinases such as cyclin-dependent kinases, polo-like kinases, and Aurora kinases being examples of key players. PAKs could therefore drive tumorigenesis by affecting cell cycle progression. PAK1 colocalizes with condensing chromatin and phosphorylates histone H3 at Ser-10 (a site in common with Aurora kinase).[60] In mitotic cells, phosphorylated PAK1 is preferentially localized on centrosomes and spindles: we have shown that the GIT1/PIX complex (but not Cdc42) plays a role in localizing PAK1 to the centrosome, where it undergoes activation without the need for small G-proteins.[12] Once activated, PAK binds to and phosphorylates two centrosome-associated kinases: Aurora kinase A on Thr-288 and Ser-342 (key sites for Aurora kinase activation) and polo-like kinase 1 on Ser-49 [VDPRp**S**RRRY].[12,61] As mitosis progresses, PAK1 moves to the spindle midbody and, finally, to the contractile ring during cytokinesis.[60]

One interesting new target for group 1 PAKs is the atypical MAPK ERK3 which has an unusual activation loop: classical MAP kinases ERK1/ERK2, JNK, p38, and ERK5 all contain the conserved T-x-Y motif in their activation loop which is dually phosphorylated by members of the MAP kinase kinases family. Two groups independently reported that PAK1/2/3 modifies ERK3 or ERK4 on Ser-189/186 [KGHLpSEGLV], the site needed for kinase activation.[62,63] This site somewhat conforms to the consensus for group 1 PAKs. The role of ERK3/4 in cell proliferation is not yet resolved, but one downstream target is MAP kinase-activated protein kinase 5 (MK5).

Both PAK1 and PAK4 have been identified as phosphorylating a RhoA GEF termed GEF-H1 at Ser-885 [PRRRpSLPAG] which generates a binding site for 14-3-3[64,65] and displaces it from microtubules. GEF-H1 (also termed Lfc) is involved in microtubule dynamics in the cell cycle,[66] and when mutated causes oncogenic transformation.[67,68] Another connection to mitotic microtubules is that Ran Ser-135 acts as a substrate for PAK4 which impedes its binding to the well-established effectors RCC1 and RanGAP1.[69]

## D. PAKs, the Cytoskeleton, and Cell Invasion

Tumor metastasis, which involves the invasion of primary tumor cells through tissue and the extracellular matrix (ECM), requires considerable cytoskeleton remodeling. Many Rho signaling cascades trigger changes in the cytoskeletal dynamics leading to cell motility.[70] Activated PAK induces loss of stress fibers and focal adhesion complexes[71] and promotes cell extension.[72] The phosphorylation targets that affect these changes to the cytoskeleton include myosin light-chain kinase (MLCK), paxillin, filamin A, cortactin, LIM kinases, and Arpc1b (as reviewed in Ref. 73). The phophorylation of the PAK partners βPIX and GIT1 Ser-517 [RQAFpSMYEP] contributes to the dynamics of the complex.[12] Both PAK1 and PAK4 are said to activate the important cofilin kinase LIMK1[74–76]: LIMK1 phosphorylate cofilin at Ser3 which prevents it from interacting with F-actin, thereby stabilizing actin filaments. PAKs activate LIMK1 by phosphorylating it at Thr-508 [KKRYpTVVGN], which is a key residue within its activation loop. The complication here is that ROCK and MRCK kinases have been invoked in the same process,[77] and further validation of cellular LIMK1 kinases is needed.

In mammalian cells, nonmuscle myosins are regulated by phosphorylation of their light chain (MLC) on Ser-19 by MLCK. The Rho target-p160 ROCK phosphorylates and inhibits the MLC phosphatase activity, leading to the accumulation of phosphorylated MLC and, thereby the increased contractility necessary for actin stress-fiber formation and cell spreading.[78] Activated PAK phosphorylates MLCK and inhibits its activity toward MLC, leading to reduced stress fibers.[79] However, MLCK has more limited roles in nonmuscle cells. Paxillin is the component of cell adhesions to which GIT1/PIX/PAK binds[80,81] and allows group 1 PAKs to locate to the leading edge of cells.[82] Paxillin is an important scaffold protein for other signaling kinases such as FAK and Src, which, in turn, act on the GIT/PIX/PAK complex.[83] We fail to observe paxillin Ser-273 [ELMApSLSDF] phosphorylation by PAK as reported,[82] consistent with the sequence of this site. Instead, we have shown Ser-273 phosphorylation (by an unknown kinase) controls nuclear–cytoplasmic shuttling of paxillin.[84]

One activity that has generated much debate is the degradation of the ECM by cancer cells via "invadopodia," as reviewed in Ref. 85. These structures correlate with high rates of cell motility and the local release of matrix metalloproteinases that facilitates degradation of the ECM. F-actin assembly at invadopodia is stimulated by Wiskott–Aldrich syndrome protein (N-WASP) through the Arp2/3 protein complex,[86] which is an actin filament nucleating and organizing regulator. PAK1 can phosphorylate the Arpc1b subunit of this complex, which stimulates Arp2/3 complex assembly and regulates the directional motility of breast cancer cells.[87] Cortactin, an F-actin binding protein, accumulates in both in lamellipodia and particularly invadopodia and can be modified by PAK1 at Ser113 in the first actin-binding repeat, which reduces actin affinity.[88] PAK1 has also been suggested to phosphorylate cortactin Ser405 and Ser418,[89] but this is inconsistent with these sites being flanked by proline residues[18] and being long established as ERK sites.[90] Thus, PAKs act downstream of many vertebrate Cdc42 and Rac1-like GTPases to affect changes in pathways needed for cytoskeletal remodeling, but very few targets are well validated because of the lack of good inhibitors.

## III. PAKs and Brain Disorders

PAK proteins and mRNAs are highly expressed in brain.[2] Forms of non-syndromic X-linked mental retardations are associated with mutation in the PAK3 or the PAK partner βPIX genes.[91–93] Thus, it is no surprise that PAK1 or PAK3 loss in mice is accompanied by abnormal synapse function and plasticity in the hippocampus.[94,95] Dominant-inhibitory (inactive) PAK1 expressed in mouse forebrain also affects synapse morphology and consolidation of long-term memory.[96] Interestingly, expression of a small PAK inhibitory peptide in the brain of a mouse modeled for Fragile-X syndrome (FXS) can rescue some aspects of the syndrome.[97] Greater spine density and elongated spines in the cortex, which are synaptic abnormalities commonly observed in FXS, are partially restored by postnatal expression of a PAK inhibitor transgene in the forebrain. The deficit in cortical long-term potentiation observed in FMR1 KO mice is fully restored by this transgene.[97] The connection between PAK1 and fragile-X proteins involves a direct interaction between the inhibitory region of PAK1 and the second KH domain of the FMR1.[20] PAK binds to FMR1 only when in an open active state and phosphorylates the fragile-X related protein (FXR1) at Ser-420, which appears to be required for its function. While FXS arises from a loss of the FMR1 protein (in males inheriting the defective X- chromosome), a rare null mutation of FMR1(I304N) fails to interact with PAKs,[20] indicating that this interaction could be critical to the function of the fragile-X complex in the normal synaptic spine.

Studies have detected changes in PAK1 and PAK3 levels in the hippocampus affected by Alzheimer's disease (AD).[93] There is suggested to be a relocalization of PAKs to the cytoskeletal fraction[98] in AD but there is no underlying biochemical understanding of this. A dominant-negative form of PAK1 sensitizes while the wild-type form protects from the effects of beta-amyloid oligomers in cultured primary neurons, but these kinds of experiments are a long way from showing that the kinase is directly implicated in the progression of AD. In case of Huntington's disease, PAK1 and the kinase adaptors PIX and GIT1 have all been independently identified as being associated with huntingtin (htt) inclusions.[80,99,100] In tissue culture models, interference with PAK1 function modestly decreases the formation of aggregates by mutant htt, while "constitutively active" PAK1 enhances the aggregation.[99] The αPIX isoform was found as an htt-interacting protein, after screening candidate proteins involved with the cytoskeleton.[100] It binds the N-terminal of both wild-type htt and mutant htt, but the latter is more prone to aggregation. These observations need to be taken with caution, as both PIX and GIT1 are sensitive to aggregate formation themselves when overexpressed (P.M. Chan and E. Manser, unpublished observations). Overall, it appears that pathological changes in the brain could be associated with both elevated and reduced functions of PAKs, and the specific role of PAK1 in these processes may be context dependent.

Finally, there is recent evidence that changes in the expression of GIT1 are associated with attention deficit hyperactivity disorder (ADHD) in humans and ADHD-like behaviors in mice.[101] As this protein is closely connected to PAK function, it would not be surprising if part of the pathology is due to a failure to properly localize the kinase.

## IV. PAKs in Pathogen Infection

One of the most interesting recent findings of PAK function regards the enterohemorrhagic *Escherichia coli* O157:H7. In the infection process, this strain uses the type III effector EspG protein to interfere with membrane trafficking at the level of the Golgi apparatus. In a yeast two-hybrid screen, two proteins, ADP-ribosylation factor (Arf1) and PAKs (1, 2, and 3), were found as its relevant host substrates.[102] Based on the structure of EspG with a small region of the PAK AID (which also binds to fragile-X proteins), the authors showed that such binding leads to kinase activation. The assembly of PAKs at this site may be accompanied by PIX and GIT1 proteins, but this was not explicitly tested or indeed mentioned.[102] As Arf1 binding to effectors is blocked by the EspG interaction, it might be that GIT1 recruitment (an ArfGAP) serves also to terminate Arf1 signaling. The kinase target(s) of PAK2 in this context has not yet been established. The use of PAKs by a bacterial pathogen clearly points to its key role in cellular events including endomembrane transport. This aspect of PAK function appears to be important for the life cycle of a number of viruses as outlined below.

The PAK-dependent signaling is implicated in the biology and pathogenicity of various viruses, as recently reviewed in Ref. 103. As PAK1 has long been associated with macropinocytosis,[104] these kinases may play a general role during infection by viruses that depend on this process. For example, siRNA-mediated PAK1 knockdown can block infection by vaccinia virus.[105] For wider discussion of macropinocytosis as a viral entry mechanism, the reader should refer to the review.[106] Two papers suggest that a target of PAK1 in macropinocytosis is

Ser-147 [TRVQpSVEQI] of "C-terminal binding protein 1 of E1A" (CtBP1) either following epidermal growth factor treatment[107] or during adenovirus uptake.[108] Indeed, PAK-dependent uptake of other viruses such as echovirus and adenovirus-3 leads to phosphorylation of CtBP1. The suggestion is that CtBP1 phosphorylation was required for fission of macropinosomes, and thus phosphorylation-defective CtBP1(S147A) can block adenovirus-3 infection.[109] The role of CtBP1 is more complicated, as it is also invoked in Golgi dynamics[110] and as a transcriptional corepressor (as reviewed in Ref. 111). A recent paper describing the plant CtBP1 suggests a Golgi-like rather than nuclear localization of the protein.[112] Thus at this stage, we do not understand how PAKs are coupled to CtBP1, nor whether these putative target proteins really have alternate functions in pinocytosis, Golgi fission, and transcriptional repression.

The Nef (negative factor) protein of HIV is a key to the pathogenesis of AIDS.[113] The Nef protein is an *N*-myristoylated protein exclusively expressed by the lentiviruses human immunodeficiency virus type 1 (HIV-1), HIV-2, and the simian immunodeficiency virus (SIV). Nef is a multifunctional protein that is expressed early upon infection of a host cell and is thought to influence virus replication through association with host signaling proteins. It binds a cellular kinase of ~68 kDa (NAK), which some time ago was suggested to be a PAK-like kinase.[114–116] Currently, a number of laboratories believe that HIV Nef does bind group 1 PAKs, with PAK2 as the favorite candidate.[117,118] How this occurs remains unclear because purified recombinant Nef and PAK proteins interact with low efficiency and PAK kinase activity is not increased in the presence of Nef (P.M. Chan and E. Manser, unpublished observations). The literature contains contradictory data on this role for HIV pathogenesis.[119,120] Perhaps Nef associates efficiently with a PAK2 only in the context of a larger complex *in vivo*,[121] which has been reconstituted somewhat using microsomal membranes.[122] This multiprotein signaling complex is suggested to include Src family kinases, Vav1, Cdc42, and PAK2[115,116,118,123] bound to lipid rafts.[118,124] The ~p80 NAK substrate has not been identified, but the tight complex between PAK2 and βPIX might suggest the latter as a substrate of Nef immune precipitates. A recent study suggests that Nef-mediated enhancement of viral replication in primary T cells is dependent on PAK2 only in the context of moderate activating stimuli.[125] PAK2 is proposed in this context to phosphorylate LIM kinase 1 at Thr-508, which can, in turn, downregulate cofilin by modifying Ser-3. *In vivo* experiments using SIV suggest that there is strong selective pressure in favor of wild-type virus that binds to PAKs.[114,126] However, other studies find that the association of Nef with PAKs is not a prerequisite for the development of AIDS.[119,120,127] In terms of an explanation for its stimulatory effect of Nef-PAK2 on viral infectivity, the kinase might enhance HIV entry across the cortical actin barrier of a host cell through PAK2-driven changes in the actin cytoskeleton.[128,129]

## V. Future Directions

At present, PAK-oriented therapy seems obvious for oncology where PAKs can contribute to progression and persistence of the disease. The potential of PAK inhibitors such as PF-3758309 for treatment of NF2 patients is quite exciting. The loss of PAK1 and PAK3 is somewhat tolerated in mice, with mild immunological and neurological defects associated with deletion of these two genes, respectively.[7,130,131] Deletion of the genes encoding PAK2 and PAK4 results in early embryonic lethality,[6] so at this stage it is not clear whether they are essential in adult tissues. Xenograph cancer models in mice suggest that PF-3758309 does not impact animal survival over the period of the experiments, though it blocks both groups of PAKs.[23] It might be that PAK inhibitors can also help to limit viral infectivity, which could be of use as a combination therapy (as e.g., in AIDS). With respect to genuine isoform-specific PAK inhibitors, targeting the kinase domain would seem to be problematic, and compensation in any case between PAKs likely. Inhibiting group I versus group II PAKs (and vice versa) should be achievable. As we understand more about PAK targets, it may be that other disease states will reveal themselves to involve misregulated PAKs.

### References

1. Zhao ZS, Manser E. PAK and other Rho-associated kinases—effectors with surprisingly diverse mechanisms of regulation. *Biochem J* 2005;**386**:201–14.
2. Manser E, Leung T, Salihuddin H, Zhao ZS, Lim L. A brain serine/threonine protein kinase activated by Cdc42 and Rac1. *Nature* 1994;**367**:40–6.
3. Bishop AL, Hall A. Rho GTPases and their effector proteins. *Biochem J* 2000;**348** (Pt. 2):241–55.
4. Dan I, Watanabe NM, Kusumi A. The Ste20 group kinases as regulators of MAP kinase cascades. *Trends Cell Biol* 2001;**11**:220–30.
5. Leberer E, Thomas DY, Whiteway M. Pheromone signalling and polarized morphogenesis in yeast. *Curr Opin Genet Dev* 1997;**7**:59–66.
6. Arias-Romero LE, Chernoff J. A tale of two Paks. *Biol Cell* 2008;**100**:97–108.
7. Asrar S, Meng Y, Zhou Z, Todorovski Z, Huang WW, Jia Z. Regulation of hippocampal long-term potentiation by p21-activated protein kinase 1 (PAK1). *Neuropharmacology* 2009;**56**:73–80.
8. Goehler H, Lalowski M, Stelzl U, Waelter S, Stroedicke M, Worm U, et al. A protein interaction network links GIT1, an enhancer of huntingtin aggregation, to Huntington's disease. *Mol Cell* 2004;**15**:853–65.
9. Sells MA, Pfaff A, Chernoff J. Temporal and spatial distribution of activated Pak1 in fibroblasts. *J Cell Biol* 2000;**151**:1449–58.
10. Turner CE, Brown MC, Perrotta JA, Riedy MC, Nikolopoulos SN, McDonald AR, et al. Paxillin LD4 motif binds PAK and PIX through a novel 95-kD ankyrin repeat, ARF-GAP protein: a role in cytoskeletal remodeling. *J Cell Biol* 1999;**145**:851–63.
11. Banerjee M, Worth D, Prowse DM, Nikolic M. Pak1 phosphorylation on t212 affects microtubules in cells undergoing mitosis. *Curr Biol* 2002;**12**:1233–9.

12. Zhao ZS, Lim JP, Ng YW, Lim L, Manser E. The GIT-associated kinase PAK targets to the centrosome and regulates Aurora-A. *Mol Cell* 2005;**20**:237–49.
13. Luo ZG, Wang Q, Zhou JZ, Wang J, Luo Z, Liu M, et al. Regulation of AChR clustering by Dishevelled interacting with MuSK and PAK1. *Neuron* 2002;**35**:489–505.
14. Phee H, Abraham RT, Weiss A. Dynamic recruitment of PAK1 to the immunological synapse is mediated by PIX independently of SLP-76 and Vav1. *Nat Immunol* 2005;**6**:608–17.
15. Dummler B, Ohshiro K, Kumar R, Field J. Pak protein kinases and their role in cancer. *Cancer Metastasis Rev* 2009;**28**:51–63.
16. Manser E, Loo TH, Koh CG, Zhao ZS, Chen XQ, Tan L, et al. PAK kinases are directly coupled to the PIX family of nucleotide exchange factors. *Mol Cell* 1998;**1**:183–92.
17. Wells CM, Jones GE. The emerging importance of group II PAKs. *Biochem J* 2010;**425**:465–73.
18. Rennefahrt UE, Deacon SW, Parker SA, Devarajan K, Beeser A, Chernoff J, et al. Specificity profiling of Pak kinases allows identification of novel phosphorylation sites. *J Biol Chem* 2007;**282**:15667–78.
19. Eswaran J, Soundararajan M, Knapp S. Targeting group II PAKs in cancer and metastasis. *Cancer Metastasis Rev* 2009;**28**:209–17.
20. Say E, Tay HG, Zhao ZS, Baskaran Y, Li R, Lim L, et al. A functional requirement for PAK1 binding to the KH(2) domain of the fragile X protein-related FXR1. *Mol Cell* 2010;**38**:236–49.
21. Deacon SW, Beeser A, Fukui JA, Rennefahrt UE, Myers C, Chernoff J, et al. An isoform-selective, small-molecule inhibitor targets the autoregulatory mechanism of p21-activated kinase. *Chem Biol* 2008;**15**:322–31.
22. Viaud J, Peterson JR. An allosteric kinase inhibitor binds the p21-activated kinase autoregulatory domain covalently. *Mol Cancer Ther* 2009;**8**:2559–65.
23. Murray BW, Guo C, Piraino J, Westwick JK, Zhang C, Lamerdin J, et al. Small-molecule p21-activated kinase inhibitor PF-3758309 is a potent inhibitor of oncogenic signaling and tumor growth. *Proc Natl Acad Sci USA* 2010;**107**:9446–51.
24. Yi C, Wilker EW, Yaffe MB, Stemmer-Rachamimov A, Kissil JL. Validation of the p21-activated kinases as targets for inhibition in neurofibromatosis type 2. *Cancer Res* 2008;**68**:7932–7.
25. Siu MK, Wong ES, Chan HY, Kong DS, Woo NW, Tam KF, et al. Differential expression and phosphorylation of Pak1 and Pak2 in ovarian cancer: effects on prognosis and cell invasion. *Int J Cancer* 2010;**127**:21–31.
26. Hirokawa Y, Tikoo A, Huynh J, Utermark T, Hanemann CO, Giovannini M, et al. A clue to the therapy of neurofibromatosis type 2: NF2/merlin is a PAK1 inhibitor. *Cancer J* 2004;**10**:20–6.
27. Bostner J, Skoog L, Fornander T, Nordenskjold B, Stal O. Estrogen receptor-alpha phosphorylation at serine 305, nuclear p21-activated kinase 1 expression, and response to tamoxifen in postmenopausal breast cancer. *Clin Cancer Res* 2010;**16**:1624–33.
28. Bostner J, Ahnstrom Waltersson M, Fornander T, Skoog L, Nordenskjold B, Stal O. Amplification of CCND1 and PAK1 as predictors of recurrence and tamoxifen resistance in postmenopausal breast cancer. *Oncogene* 2007;**26**:6997–7005.
29. Kichina JV, Goc A, Al-Husein B, Somanath PR, Kandel ES. PAK1 as a therapeutic target. *Expert Opin Ther Targets* 2010;**14**:703–25.
30. Ong CC, Jubb AM, Haverty PM, Zhou W, Tran V, Truong T, et al. Targeting p21-activated kinase 1 (PAK1) to induce apoptosis of tumor cells. *Proc Natl Acad Sci USA* 2011;**108**:7177–82.
31. Liu F, Li X, Wang C, Cai X, Du Z, Xu H, et al. Downregulation of p21-activated kinase-1 inhibits the growth of gastric cancer cells involving cyclin B1. *Int J Cancer* 2009;**125**:2511–9.
32. Zhu G, Wang Y, Huang B, Liang J, Ding Y, Xu A, et al. A Rac1/PAK1 cascade controls beta-catenin activation in colon cancer cells. *Oncogene* 2011; in press, doi: 10.1038/onc.2011.294.

33. Tang Y, Chen Z, Ambrose D, Liu J, Gibbs JB, Chernoff J, et al. Kinase-deficient Pak1 mutants inhibit Ras transformation of Rat-1 fibroblasts. *Mol Cell Biol* 1997;**17**:4454–64.
34. Somanath PR, Vijai J, Kichina JV, Byzova T, Kandel ES. The role of PAK-1 in activation of MAP kinase cascade and oncogenic transformation by Akt. *Oncogene* 2009;**28**:2365–9.
35. Alavi A, Hood JD, Frausto R, Stupack DG, Cheresh DA. Role of Raf in vascular protection from distinct apoptotic stimuli. *Science* 2003;**301**:94–6.
36. Chaudhary A, King WG, Mattaliano MD, Frost JA, Diaz B, Morrison DK, et al. Phosphatidylinositol 3-kinase regulates Raf1 through Pak phosphorylation of serine 338. *Curr Biol* 2000;**10**:551–4.
37. Frost JA, Steen H, Shapiro P, Lewis T, Ahn N, Shaw PE, et al. Cross-cascade activation of ERKs and ternary complex factors by Rho family proteins. *EMBO J* 1997;**16**:6426–38.
38. Huang Z, Traugh JA, Bishop JM. Negative control of the Myc protein by the stress-responsive kinase Pak2. *Mol Cell Biol* 2004;**24**:1582–94.
39. Sauzeau V, Berenjeno IM, Citterio C, Bustelo XR. A transcriptional cross-talk between RhoA and c-Myc inhibits the RhoA/Rock-dependent cytoskeleton. *Oncogene* 2010;**29**:3781–92.
40. Jung JH, Pendergast AM, Zipfel PA, Traugh JA. Phosphorylation of c-Abl by protein kinase Pak2 regulates differential binding of ABI2 and CRK. *Biochemistry* 2008;**47**:1094–104.
41. Vadlamudi RK, Adam L, Wang RA, Mandal M, Nguyen D, Sahin A, et al. Regulatable expression of p21-activated kinase-1 promotes anchorage-independent growth and abnormal organization of mitotic spindles in human epithelial breast cancer cells. *J Biol Chem* 2000;**275**:36238–44.
42. Wang RA, Mazumdar A, Vadlamudi RK, Kumar R. P21-activated kinase-1 phosphorylates and transactivates estrogen receptor-alpha and promotes hyperplasia in mammary epithelium. *EMBO J* 2002;**21**:5437–47.
43. Wang RA, Zhang H, Balasenthil S, Medina D, Kumar R. PAK1 hyperactivation is sufficient for mammary gland tumor formation. *Oncogene* 2006;**25**:2931–6.
44. Ng Y-W, Raghunathan D, Chan PM, Baskaran Y, Smith DJ, Lee C-H, et al. Why an A-loop phospho-mimetic fails to activate PAK1: understanding an inaccessible kinase state by molecular dynamics simulations. *Structure* 2010;**18**:879–90.
45. Callow MG, Clairvoyant F, Zhu S, Schryver B, Whyte DB, Bischoff JR, et al. Requirement for PAK4 in the anchorage-independent growth of human cancer cell lines. *J Biol Chem* 2002;**277**:550–8.
46. Greenman C, Stephens P, Smith R, Dalgliesh GL, Hunter C, Bignell G, et al. Patterns of somatic mutation in human cancer genomes. *Nature* 2007;**446**:153–8.
47. Chen S, Auletta T, Dovirak O, Hutter C, Kuntz K, El-ftesi S, et al. Copy number alterations in pancreatic cancer identify recurrent PAK4 amplification. *Cancer Biol Ther* 2008;**7**:1793–802.
48. Kimmelman AC, Hezel AF, Aguirre AJ, Zheng H, Paik JH, Ying H, et al. Genomic alterations link Rho family of GTPases to the highly invasive phenotype of pancreas cancer. *Proc Natl Acad Sci USA* 2008;**105**:19372–7.
49. Parsons M, Monypenny J, Ameer-Beg SM, Millard TH, Machesky LM, Peter M, et al. Spatially distinct binding of Cdc42 to PAK1 and N-WASP in breast carcinoma cells. *Mol Cell Biol* 2005;**25**:1680–95.
50. Marlin JW, Eaton A, Montano GT, Chang YW, Jakobi R. Elevated p21-activated kinase 2 activity results in anchorage-independent growth and resistance to anticancer drug-induced cell death. *Neoplasia* 2009;**11**:286–97.
51. Deacon K, Mistry P, Chernoff J, Blank JL, Patel R. p38 Mitogen-activated protein kinase mediates cell death and p21-activated kinase mediates cell survival during chemotherapeutic drug-induced mitotic arrest. *Mol Biol Cell* 2003;**14**:2071–87.
52. Jakobi R, Moertl E, Koeppel MA. p21-Activated protein kinase gamma-PAK suppresses programmed cell death of BALB3T3 fibroblasts. *J Biol Chem* 2001;**276**:16624–34.

53. Menard RE, Jovanovski AP, Mattingly RR. Active p21-activated kinase 1 rescues MCF10A breast epithelial cells from undergoing anoikis. *Neoplasia* 2005;**7**:638–45.
54. Schurmann A, Mooney AF, Sanders LC, Sells MA, Wang HG, Reed JC, et al. p21-Activated kinase 1 phosphorylates the death agonist bad and protects cells from apoptosis. *Mol Cell Biol* 2000;**20**:453–61.
55. Jin S, Zhuo Y, Guo W, Field J. p21-Activated Kinase 1 (Pak1)-dependent phosphorylation of Raf-1 regulates its mitochondrial localization, phosphorylation of BAD, and Bcl-2 association. *J Biol Chem* 2005;**280**:24698–705.
56. Friedland JC, Lakins JN, Kazanietz MG, Chernoff J, Boettiger D, Weaver VM. Alpha6beta4 integrin activates Rac-dependent p21-activated kinase 1 to drive NF-kappaB-dependent resistance to apoptosis in 3D mammary acini. *J Cell Sci* 2007;**120**:3700–12.
57. Gnesutta N, Qu J, Minden A. The serine/threonine kinase PAK4 prevents caspase activation and protects cells from apoptosis. *J Biol Chem* 2001;**276**:14414–9.
58. Gnesutta N, Minden A. Death receptor-induced activation of initiator caspase 8 is antagonized by serine/threonine kinase PAK4. *Mol Cell Biol* 2003;**23**:7838–48.
59. Lu Y, Pan ZZ, Devaux Y, Ray P. p21-Activated protein kinase 4 (PAK4) interacts with the keratinocyte growth factor receptor and participates in keratinocyte growth factor-mediated inhibition of oxidant-induced cell death. *J Biol Chem* 2003;**278**:10374–80.
60. Li F, Adam L, Vadlamudi RK, Zhou H, Sen S, Chernoff J, et al. p21-Activated kinase 1 interacts with and phosphorylates histone H3 in breast cancer cells. *EMBO Rep* 2002;**3**:767–73.
61. Maroto B, Ye MB, von Lohneysen K, Schnelzer A, Knaus UG. P21-Activated kinase is required for mitotic progression and regulates Plk1. *Oncogene* 2008;**27**:4900–8.
62. De la Mota-Peynado A, Chernoff J, Beeser A. Identification of the atypical MAPK Erk3 as a novel substrate for p21-activated kinase (Pak) activity. *J Biol Chem* 2011;**286**:13603–11.
63. Deleris P, Trost M, Topisirovic I, Tanguay PL, Borden KL, Thibault P, et al. Activation loop phosphorylation of ERK3/ERK4 by group I p21-activated kinases (PAKs) defines a novel PAK-ERK3/4-MAPK-activated protein kinase 5 signaling pathway. *J Biol Chem* 2011;**286**:6470–8.
64. Callow MG, Zozulya S, Gishizky ML, Jallal B, Smeal T. PAK4 mediates morphological changes through the regulation of GEF-H1. *J Cell Sci* 2005;**118**:1861–72.
65. Zenke FT, Krendel M, DerMardirossian C, King CC, Bohl BP, Bokoch GM. p21-Activated kinase 1 phosphorylates and regulates 14-3-3 binding to GEF-H1, a microtubule-localized Rho exchange factor. *J Biol Chem* 2004;**279**:18392–400.
66. Bakal CJ, Finan D, LaRose J, Wells CD, Gish G, Kulkarni S, et al. The Rho GTP exchange factor Lfc promotes spindle assembly in early mitosis. *Proc Natl Acad Sci USA* 2005;**102**:9529–34.
67. Krendel M, Zenke FT, Bokoch GM. Nucleotide exchange factor GEF-H1 mediates cross-talk between microtubules and the actin cytoskeleton. *Nat Cell Biol* 2002;**4**:294–301.
68. Brecht M, Steenvoorden AC, Collard JG, Luf S, Erz D, Bartram CR, et al. Activation of gef-h1, a guanine nucleotide exchange factor for RhoA, by DNA transfection. *Int J Cancer* 2005;**113**:533–40.
69. Bompard G, Rabeharivelo G, Frank M, Cau J, Delsert C, Morin N. Subgroup II PAK-mediated phosphorylation regulates Ran activity during mitosis. *J Cell Biol* 2010;**190**:807–22.
70. Sit ST, Manser E. Rho GTPases and their role in organizing the actin cytoskeleton. *J Cell Sci* 2011;**124**:679–83.
71. Manser E, Huang HY, Loo TH, Chen XQ, Dong JM, Leung T, et al. Expression of constitutively active alpha-PAK reveals effects of the kinase on actin and focal complexes. *Mol Cell Biol* 1997;**17**:1129–43.
72. Sells MA, Knaus UG, Bagrodia S, Ambrose DM, Bokoch GM, Chernoff J. Human p21-activated kinase (Pak1) regulates actin organization in mammalian cells. *Curr Biol* 1997;**7**:202–10.

73. Kumar R, Gururaj AE, Barnes CJ. p21-Activated kinases in cancer. *Nat Rev Cancer* 2006;**6**:459–71.
74. Edwards DC, Sanders LC, Bokoch GM, Gill GN. Activation of LIM-kinase by Pak1 couples Rac/Cdc42 GTPase signalling to actin cytoskeletal dynamics. *Nat Cell Biol* 1999;**1**:253–9.
75. Dan C, Kelly A, Bernard O, Minden A. Cytoskeletal changes regulated by the PAK4 serine/threonine kinase are mediated by LIM kinase 1 and cofilin. *J Biol Chem* 2001;**276**:32115–21.
76. Ahmed T, Shea K, Masters JR, Jones GE, Wells CM. A PAK4-LIMK1 pathway drives prostate cancer cell migration downstream of HGF. *Cell Signal* 2008;**20**:1320–8.
77. Scott RW, Olson MF. LIM kinases: function, regulation and association with human disease. *J Mol Med (Berl)* 2007;**85**:555–68.
78. Totsukawa G, Wu Y, Sasaki Y, Hartshorne DJ, Yamakita Y, Yamashiro S, et al. Distinct roles of MLCK and ROCK in the regulation of membrane protrusions and focal adhesion dynamics during cell migration of fibroblasts. *J Cell Biol* 2004;**164**:427–39.
79. Chew TL, Masaracchia RA, Goeckeler ZM, Wysolmerski RB. Phosphorylation of non-muscle myosin II regulatory light chain by p21-activated kinase (gamma-PAK). *J Muscle Res Cell Motil* 1998;**19**:839–54.
80. Hoefen RJ, Berk BC. The multifunctional GIT family of proteins. *J Cell Sci* 2006;**119**:1469–75.
81. Zhao ZS, Manser E, Loo TH, Lim L. Coupling of PAK-interacting exchange factor PIX to GIT1 promotes focal complex disassembly. *Mol Cell Biol* 2000;**20**:6354–63.
82. Nayal A, Webb DJ, Brown CM, Schaefer EM, Vicente-Manzanares M, Horwitz AR. Paxillin phosphorylation at Ser273 localizes a GIT1-PIX-PAK complex and regulates adhesion and protrusion dynamics. *J Cell Biol* 2006;**173**:587–9.
83. Brown MC, West KA, Turner CE. Paxillin-dependent paxillin kinase linker and p21-activated kinase localization to focal adhesions involves a multistep activation pathway. *Mol Biol Cell* 2002;**13**:1550–65.
84. Dong JM, Lau LS, Ng YW, Lim L, Manser E. Paxillin nuclear-cytoplasmic localization is regulated by phosphorylation of the LD4 motif: evidence that nuclear paxillin promotes cell proliferation. *Biochem J* 2009;**418**:173–84.
85. Murphy DA, Courtneidge SA. The 'ins' and 'outs' of podosomes and invadopodia: characteristics, formation and function. *Nat Rev Mol Cell Biol* 2011;**12**:413–26.
86. Oser M, Yamaguchi H, Mader CC, Bravo-Cordero JJ, Arias M, Chen X, et al. Cortactin regulates cofilin and N-WASp activities to control the stages of invadopodium assembly and maturation. *J Cell Biol* 2009;**186**:571–87.
87. Vadlamudi RK, Li F, Barnes CJ, Bagheri-Yarmand R, Kumar R. p41-Arc subunit of human Arp2/3 complex is a p21-activated kinase-1-interacting substrate. *EMBO Rep* 2004;**5**:154–60.
88. Webb BA, Zhou S, Eves R, Shen L, Jia L, Mak AS. Phosphorylation of cortactin by p21-activated kinase. *Arch Biochem Biophys* 2006;**456**:183–93.
89. Grassart A, Meas-Yedid V, Dufour A, Olivo-Marin JC, Dautry-Varsat A, Sauvonnet N. Pak1 phosphorylation enhances cortactin-N-WASP interaction in clathrin-caveolin-independent endocytosis. *Traffic* 2010;**11**:1079–91.
90. Davis RJ. The mitogen-activated protein kinase signal transduction pathway. *J Biol Chem* 1993;**268**:14553–6.
91. McPhie DL, Coopersmith R, Hines-Peralta A, Chen Y, Ivins KJ, Manly SP, et al. DNA synthesis and neuronal apoptosis caused by familial Alzheimer disease mutants of the amyloid precursor protein are mediated by the p21 activated kinase PAK3. *J Neurosci* 2003;**23**:6914–27.
92. Allen KM, Gleeson JG, Bagrodia S, Partington MW, MacMillan JC, Cerione RA, et al. PAK3 mutation in nonsyndromic X-linked mental retardation. *Nat Genet* 1998;**20**:25–30.
93. Zhao L, Ma QL, Calon F, Harris-White ME, Yang F, Lim GP, et al. Role of p21-activated kinase pathway defects in the cognitive deficits of Alzheimer disease. *Nat Neurosci* 2006;**9**:234–42.

94. Boda B, Alberi S, Nikonenko I, Node-Langlois R, Jourdain P, Moosmayer M, et al. The mental retardation protein PAK3 contributes to synapse formation and plasticity in hippocampus. *J Neurosci* 2004;**24**:10816–25.
95. Huang W, Zhou Z, Asrar S, Henkelman M, Xie W, Jia Z. p21-activated kinases 1 and 3 control brain size through coordinating neuronal complexity and synaptic properties. *Mol Cell Biol* 2011;**31**:388–403.
96. Hayashi ML, Choi SY, Rao BS, Jung HY, Lee HK, Zhang D, et al. Altered cortical synaptic morphology and impaired memory consolidation in forebrain-specific dominant-negative PAK transgenic mice. *Neuron* 2004;**42**:773–87.
97. Hayashi ML, Rao BS, Seo JS, Choi HS, Dolan BM, Choi SY, et al. Inhibition of p21-activated kinase rescues symptoms of fragile X syndrome in mice. *Proc Natl Acad Sci USA* 2007;**104**:11489–94.
98. Ma QL, Yang F, Calon F, Ubeda OJ, Hansen JE, Weisbart RH, et al. p21-Activated kinase-aberrant activation and translocation in Alzheimer disease pathogenesis. *J Biol Chem* 2008;**283**:14132–43.
99. Luo S, Mizuta H, Rubinsztein DC. p21-activated kinase 1 promotes soluble mutant huntingtin self-interaction and enhances toxicity. *Hum Mol Genet* 2008;**17**:895–905.
100. Eriguchi M, Mizuta H, Luo S, Kuroda Y, Hara H, Rubinsztein DC. Alpha Pix enhances mutant huntingtin aggregation. *J Neurol Sci* 2010;**290**:80–5.
101. Won H, Mah W, Kim E, Kim JW, Hahm EK, Kim MH, et al. GIT1 is associated with ADHD in humans and ADHD-like behaviors in mice. *Nat Med* 2011;**17**:566–72.
102. Selyunin AS, Sutton SE, Weigele BA, Reddick LE, Orchard RC, Bresson SM, et al. The assembly of a GTPase-kinase signalling complex by a bacterial catalytic scaffold. *Nature* 2011;**469**:107–11.
103. Van den Broeke C, Radu M, Chernoff J, Favoreel HW. An emerging role for p21-activated kinases (Paks) in viral infections. *Trends Cell Biol* 2010;**20**:160–9.
104. Dharmawardhane S, Schurmann A, Sells MA, Chernoff J, Schmid SL, Bokoch GM. Regulation of macropinocytosis by p21-activated kinase-1. *Mol Biol Cell* 2000;**11**:3341–52.
105. Mercer J, Helenius A. Vaccinia virus uses macropinocytosis and apoptotic mimicry to enter host cells. *Science* 2008;**320**:531–5.
106. Mercer J, Helenius A. Virus entry by macropinocytosis. *Nat Cell Biol* 2009;**11**:510–20.
107. Liberali P, Kakkonen E, Turacchio G, Valente C, Spaar A, Perinetti G, et al. The closure of Pak1-dependent macropinosomes requires the phosphorylation of CtBP1/BARS. *EMBO J* 2008;**27**:970–81.
108. Amstutz B, Gastaldelli M, Kalin S, Imelli N, Boucke K, Wandeler E, et al. Subversion of CtBP1-controlled macropinocytosis by human adenovirus serotype 3. *EMBO J* 2008;**27**:956–69.
109. Kalin S, Amstutz B, Gastaldelli M, Wolfrum N, Boucke K, Havenga M, et al. Macropinocytotic uptake and infection of human epithelial cells with species B2 adenovirus type 35. *J Virol* 2010;**84**:5336–50.
110. Corda D, Colanzi A, Luini A. The multiple activities of CtBP/BARS proteins: the Golgi view. *Trends Cell Biol* 2006;**16**:167–73.
111. Chinnadurai G. The transcriptional corepressor CtBP: a foe of multiple tumor suppressors. *Cancer Res* 2009;**69**:731–4.
112. Minamisawa N, Sato M, Cho KH, Ueno H, Takechi K, Kajikawa M, et al. ANGUSTIFOLIA, a plant homolog of CtBP/BARS, functions outside the nucleus. *Plant J* 2011;**68**:788–99.
113. Kestler HW 3rd, Ringler DJ, Mori K, Panicali DL, Sehgal PK, Daniel MD, et al. Importance of the nef gene for maintenance of high virus loads and for development of AIDS. *Cell* 1991;**65**:651–62.
114. Sawai ET, Khan IH, Montbriand PM, Peterlin BM, Cheng-Mayer C, Luciw PA. Activation of PAK by HIV and SIV Nef: importance for AIDS in rhesus macaques. *Curr Biol* 1996;**6**:1519–27.

115. Nunn MF, Marsh JW. Human immunodeficiency virus type 1 Nef associates with a member of the p21-activated kinase family. *J Virol* 1996;**70**:6157–61.
116. Lu X, Wu X, Plemenitas A, Yu H, Sawai ET, Abo A, et al. CDC42 and Rac1 are implicated in the activation of the Nef-associated kinase and replication of HIV-1. *Curr Biol* 1996;**6**:1677–84.
117. Nguyen DG, Wolff KC, Yin H, Caldwell JS, Kuhen KL. "UnPAKing" human immunodeficiency virus (HIV) replication: using small interfering RNA screening to identify novel cofactors and elucidate the role of group I PAKs in HIV infection. *J Virol* 2006;**80**:130–7.
118. Rauch S, Pulkkinen K, Saksela K, Fackler OT. Human immunodeficiency virus type 1 Nef recruits the guanine exchange factor Vav1 via an unexpected interface into plasma membrane microdomains for association with p21-activated kinase 2 activity. *J Virol* 2008;**82**:2918–29.
119. Carl S, Iafrate AJ, Lang SM, Stolte N, Stahl-Hennig C, Matz-Rensing K, et al. Simian immunodeficiency virus containing mutations in N-terminal tyrosine residues and in the PxxP motif in Nef replicates efficiently in rhesus macaques. *J Virol* 2000;**74**:4155–64.
120. Schindler M, Rajan D, Specht A, Ritter C, Pulkkinen K, Saksela K, et al. Association of Nef with p21-activated kinase 2 is dispensable for efficient human immunodeficiency virus type 1 replication and cytopathicity in ex vivo-infected human lymphoid tissue. *J Virol* 2007;**81**:13005–14.
121. Fackler OT, Baur AS. Live and let die: Nef functions beyond HIV replication. *Immunity* 2002;**16**:493–7.
122. Raney A, Kuo LS, Baugh LL, Foster JL, Garcia JV. Reconstitution and molecular analysis of an active human immunodeficiency virus type 1 Nef/p21-activated kinase 2 complex. *J Virol* 2005;**79**:12732–41.
123. Linnemann T, Zheng YH, Mandic R, Peterlin BM. Interaction between Nef and phosphatidylinositol-3-kinase leads to activation of p21-activated kinase and increased production of HIV. *Virology* 2002;**294**:246–55.
124. Krautkramer E, Giese SI, Gasteier JE, Muranyi W, Fackler OT. Human immunodeficiency virus type 1 Nef activates p21-activated kinase via recruitment into lipid rafts. *J Virol* 2004;**78**:4085–97.
125. Olivieri KC, Mukerji J, Gabuzda D. Nef-mediated enhancement of cellular activation and human immunodeficiency virus type 1 replication in primary T cells is dependent on association with p21-activated kinase 2. *Retrovirology* 2011;**8**:64.
126. Khan IH, Sawai ET, Antonio E, Weber CJ, Mandell CP, Montbriand P, et al. Role of the SH3-ligand domain of simian immunodeficiency virus Nef in interaction with Nef-associated kinase and simian AIDS in rhesus macaques. *J Virol* 1998;**72**:5820–30.
127. Lang SM, Iafrate AJ, Stahl-Hennig C, Kuhn EM, Nisslein T, Kaup FJ, et al. Association of simian immunodeficiency virus Nef with cellular serine/threonine kinases is dispensable for the development of AIDS in rhesus macaques. *Nat Med* 1997;**3**:860–5.
128. Fackler OT, Luo W, Geyer M, Alberts AS, Peterlin BM. Activation of Vav by Nef induces cytoskeletal rearrangements and downstream effector functions. *Mol Cell* 1999;**3**:729–39.
129. Campbell EM, Nunez R, Hope TJ. Disruption of the actin cytoskeleton can complement the ability of Nef to enhance human immunodeficiency virus type 1 infectivity. *J Virol* 2004;**78**:5745–55.
130. Allen JD, Jaffer ZM, Park SJ, Burgin S, Hofmann C, Sells MA, et al. p21-activated kinase regulates mast cell degranulation via effects on calcium mobilization and cytoskeletal dynamics. *Blood* 2009;**113**:2695–705.
131. Meng J, Meng Y, Hanna A, Janus C, Jia Z. Abnormal long-lasting synaptic plasticity and cognition in mice lacking the mental retardation gene Pak3. *J Neurosci* 2005;**25**:6641–50.

# Phosphoproteins in Stress-Induced Disease

Lucy E. Dalton, Eleanor Healey, James Irving, and Stefan J. Marciniak

*Division of Respiratory Medicine, Department of Medicine, University of Cambridge, Cambridge Institute for Medical Research, Cambridge, United Kingdom*



The integrated stress response (ISR) is an evolutionarily conserved homeostatic program activated by specific pathological states. These include amino acid deprivation, viral infection, iron deficiency, and the misfolding of proteins within the endoplasmic reticulum (ER), the so-called ER stress. Although apparently disparate, each of these stresses induces phosphorylation of a translation initiation factor, eIF2α, to attenuate new protein translation while simultaneously triggering a transcriptional program. This is achieved by four homologous stress-sensing kinases: GCN2, PKR, HRI, and PERK. In addition to these kinases, mammals possess two specific eIF2α phosphatases, GADD34 and CReP, which play crucial roles in the recovery of protein synthesis following the initial insult. They are not only important in embryonic development but also appear to play important roles in disease, particularly cancer. In this chapter, we discuss each of the eIF2α kinases, in turn, with particular emphasis on their regulation and the new insights provided by recent structural studies. We also discuss the potential for developing novel drug therapies that target the ISR.

Progress in Molecular Biology
and Translational Science, Vol. 106
DOI: 10.1016/B978-0-12-396456-4.00003-1


1877-1173/12 $35.00

# I. Introduction

Approximately one-third of all proteins undergo phosphorylation. It is therefore unsurprising that protein kinases and phosphatases are important in many signaling pathways initiated by cellular stress. For the purposes of this chapter, we shall focus on aspects of cellular stress relevant to human diseases caused by protein misfolding, metabolic disturbances, and infection. This is in part to maintain a manageable scope, but also because recent advances in these areas have provided fascinating insights into disease pathogenesis that might offer novel therapeutic targets.

Early in the evolution of eukaryotes, a kinase evolved that enabled protein translation rates to be linked with the availability of amino acids. This kinase is general control nondepressible-2 (GCN2) and it has been conserved in all eukaryotes.[1–3] Its target, the alpha subunit of eukaryotic translation initiation factor 2 (eIF2α), is required for the recruitment of the methionyl-tRNA to the 40S ribosome at the beginning of protein translation.[4] EIF2α is inhibited by this phosphorylation, bringing about a general attenuation of protein translation (see Chapter 4). As eukaryotes diversified, in particular with the dawning of multicellularity, this kinase gave rise to a family of related eIF2α kinases, which in humans comprises four members: GCN2,[5,6] PKR,[7,8] HRI,[9] and PKR-like ER kinase (PERK)[10,11] (Fig. 1). Each is heavily phosphorylated and this is central to modulating their activities. Moreover, each has been implicated in either the pathogenesis of, or the response to, human disease. These kinases respond to diverse and distinct stresses: GCN2 to amino acid starvation, PKR to viral infection, HRI to iron deficiency, and PERK to protein misfolding in the endoplasmic reticulum (ER), the so-called ER stress. This ability to regulate protein translation in response to many different stresses has led to the pathway being named the "integrated stress response" (ISR).[15]

# II. GCN2

Although yeast can synthesize all 20 amino acids from other carbon sources, they remain responsive to the amino acid content of their environment. When amino acid levels are limiting, yeast will attenuate protein synthesis. Genetic studies identified the kinase gcn2p to be the master regulator of this response[1,16] through its activation by the uncharged tRNAs that accumulate in cells deficient in amino acids.[17] Uncharged tRNAs bind to the histidyl-tRNA synthetase-related domain of gcn2p, causing a reorientation of the protein dimer from an inactive antiparallel state to an active parallel configuration (Fig. 2).[18,19] Phosphorylation of eIF2α follows and protein synthesis is attenuated.

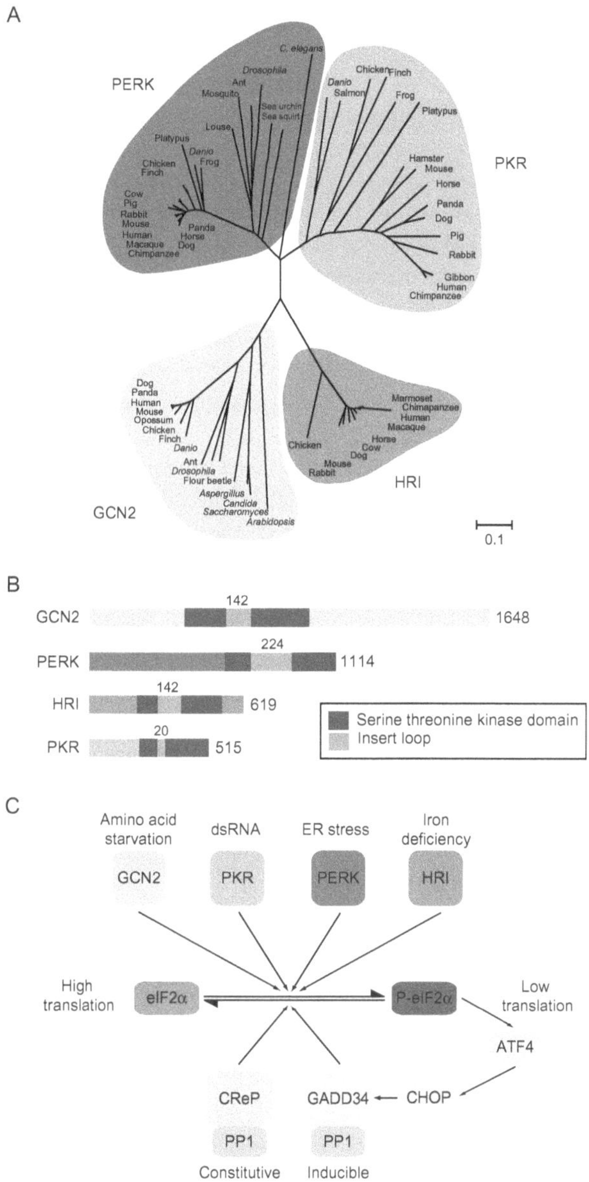


FIG. 1. The eIF2α kinases. (A) Evolutionary relationships of the eIF2α kinases. The evolutionary history was inferred using the neighbor-joining method.[12] The optimal tree with the sum of branch length=8.94 is drawn to scale, with branch lengths in the same units as those of the evolutionary distances used to infer the phylogenetic tree. The evolutionary distances were computed using the Poisson correction method[13] and are in the units of the number of amino acid

In metazoa, GCN2 has retained its role in the response to amino acid starvation. For example, in mice fed a diet lacking in the essential amino acid leucine, GCN2 is activated in the liver to inhibit protein synthesis and prevent loss of body muscle mass.[20] Remarkably, this ancient response also regulates the behavior of mammals, enabling them to choose a balanced amino acid diet over an unbalanced diet. When a rodent is offered a diet deficient in an essential amino acid, the limiting amino acid is sensed in the anterior piriform cortex of the brain causing the animal to reject the food in favor of more balanced fare.[21] In contrast, *Gcn2−/−* mice fail to make this distinction.[22,23] The response is triggered when amino acid deficiency causes the accumulation of uncharged tRNA in neurons of the anterior piriform cortex, where phosphorylation of eIF2α can be detected within 20 min of eating. This therefore represents a rare example of a behavioral program being triggered by a defined phosphorylation event.

GCN2 can also respond to local amino acid concentrations, enabling cells to adjust their enzyme activities accordingly. The diffusible second messenger nitric oxide (NO) is generated from the amino acid arginine by one of the three NO synthases. Its generation by the endothelial and neuronal NO synthases (eNOS and nNOS) for intercellular communication is tightly regulated by intracellular calcium and protein phosphorylation. In contrast, the transcription of inducible NO synthase (iNOS) is induced by inflammatory signals to generate high levels of NO that are toxic to bacterial pathogens through macromolecule oxidation.[24] However, this toxicity can contribute to the pathology of neurodegenerative diseases including multiple sclerosis, Huntington's disease, Alzheimer's disease, Parkinson's disease, and stroke.[25] Consequently, the regulation of iNOS in the brain has been studied intensively. In astrocytes, the activity of iNOS is determined by the transport of its substrate, arginine, into the cell, but the mechanism underlying this was unclear because intracellular arginine levels almost always exceed the $K_m$ of the enzyme: the so-called arginine paradox. The resolution of

substitutions per site. All positions containing gaps and missing data were eliminated from the dataset (complete deletion option). There were a total of 211 positions in the final dataset. Phylogenetic analyses were conducted in MEGA4.[14] (B) Representation of the eIF2α kinases with their serine–threonine kinase domains indicated in dark gray and their insert loop in light gray. The protein length is indicated on the right of each protein and the insert loop size is given above each protein. (C) The integrated stress response (ISR). Each eIF2α kinase is activated by a distinct stress leading to eIF2α phosphorylation. In its unphosphorylated form, eIF2α efficiently supports new protein translation, but once phosphorylated inhibits most translation. Paradoxically, ATF4 translation increases during eIF2α phosphorylation. ATF4 transactivates many genes of the ISR, including the transcription factor *CHOP*. One of the target genes of CHOP is *GADD34*, which encodes a protein phosphatase 1 (PP1) regulatory subunit causing eIF2α dephosphorylation and the recovery of protein synthesis. The constitutively expressed PP1 subunit, CReP, also slowly dephosphorylates eIF2α. (See Color Insert.)

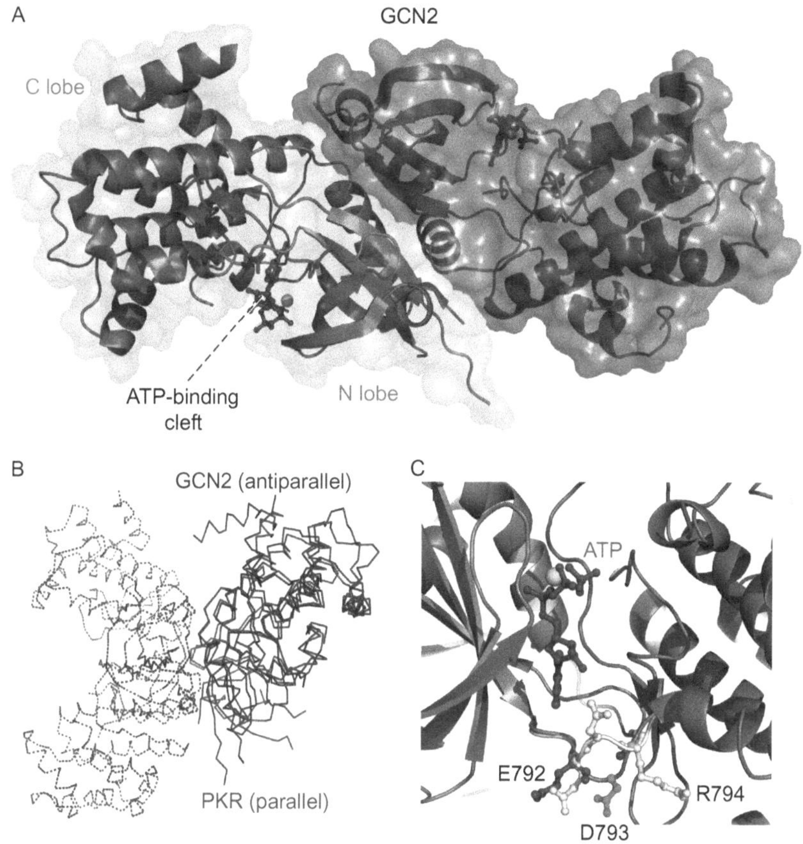


FIG. 2. GCN2 structure. (A) The configuration of the GCN2 dimer is shown (PDB accession 1ZYC), with different surface shades used to distinguish the two monomers and contrasting colors used to indicate the N and C lobes. An ATP analogue is shown in the red ball-and-stick diagram. (B) The relative orientation of the monomers within the dimer differs for the structures of PKR (red) and GCN2 (blue). (C) The ATP-binding cleft is partly occluded in the structure of nonactivated GCN2 (1ZYC) by the side- and main chain of D793. In the activated variant R794G (accession 1ZY5), a switch in backbone conformation is observed, which believed to reflect a mechanism by which binding of ATP is regulated. Figures were prepared using PyMol (The PyMOL Molecular Graphics System, Version 1.3, Schrödinger, LLC). (See Color Insert.)

this paradox appears to involve amino acid sensing by GCN2. Although NO can be generated in cells depleted of arginine, this has been shown to cause higher levels of superoxide formation with toxic consequences.[26] A reduction in intracellular arginine increases the level of uncharged tRNA causing GCN2 activation that represses iNOS translation, thus protecting the cell.[27]

The effects of eIF2α phosphorylation are not restricted to inhibiting protein translation, as it can also activate a transcriptional program. In yeast, translation attenuation induced by gcn2p leads to increased translation of the transcription factor gcn4.[1,2,16,28,29] This then transactivates the genes involved in amino acid sufficiency, thereby promoting growth under conditions of nutrient limitation.

The transcriptional consequences of eIF2α phosphorylation in humans can be illustrated by a family of autosomal recessive white matter hypomyelination disorders, termed CACH/VWM leukodystrophies. They involve the progressive loss of mental and motor faculties due to deterioration of brain white matter. In all cases, the causative mutations have been found within subunits of an eIF2α-interacting protein, eIF2B.[30–33] As described in Chapter 4, eIF2 undergoes rounds of GTP binding, hydrolysis, and guanine nucleotide exchange as it recruits methionyl-tRNA to the ribosome. Only in its GTP-bound state can eIF2 perform this function and so a guanine nucleotide exchange factor (GEF) is required to replace GDP with GTP after each hydrolysis. In mammals, the GEF is called eIF2B. Once phosphorylated, eIF2α binds avidly to this GEF, inhibiting further exchange.[34,35] Thus, phosphorylation of a small pool of eIF2α can inhibit most GEF activity. The mutations that cause the CACH/VWM family of diseases interfere with the GEF activity of eIF2B, but rather than affecting global translation rates during stress, these mutations increase signaling via the transcription factor ATF4.[33] Like yeast gcn4p, mammalian ATF4 is a transcription factor whose translation is enhanced when active eIF2 complex levels are limiting. The progressive loss of brain white matter is believed to result from the resulting aberrant stress signaling caused by inappropriately high ATF4 protein.

While most RNAs are translated less efficiently under conditions where eIF2α is phosphorylated, a small number including *ATF4*, *CHOP*, and *GADD34* are translated more efficiently.[15,36–42] This is caused by upstream open reading frames (uORFs) that are located in their mRNAs 5' to the initiation codon of each true coding sequence. Of these, the most well-characterized example in mammals is *ATF4*.[42] This transcript has two uORFs, the second of which overlaps out of frame with the true *ATF* coding sequence. In unstressed conditions, ribosomes scan along the mRNA translating uORF1 and then recapacitate by rebinding the eIF2/GTP/methionyl-tRNA ternary complex in time to translate inhibitory uORF2. During stress, limiting levels of eIF2 ternary complex lead to a delay in recapacitation of these scanning ribosomes, such that they fail to reinitiate at uORF2 but instead scan to the *ATF4* initiation codon. By then, a proportion would have reacquired the ternary complex, allowing translation of the active transcription factor. Similar uORFs have also been identified within the 5'regions of the mRNAs encoding both CHOP[36] and GADD34.[37] Thus, eIF2α phosphorylation not only reduces translation rates but also participates in the transcriptional reprogramming of the cell which helps it to adapt to the current stress. It is interesting that each of these three uORF-dependent products constitutes a pathway that is essential for the recovery of protein translation and follows an eIF2α phosphorylation event. This pathway will be explored in more detail later.

## III. HRI

Numerous proteins involved in oxygen metabolism contain stoichiometric amounts iron in the form of heme (Fe(II)-protoporphyrin IX complex). Under conditions of limiting heme, these proteins fail to incorporate this prosthetic group and are therefore unable to fold correctly. Should the apoproteins accumulate, they would go on to aggregate and so it is frequently advantageous for the cell to match protein synthesis with heme availability. This balance is achieved by the activity of another eIF2α kinase, the heme-regulated inhibitor of protein synthesis (HRI) (Fig. 1). This kinase is widely expressed, but its activity has been most thoroughly studied in erythroid precursors in which it plays a central role in matching globin synthesis to iron availability.[43,44] In the absence of HRI, a lack of heme due either to dietary deficiency of iron or to inborn defects of heme metabolism leads to the accumulation of globin chains that aggregate, causing accelerated apoptosis within the bone marrow.[44] Interestingly, the phenotype of β-thalassemia caused by mutations of the β-globin gene is also worsened by HRI deficiency, indicating that the heme levels are not the sole trigger for this kinase.[45] Indeed, it has been suggested that HRI is able to respond to a wide repertoire of cytosolic stresses other than heme levels, including cytosolic protein denaturation, oxidative stress, and osmotic stress.[43]

The HRI protein comprises a heme-binding N-terminal domain and an eIF2α kinase domain[46–48] (Fig. 1). When the N-terminal domain is deleted, heme sensing is markedly impaired but not completely abolished. Heme binds to the N-terminal domain with a 1:1 stoichiometry and, although the precise mechanism remains unknown, this leads to inhibition of eIF2α kinase activity.[48,49] When heme levels are low, heme rapidly dissociates from HRI, allowing autophosphorylation and kinase activation.[49,50] This activation involves multiple autophosphorylation events, but only phosphorylation of threonine 485 within the kinase activation loop has been shown to be absolutely essential for eIF2α kinase activity.[47] Interestingly, some degree of autophosphorylation appears to be necessary to maintain solubility of the inactive kinase even in the presence of heme.[50]

The interaction between the heme-containing N-terminal domain of HRI with its catalytic domain appears to be modulated by diffusible factors *in vitro*, and this might be responsible for some aspects of stress sensing *in vivo*.[51]

For example, NO can disrupt the interaction between the N-terminal domain and the kinase domain, leading to kinase activation. In contrast, carbon monoxide stabilizes this inhibitory interaction. Although this molecular mechanism requires further study, it is tempting to speculate that heme serves as a sensor of other diffusible cytosolic toxins. For example, arsenite is an important environmental toxin that has multiple deleterious effects of cell metabolism, notably oxidative damage.[43,52] It appears that HRI is responsible for a prosurvival response to arsenite that is mediated by eIF2α phosphorylation.[52] This protection involves translational attenuation and the formation of stress granules, which are cytosolic aggregates of stalled mRNA preinitiation complexes, accompanied by the induction of multiple antioxidant components of the ISR.

## IV. PKR

Viral infection is responsible for an enormous burden of morbidity and mortality. Constantly, viruses infect cells and hijack their protein synthetic machinery to manufacture new viral particles. With some important exceptions, such infections are usually self-limiting. This is due in large part to an efficient innate immune response that limits cell-to-cell spread of cytolytic viruses, part of which is triggered by the presence of viral double-stranded RNA (dsRNA).[7,53,54] The genome of some viruses is composed of dsRNA (e.g., rotavirus), while some single-stranded RNA viruses generate dsRNA during their replication (e.g., HCV). Even some DNA viruses (e.g., EBV) can generate large dsRNAs during the transcription of their genome. This dsRNA can induce interferon either through direct recognition by cell surface Toll-like receptors on plasmacytoid dendritic cells or through interaction with the cytosolic protein RIG-I.[55] This releases the interferon and activates surface receptors on neighboring cells, triggering the induction of hundreds of genes, among which is the eIF2α kinase PKR.[53]

It was recognized in 1972 that interferon inhibits protein translation in virus-infected cells.[56] Subsequently, this was shown to be caused by the phosphorylation of eIF2α by PKR.[53] Although PKR is expressed constitutively at low levels in most mammalian tissues, it is strongly induced upon interferon stimulation.[8,57] It consists of two tandem copies of dsRNA-binding motifs at its N-terminus and a C-terminal kinase domain separated by a 90-amino acid linker (Fig. 3). These dsRNA-binding motifs serve as an autoinhibitory domain in uninfected cells, causing PKR to exist in a latent, monomeric state.[59] On binding dsRNA, these motifs release the kinase domain, allowing its homodimerization and autophosphorlyation, leading to full activation.[60] However, in the absence of dsRNA, other signals appear to be able to activate PKR via a protein called PACT (RAX in mice) which directly binds to the kinase domain

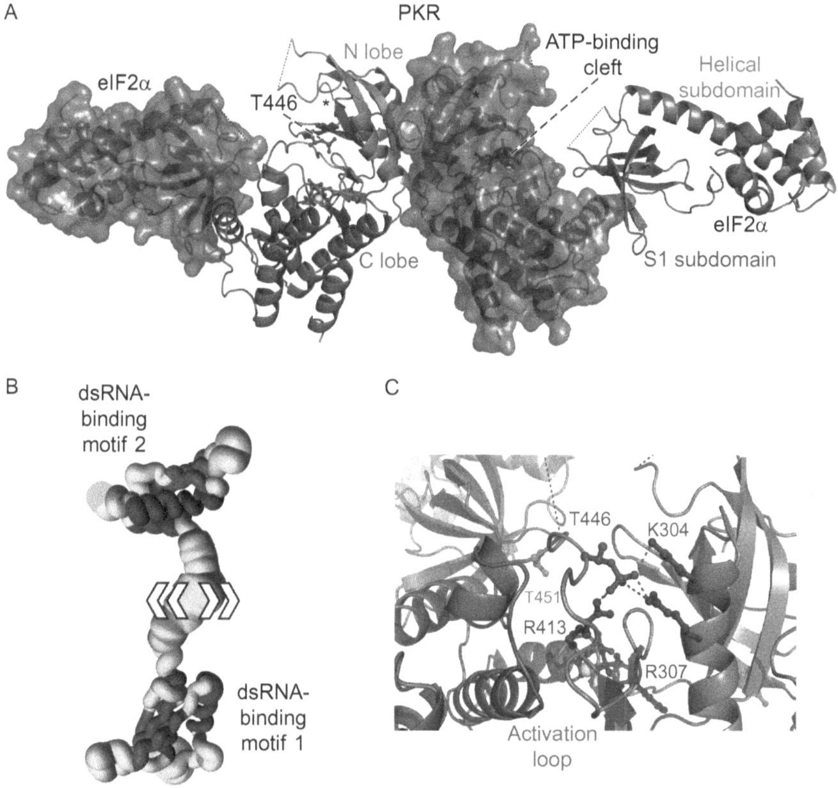


FIG. 3. PKR structure. (A) A back-to-back dimer of the PKR kinase domain is shown (accession 2A1A) in complex with two molecules of eIF2α. The activation loop bearing phosphothreonine 446 is in orange, and the position of the ATP-binding cleft is indicated based on 2A19. Asterisks indicate the site of the N-termini of this structure; the dsRNA-binding domains that precede this point in the structure have been partially characterized by NMR (B). (B) A "putty" representation of the dsRNA-binding domains of PKR; the width of the sausage indicates the extent to which a region is flexible/disordered. The chevrons indicate the direction of the double helix, with the stretch of amino acids connecting the two binding motifs believed to extend in the direction of the major groove. (C) The positively charged residues coordinating the phosphothreonine T446 of the activation loop are shown. Figures were prepared using PyMol and MolMol 2K,2 (MOLMOL: a program for display and analysis of macromolecular structures[58]). (See Color Insert.)

and displaces the inhibitory dsRNA-binding motifs.[61–63] These non-dsRNA signals include some forms of growth factor deprivation, and treatment with toxins that induce ER stress (thapsigargin), oxidative stress (hydrogen peroxide), or otherwise trigger cell death (ceramide).[62,64] PACT is itself phosphorylated on two serine residues, serine 246 and serine 287, that promote its binding and activation of PKR.[61,63,65,66] This appears to be regulated, as phosphorylation of these sites increases during stress, although the relevant signaling kinase remains unknown.

As with all members of this kinase family, phosphorylation of eIF2α by PKR leads to translation attenuation and the induction of the ISR transcriptional program. However, PKR has been shown to interact with additional targets involved in regulating inflammation. NFκB is a proinflammatory transcription factor that is held inactive in the cytosol of unstressed cells through binding to its inhibitor IκB. When an IκB kinase (IKK) phosphorylates IκB, the inhibitory subunit is targeted for degradation, allowing NFκB to migrate into the nucleus where it transactivates the genes involved in the inflammatory response. Dimerization of PKR enables it to bind the scaffolding proteins TRAF2 and TRAF5 and thereby activates NFκB through an interaction with IKK.[67–69] It is not clear whether the kinase activity of PKR is absolutely required for the stimulation of IKK, since only the N-terminus is necessary to restore NFκB activation in *Pkr−/−* fibroblasts.[70,71] However, inactive full-length PKR mutants are incapable of restoring IKK activation in *Pkr−/−* cells.[72] This suggests that, in the context of the whole protein, kinase activity is required, perhaps in order to reorganize the protein structure to enable TRAF interaction.

The structure of the PKR dsRNA-binding domain has been solved by nuclear magnetic resonance and found to consist of two identical α-β–β-α folds separated by a random coil linker of 20 amino acids[73,74] (Fig. 3A). This generates the flexibility required to wrap around dsRNA molecules for optimal protein–RNA interactions. After RNA binding, PKR undergoes a number of conformational changes which lead to dimerization and autophosphorylation on at least 15 threonine and serine residues.[75–77] Mutation studies suggest that phosphorylation of residues threonine 446 and threonine 451 in the kinase activation loop of PKR is required for its catalytic activity.[76–79] However, the crystal structure of PKR revealed that only threonine 446 was phosphorylated stoichiometrically[78] (Fig. 3). An analogous site is also autophosphorylated in GCN2, PERK, and HRI, suggesting that it is crucial in kinase activation.[47,76,80] In addition, many phosphorylation sites lie outside the activation loop, although only threonine 258, which resides in the linker region between the dsRNA-binding motifs and the kinase domain, has been shown definitively to be of functional importance.[75] PKR can also autophosphorylate three tyrosine residues, tyrosine 101, tyrosine 162, and tyrosine 293, both *in vitro* and *in vivo*, and this appears important for its kinase activity.[81] As other members of the family also autophosphorylate tyrosine residues, the eIF2α kinases should more accurately be considered dual-specificity kinases.[80,82] Indeed, when serine 51 of eIF2α was mutated to tyrosine, PKR and HRI were still able to phosphorylate this site and regulate gcn4 expression in yeast.[83]

In response to the selection pressure imposed by PKR, many viruses have evolved mechanisms to evade its antiviral effects. These include direct inhibition or degradation of the kinase or dephosphorylation of its target. A simple mechanism to prevent PKR activation is to mask their viral dsRNA, for example, E3L from vaccinia,[84,85] NS1 from influenza,[86,87] and σ3 from reovirus.[88] In contrast, the HSV-1 protein US11 directly binds PKR to prevent its activation[89] and NS5A from HCV blocks PKR dimerization.[90] The HIV protein Tat, which is involved in viral gene transactivation, and another vaccinia protein K3L act as eIF2α analogues competitively inhibiting PKR.[85,91–93] It is not clear whether endogenous PKR inhibitors modulate its function. Many reports have described inhibitory interactions between p58IPK and both PKR and PERK.[94,95] However, recent reports have suggested that p58IPK is predominantly located within the ER lumen where it functions as a DNAJ cochaperone, rather than in the cytosol as a kinase inhibitor.[96]

Another mechanism employed by viruses to evade PKR that does have a direct endogenous analogue is target dephosphorylation. The herpes viral gene ICP34.5 encodes a protein that directs mammalian protein phosphatase 1 (PP1) specifically to phosphorylayted eIF2α.[97] ICP34.5 is, in fact, homologous to two endogenous PP1 regulatory subunits, GADD34 and CReP, that cause eIF2α dephosphorylation.[98,99] After the function of ICP34.5 had first been elucidated in virally infected cells, screens for mammalian suppressors of the ER stress response identified GADD34 and then CReP.[98,99] GADD34 will be discussed in more detail later.

## V. PERK

The ER is the site of folding and maturation of transmembrane and secreted proteins.[100] Maintaining the balance between the load of immature ER client proteins and the folding capacity of the organelle is vital for the health of a cell. Consequently, homeostatic mechanisms have evolved to ensure that the folding machinery of the ER can rapidly adapt to fluctuations in the rate at which newly synthesized proteins enter its lumen. The intracellular signaling that mediates this regulation is termed the "unfolded protein response" (UPR). If the load of unfolded client proteins within the ER exceeds its capacity to fold them, the cell is said to experience ER stress. Three ER transmembrane molecules sense this: inositol-requiring protein-1 (IRE1), activating transcription factor-6 (ATF6), and PERK. These ER sensors monitor the protein folding status of the ER lumen and report this to the cytoplasm. IRE1 and ATF6 regulate UPR gene transcription and have been reviewed elsewhere.[100] In contrast, PERK serves both to attenuate new protein translation, thus reducing the translocation of newly synthesized proteins into the ER, and via enhanced translation of *ATF4* to invoke the transcriptional program of the ISR.[11,40,101,102]

PERK is a type-1 ER membrane protein, first identified through its homology both to the eIF2α kinases and to the ER stress sensor IRE1.[11] It consists of an ER luminal stress sensor domain that is structurally and functionally similar to that of IRE1 and a cytoplasmic eIF2α kinase domain homologous to GCN2, HRI, and PKR.[11,103] Under unstressed conditions, PERK exists in an inactive monomeric state. Upon ER stress, it clusters in the plane of the ER membrane to form oligomers.[103] The close proximity of the cytoplasmic kinase domains of PERK allows their *trans*-autophoshorylation at multiple sites including the kinase activation loop and an insert loop that is a hallmark of the eIF2α kinase family (Fig. 1).[80,82]

The precise mechanism by which ER stress induces these stable clusters remains a subject of debate. The model that has held sway for much of the past decade stemmed from the observation that, in unstressed conditions, the ER-resident chaperone B-cell immunoglobulin heavy chain-binding protein (BiP) is found bound to the luminal domains of both PERK and IRE1, but dissociates during ER stress.[103] It is believed that this detachment is mediated by sequestration of BiP by incompletely folded and misfolded ER client proteins. The model suggests that the binding of BiP acts to prevent interaction between the luminal domains of PERK molecules and so its dissociation frees the kinase to oligomerize. Although supported by considerable data, this model may prove to be only one part of PERK's regulatory mechanism. Recent studies have suggested an additional mechanism by which IRE1, and by extension PERK, might sense the misfolded protein. Mutations within the BiP-binding domain of yeast ire1, which one might expect should render it constitutively active, instead leave the cells still able to respond to ER stress.[104,105] Moreover, the crystal structure of the luminal domain of yeast ire1p revealed a hydrophobic groove that shares characteristics with that of the major histocompatibility complex, suggesting that peptides might bind to it directly.[106] This region of the IRE1 luminal domain can inhibit protein aggregation *in vitro*, supporting such a direct interaction.[107] This has led some to propose a second model of ER stress sensing in which direct interaction between misfolded proteins and the luminal domains of PERK or IRE1 causes their oligomerization. Clearly, these two models are not mutually exclusive and it may yet prove that ER stress sensing combines elements of both in order to allow subtle regulation of PERK.

When the PERK kinase domain is expressed in bacteria as a GST fusion protein, it can be isolated fully autophosphorylated and active.[108] GST tags are frequently used to enable affinity purification of recombinant proteins, but in the case of PERK, GST serves an additional role. GST forms homodimers and so recombinant fusion proteins containing GST will also be dimeric. This is responsible for the activity of recombinant PERK kinase domains, as the use of alternative monomeric tags, for example, His6, leads to the generation of inactive and often insoluble aggregates of unphosphorylated PERK (unpublished observations). This is remarkable, as bacterial expression of a recombinant kinase dead mutant of PERK, K618R, with an inactivating mutation in the protein's ATP-binding pocket remains unphosphorylated yet folds correctly, is soluble, and can easily be purified.[108] Moreover, if recombinant PERK is dephosphorylated *in vitro* using lambda phosphatase, only 50% can then undergo re-autophosphorylation on the addition of ATP.[108] The remaining material forms aggregates, suggesting that the transition between the active and inactive states of PERK may require additional as-yet unidentified factors. The ability to rephosphorylate PERK *in vitro* demonstrates that dimerization is sufficient to activate the kinase. This is well illustrated by the FV2E–PERK construct generated in the lab of David Ron.[109] FV2E–PERK is a chimeric protein that has the luminal ER stress-sensing domain of PERK replaced by an artificial drug-inducible dimerization domain. It is inactive when expressed in mammalian cells but rapidly gets activated upon addition of the dimerization compound AP20187. A crystal structure of the PERK kinase domain in its active conformation has recently been solved.[110] This has shown that PERK also assumes a similar back-to-back N-lobe dimer conformation as had been shown previously for both GCN2 and PKR.

PERK plays an important role in cellular survival under conditions of ER stress, particularly in secretory cells, owing to the high flux of new proteins passing through their ER. A good example of this is the pancreatic β-cell, which in addition to a heavy client protein load has to cope with dramatic swings in protein flux.[111] When circulating glucose levels are low, insulin is not required and so β-cells discontinue hormone production. However, following ingestion of a carbohydrate meal, the rise in circulating glucose triggers both the exocytosis of preexisting insulin and increased proinsulin biosynthesis. The importance of PERK in this response was first recognized in the rare human disease Wolcott–Rallison syndrome. Wolcott and Rallison had described two siblings who developed early onset insulin-dependent diabetes, bone developmental defects, and growth retardation.[112] Subsequent studies mapped the causative mutation to 2p12 in two further families.[113] Candidate gene analysis then identified mutations in the *PERK* gene, *EIF2AK3*. One of these families harbored an insertional mutation causing a frame shift that truncated the protein prior to its catalytic domain. The second family had a missense mutation within the catalytic domain impairing kinase activity. Since then, multiple disease-causing mutations have been identified in the *PERK* gene, each causing loss of catalytic function.[114]

*Perk−/−* mice recapitulate many of the phenotypic features of the Wolcott and Rallison syndrome.[102] Although *Perk−/−* mice appear normal in the neonatal period, shortly thereafter they develop hyperglycemia due to reduced insulin levels. Islets from neonatal mutant mice develop normally for the first 12 days, but between postnatal days 21 and 42, all insulin-secreting β-cells die by apoptosis. Increased signaling via the parallel UPR pathways and the development of a grossly dilated ER suggest that PERK is essential for β-cells to cope with physiological ER stress arising from day to day insulin synthesis.[11] No mutations in other eIF2α kinases have yet been found to cause diabetes, highlighting the importance of the ER stress signaling in this disease.[6,44,115]

The relevance of ER stress to diabetes has been explored by mutating the phosphorylation site of eIF2α from serine to alanine,[116] which in the mouse uncouples protein translation rates from PERK activation. This $eIF2\alpha^{S51A}$ mutant protein functions normally in the eIF2 complex but is insensitive to the eIF2α kinases. Homozygous $eIF2\alpha^{S51A}$ animals are born with severe β-cell deficiency and die within 24h. In contrast, $eIF2\alpha^{S51A}$ heterozygotes, with one normal allele, appear normal in many respects. They are, however, hypersensitive to nutrient excess.[117] When they are fed a high-fat diet, they develop obesity because of a failure to increase energy expenditure appropriately and suffer failure of glucose homeostasis. Unlike the *Perk−/−* animals that rapidly lose the capacity to make insulin, $eIF2\alpha^{S51A}$ heterozygotes develop hyperinsulinemia and hyperleptinemia, which are features of type II diabetes and the so-called metabolic syndrome that is becoming prevalent in affluent (over-fed) countries. Isolated islets from these mice show increased basal insulin secretion but reduced stimulated insulin secretion. Eventually, prolonged high-fat feeding leads to distension of the ER by misfolded proteins (including proinsulin). This demonstrates that eIF2α signaling plays an important role in helping β-cells survive prolonged increased demand and, moreover, shows that subtle perturbations of eIF2α phosphorylation can have detrimental effects on peripheral tissue insulin sensitivity.

## VI. ISR Target Genes

We have already discussed how the eIF2α kinases regulate both the rate of protein translation and also a transcriptional program governed by the transcription factor ATF4. Among the target genes of ATF4 is *CHOP*, a transcription factor that has been implicated in mediating the lethal effects of PERK signaling.[118–120] *Chop* deletion significantly delays the onset of diabetes in mice harboring the Akita insulin mutation (C96Y) in the insulin-2 gene.[121] Unlike humans, mice possess two insulin genes that are functionally redundant.[122] However, the Akita mutation causes a codominant form of diabetes suggesting a toxic gain-of-function.[123,124] The mutation causes substitution of a conserved cysteine residue required for the formation of an intramolecular disulfide bond in proinsulin.[125,126] This mutant proinsulin fails to fold correctly and so is degraded via ER-associated degradation by the proteasome.[123,127] Further, wild-type proinsulin becomes trapped in complexes with misfolded Akita proinsulin impairing its secretion.[128] Recently, a human mutation homologous to that in the Akita mouse has been identified as a rare cause of neonatal diabetes in humans.[129] Indeed, many human insulin mutations and at least one further spontaneous mouse mutation (Munich) have now been described as causing diabetes by proinsulin misfolding and UPR activation.[129,130]

Pancreatic β-cells are also protected from other insults by *Chop* deletion. For example, *Chop*−/− mice are resistant to cytokine-induced β-cell death in which NO triggers ER stress by the depletion of ER calcium stores.[131] When we attempted to identify target genes of CHOP that mediate its apparent toxicity, we obtained surprising results. Rather than activating a clear set of proapoptotic genes, CHOP was shown to induce genes that defend ongoing protein secretion during ER stress, one of which is *Gadd34*.[132] GADD34 had been thought to mediate prosurvival effects of UPR signaling, but we found that this gene was responsible for some of the apparent toxicity of CHOP.[132] This makes sense only when one considers that during ER stress, a cell must balance its need to protect its ER from excess load against the need to fulfill its secretory duties. It is likely that, on occasions, the organism may benefit from continued protein secretion even at the expense of impairing the viability of some of its secretory cells. The apparent protective effect of *Chop* deletion would reflect a resetting of this balance in favor of reduced ER load (Fig. 4). The prediction of such a model is that *Gadd34* mutation would have a similar protective effect, and this indeed is the case.[132] Mice in which the *Gadd34* gene is inactivated are outwardly healthy and, when treated with toxins that cause ER stress, suffer less tissue damage.

Given that GADD34 enhances the toxicity of ER stress and that animals remain healthy despite GADD34 inactivation, we have previously speculated that GADD34 inhibition might be of therapeutic benefit in diseases involving ER stress.[111] A screen for small molecules that ameliorate the toxicity of ER stress identified salubrinal, a compound that prevented eIF2α dephosphorylation in HSV-infected cells.[134] It is unclear whether salubrinal acts directly or selectively against eIF2α phosphatases, as *in vitro* we have not yet observed inhibition of purified GADD34-PP1 holoenzyme by this compound (unpublished observations). But, as many drugs require metabolism or the presence of cellular cofactors to inhibit their target, salubrinal may still represent the prototype for a new class of GADD34 inhibitor. Undoubtedly, any therapeutic inhibition of the eIF2α phosphatases would require careful titration. Indeed, in primary β-cell cultures, salubrinal itself can cause toxicity through eIF2α hyperphosphorylation leading to apoptosis.[135]

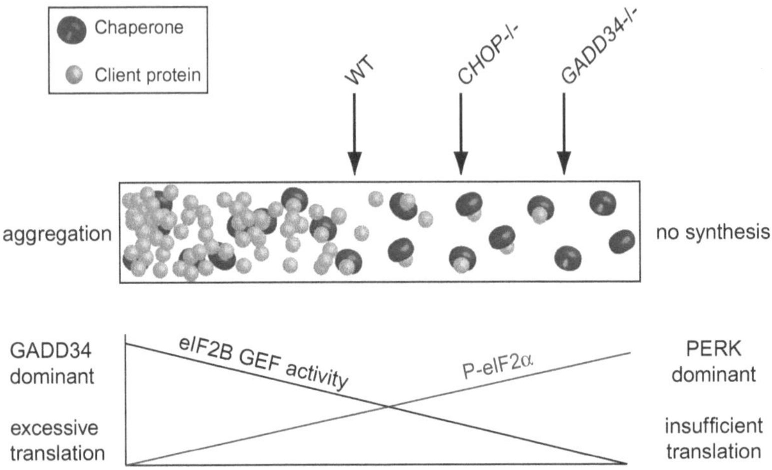


FIG. 4. Failure of ER homeostasis. The ratio of client protein within the ER (gray spheres) to the availability of chaperones (red caps) varies depending upon the protein translation rates. When new protein biosynthesis exceeds the capacity of ER chaperones to fold them, aggregation ensues. PERK promotes eIF2α phosphorylation, thus inhibiting protein translation, which is represented by a rightward shift in this figure. GADD34 induction, through eIF2α dephosphorylation and disinhibition of eIF2B GEF activity, promotes protein translation causing a leftward shift and protein aggregation. *CHOP−/−* and *GADD34* mutant cells experience lower levels of new protein aggregation because reduced new protein synthesis defends the reserve of ER chaperones (adapted with permission from Ref. 133). (See Color Insert.)

The regulation of the eIF2α phosphatases deserves further study. GADD34 has been shown to be the substrate of protein kinases that affect its toxicity in other settings, and it has been shown to interact with many other proteins (see below).[136–143] It is interesting therefore that protein synthesis in β-cells appears to be enhanced by increased glucose levels via a rapid dephosphorylation of eIF2α and that salubrinal treatment is able to blunt this response.[144] Conversely, new agents that enhance insulin biosynthesis may function by augmenting the translational recovery caused by GADD34. Incretins are a family of hormones that stimulate insulin secretion in response to nutrient entry into the gut.[145] Their actions are mediated, at least in part, by the glucagon-like peptide-1 receptor (GLP1-R). Exendin-4 is an agonist at this receptor and has been shown to enhance insulin synthesis and secretion while promoting β-cell survival.[146–148] Remarkably, these actions appear to involve induction of ATF4 and increased GADD34 expression.[149] The increased levels of ATF4 and early reversal of translational attenuation appear to allow both increased insulin synthesis while promoting β-cell survival during ER stress. An independent study showed exendin-4 to promote cellular survival by reducing levels of ER stress.[150] These encouraging findings suggest that, through their effects on the ER, GLP1-R agonists may prove useful in treating type II diabetes by augmenting insulin secretion and opposing the loss of β-cell mass.

Although it is clear that GADD34 is a phosphoprotein, the regulatory significance of this remains unclear.[151] Deletion studies have mapped some of these phosphorylation sites to the central portion of GADD34 composed of four PEST domains.[151] We have identified a number of these by mass spectroscopy and determined them to be predominantly on serine and threonine residues (unpublished data L Dalton). PEST motifs have been associated with rapid turnover of other short-lived proteins such as IκBβ and ABCA1, and their propensity to promote protein degradation can be modified by phosphorylation.[152–154] It is therefore tempting to speculate that phosphorylation of the PEST domains of GADD34 might regulate its rapid turnover. To date, only a single study has addressed the functional consequences of GADD34 phosphorylation. It was found that in a yeast-two hybrid system, GADD34 could interact with Lyn kinase,[140] a member of the src family of protein tyrosine kinases implicated in oncogenesis.[155] It was noted that GADD34 undergoes tyrosine phosphorylation in cells expressing Lyn and that this enhanced GADD34-mediated toxicity during genotoxic stress.[140]

GADD34 has been reported to participate in a number of protein–protein interactions (with BAG-1, Smad7, hSNF5/INI1, inhibitor 1, Lyn, KIF3A, HRX, Translin, TGFβ-receptor)[136,137,139,141–143,156]; however, their functional significance remains uncertain. It is worth noting that while deletion of *Gadd34* is well tolerated in mice, deletion of the constitutively expressed eIF2α phosphatase regulatory subunit *CReP* leads to growth retardation.[157] Interestingly, these mice also exhibit impaired erythropoiesis, which may relate to hyperactivity of HRI. However, when *Gadd34* and *CReP* were deleted simultaneously, the animals died at an early embryonic stage. Importantly, when the compound double knockout was crossed with the eIF2α$^{S51A}$ mouse, which is resistant to eIF2α kinases, the embryonic lethality was rescued. This strongly suggests that the predominant function of these phosphatases, at least during embryogenesis, is the dephosphorylation of eIF2α. This brings into question the significance of the other proposed GADD34 protein interactions.

Early reports implicated GADD34 in tumor suppression because of its growth arrest and proapoptotic properties.[158–161] Attempts have been made to harness this antitumor effect in the treatment of cancer. Concurrent treatment with chemotherapy and radiotherapy has been shown to upregulate GADD34 expression in cancer cells and to promote increased oncolytic viral replication, presumably through antagonizing the antiviral effects of PKR.[162,163] This upregulation of GADD34 potentiates cell killing by the virus even in highly resistant tumors such as malignant mesothelioma.[163] Conversely, loss of

GADD34 may play a role in cancer cell survival or even in cell transformation. During an effort to identify genes involved in tumor progression, an N-terminal truncated sequence of GADD34 was identified by subtractive hybridization from a rat embryonic cell line.[160] This transcript, which was named progression elevated gene-3 (PEG-3), was found to enhance tumor aggressiveness and promote angiogenesis *in vivo*.[164] Similar truncating mutations of GADD34 are frequent occurrences during transformation of rat cell lines.[165] The mechanism of the tumor-promoting activity is unclear but might plausibly involve a dominant negative interaction with endogenous GADD34. This observation has led to a number of studies using the viral delivery of constructs driven by the PEG-3/GADD34 promoter to target malignant tissue either with cytotoxins or with imaging reporters.[166,167]

## VII. ISR and Cell Growth

When a yeast cell experiences ER stress, its UPR is activated exclusively via ire1p because, in single-celled eukaryotes, eIF2α phosphorylation is restricted to gcn2p governed by amino acid sufficiency. A recent report has shown that in budding yeast at least, asymmetric inheritance of the ER can protect the mother cell from death at the expense of generating an unviable daughter cell.[168] This mechanism is not available to multicellular organisms that are unlikely to tolerate the accumulation of defective daughter cells. Instead, in metazoa, it appears that phosphorylation of eIF2α triggers cell cycle arrest at multiple levels to prevent the generation of daughter cells with stressed ER, which has relevance for many human diseases including infection and cancer. Solid tumors deficient in PERK grow poorly in hypoxic conditions thus limiting their size, supporting a role for the ISR in enabling tissues to match cell proliferation with oxygen and nutrient supply.[169] *Perk−/−* islets, for example, have been shown to develop fewer and smaller insulinomas when made to express SV40 Large T antigen owing to reduced β-cell proliferation and reduced tumor vascularity.[170] It appears that regulating the rate of protein translation in response to ambient nutrient levels is an important factor in determining cell growth and helps to integrate signaling from other stress-sensitive kinases such as AMPK and mTOR.[171]

PERK activation has been shown to cause loss of cyclin D1 resulting in G1 cell cycle arrest during ER stress, which may reflect inhibition of cyclin D1 translation or its increased proteasomal degradation.[172–174] In addition, islets from neonatal *Perk−/−* mice show a deficit of gene products involved in G2-M cell cycle transition.[175] A number of reports have identified changes in p53 phosphorylation and its subcellular localization during ER stress, which may mediate some of these effects.[176–180] Impaired ribosomal biogenesis leads to G1 cell cycle arrest through an inhibitory interaction between ribosomal proteins with MDM2, an E3 ligase of p53.[181–185] Recently, ribosomal RNA synthesis has been shown to be repressed rapidly upon eIF2α phosphorylation by PERK,[186] and ER stress has been shown to promote ribosomal protein interaction with MDM2 causing p53 stabilization.[179] However, ER stress has also been shown to induce glycogen synthase-3β (GSK-3β) kinase activity in a PERK-dependent manner causing p53 destabilization.[176,180] We recently conducted a forward genetic screen for modifiers of PERK function and identified checkpoint kinase 1 (CHK1).[187] We found that, during ER stress, CHK1 is rapidly activated and mediates a G2/M cell cycle delay. This is intriguing since CHK1 is well known to regulate cell cycle progression in the context of DNA damage.[188–191] Moreover, CHK1 has generated much interest as a potential therapeutic target. Many anticancer therapies induce DNA damage as a means to cause selective tumor killing. If cells undergo mitosis without first repairing their damaged DNA, they die through the process of mitotic catastrophe. The protein p53 is responsible in large part for mediating the cell cycle arrest that prevents mitotic catastrophe, but approximately 50% of all cancers are effectively p53 null. Nevertheless, they can respond to DNA damage by inducing G2 arrest via CHK1. It has been proposed that CHK1 inhibitors will therefore sensitize cancer cells to DNA damaging agents. But the recent finding that CHK1 is activated during ER stress by eIF2α phosphorylation may complicate this approach. In the context of DNA damage, CHK1 clearly has antiapoptotic effects for the cancer cell, but we observed reduced ER stress-induced cell death in cells depleted of CHK1 by RNAi.[187] As ER stress is commonly found within the hypoxic cores of solid tumors, it remains to be seen whether CHK1 antagonism will have a net pro- or antiapoptotic effect in this setting. Independently, a second group has identified an ER-stress-induced G2 checkpoint but found altered p53 isoform translation to be involved.[192] It remains to be determined whether both mechanisms are required or indeed if they comprise part of the same pathway.

## VIII. Structural Considerations

Modest sequence conservation within the eIF2α kinase domain makes it possible to extend observations from the crystal structures of GCN2[193] and PKR[78] to other members of the family. Both GCN2 and PKR possess a typical bi-lobal kinase domain structure with an ATP-binding cleft situated between the two lobes in a hinge region (Figs. 2A and 3A).[193] The smaller N-terminal domain takes the form of a twisted β-sheet with the large "variable insert" loop between adjacent strands of a β-sheet (termed β4–β5), while the larger

C-terminal lobe is made up of mainly α-helices and contains the eIF2α binding site. Indeed, the recent crystal structure of the PERK kinase domain that was solved to 2.8Å resolution shows many of the features that had been predicted from the PKR and GCN2 structures.[110]

Sequence alignments indicate that, in addition to the conserved kinase domain, most members of this family contain the insert loop.[108,194] This region displays only limited conservation between orthologues of each family member between species, apart from overall length, high acidity, and the presence of multiple phosphorylation sites (Fig. 1).[195] Further, paralogues within the same species exhibit very different insert loop lengths, with PERK having the longest at 224 residues.[80] The insert domain appears to be involved, at least in the case of PERK, in an unusual interaction with its substrate eIF2α. The *in vitro* affinity of PERK for eIF2α, but not its catalytic activity, is substantially affected by deleting large lengths of the insert loop, and the presence of an intact insert loop is important *in vivo* for efficient eIF2α phosphorylation.[108] *In vitro* studies suggest that the insert loop, when phosphorylated, increases the affinity of the PERK for eIF2α. Trans-autophosphorylated PERK will bind eIF2α and can deplete the eIF2 complex completely from cell lysates; however, inactive or dephosphorylated PERK fails to interact efficiently with its substrate. This is thought to involve an ionic interaction between the acidic (negatively charged) loop and a patch of basic (positively charged) residues on eIF2α adjacent to its phosphoacceptor serine 51.[108] In support of this, phosphomimetic mutations of the insert loop of inactive K618R PERK enable recruitment of eIF2α to be restored. The crystal structure obtained for GCN2 is widely considered to represent the inactive form of the kinase and therefore, in conjunction with eIF2α-bound PKR, can provide insight into the conformational changes required for activation of the kinase[193] (Fig. 2A). However, it is worth noting that GCN2 is the only member of the eIF2α kinase family that exists as a constitutive dimer[18,19] and so some caution is required when interpreting structural predications based on it. Comparisons between the apo- and ATP-bound forms of GCN2 indicate that, in its inactive form, GCN2 exhibits a closed bi-lobal structure that requires the opening of the hinge region to allow ATP binding. A hydrogen bond between arginine 794 and glutamate 792 locks the hinge and forces aspargine 793 into a position that partially blocks the ATP-binding cleft[193] (Fig. 2C). A conformational change would be required to release this autoinhibition and, in the case of GCN2, this is brought about by the binding of uncharged tRNAs.

In addition to the conformational change at its hinge region, autophosphorylation of the GCN2 activation loop at threonine 887 and threonine 882 is required for maximal kinase activity. Typically, anchoring the α- and β-phosphates of ATP in the correct position in the binding cleft is essential for kinase activity, and this is ordinarily achieved by coordination of a conserved lysine residue in sheet β3.[193] This lysine is stabilized by a classical ion-pair interaction with a glutamic acid residue. However, in GCN2, this lysine 628 is separated from glutamate 643 by approximately 11 Å, and so this glutamate instead forms a salt bridge with arginine 834. It has been proposed that, upon GCN2 autophosphorylation on threonines 887 and 882, arginine 834 interacts preferentially with the phosphothreonines. This allows glutamate 643 to swing into the catalytic site, with a corresponding conformational change of helix-αC, to form the ion pair with lysine 628 which facilitates ATP positioning.

What cannot be gained from the GCN2 structure is an appreciation of how dimerization regulates the activity of PERK since GCN2 is a constitutive dimer. The high-resolution crystal structure of PKR, however, might provide us with this information. While oligomerization commonly contributes to catalytic activation by facilitating autophosphorylation, in the case of PKR, catalytic domain dimerization plays a far more pivotal role and is required for substrate recognition. All the structural determinants for eIF2α binding are contained within the catalytic domain.

The contact surface between eIF2α and its binding partners is highly conserved, whereas the reciprocal surface on PKR is poorly conserved across the eIF2α-kinase family.[196] When compared with determined protein kinase structures, only helix-αG was significantly displaced from its canonical position in the PKR and GCN2 structures. The eIF2α/PKR binding interface spans approximately 1200 $Å^2$ and is centered on helix-αG.[196] The noncanonical positioning of the αG-helix optimally positions eIF2α toward the catalytic cleft of PKR, whereas a traditional positioning of the αG-helix would result in steric clashes and too large a gap between the phosphoacceptor binding site and the serine 51 phosphotarget to allow catalysis to occur. This may be a common feature of the eIF2α kinases, as the recently published structure of PERK has also demonstrated displacement of helix-αG into this nonclassical location.[110]

Dimerization is important in promoting binding to eIF2α and regulating the catalytic activity of PKR. The crystal structure reveals a specific antiparallel N lobe-to-N lobe dimer form. Important residues at the dimer interface are arginine 262 and asparagines 266, which form a salt bridge, and asparagine 289, tyrosine 293, and tyrosine 323 which form a hydrogen-bond triad; these residues are conserved across all members of the eIF2α kinase family.[78] In many kinases, dimerization serves only to facilitate autophosphorylation by bringing the two monomers together; however, in the case of PKR, a specific dimer orientation between the two monomers is required for full catalytic activity. Mutations of arginine 262 and asparagine 266 to give opposing charges (e.g., R262D and D266R) impaired PKR autophosphorylation and eIF2α phosphorylation.[197] However, when both mutations were introduced into the same PKR molecule, which would be predicted to restore the salt bridge, but with the opposite polarity, PKR function was restored.[197] As the residues forming these contacts are conserved across the entire family, we can infer that precise orientation of the PERK dimer is also required for catalytic function.

A mutation, R587Q in PERK, equivalent to mutating arginine 262 in PKR has been identified in a patient with the Wolcott–Rallison syndrome.[113] The structural studies of PKR suggest that this mutation disrupts the vital interlobe salt bridge and that the resultant defect in dimerization may be the root cause of the disease. In addition, the same mutational studies described above for PKR were also conducted for the equivalent residues in PERK: arginine 599 and glutamate 603. The single mutations blocked autophosphorylation, activation, and eIF2α phosphorylation, whereas the double mutation restored the salt bridge and rescued PERK function. These data highlight the importance of the precise N lobe-to-N lobe mode of kinase domain dimerization for PERK activation and autophosphorylation.[197]

Aside from dimerization, another central point of PKR catalytic regulation is autophosphorylation of conserved residues in its activation segment. Threonine 446 phosphorylation, in particular, is necessary for kinase function, with the equivalent residue in PERK being threonine 980. Indeed, in the recent crystal structure of PERK, this residue was shown to be phosphorylated stoichiometrically.[110] Mutation of conserved dimer interface residues in PKR blocked autophosphorylation on threonine 446, indicating that catalytic domain dimerization promotes activation segment autophosphorylation.[78] In turn, activation segment phosphorylation is essential for efficient PKR catalytic function, eIF2α substrate recognition, and stabilization of the dimer configuration. This was shown by a T446A mutation preventing eIF2α recognition and phosphorylation, as well as by a mutation blocking PKR autophosphorylation also impairing dimerization, respectively.

The mechanics of autophosphorylation has been a source of debate. It has been unclear, for example, whether this occurs in a *cis* or in a *trans* manner within the dimer. Interestingly, the back-to-back arrangement of the two monomers in the PKR dimer means that each activation segment is inaccessible to the active site of the other protomer. Hence phosphorylation must occur either in *cis* or through dimer–dimer/dimer–monomer interactions. We have observed that a kinase-inactivation mutant of PERK is unable to autophosphorylate and therefore cannot form dimers, nor is it phosphorylated when mixed with active PERK-WT dimers (unpublished observations SJ Marciniak). This might suggest that dimers of PERK cannot phosphorylate other dimers, indicating that PERK may *cis*-phosphorylate its activation segment.

## IX. Concluding Remarks

In summary, phosphorylation of eIF2α is an ancient mechanism that allows stress to be linked directly to the rate of protein translation and indirectly to an adaptive transcriptional program. The response it governs helps to adapt cells to many challenges including nutrient deprivation, protein misfolding, and viral infection. This crucial role means that its efficient functioning is necessary if an organism is to remain healthy; on the contrary, defects in this pathway are linked strongly with disease. Now that our understanding of the protein interactions involved has reached extraordinary molecular detail, it is likely that components of the ISR will become important targets for the development of novel therapeutics.

### References


1. Tzamarias D, Thireos G. Evidence that the GCN2 protein kinase regulates reinitiation by yeast ribosomes. *EMBO J* 1988;**7**:3547–51.
2. Dever TE, Feng L, Wek RC, Cigan AM, Donahue TF, Hinnebusch AG. Phosphorylation of initiation factor 2 alpha by protein kinase GCN2 mediates gene-specific translational control of GCN4 in yeast. *Cell* 1992;**68**:585–96.
3. Dever TE, Chen JJ, Barber GN, Cigan AM, Feng L, Donahue TF, et al. Mammalian eukaryotic initiation factor 2 alpha kinases functionally substitute for GCN2 protein kinase in the GCN4 translational control mechanism of yeast. *Proc Natl Acad Sci USA* 1993;**90**:4616–20.
4. Hinnebusch AG. Mechanism and regulation of initiator methionyl-tRNA binding to ribosomes. In: Sonenberg N, Hershey JWB, Mathews MB, editors. *Translational control of gene expression*. Cold Spring Harbor: CSHL Press; 2000. pp. 185–243.
5. Berlanga JJ, Santoyo J, De Haro C. Characterization of a mammalian homolog of the GCN2 eukaryotic initiation factor 2alpha kinase. *Eur J Biochem* 1999;**265**:754–62.
6. Zhang P, McGrath BC, Reinert J, Olsen DS, Lei L, Gill S, et al. The GCN2 eIF2alpha kinase is required for adaptation to amino acid deprivation in mice. *Mol Cell Biol* 2002;**22**:6681–8.
7. Kostura M, Mathews MB. Purification and activation of the double-stranded RNA-dependent eIF-2 kinase DAI. *Mol Cell Biol* 1989;**9**:1576–86.
8. Meurs E, Chong K, Galabru J, Thomas NS, Kerr IM, Williams BR, et al. Molecular cloning and characterization of the human double-stranded RNA-activated protein kinase induced by interferon. *Cell* 1990;**62**:379–90.
9. Chen JJ, Throop MS, Gehrke L, Kuo I, Pal JK, Brodsky M, et al. Cloning of the cDNA of the heme-regulated eukaryotic initiation factor 2 alpha (eIF-2 alpha) kinase of rabbit reticulocytes: homology to yeast GCN2 protein kinase and human double-stranded-RNA-dependent eIF-2 alpha kinase. *Proc Natl Acad Sci USA* 1991;**88**:7729–33.
10. Shi Y, Vattem KM, Sood R, An J, Liang J, Stramm L, et al. Identification and characterization of pancreatic eukaryotic initiation factor 2 alpha-subunit kinase, PEK, involved in translational control. *Mol Cell Biol* 1998;**18**:7499–509.
11. Harding H, Zhang Y, Ron D. Translation and protein folding are coupled by an endoplasmic reticulum resident kinase. *Nature* 1999;**397**:271–4.

12. Saitou N, Nei M. The neighbor-joining method: a new method for reconstructing phylogenetic trees. *Mol Biol Evol* 1987;**4**:406–25.
13. Zuckerkandl E, Pauling L. Evolutionary divergence and convergence in proteins. In: Bryson V, Vogel HJ, editors. *Evolving genes and proteins*. New York: Academic Press; 1965. pp. 97–166.
14. Tamura K, Dudley J, Nei M, Kumar S. MEGA4: Molecular evolutionary genetics analysis (MEGA) software version 4.0. *Mol Biol Evol* 2007;**24**:1596–9.
15. Harding H, Zhang Y, Zeng H, Novoa I, Lu P, Calfon M, et al. An integrated stress response regulates amino acid metabolism and resistance to oxidative stress. *Mol Cell* 2003;**11**:619–33.
16. Tzamarias D, Roussou I, Thireos G. Coupling of GCN4 mRNA translational activation with decreased rates of polypeptide chain initiation. *Cell* 1989;**57**:947–54.
17. Wek SA, Zhu S, Wek RC. The histidyl-tRNA synthetase-related sequence in the eIF-2 alpha protein kinase GCN2 interacts with tRNA and is required for activation in response to starvation for different amino acids. *Mol Cell Biol* 1995;**15**:4497–506.
18. Dong J, Qiu H, Garcia-Barrio M, Anderson J, Hinnebusch AG. Uncharged tRNA activates GCN2 by displacing the protein kinase moiety from a bipartite tRNA-binding domain. *Mol Cell* 2000;**6**:269–79.
19. Qiu H, Dong J, Hu C, Francklyn CS, Hinnebusch AG. The tRNA-binding moiety in GCN2 contains a dimerization domain that interacts with the kinase domain and is required for tRNA binding and kinase activation. *EMBO J* 2001;**20**:1425–38.
20. Anthony TG, McDaniel BJ, Byerley RL, McGrath BC, Cavener DR, McNurlan MA, et al. Preservation of liver protein synthesis during dietary leucine deprivation occurs at the expense of skeletal muscle mass in mice deleted for eIF2 kinase GCN2. *J Biol Chem* 2004;**279**:36553–61.
21. Gietzen DW, Ross CM, Hao S, Sharp JW. Phosphorylation of eIF2alpha is involved in the signaling of indispensable amino acid deficiency in the anterior piriform cortex of the brain in rats. *J Nutr* 2004;**134**:717–23.
22. Hao S, Sharp JW, Ross-Inta CM, McDaniel BJ, Anthony TG, Wek RC, et al. Uncharged tRNA and sensing of amino acid deficiency in mammalian piriform cortex. *Science* 2005;**307**:1776–8.
23. Maurin AC, Jousse C, Averous J, Parry L, Bruhat A, Cherasse Y, et al. The GCN2 kinase biases feeding behavior to maintain amino acid homeostasis in omnivores. *Cell Metab* 2005;**1**:273–7.
24. Chan ED, Chan J, Schluger NW. What is the role of nitric oxide in murine and human host defense against tuberculosis. *Am J Respir Cell Mol Biol* 2001;**25**:606–12.
25. Szabo C, Ischiropoulos H, Radi R. Peroxynitrite: biochemistry, pathophysiology and development of therapeutics. *Nat Rev Drug Discov* 2007;**6**:662–80.
26. Xia Y, Dawson VL, Dawson TM, Snyder SH, Zweier JL. Nitric oxide synthase generates superoxide and nitric oxide in arginine-depleted cells leading to peroxynitrite-mediated cellular injury. *Proc Natl Acad Sci USA* 1996;**93**:6770–4.
27. Lee J, Ryu H, Ferrante RJ, Morris Jr. SM, Ratan RR. Translational control of inducible nitric oxide synthase expression by arginine can explain the arginine paradox. *Proc Natl Acad Sci USA* 2003;**100**:4843–8.
28. Mueller PP, Hinnebusch AG. Multiple upstream AUG codons mediate translational control of GCN4. *Cell* 1986;**45**:201–7.
29. Abastado JP, Miller PF, Jackson BM, Hinnebusch AG. Suppression of ribosomal reinitiation at upstream open reading frames in amino acid-starved cells forms the basis for GCN4 translational control. *Mol Cell Biol* 1991;**11**:486–96.
30. Fogli A, Wong K, Eymard-Pierre E, Wenger J, Bouffard JP, Goldin E, et al. Cree leukoencephalopathy and CACH/VWM disease are allelic at the EIF2B5 locus. *Ann Neurol* 2002;**52**:506–10.

31. Fogli A, Schiffmann R, Bertini E, Ughetto S, Combes P, Eymard-Pierre E, et al. The effect of genotype on the natural history of eIF2B-related leukodystrophies. *Neurology* 2004;**62**:1509–17.
32. Richardson JP, Mohammad SS, Pavitt GD. Mutations causing childhood ataxia with central nervous system hypomyelination reduce eukaryotic initiation factor 2B complex formation and activity. *Mol Cell Biol* 2004;**24**:2352–63.
33. Kantor L, Harding HP, Ron D, Schiffmann R, Kaneski CR, Kimball SR, et al. Heightened stress response in primary fibroblasts expressing mutant eIF2B genes from CACH/VWM leukodystrophy patients. *Hum Genet* 2005;**118**:99–106.
34. Siekierka J, Manne V, Ochoa S. Mechanism of translational control by partial phosphorylation of the alpha subunit of eukaryotic initiation factor 2. *Proc Natl Acad Sci USA* 1984;**81**:352–6.
35. Matts RL, Levin DH, London IM. Effect of phosphorylation of the alpha-subunit of eukaryotic initiation factor 2 on the function of reversing factor in the initiation of protein synthesis. *Proc Natl Acad Sci USA* 1983;**80**:2559–63.
36. Palam LR, Baird TD, Wek RC. Phosphorylation of eIF2 facilitates ribosomal bypass of an inhibitory upstream ORF to enhance CHOP translation. *J Biol Chem* 2011;**286**:10939.
37. Lee YY, Cevallos RC, Jan E. An upstream open reading frame regulates translation of GADD34 during cellular stresses that induce eIF2alpha phosphorylation. *J Biol Chem* 2009;**284**:6661–73.
38. Mueller PP, Jackson BM, Miller PF, Hinnebusch AG. The first and fourth upstream open reading frames in GCN4 mRNA have similar initiation efficiencies but respond differently in translational control to change in length and sequence. *Mol Cell Biol* 1988;**8**:5439–47.
39. Mueller PP, Harashima S, Hinnebusch AG. A segment of GCN4 mRNA containing the upstream AUG codons confers translational control upon a heterologous yeast transcript. *Proc Natl Acad Sci USA* 1987;**84**:2863–7.
40. Harding H, Novoa I, Zhang Y, Zeng H, Wek RC, Schapira M, et al. Regulated translation initiation controls stress-induced gene expression in mammalian cells. *Mol Cell* 2000;**6**:1099–108.
41. Ma Y, Brewer JW, Diehl JA, Hendershot LM. Two distinct stress signaling pathways converge upon the CHOP promoter during the mammalian unfolded protein response. *J Mol Biol* 2002;**318**:1351–65.
42. Lu PD, Harding HP, Ron D. Translation re-initiation at alternative open reading frames regulates gene expression in an integrated stress response. *J Cell Biol* 2004;**167**:27–33.
43. Lu L, Han AP, Chen JJ. Translation initiation control by heme-regulated eukaryotic initiation factor 2alpha kinase in erythroid cells under cytoplasmic stresses. *Mol Cell Biol* 2001;**21**:7971–80.
44. Han AP, Yu C, Lu L, Fujiwara Y, Browne C, Chin G, et al. Heme-regulated eIF2alpha kinase (HRI) is required for translational regulation and survival of erythroid precursors in iron deficiency. *EMBO J* 2001;**20**:6909–18.
45. Han AP, Fleming MD, Chen JJ. Heme-regulated eIF2alpha kinase modifies the phenotypic severity of murine models of erythropoietic protoporphyria and beta-thalassemia. *J Clin Invest* 2005;**115**:1562–70.
46. Rafie-Kolpin M, Chefalo PJ, Hussain Z, Hahn J, Uma S, Matts RL, et al. Two heme-binding domains of heme-regulated eukaryotic initiation factor-2alpha kinase. N terminus and kinase insertion. *J Biol Chem* 2000;**275**:5171–8.
47. Rafie-Kolpin M, Han AP, Chen JJ. Autophosphorylation of threonine 485 in the activation loop is essential for attaining eIF2alpha kinase activity of HRI. *Biochemistry* 2003;**42**:6536–44.
48. Miksanova M, Igarashi J, Minami M, Sagami I, Yamauchi S, Kurokawa H, et al. Characterization of heme-regulated eIF2alpha kinase: roles of the N-terminal domain in the oligomeric state, heme binding, catalysis, and inhibition. *Biochemistry* 2006;**45**:9894–905.
49. Chefalo PJ, Oh J, Rafie-Kolpin M, Kan B, Chen JJ. Heme-regulated eIF-2alpha kinase purifies as a hemoprotein. *Eur J Biochem* 1998;**258**:820–30.

50. Bauer BN, Rafie-Kolpin M, Lu L, Han A, Chen JJ. Multiple autophosphorylation is essential for the formation of the active and stable homodimer of heme-regulated eIF2alpha kinase. *Biochemistry* 2001;**40**:11543–51.
51. Yun BG, Matts JA, Matts RL. Interdomain interactions regulate the activation of the heme-regulated eIF 2 alpha kinase. *Biochim Biophys Acta* 2005;**1725**:174–81.
52. McEwen E, Kedersha N, Song B, Scheuner D, Gilks N, Han A, et al. Heme-regulated inhibitor kinase-mediated phosphorylation of eukaryotic translation initiation factor 2 inhibits translation, induces stress granule formation, and mediates survival upon arsenite exposure. *J Biol Chem* 2005;**280**:16925–33.
53. Berry MJ, Knutson GS, Lasky SR, Munemitsu SM, Samuel CE. Mechanism of interferon action. Purification and substrate specificities of the double-stranded RNA-dependent protein kinase from untreated and interferon-treated mouse fibroblasts. *J Biol Chem* 1985;**260**:11240–7.
54. Galabru J, Hovanessian A. Autophosphorylation of the protein kinase dependent on double-stranded RNA. *J Biol Chem* 1987;**262**:15538–44.
55. Kato H, Sato S, Yoneyama M, Yamamoto M, Uematsu S, Matsui K, et al. Cell type-specific involvement of RIG-I in antiviral response. *Immunity* 2005;**23**:19–28.
56. Metz DH, Esteban M. Interferon inhibits viral protein synthesis in L cells infected with vaccinia virus. *Nature* 1972;**238**:385–8.
57. Balachandran S, Roberts PC, Brown LE, Truong H, Pattnaik AK, Archer DR, et al. Essential role for the dsRNA-dependent protein kinase PKR in innate immunity to viral infection. *Immunity* 2000;**13**:129–41.
58. Koradi R, Billeter M, Wuthrich K. MOLMOL: a program for display and analysis of macromolecular structures. *J Mol Graph* 1996;**14**:51–5. [29–32].
59. Nanduri S, Carpick BW, Yang Y, Williams BR, Qin J. Structure of the double-stranded RNA-binding domain of the protein kinase PKR reveals the molecular basis of its dsRNA-mediated activation. *EMBO J* 1998;**17**:5458–65.
60. Sadler AJ, Williams BR. Structure and function of the protein kinase R. *Curr Top Microbiol Immunol* 2007;**316**:253–92.
61. Patel CV, Handy I, Goldsmith T, Patel RC. PACT, a stress-modulated cellular activator of interferon-induced double-stranded RNA-activated protein kinase, PKR. *J Biol Chem* 2000;**275**:37993–8.
62. Ito T, Yang M, May WS. RAX, a cellular activator for double-stranded RNA-dependent protein kinase during stress signaling. *J Biol Chem* 1999;**274**:15427–32.
63. Patel RC, Sen GC. PACT, a protein activator of the interferon-induced protein kinase, PKR. *EMBO J* 1998;**17**:4379–90.
64. Ruvolo PP, Gao F, Blalock WL, Deng X, May WS. Ceramide regulates protein synthesis by a novel mechanism involving the cellular PKR activator RAX. *J Biol Chem* 2001;**276**:11754–8.
65. Peters GA, Dickerman B, Sen GC. Biochemical analysis of PKR activation by PACT. *Biochemistry* 2009;**48**:7441–7.
66. Peters GA, Hartmann R, Qin J, Sen GC. Modular structure of PACT: distinct domains for binding and activating PKR. *Mol Cell Biol* 2001;**21**:1908–20.
67. Zamanian-Daryoush M, Mogensen TH, DiDonato JA, Williams BR. NF-kappaB activation by double-stranded-RNA-activated protein kinase (PKR) is mediated through NF-kappaB-inducing kinase and IkappaB kinase. *Mol Cell Biol* 2000;**20**:1278–90.
68. Gil J, Garcia MA, Gomez-Puertas P, Guerra S, Rullas J, Nakano H, et al. TRAF family proteins link PKR with NF-kappa B activation. *Mol Cell Biol* 2004;**24**:4502–12.
69. Takada Y, Ichikawa H, Pataer A, Swisher S, Aggarwal BB. Genetic deletion of PKR abrogates TNF-induced activation of IkappaBalpha kinase, JNK, Akt and cell proliferation but potentiates p44/p42 MAPK and p38 MAPK activation. *Oncogene* 2007;**26**:1201–12.

70. Bonnet MC, Daurat C, Ottone C, Meurs EF. The N-terminus of PKR is responsible for the activation of the NF-kappaB signaling pathway by interacting with the IKK complex. *Cell Signal* 2006;**18**:1865–75.
71. Bonnet MC, Weil R, Dam E, Hovanessian AG, Meurs EF. PKR stimulates NF-kappaB irrespective of its kinase function by interacting with the IkappaB kinase complex. *Mol Cell Biol* 2000;**20**:4532–42.
72. Gil J, Rullas J, Garcia MA, Alcami J, Esteban M. The catalytic activity of dsRNA-dependent protein kinase, PKR, is required for NF-kappaB activation. *Oncogene* 2001;**20**:385–94.
73. Nanduri S, Rahman F, Williams BR, Qin J. A dynamically tuned double-stranded RNA binding mechanism for the activation of antiviral kinase PKR. *EMBO J* 2000;**19**:5567–74.
74. Bycroft M, Grunert S, Murzin AG, Proctor M, St Johnston D. NMR solution structure of a dsRNA binding domain from Drosophila staufen protein reveals homology to the N-terminal domain of ribosomal protein S5. *EMBO J* 1995;**14**:3563–71.
75. Taylor DR, Lee SB, Romano PR, Marshak DR, Hinnebusch AG, Esteban M, et al. Autophosphorylation sites participate in the activation of the double-stranded-RNA-activated protein kinase PKR. *Mol Cell Biol* 1996;**16**:6295–302.
76. Romano PR, Garcia-Barrio MT, Zhang X, Wang Q, Taylor DR, Zhang F, et al. Autophosphorylation in the activation loop is required for full kinase activity in vivo of human and yeast eukaryotic initiation factor 2alpha kinases PKR and GCN2. *Mol Cell Biol* 1998;**18**:2282–97.
77. Zhang F, Romano PR, Nagamura-Inoue T, Tian B, Dever TE, Mathews MB, et al. Binding of double-stranded RNA to protein kinase PKR is required for dimerization and promotes critical autophosphorylation events in the activation loop. *J Biol Chem* 2001;**276**:24946–58.
78. Dey M, Cao C, Dar AC, Tamura T, Ozato K, Sicheri F, et al. Mechanistic link between PKR dimerization, autophosphorylation, and eIF2alpha substrate recognition. *Cell* 2005;**122**:901–13.
79. Zhang X, Herring CJ, Romano PR, Szczepanowska J, Brzeska H, Hinnebusch AG, et al. Identification of phosphorylation sites in proteins separated by polyacrylamide gel electrophoresis. *Anal Chem* 1998;**70**:2050–9.
80. Ma Y, Lu Y, Zeng H, Ron D, Mo W, Neubert TA. Characterization of phosphopeptides from protein digests using matrix-assisted laser desorption/ionization time-of-flight mass spectrometry and nanoelectrospray quadrupole time-of-flight mass spectrometry. *Rapid Commun Mass Spectrom* 2001;**15**:1693–700.
81. Su Q, Wang S, Baltzis D, Qu LK, Wong AH, Koromilas AE. Tyrosine phosphorylation acts as a molecular switch to full-scale activation of the eIF2alpha RNA-dependent protein kinase. *Proc Natl Acad Sci USA* 2006;**103**:63–8.
82. Su Q, Wang S, Gao HQ, Kazemi S, Harding HP, Ron D, et al. Modulation of the eukaryotic initiation factor 2 alpha-subunit kinase PERK by tyrosine phosphorylation. *J Biol Chem* 2008;**283**:469–75.
83. Lu J, O'Hara EB, Trieselmann BA, Romano PR, Dever TE. The interferon-induced double-stranded RNA-activated protein kinase PKR will phosphorylate serine, threonine, or tyrosine at residue 51 in eukaryotic initiation factor 2alpha. *J Biol Chem* 1999;**274**:32198–203.
84. Langland JO, Jacobs BL. Inhibition of PKR by vaccinia virus: role of the N- and C-terminal domains of E3L. *Virology* 2004;**324**:419–29.
85. Davies MV, Chang HW, Jacobs BL, Kaufman RJ. The E3L and K3L vaccinia virus gene products stimulate translation through inhibition of the double-stranded RNA-dependent protein kinase by different mechanisms. *J Virol* 1993;**67**:1688–92.
86. Li S, Min JY, Krug RM, Sen GC. Binding of the influenza A virus NS1 protein to PKR mediates the inhibition of its activation by either PACT or double-stranded RNA. *Virology* 2006;**349**:13–21.

87. Lu Y, Wambach M, Katze MG, Krug RM. Binding of the influenza virus NS1 protein to double-stranded RNA inhibits the activation of the protein kinase that phosphorylates the elF-2 translation initiation factor. *Virology* 1995;**214**:222–8.
88. Yue Z, Shatkin AJ. Double-stranded RNA-dependent protein kinase (PKR) is regulated by reovirus structural proteins. *Virology* 1997;**234**:364–71.
89. Poppers J, Mulvey M, Khoo D, Mohr I. Inhibition of PKR activation by the proline-rich RNA binding domain of the herpes simplex virus type 1 Us11 protein. *J Virol* 2000;**74**:11215–21.
90. He Y, Tan SL, Tareen SU, Vijaysri S, Langland JO, Jacobs BL, et al. Regulation of mRNA translation and cellular signaling by hepatitis C virus nonstructural protein NS5A. *J Virol* 2001;**75**:5090–8.
91. Brand SR, Kobayashi R, Mathews MB. The Tat protein of human immunodeficiency virus type 1 is a substrate and inhibitor of the interferon-induced, virally activated protein kinase, PKR. *J Biol Chem* 1997;**272**:8388–95.
92. Davies MV, Elroy-Stein O, Jagus R, Moss B, Kaufman RJ. The vaccinia virus K3L gene product potentiates translation by inhibiting double-stranded-RNA-activated protein kinase and phosphorylation of the alpha subunit of eukaryotic initiation factor 2. *J Virol* 1992;**66**:1943–50.
93. Carroll K, Elroy-Stein O, Moss B, Jagus R. Recombinant vaccinia virus K3L gene product prevents activation of double-stranded RNA-dependent, initiation factor 2 alpha-specific protein kinase. *J Biol Chem* 1993;**268**:12837–42.
94. Gale Jr. M, Tan SL, Wambach M, Katze MG. Interaction of the interferon-induced PKR protein kinase with inhibitory proteins P58IPK and vaccinia virus K3L is mediated by unique domains: implications for kinase regulation. *Mol Cell Biol* 1996;**16**:4172–81.
95. Van Huizen R, Martindale JL, Gorospe M, Holbrook NJ. P58IPK, a novel endoplasmic reticulum stress-inducible protein and potential negative regulator of eIF2alpha signaling. *J Biol Chem* 2003;**278**:15558–64.
96. Rutkowski DT, Kang SW, Goodman AG, Garrison JL, Taunton J, Katze MG, et al. The role of p58IPK in protecting the stressed endoplasmic reticulum. *Mol Biol Cell* 2007;**18**:3681–91.
97. Brown SM, MacLean AR, McKie EA, Harland J. The herpes simplex virus virulence factor ICP34.5 and the cellular protein MyD116 complex with proliferating cell nuclear antigen through the 63-amino-acid domain conserved in ICP34.5, MyD116, and GADD34. *J Virol* 1997;**71**:9442–9.
98. Jousse C, Oyadomari S, Novoa I, Lu PD, Zhang Y, Harding HP, et al. Inhibition of a constitutive translation initiation factor 2a phosphatase, CReP, promotes survival of stressed cells. *J Cell Biol* 2003;**163**:767–75.
99. Novoa I, Zeng H, Harding H, Ron D. Feedback inhibition of the unfolded protein response by GADD34-mediated dephosphorylation of eIF2a. *J Cell Biol* 2001;**153**:1011–22.
100. Ron D, Walter P. Signal integration in the endoplasmic reticulum unfolded protein response. *Nat Rev Mol Cell Biol* 2007;**8**:519–29.
101. Harding H, Zhang Y, Bertolotti A, Zeng H, Ron D. *Perk* is essential for translational regulation and cell survival during the unfolded protein response. *Mol Cell* 2000;**5**:897–904.
102. Harding H, Zeng H, Zhang Y, Jungreis R, Chung P, Plesken H, et al. Diabetes Mellitus and excocrine pancreatic dysfunction in Perk-/- mice reveals a role for translational control in survival of secretory cells. *Mol Cell* 2001;**7**:1153–63.
103. Bertolotti A, Zhang Y, Hendershot L, Harding H, Ron D. Dynamic interaction of BiP and the ER stress transducers in the unfolded protein response. *Nat Cell Biol* 2000;**2**:326–32.
104. Kimata Y, Oikawa D, Shimizu Y, Ishiwata-Kimata Y, Kohno K. A role for BiP as an adjustor for the endoplasmic reticulum stress-sensing protein Ire1. *J Cell Biol* 2004;**167**:445–56.
105. Oikawa D, Kimata Y, Kohno K. Self-association and BiP dissociation are not sufficient for activation of the ER stress sensor Ire1. *J Cell Sci* 2007;**120**:1681–8.

106. Credle JJ, Finer-Moore JS, Papa FR, Stroud RM, Walter P. Inaugural Article: On the mechanism of sensing unfolded protein in the endoplasmic reticulum. *Proc Natl Acad Sci USA* 2005;**102**:18773–84.
107. Kimata Y, Ishiwata-Kimata Y, Ito T, Hirata A, Suzuki T, Oikawa D, et al. Two regulatory steps of ER-stress sensor Ire1 involving its cluster formation and interaction with unfolded proteins. *J Cell Biol* 2007;**179**:75–86.
108. Marciniak SJ, Garcia-Bonilla L, Hu J, Harding HP, Ron D. Activation-dependent substrate recruitment by the eukaryotic translation initiation factor 2 kinase PERK. *J Cell Biol* 2006;**172**:201–9.
109. Lu PD, Jousse C, Marciniak SJ, Zhang Y, Novoa I, Scheuner D, et al. Cytoprotection by pre-emptive conditional phosphorylation of translation initiation factor 2. *EMBO J* 2004;**23**:169–79.
110. Cui W, Li J, Ron D, Sha B. The structure of the PERK kinase domain suggests the mechanism for its activation. *Acta Crystallogr D Biol Crystallogr* 2011;**67**:423–8.
111. Thomas SE, Dalton LE, Daly ML, Malzer E, Marciniak SJ. Diabetes as a disease of endoplasmic reticulum stress. *Diabetes Metab Res Rev* 2010;**26**:611–21.
112. Wolcott C, Rallison M. Infancy-onset diabetes mellitus and multiple epiphyseal dysplasia. *J Pediatrics* 1972;**80**:292–7.
113. Delepine M, Nicolino M, Barrett T, Golamaully M, Lathrop GM, Julier C. EIF2AK3, encoding translation initiation factor 2-alpha kinase 3, is mutated in patients with Wolcott-Rallison syndrome. *Nat Genet* 2000;**25**:406–9.
114. Rubio-Cabezas O, Patch AM, Minton JA, Flanagan SE, Edghill EL, Hussain K, et al. Wolcott-Rallison syndrome is the most common genetic cause of permanent neonatal diabetes in consanguineous families. *J Clin Endocrinol Metab* 2009;**94**:4162–70.
115. Yang YL, Reis LF, Pavlovic J, Aguzzi A, Schafer R, Kumar A, et al. Deficient signaling in mice devoid of double-stranded RNA-dependent protein kinase. *EMBO J* 1995;**14**:6095–106.
116. Scheuner D, Song B, McEwen E, Gillespie P, Saunders T, Bonner-Weir S, et al. Translational control is required for the unfolded protein response and in-vivo glucose homeostasis. *Mol Cell* 2001;**7**:1165–76.
117. Scheuner D, Mierde DV, Song B, Flamez D, Creemers JW, Tsukamoto K, et al. Control of mRNA translation preserves endoplasmic reticulum function in beta cells and maintains glucose homeostasis. *Nat Med* 2005;**11**:757–64.
118. Wang X-Z, Lawson B, Brewer J, Zinszner H, Sanjay A, Mi L, et al. Signals from the stressed endoplasmic reticulum induce C/EBP homologous protein (CHOP/GADD153). *Mol Cell Biol* 1996;**16**:4273–80.
119. Zinszner H, Kuroda M, Wang X, Batchvarova N, Lightfoot RT, Remotti H, et al. CHOP is implicated in programmed cell death in response to impaired function of the endoplasmic reticulum. *Genes Dev* 1998;**12**:982–95.
120. Jousse C, Bruhat A, Harding HP, Ferrara M, Ron D, Fafournoux P. Amino acid limitation regulates CHOP expression through a specific pathway independent of the unfolded protein response. *FEBS Lett* 1999;**448**:211–6.
121. Oyadomari S, Koizumi A, Takeda K, Gotoh T, Akira S, Araki E, et al. Targeted disruption of the Chop gene delays endoplasmic reticulum stress-mediated diabetes. *J Clin Invest* 2002;**109**:525–32.
122. Leroux L, Desbois P, Lamotte L, Duvillie B, Cordonnier N, Jackerott M, et al. Compensatory responses in mice carrying a null mutation for Ins1 or Ins2. *Diabetes* 2001;**50**(Suppl. 1): S150–3.
123. Wang J, Takeuchi T, Tanaka S, Kubo SK, Kayo T, Lu D, et al. A mutation in the insulin 2 gene induces diabetes with severe pancreatic beta-cell dysfunction in the Mody mouse. *J Clin Invest* 1999;**103**:27–37.

124. Izumi T, Yokota-Hashimoto H, Zhao S, Wang J, Halban PA, Takeuchi T. Dominant negative pathogenesis by mutant proinsulin in the Akita diabetic mouse. *Diabetes* 2003;**52**:409–16.
125. Nozaki J, Kubota H, Yoshida H, Naitoh M, Goji J, Yoshinaga T, et al. The endoplasmic reticulum stress response is stimulated through the continuous activation of transcription factors ATF6 and XBP1 in Ins2+/Akita pancreatic beta cells. *Genes Cells* 2004;**9**:261–70.
126. Liu M, Li Y, Cavener D, Arvan P. Proinsulin disulfide maturation and misfolding in the endoplasmic reticulum. *J Biol Chem* 2005;**280**:13209–12.
127. Allen JR, Nguyen LX, Sargent KE, Lipson KL, Hackett A, Urano F. High ER stress in beta-cells stimulates intracellular degradation of misfolded insulin. *Biochem Biophys Res Commun* 2004;**324**:166–70.
128. Liu M, Hodish I, Rhodes CJ, Arvan P. Proinsulin maturation, misfolding, and proteotoxicity. *Proc Natl Acad Sci USA* 2007;**104**:15841–6.
129. Stoy J, Edghill EL, Flanagan SE, Ye H, Paz VP, Pluzhnikov A, et al. Insulin gene mutations as a cause of permanent neonatal diabetes. *Proc Natl Acad Sci USA* 2007;**104**:15040–4.
130. Herbach N, Rathkolb B, Kemter E, Pichl L, Klaften M, de Angelis MH, et al. Dominant-negative effects of a novel mutated Ins2 allele causes early-onset diabetes and severe beta-cell loss in Munich Ins2C95S mutant mice. *Diabetes* 2007;**56**:1268–76.
131. Oyadomari S, Takeda K, Takiguchi M, Gotoh T, Matsumoto M, Wada I, et al. Nitric oxide-induced apoptosis in pancreatic b-cells is mediated by the endoplasmic reticulum stress pathway. *Proc Natl Acad Sci USA* 2001;**98**:10845–50.
132. Marciniak SJ, Yun CY, Oyadomari S, Novoa I, Zhang Y, Jungreis R, et al. CHOP induces death by promoting protein synthesis and oxidation in the stressed endoplasmic reticulum. *Genes Dev* 2004;**18**:3066–77.
133. Marciniak SJ, Ron D. Endoplasmic reticulum signalling in disease. *Physiol Rev* 2006;**86**:1133–49.
134. Boyce M, Bryant KF, Jousse C, Long K, Harding HP, Scheuner D, et al. A Selective inhibitor of eIF2a dephosphorylation protects cells from ER stress. *Science* 2005;**307**:935–9.
135. Cnop M, Ladriere L, Hekerman P, Ortis F, Cardozo AK, Dogusan Z, et al. Selective inhibition of eukaryotic translation initiation factor 2 alpha dephosphorylation potentiates fatty acid-induced endoplasmic reticulum stress and causes pancreatic beta-cell dysfunction and apoptosis. *J Biol Chem* 2007;**282**:3989–97.
136. Adler HT, Chinery R, Wu DY, Kussick SJ, Payne JM, Fornace Jr. AJ, et al. Leukemic HRX fusion proteins inhibit GADD34-induced apoptosis and associate with the GADD34 and hSNF5/INI1 proteins. *Mol Cell Biol* 1999;**19**:7050–60.
137. Hasegawa T, Isobe K. Evidence for the interaction between Translin and GADD34 in mammalian cells. *Biochim Biophys Acta* 1999;**1428**:161–8.
138. Hasegawa T, Xiao H, Hamajima F, Isobe K. Interaction between DNA-damage protein GADD34 and a new member of the Hsp40 family of heat shock proteins that is induced by a DNA-damaging reagent. *Biochem J* 2000;**352**:795–800.
139. Hasegawa T, Yagi A, Isobe K. Interaction between GADD34 and kinesin superfamily, KIF3A. *Biochem Biophys Res Commun* 2000;**267**:593–6.
140. Grishin AV, Azhipa O, Semenov I, Corey SJ. Interaction between growth arrest-DNA damage protein 34 and Src kinase Lyn negatively regulates genotoxic apoptosis. *Proc Natl Acad Sci USA* 2001;**98**:10172–7.
141. Wu DY, Tkachuck DC, Roberson RS, Schubach WH. The human SNF5/INI1 protein facilitates the function of the growth arrest and DNA damage-inducible protein (GADD34) and modulates GADD34-bound protein phosphatase-1 activity. *J Biol Chem* 2002;**277**:27706–15.
142. Hung WJ, Roberson RS, Taft J, Wu DY. Human BAG-1 proteins bind to the cellular stress response protein GADD34 and interfere with GADD34 functions. *Mol Cell Biol* 2003;**23**:3477–86.

143. Shi W, Sun C, He B, Xiong W, Shi X, Yao D, et al. GADD34-PP1c recruited by Smad7 dephosphorylates TGF{beta} type I receptor. *J Cell Biol* 2004;**164**:291–300.
144. Vander Mierde D, Scheuner D, Quintens R, Patel R, Song B, Tsukamoto K, et al. Glucose activates a protein phosphatase-1-mediated signaling pathway to enhance overall translation in pancreatic beta-cells. *Endocrinology* 2007;**148**:609–17.
145. Drucker DJ. The biology of incretin hormones. *Cell Metab* 2006;**3**:153–65.
146. Drucker DJ. Glucagon-like peptide-1 and the islet beta-cell: augmentation of cell proliferation and inhibition of apoptosis. *Endocrinology* 2003;**144**:5145–8.
147. Kim JG, Baggio LL, Bridon DP, Castaigne JP, Robitaille MF, Jette L, et al. Development and characterization of a glucagon-like peptide 1-albumin conjugate: the ability to activate the glucagon-like peptide 1 receptor in vivo. *Diabetes* 2003;**52**:751–9.
148. Li Y, Hansotia T, Yusta B, Ris F, Halban PA, Drucker DJ. Glucagon-like peptide-1 receptor signaling modulates beta cell apoptosis. *J Biol Chem* 2003;**278**:471–8.
149. Yusta B, Baggio LL, Estall JL, Koehler JA, Holland DP, Li H, et al. GLP-1 receptor activation improves beta cell function and survival following induction of endoplasmic reticulum stress. *Cell Metab* 2006;**4**:391–406.
150. Tsunekawa S, Yamamoto N, Tsukamoto K, Itoh Y, Kaneko Y, Kimura T, et al. Protection of pancreatic beta-cells by exendin-4 may involve the reduction of endoplasmic reticulum stress; in vivo and in vitro studies. *J Endocrinol* 2007;**193**:65–74.
151. Brush MH, Shenolikar S. Control of cellular GADD34 levels by the 26S proteasome. *Mol Cell Biol* 2008;**28**:6989–7000.
152. Mathes E, O'Dea EL, Hoffmann A, Ghosh G. NF-kappaB dictates the degradation pathway of IkappaBalpha. *EMBO J* 2008;**27**:1357–67.
153. Chu ZL, McKinsey TA, Liu L, Qi X, Ballard DW. Basal phosphorylation of the PEST domain in the I(kappa)B(beta) regulates its functional interaction with the c-rel proto-oncogene product. *Mol Cell Biol* 1996;**16**:5974–84.
154. Martinez LO, Agerholm-Larsen B, Wang N, Chen W, Tall AR. Phosphorylation of a pest sequence in ABCA1 promotes calpain degradation and is reversed by ApoA-I. *J Biol Chem* 2003;**278**:37368–74.
155. Stoeckle MY, Sugano S, Hampe A, Vashistha A, Pellman D, Hanafusa H. 78-kilodalton glucose-regulated protein is induced in Rous sarcoma virus-transformed cells independently of glucose deprivation. *Mol Cell Biol* 1988;**8**:2675–80.
156. Connor JH, Weiser DC, Li S, Hallenbeck JM, Shenolikar S. Growth arrest and dna damage-inducible protein gadd34 assembles a novel signaling complex containing protein phosphatase 1 and inhibitor 1. *Mol Cell Biol* 2001;**21**:6841–50.
157. Harding HP, Zhang Y, Scheuner D, Chen JJ, Kaufman RJ, Ron D. Ppp1r15 gene knockout reveals an essential role for translation initiation factor 2 alpha (eIF2alpha) dephosphorylation in mammalian development. *Proc Natl Acad Sci USA* 2009;**106**:1832–7.
158. Zhan Q, Lord KA, Alamo Jr. I, Hollander MC, Carrier F, Ron D, et al. The gadd and MyD genes define a novel set of mammalian genes encoding acidic proteins that synergistically suppress cell growth. *Mol Cell Biol* 1994;**14**:2361–71.
159. Fornace AJ, Neibert DW, Hollander MC, Luethy JD, Papathanasiou M, Fragoli J, et al. Mammalian genes coordinately regulated by growth arrest signals and DNA-damaging agents. *Mol Cell Biol* 1989;**9**:4196–203.
160. Su ZZ, Shi Y, Fisher PB. Subtraction hybridization identifies a transformation progression-associated gene PEG-3 with sequence homology to a growth arrest and DNA damage-inducible gene. *Proc Natl Acad Sci USA* 1997;**94**:9125–30.
161. Hollander MC, Zhan Q, Bae I, Fornace Jr. AJ. Mammalian GADD34, an apoptosis- and DNA damage-inducible gene. *J Biol Chem* 1997;**272**:13731–7.

162. Adusumilli PS, Chan MK, Hezel M, Yu Z, Stiles BM, Chou TC, et al. Radiation-induced cellular DNA damage repair response enhances viral gene therapy efficacy in the treatment of malignant pleural mesothelioma. *Ann Surg Oncol* 2007;**14**:258–69.
163. Adusumilli PS, Chan MK, Chun YS, Hezel M, Chou TC, Rusch VW, et al. Cisplatin-induced GADD34 upregulation potentiates oncolytic viral therapy in the treatment of malignant pleural mesothelioma. *Cancer Biol Ther* 2006;**5**:48–53.
164. Su ZZ, Goldstein NI, Jiang H, Wang MN, Duigou GJ, Young CS, et al. PEG-3, a nontransforming cancer progression gene, is a positive regulator of cancer aggressiveness and angiogenesis. *Proc Natl Acad Sci USA* 1999;**96**:15115–20.
165. Su ZZ, Emdad L, Sarkar D, Randolph A, Valerie K, Yacoub A, et al. Potential molecular mechanism for rodent tumorigenesis: mutational generation of Progression Elevated Gene-3 (PEG-3). *Oncogene* 2005;**24**:2247–55.
166. Su ZZ, Sarkar D, Emdad L, Duigou GJ, Young CS, Ware J, et al. Targeting gene expression selectively in cancer cells by using the progression-elevated gene-3 promoter. *Proc Natl Acad Sci USA* 2005;**102**:1059–64.
167. Bhang HE, Gabrielson KL, Laterra J, Fisher PB, Pomper MG. Tumor-specific imaging through progression elevated gene-3 promoter-driven gene expression. *Nat Med* 2010;**17**:123–9.
168. Babour A, Bicknell AA, Tourtellotte J, Niwa M. A surveillance pathway monitors the fitness of the endoplasmic reticulum to control its inheritance. *Cell* 2010;**142**:256–69.
169. Bi M, Naczki C, Koritzinsky M, Fels D, Blais J, Hu N, et al. ER stress-regulated translation increases tolerance to extreme hypoxia and promotes tumor growth. *EMBO J* 2005;**24**:3470–81.
170. Gupta S, McGrath B, Cavener DR. PERK regulates the proliferation and development of insulin-secreting beta-cell tumors in the endocrine pancreas of mice. *PLoS One* 2009;**4**:e8008.
171. Liu L, Cash TP, Jones RG, Keith B, Thompson CB, Simon MC. Hypoxia-induced energy stress regulates mRNA translation and cell growth. *Mol Cell* 2006;**21**:521–31.
172. Brewer JW, Hendershot LM, Sherr CJ, Diehl JA. Mammalian unfolded protein response inhibits cyclin D1 translation and cell-cycle progression. *Proc Natl Acad Sci USA* 1999;**96**:8505–10.
173. Brewer JW, Diehl JA. PERK mediates cell-cycle exit during the mammalian unfolded protein response. *Proc Natl Acad Sci USA* 2000;**97**:12625–30.
174. Raven JF, Baltzis D, Wang S, Mounir Z, Papadakis AI, Gao HQ, et al. PKR and PKR-like endoplasmic reticulum kinase induce the proteasome-dependent degradation of cyclin D1 via a mechanism requiring eukaryotic initiation factor 2alpha phosphorylation. *J Biol Chem* 2008;**283**:3097–108.
175. Zhang W, Feng D, Li Y, Iida K, McGrath B, Cavener DR. PERK EIF2AK3 control of pancreatic beta cell differentiation and proliferation is required for postnatal glucose homeostasis. *Cell Metab* 2006;**4**:491–7.
176. Qu L, Huang S, Baltzis D, Rivas-Estilla AM, Pluquet O, Hatzoglou M, et al. Endoplasmic reticulum stress induces p53 cytoplasmic localization and prevents p53-dependent apoptosis by a pathway involving glycogen synthase kinase-3beta. *Genes Dev* 2004;**18**:261–77.
177. Pluquet O, Qu LK, Baltzis D, Koromilas AE. Endoplasmic reticulum stress accelerates p53 degradation by the cooperative actions of Hdm2 and glycogen synthase kinase 3beta. *Mol Cell Biol* 2005;**25**:9392–405.
178. Li J, Lee B, Lee AS. Endoplasmic reticulum stress-induced apoptosis: multiple pathways and activation of p53-up-regulated modulator of apoptosis (PUMA) and NOXA by p53. *J Biol Chem* 2006;**281**:7260–70.
179. Zhang F, Hamanaka RB, Bobrovnikova-Marjon E, Gordan JD, Dai MS, Lu H, et al. Ribosomal stress couples the unfolded protein response to p53-dependent cell cycle arrest. *J Biol Chem* 2006;**281**:30036–45.

180. Baltzis D, Pluquet O, Papadakis AI, Kazemi S, Qu LK, Koromilas AE. The eIF2alpha kinases PERK and PKR activate glycogen synthase kinase 3 to promote the proteasomal degradation of p53. *J Biol Chem* 2007;**282**:31675–87.
181. Lohrum MA, Ludwig RL, Kubbutat MH, Hanlon M, Vousden KH. Regulation of HDM2 activity by the ribosomal protein L11. *Cancer Cell* 2003;**3**:577–87.
182. Zhang Y, Wolf GW, Bhat K, Jin A, Allio T, Burkhart WA, et al. Ribosomal protein L11 negatively regulates oncoprotein MDM2 and mediates a p53-dependent ribosomal-stress checkpoint pathway. *Mol Cell Biol* 2003;**23**:8902–12.
183. Dai MS, Lu H. Inhibition of MDM2-mediated p53 ubiquitination and degradation by ribosomal protein L5. *J Biol Chem* 2004;**279**:44475–82.
184. Dai MS, Zeng SX, Jin Y, Sun XX, David L, Lu H. Ribosomal protein L23 activates p53 by inhibiting MDM2 function in response to ribosomal perturbation but not to translation inhibition. *Mol Cell Biol* 2004;**24**:7654–68.
185. Jin A, Itahana K, O'Keefe K, Zhang Y. Inhibition of HDM2 and activation of p53 by ribosomal protein L23. *Mol Cell Biol* 2004;**24**:7669–80.
186. DuRose JB, Scheuner D, Kaufman RJ, Rothblum LI, Niwa M. Phosphorylation of eukaryotic translation initiation factor 2alpha coordinates rRNA transcription and translation inhibition during endoplasmic reticulum stress. *Mol Cell Biol* 2009;**29**:4295–307.
187. Malzer E, Daly ML, Moloney A, Sendall TJ, Thomas SE, Ryder E, et al. Impaired tissue growth is mediated by checkpoint kinase 1 (CHK1) in the integrated stress response. *J Cell Sci* 2010;**123**:2892–900.
188. Sanchez Y, Wong C, Thoma RS, Richman R, Wu Z, Piwnica-Worms H, et al. Conservation of the Chk1 checkpoint pathway in mammals: linkage of DNA damage to Cdk regulation through Cdc25. *Science* 1997;**277**:1497–501.
189. Liu Q, Guntuku S, Cui XS, Matsuoka S, Cortez D, Tamai K, et al. Chk1 is an essential kinase that is regulated by Atr and required for the G(2)/M DNA damage checkpoint. *Genes Dev* 2000;**14**:1448–59.
190. Shieh SY, Ahn J, Tamai K, Taya Y, Prives C. The human homologs of checkpoint kinases Chk1 and Cds1 (Chk2) phosphorylate p53 at multiple DNA damage-inducible sites. *Genes Dev* 2000;**14**:289–300.
191. Takai H, Tominaga K, Motoyama N, Minamishima YA, Nagahama H, Tsukiyama T, et al. Aberrant cell cycle checkpoint function and early embryonic death in Chk1(-/-) mice. *Genes Dev* 2000;**14**:1439–47.
192. Bourougaa K, Naski N, Boularan C, Mlynarczyk C, Candeias MM, Marullo S, et al. Endoplasmic reticulum stress induces G2 cell-cycle arrest via mRNA translation of the p53 isoform p53/47. *Mol Cell* 2010;**38**:78–88.
193. Padyana AK, Qiu H, Roll-Mecak A, Hinnebusch AG, Burley SK. Structural basis for autoinhibition and mutational activation of eukaryotic initiation factor 2alpha protein kinase GCN2. *J Biol Chem* 2005;**280**:29289–99.
194. Lageix S, Lanet E, Pouch-Pelissier MN, Espagnol MC, Robaglia C, Deragon JM, et al. Arabidopsis eIF2alpha kinase GCN2 is essential for growth in stress conditions and is activated by wounding. *BMC Plant Biol* 2008;**8**:134.
195. Hanks SK, Hunter T. Protein kinases 6. The eukaryotic protein kinase superfamily: kinase (catalytic) domain structure and classification. *FASEB J* 1995;**9**:576–96.
196. Dey M, Trieselmann B, Locke EG, Lu J, Cao C, Dar AC, et al. PKR and GCN2 kinases and guanine nucleotide exchange factor eukaryotic translation initiation factor 2B (eIF2B) recognize overlapping surfaces on eIF2alpha. *Mol Cell Biol* 2005;**25**:3063–75.
197. Dey M, Cao C, Sicheri F, Dever TE. Conserved intermolecular salt bridge required for activation of protein kinases PKR, GCN2, and PERK. *J Biol Chem* 2007;**282**:6653–60.

# HijAkt: The PI3K/Akt Pathway in Virus Replication and Pathogenesis

Ewan F. Dunn and John H. Connor

*Department of Microbiology, Boston University School of Medicine, Boston, Massachusetts, USA*



As obligate parasites of cellular processes, viruses must take over cellular macromolecular machinery. It is also becoming clear that viruses routinely control intracellular signaling pathways through the direct or indirect control of kinases and phosphatases. This control of cellular phosphoproteins is important to promote a variety of viral processes, from control of entry to nuclear function to the stimulation of viral protein synthesis. This review focuses on the takeover of the cellular phosphatidylinositol-3-kinase (PI3K)/Akt signaling pathway by a variety of retroviruses, DNA viruses, and RNA viruses, highlighting the functions ascribed to virus activation of PI3K and Akt activity. This review also describes the role that the PI3K/Akt pathway plays in the host response, noting that it that can trigger anti- as well as proviral functions.

**Abbreviations:** 4E-BP1, eIF4E binding protein 1; CMV, cytomegalovirus; DNA, deoxyribonucleic acid; EGFR, epidermal growth factor receptor; FAK, focal adhesion kinase; FOXO1, forkhead box O1; HSV1, herpes simplex virus 1; IL-1, interleukin-1; IL-6, interleukin-6; IL-8, interleukin-8; JEV, Japanese encephalitis virus; mTOR, mammalian target of rapamycin; mTORC2, mTOR complex 2; NF-κB, nuclear factor kappa B; NSP1, nonstructural protein 1; P70S6k, p70 ribosomal protein S6 kinase; PI3K, phosphatidyl inositol 3 kinase; PP2A, protein phosphatase 2A; RNA, ribonucleic acid; rpS6, ribosomal protein S6; RSV, respiratory syncytial virus; RTK, receptor tyrosine kinase; SARS, severe acute respiratory syndrome; SV40, simian virus 40.

Progress in Molecular Biology and Translational Science, Vol. 106
DOI: 10.1016/B978-0-12-396456-4.00002-X


1877-1173/12 $35.00

# I. Introduction

Viral disease is one of the most significant human health concerns in modern medicine, manifesting as a broad range of acute infections, chronic diseases, and virus-associated malignancies. Viruses such as human immunodeficiency virus (HIV), hepatitis C, influenza, rotavirus, chikungunya, and dengue affect millions of people each year worldwide. The emergence of novel strains of H1N1 flu[1] and previously unrecognized pathogens such as severe acute respiratory syndrome (SARS) coronavirus[2] and SFTS bunyavirus[3,4] has highlighted the threats of viral emergence and pandemic with which humanity regularly contends. This, along with the emergence of antiviral resistance among viruses such as influenza, highlights the need to better understand virus replication and the virus–host interaction to identify critical restriction points that can be targets for antiviral drug development and vaccine generation.

One controlling aspect of viral pathogenesis and replication is their dependence on the cells that they infect. Unlike other microbial pathogens such as bacteria and fungi which are capable of autonomous growth, viruses are completely dependent on cellular processes for replication. It has long been recognized that viruses are dependent upon the use of cellular machinery for protein synthesis: nucleotide and protein building blocks, ribonucleic acid (RNA) and deoxyribonucleic acid (DNA) polymerases, and membranes. More recently, it has become clear that viruses also rely upon intracellular communication.[5,6] Particularly, viruses appear to be adept at capturing protein phosphorylation cascades for their own use.

Most, if not all, viruses require protein phosphorylation in order to replicate. This would initially suggest that viruses would gain a replication edge by generally promoting protein phosphorylation. However, simply promoting all cellular protein phosphorylation is not likely to be advantageous to viruses, as the activity of some host kinases (and the resultant phosphorylation events) are known to antagonize viral replication. Thus, a nonspecific increase in phosphorylation would trigger an antiviral response. Viruses have a need for specificity in activating host kinases and phosphatases. Simply stated, viruses need to activate signaling pathways that promote their replication[7–9] while blocking or avoiding the activation of antiviral signaling pathways.[10]

Controlling kinase activity can promote protein translation,[11] increase metabolic activity, drive cell division, and inhibit cellular apoptosis. All of these are advantageous for virus replication. In addition, many viruses produce virus proteins that require direct phosphorylation in order to properly function but make no kinase that will carry out this phosphorylation.[9,10] This dictates that they must recruit kinase activity from their host.

It is therefore not surprising that viruses have developed mechanisms to control cellular signaling through regulating kinase and phosphatase signaling cascades. What is surprising is the commonality of host pathways that are targeted. In theory, any number of kinases might be important for the replication of any particular virus, but this possibility of each virus using different kinases is not actually observed. Rather, as is emphasized below, one signaling pathway has emerged as a common player in the replication of many different viruses—the phosphatidylinositol-3-kinase (PI3K)/Akt phosphorylation cascade.

The repeated identification of the PI3K/Akt pathway as a central player in virus replication has underscored the importance of this pathway as a central regulator of cell health and metabolism. As a central regulator of multiple cellar processes that control translation, metabolism, and cell death, active PI3K/Akt signaling can fulfill many viral "needs." This is not to suggest that this signaling pathway only provides positive feedback for the virus. Studies of virus interaction with this pathway have also shown that the pathway is not simply a "proviral" kinase when activated, but that it also acts as part of the host response to viral infection. This suggests that PI3K/Akt signaling represents a two-faced player in interactions between the virus and cell, able to both promote viral replication and be an active part of the immune response that will eventually quash virus replication. This review will describe the different approaches that viruses have taken to dominate the PI3K/Akt signaling pathway. The viruses chosen in this review highlight the varied manner in which this pathway is manipulated and turned to the advantage of the invader, while also pointing out how the pathway also plays a role in host defense.

## II. Overview of the PI3K/Akt Signaling Pathway

The PI3K/Akt signal transduction cascade is a classical phosphorylation cascade that utilizes tyrosine, lipid, and serine–threonine phosphorylation to transduce external signals to internal responses (see Fig. 1 for a schematic diagram). Normally, signaling through this pathway is initiated by the stimulation of a receptor tyrosine kinase (RTK) by a cytokine or a growth factor at the cell surface. Activation of the RTK recruits and activates PI3K. PI3K is a heterodimer consisting of a p110 catalytic and a p85 regulatory subunit.[12] PI3K is responsible for converting phosphatidylinositol 4,5-bisphosphate (PIP2) to 3,4,5-triphosphorylated phosphoinositide (PIP3). PIP3-rich membrane domains serve as docking sites for proteins that contain a pleckstrin homology domain such as Akt.

The recruitment of inactive Akt protein to PIP3-rich areas of the plasma membrane results in a conformational change that exposes the activation loop of Akt.[13] Akt's activating kinase, phosphoinositide-dependent protein kinase

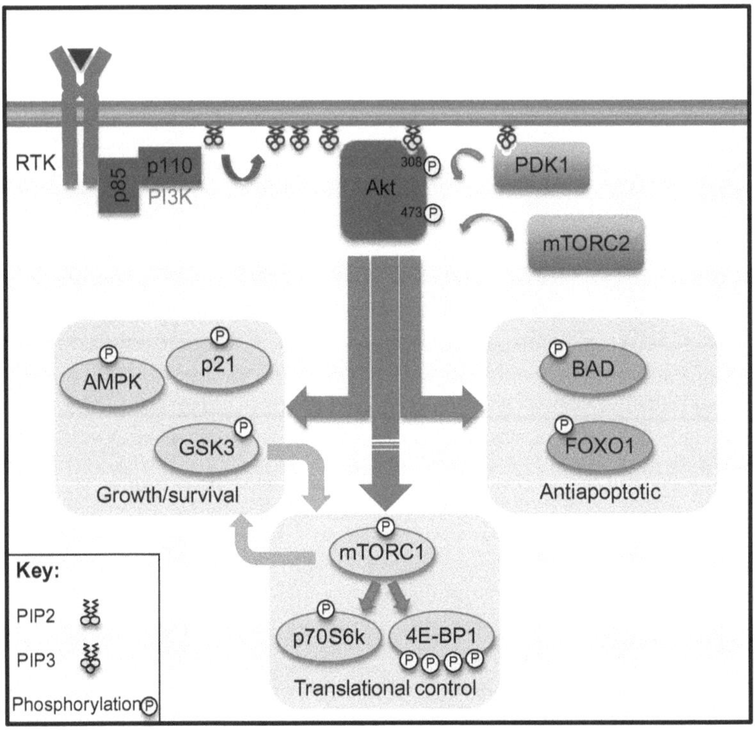


FIG. 1. Overview of the PI3K/Akt pathway. The diagram illustrates the two subunits of PI3K (p85 and p110) bound to an activated receptor tyrosine kinase (classical examples are the epidermal growth factor and the insulin receptor). Active PI3K catalyzes the addition of a single phosphate to the 3 position of PIP4,5-bisphosphate to make PIP3. This triphosphorylated inositol serves as an affinity ligand for the pleckstrin homology domains of the Akt kinase PDK1 and Akt, with the binding of PIP3 by the Akt PH domain serving to uncover the activation domain of Akt. Phosphorylation by PDK1 occurs at position T308, and a second activating phosphorylation occurs at S473. Active Akt will phosphorylate various downstream protein targets that control cell growth (AMPK, p21, and GSK3 are illustrated) and translational control (mTOR and its downstream effectors p70S6K and 4E-BP1 are illustrated) and act to suppress apoptosis (inhibitory phosphorylation of BAD and FOXO1 are illustrated). (See Color Insert.)

(PDK1)[14], is also recruited to PIP3 microdomains. PDK1 phosphorylates Akt on threonine 308 (Thr308) of the exposed activation loop, activating Akt and leading to a second phosphorylation of Akt at serine 473 (Ser473) by a kinase presumed to be mTORC2 that further potentiates kinase activity.[13,15–17]

Activated Akt can control the signaling of several key pathways supporting different types of cellular functions (Fig. 1). The PI3/Akt signaling pathway promotes cell growth, cell survival, and tumorigenesis through the phosphorylation and inactivation of cell cycle kinase inhibitors such as p21 and the inactivation of transcription factors that can inhibit cell cycle progression and promote apoptosis such as the forkhead family of proteins (e.g., FOXO1[18]). Akt also can block the actions of metabolically repressive kinases such as AMPK[19] and so can drive the up regulation of metabolic activity in the cell.[20]

Akt also promotes cellular translation through GSK3 phosphorylation and the activation of mammalian target of rapamycin complex 1 (mTORC1), a kinase complex that activates ribosomal S6 kinase and inhibits the translational repressor eIF4E-BP1 (4E-BP1).[21] Akt activation of mTORC1 is indirect. Akt phosphorylates and represses the hamartin and tuberin complex (TSC1 and TSC2[22]). Inhibition of TSC1/TSC2 allows Rheb-mediated activation of mTORC1 complex kinase activity. MTORC1 phosphorylates and activates the ribosomal protein S6 kinase (p70S6K) and inactivates the translation suppressor 4EBP1, leading to increased translation.

Akt activity antagonizes apoptotic signaling. This is accomplished through direct actions such as inhibiting pro-apoptotic factors (e.g., phosphorylation and inhibition of BAD,[23] a pro-apoptotic member of the BCL-2 family) and through more indirect actions, such as the activation of the transcription factor nuclear factor kappa B (NF-κB)[24] and, as mentioned above, the inactivation of FOXO1. With actions that influence multiple phosphoprotein cascades involved in different core aspects of cell function, PI3K/Akt signaling pathway is a prime candidate for a "hub" kinase whose activation can have a diverse set of consequences, all of which are centered on promoting cell survival and increasing the metabolic capacity of the cell.

## III. Virus Modulation of PI3K/Akt Signaling

With the central nature of the PI3K/Akt pathway in controlling cell functions, it stands to reason that the control of such a signaling "hub" would greatly benefit an invading virus that is dependent upon cellular functions. Extensive research supports this hypothesis, finding viral domination of this pathway in various contexts.[5,25,26] To illustrate the different ways in which viruses attack and control the pathway, this review will explain different examples of how viruses co-opt the host signaling apparatus through the PI3K/Akt pathway for selfish gain. The review will also highlight how viral activation of this pathway can contribute to the host's response to viral infection. This review will classify virus interactions with the PI3K/Akt pathway based on the aspect(s) of PI3K/Akt signaling from which each virus benefits. This approach is complementary to other reviews which have taken the approach of classifying the interaction based on virus families.[5,26]

The first examples presented here highlight viruses that are suggested to have "simple" interaction paradigms. For the purpose of this review, a "simple" interaction is one where the virus or virus family appears to activate the PI3K/Akt pathway to benefit from just one aspect of its cellular control, such as the promotion of cell cycle progression, or the inhibition of apoptosis. Following this, examples of viruses that benefit from several of the outcomes of PI3K/Akt signaling will be discussed.

## A. Simple Paradigms of Viruses That Interact with PI3K/Akt

### 1. Viral Proteins That use Molecular Mimicry to Activate the PI3K/Akt Signaling Axis

Several viral proteins are known to directly activate the PI3K/Akt signaling axis through molecular mimicry or replacement (Fig. 2). These viral proteins helped in elucidating the initial understanding of the PI3K/Akt signaling cascade. The paradigm example v-akt was originally described as a retroviral oncogene encoded by the AKT8 mouse transforming retrovirus.[27] The kinase was described as a fusion protein with the viral gag (matrix protein). It was quickly recognized that the protein had a cellular homolog,[28] which was named c-akt. This protein was later determined to be the same as the cellular kinases that had been characterized under the names of PKB[29] and RAC-PK.[30] Though the role of v-akt in the life cycle of the AKT8 virus was never established, v-akt is capable of transforming cells, suggesting that the virus utilized the v-akt protein to drive cell cycle progression. The finding that a cellular kinase had been essentially directly incorporated into the viral genome as a fusion protein with the viral gag protein was the first demonstration of the importance of this kinase for viral replication.

### 2. Oncoretrovirus Control of PI3K/Akt Signaling

The control of the PI3K/Akt signaling pathway by molecular replacement extends to several other viruses, notably avian sarcoma viruses. The AV16 sarcoma virus encodes for a PI3K homolog, named vP3K,[31] which encodes a homolog of the cellular catalytic subunit of PI3K (p110). Like v-Akt, vP3K acts to provide constitutively active PI4,5-bisphosphate kinase activity without the need for the p85 regulatory subunit, generating high levels of PIP3. This serves to activate Akt,[32] which phosphorylates and inactivates the FOXO1 transcription repressor[33] and promotes cellular translation through the phosphorylation of 4E-BP1.[34] These actions remove a cell-cycle brake, allowing proliferative signals to dominate.

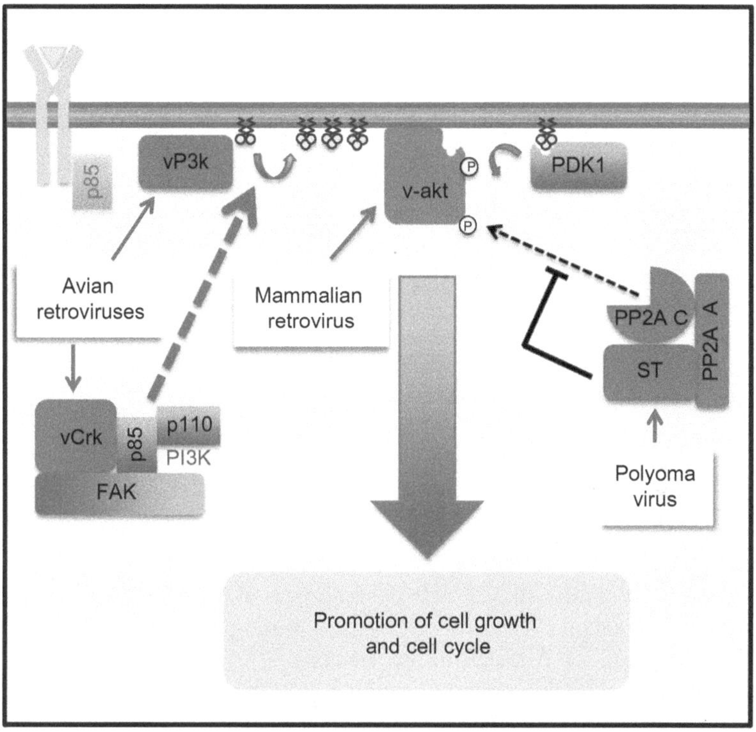


FIG. 2. Viral mimicry of PI3K pathway members. The diagram illustrates steps at which viral proteins either replace or actively modulate Akt signaling. Shown are the vPI3K (p110 subunit) produced by the avian retrovirus A16, and the vAkt protein produced by the mammalian AKT8 retrovirus which directly activates Akt signaling. Also shown is vCrk, which stimulates Akt signaling in a focal adhesion kinase (FAK)-dependent manner. Illustrated on the right-hand side of the figure is the SV40 small t (smt) protein. The smt protein binds the PP2A/C heterodimer and inactivates Akt dephosphorylation, providing an indirect but potent activation of Akt kinase activity. The actions of all these proteins can drive cell proliferation in a manner that is dependent on Akt-phosphorylated proteins. (See Color Insert.)

A slightly different approach to activation of the PI3K/Akt pathway activation is taken by the CT10 and AV1 viruses. These viruses each encode a homolog to the cellular Crk protein.[35–37] vCrk activates the PI3K/Akt signaling pathway through the recruitment of PI3K to the focal adhesion kinase.[38] This activity leads to the transformation of vCrk-infected cells[39] presumably by mechanisms similar to those described for vP3K (activation of cellular translation and inactivation of transcription factors such as FOXO1).

In all of these examples, a consistent theme is evident—retroviruses have repeatedly captured host proteins that can activate the PI3K/Akt pathway. It is thought that the cell proliferation induced by these proteins is critical for the replication of these viruses, as the breakdown of the nuclear envelope during the cell cycle gives these retroviruses access to cellular DNA and thereby allows the virus to integrate within this DNA. Further, proliferating cells are more likely to activate retroviral gene expression to stimulate virus production. However, this may not be the only role for the PI3K/Akt pathway in the retrovirus life cycle, and more recent research into viruses, such as HIV, suggests that the virus may benefit from inducing Akt activity to prevent premature apoptosis.[40]

## B. Potentiation of Akt Signaling Through the Direct Control of Phosphatase Activity

### 1. Polyomavirus Control of Akt Signaling

Viruses have also adopted approaches of inducing PI3K/Akt pathway signaling by attacking phosphatase activity. As is true in most kinase cascades, the activity of Akt signaling can be decreased through phosphatase-mediated dephosphorylation of Akt activation domains. Conversely, kinase signaling can be potentiated by decreasing phosphatase activity. For PI3K/Akt signaling, the cellular phosphatase PP2A is at least partially responsible for the dephosphorylation and inactivation of Akt.[41,42] By decreasing the activity of PP2A, more Akt will remain in the active state. This approach to activating Akt signaling has been adopted by the genus *Polyomaviridae*. Simian virus 40 (SV40) is the prototype example of a polyomavirus that stimulates Akt activation by inhibiting the dephosphorylation of Akt.[43] To do this, the SV40 small t antigen of SV40 (SV40ST) binds the cellular phosphatase PP2A that is responsible for a large fraction of Akt phosphatase activity.[44] PP2A is a heterotrimeric protein comprising a scaffolding subunit (A), a catalytic subunit (C), and a regulatory subunit (B). There are many variants of the B subunit, and each is believed to be responsible for directing the phosphatase catalytic subunit directly to a single substrate (such as Akt) or set of substrates. SV40ST displaces the B subunit from the A and C PP2A subunits, inhibiting much of the phosphatase activity which would normally dampen or reverse Akt and mTOR activation (see Fig. 2). The resulting potentiation of Akt signaling helps drive cell proliferation, moving the cell into S phase, where virus DNA synthesis is initiated.

This modulation of PP2A to alter PI3K/Akt signaling is utilized by other viruses in the polyomavirus genus as well, including BK virus,[45] mouse polyoma virus,[46] and JC virus.[47] In addition, polyomaviruses can indirectly stimulate the PI3K/Akt pathway through additional mechanisms,[48–50] indicating that these viruses attack the PI3K/Akt pathway at multiple points to ensure pathway activation.

## 2. Papillomavirus Control of Akt Signaling

A similar mechanism of activating Akt signaling (inhibition of PP2A activity that increases Akt activation) has been proposed for the E7 oncoprotein of papillomavirus HPV.[51] The HPV E7 protein is critical for the maintenance of cellular DNA synthesis in HPV-infected cells,[52] which is a necessity, as HPV requires the host cell DNA synthesis machinery for its replication.[53] Thus, activation of Akt to remove brakes on cell DNA synthesis (i.e., FOXO) is of clear value to virus replication. This action is also important in the cases where the E7 gene is incorporated into host DNA (not part of the normal virus life cycle). Under these circumstances, Akt signaling may promote malignant transformation,[51] thereby underscoring a potential link between viral control of cellular phosphorylation for its own means and the induction of cancer.

## 3. Virus Activation of Akt to Promote an Antiapoptotic Signal

In the examples presented above, virus takeover of the Akt signaling pathway was associated with the induction of cellular proliferation. In many examples, though, viruses that activate the PI3K/Akt pathway do not benefit from the promotion of cellular proliferation or cellular transformation. For many viruses, the PI3K/Akt pathway can provide the short-term benefit of keeping the infected cell alive through the suppression of apoptotic signals. The critical nature of PI3K/Akt-mediated suppression of apoptosis is best illustrated by viruses that cause acute infections, such as the RNA viruses.

For most RNA viruses, their life cycle is carried out in the cytoplasm of their host cell. These viruses therefore have minimal requirements for nuclear components or cell cycle progression, and in many cases block nuclear functions and cell-cycle progression.[54–56] Thus, PI3K/Akt-mediated promotion of cell division or alteration of transcription does not play an essential role in virus replication. Instead, phosphorylation of cytoplasmic proteins in order to delay apoptosis keeps the host cell alive until viral progeny can be produced.

## 4. Activation of Akt Signaling by Paramyxoviruses

The trigger for PI3K/Akt activation in infection with paramyxoviruses often occurs early in infection. As an example, respiratory syncytial virus (RSV), a significant pediatric pathogen, activates PI3K/Akt signaling at the level of PI3K activation.[57] Pathway stimulation appears to involve the activation of sphingosine kinase, which, in turn, leads to the phosphorylation and activation of Akt through a mechanism that may involve the extracellular signaling of sphingosine 1 phosphate.[58] Akt activity is critical for delaying cellular apoptosis during RSV infection.[57]

Similarly, the genetically related Sendai virus, human parainfluenza virus 5 (HPIV5), and HPIV3 have also been shown[59] or suggested[59,60] to activate PI3K and Akt early in infection, an action associated with delaying apoptosis and potentially activating the viral polymerase. For all of these viruses, signaling through the Akt pathway allows the completion of the viral replication cycle. Inhibition of PI3K or Akt signaling leads to faster apoptosis[59] and, in some reports, reduced virus growth.[60] In the case of Sendai virus, Akt activation may also play a role in the ability of this virus to establish a persistent infection in cells in tissue culture.[59]

## 5. Activation of Akt Signaling by Picornaviruses

Viruses in the picornavirus family are also reported to activate PI3K/Akt signaling. For both poliovirus and the common-cold-causing rhinovirus, viral attachment and entry serve as the trigger for PI3K/Akt activity.[61] For poliovirus, the exact trigger is unknown; for rhinovirus, binding of the virus to its receptor recruits the tyrosine kinase Syk, which interacts directly with the p85 subunit of PI3K and activates PI3K/Akt signaling.[62,63] It is likely that the activation of Akt is important for avoiding cell death in the case of both viruses. An important role for Akt in forestalling apoptosis has been shown in the case of poliovirus, where the Akt phosphorylation of Ask1 limits the activity of its downstream effector Jnk1 and thereby decreases apoptotic signals.

Cardioviruses also appear to manipulate Akt activity to limit apoptosis in infected cells. For both encephalomyocarditis virus (EMCV) and coxsackievirus, viral infection is associated with the stimulation of PI3K/Akt activity.[64] Unlike poliovirus and rhinovirus infection, the activation of PI3K does not appear to be entry dependent.[64] Activation is important for suppression of apoptosis and promotion of virus replication. For coxsackievirus, Akt activation results in the stimulation of NF-κB activity, which promotes cell survival.[65]

## 6. Activation of Akt Signaling by Reoviruses

PI3K/Akt activation to promote the survival of infected cells is also seen during infection of cells with double-stranded RNA viruses of the reovirus family. As an example, rotavirus stimulates PI3K/Akt activity within the first 2 h of infection.[66] Through the use of both mutant virus and viral protein overexpression approaches, it has been shown that Akt activation is carried out by the rotavirus nonstructural protein 1 (NSP1).[67] The activation of Akt by NSP1 can occur even in the absence of other viral components and involves a direct interaction between NSP1 and PI3K (see Fig. 3), suggesting that the protein alters kinase activity. The mechanism of this targeting is currently unknown.

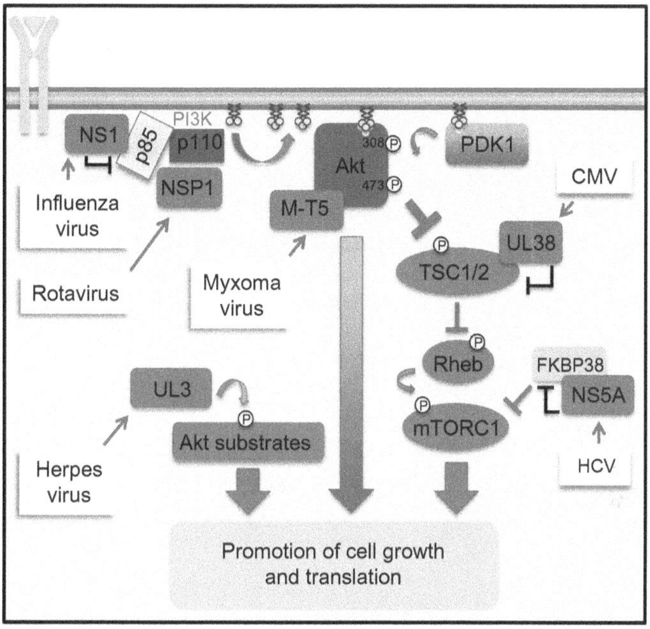


FIG. 3. Virus proteins that directly interact with the Akt signaling pathway. Illustrated on the left-hand side of the diagram are activation of PI3K by the influenza NS1 protein through direct interaction with the p85 subunit and the activation of Akt by the rotavirus NSP1 protein. Illustrated in the center of the image are the myxomavirus Akt activator M-T5 and the herpes virus kinase UL3, a viral ortholog of the Akt kinase. Illustrated on the right-hand side of the diagram is action of the hepatitis C protein NS5A to bind FKBP38, relieving an inhibitory regulation to activate mTOR. Also illustrated is action of UL38, which inhibits TSC2, also relieving an inhibitory regulation to activate mTOR. (See Color Insert.)

Activation of Akt has also been suggested for the avian reovirus A1133.[68] The activation of Akt in rotavirus-infected cells suppresses apoptosis to allow the completion of the virus life cycle.[69] Akt signaling in infected cells is also required for the expression of integrin proteins[66] which can strongly promote both survival signaling and virus attachment, suggesting that Akt activation may send survival signals and promote viral attachment in infected cells.

## 7. Flavivirus Activation of PI3K/Akt Signaling

Like many viruses, the flaviviruses dengue virus and Japanese encephalitis virus[70] stimulate PI3K/Akt activation within the first few hours of infection. For these viruses, the mechanism of activation is unknown, but activation appears to be transient[70] and curiously does not seem to impact virus replication in any fashion. Hepatitis C, another flavivirus, has not been reported to activate either PI3K or Akt but has been shown to promote cell survival through the activation of the downstream effector, mTOR kinase.[71] This activation is facilitated by the viral nonstructural protein NS5A, which has the capacity to bind directly to FKBP38, a negative regulator of mTOR[72] (see Fig. 3). The association of FKBP38 with mTORC1 is blocked by NS5A binding, thereby stimulating mTORC1 activity, promoting translation, and limiting apoptosis. This mechanism of targeting mTORC1 by relieving an inhibitor action is seen in the actions of a cytomegalovirus (CMV) virus protein, UL38 (see below).

## C. Viruses That Rely on Multiple Signaling Endpoints of the PI3K/Akt Pathway

As described in the introduction, the PI3K/Akt pathway acts as a signaling "hub" triggering a branching phosphorylation cascade that alters multiple cellular processes. The viruses described above appear to benefit primarily from one aspect of "hub" activation or one subset of the phosphoproteins. This is not true of all viruses. Some viruses benefit from more than one signaling arm of the PI3K/Akt pathway. Three examples of this more complex virus/host interaction illustrate this well: the domination of signaling by herpes viruses, poxviruses, and the influenza virus.

### 1. Herpes Viruses Benefit from the Stimulation of Translation Through PI3K/Akt Signaling

The herpes virus family consists of several viruses that have each taken differing approaches to achieve the same result: preservation of a favorable environment for both viral transcription and viral translation. The life cycle of these viruses is particularly pertinent to their interaction with the PI3K/Akt pathway, as herpes viruses have both a lytic phase, where they actively replicate in cells, and a latent phase, where they are maintained as episomes in the host genome. This dual life cycle means the virus requires different cell environments at different times. During latent infection, the virus requires little from the host, while during the lytic phase, many host factors and a high metabolic throughput are desirable.

### 2. Epstein-Barr Virus Activation of PI3K/Akt Signaling

One example of how herpes viruses interact with PI3K/Akt signaling is the Epstein–Barr virus (EBV). EBV is associated with Burkitt lymphoma/leukemia and other lymphoproliferative disorders. EBV establishes a latent stage in which most viral proteins are not produced and replication is dormant. In this phase, virus-infected cells produce a latency protein LMP2A. LMP2A promotes Akt phosphorylation, provides a general antiapoptotic signal, and also promotes c-MYC translation through translational upregulation caused by mTOR activation.[73] The development of EBV-driven Burkitt lymphoma/leukemia is dependent upon the constitutive expression of c-MYC, which modulates the expression of target genes that encode many cellular processes including cell growth, division, and apoptosis. LMP2A protects cells that express c-MYC from apoptosis through the upregulation of antiapoptotic genes.[74]

LMP1, another latency-associated protein produced by the EBV genome, also activates the Akt pathway.[75] For LMP1, the activation of Akt activity is associated with a nuclear function—the inhibition of DNA damage repair—and not protein synthesis promotion. This suggests the LMP1 and LMP2 proteins of EBV target different pools of PI3K and Akt to control the function of this signaling pathway in different cellular compartments.

The activity of Akt is also targeted in the lytic phase of EBV replication. Expression of BRLF1, a protein that induces lytic replication of EBV, results in the activation of Akt in overexpression studies, suggesting that pathway PI3K activity is an important signal for the emergence of the virus from latency.[76] Consistent with this, inhibition of PI3K signaling blocks EBV reactivation from a latent state.

This dual use of the PI3K/Akt signaling for both latent and lytic replication reveals an interesting paradigm for cellular signaling in virus infection. Akt activity appears to be a basic requirement for the survival of EBV-infected cells, but cannot in itself be a trigger for reactivation. Akt activity is much more likely to be a basal maintenance factor. Further research will undoubtedly uncover true triggers that cooperate with the PI3K/Akt pathway to drive reactivation. In the interim, the apparent requirement for PI3K/Akt signaling to maintain a latent state and help drive reactivation has identified an Achilles heel of EBV-infected cells. Targeting EBV-infected cells for death by adding inhibitors of this pathway are showing initial promise as a way to inhibit EBV +tumor growth.[77]

### 3. Herpes Virus Activation of PI3K/Akt Signaling

Unlike EBV, the activation of the PI3K/Akt phosphorylation cascade in herpes simplex virus 1 infection is transient.[78] The activation has been proposed to be due to the actions of the viral VP11/12 protein,[79] though virus binding to the cell surface has also been proposed as the activating trigger.[80] Despite the transient nature of Akt activation,[78] downstream effectors of Akt such as 4E-BP1 remain phosphorylated throughout virus infection.[81] The preservation of the phosphorylated state appears to be the responsibility of the viral kinase UL3. UL3 is capable of phosphorylating multiple Akt substrates including TSC2, GSK3, and the transcription factor FOXO. UL3 appears to stimulate the assembly of translation initiation factor complexes that drive the translation of viral mRNAs, though additional actions seem likely[81] as the phosphorylation of multiple Akt substrates suggest UL3's kinase activity impacts multiple signaling pathways (Fig. 3). This host kinase mimicry is functionally analogous to that seen for the retroviruses described above, where a virus acquires or evolves a kinase that phosphorylates the same substrates as Akt. The HSV example is distinct in that UL3 is not homologous to Akt at a protein sequence level.

### 4. Cytomegalovirus Activation of PI3K/Akt Signaling

Unlike EBV and HSV, CMV is a herpes virus family member that has not been shown to directly or indirectly activate PI3K or Akt. Instead, CMV targets the downstream effector mTOR through a viral protein/host protein interaction strategy. CMV encodes a protein, named UL38,[82] that interacts directly with the tuberous sclerosis protein 2 (TSC2), a protein that is involved in suppressing mTOR activity. Binding of UL38 acts to "silence" the TSC2 inhibition of mTOR activity (Fig. 3), resulting in an increase in mTOR substrate phosphorylation.[82,83] This action phenocopies Akt activation, as Akt will directly phosphorylate and inhibit TSC2. By producing UL38, CMV replaces this function of Akt activation. Further, the UL38 inactivation of TSC2 also blocks the ability of the metabolic sensor AMPK to modulate mTOR activity by activating TSC2. In CMV-infected cells, expression of UL38 acts to preserve the mTOR arm of the PI3K/Akt pathway, an action that promotes viral translation and replication.

Similar to the multifactor targeting of the PI3K/Akt pathway seen in EBV-infected cells, it has also been reported that CMV-infected cells show a change in the substrate specificities of the mTOR complexes mTORC1 and mTORC2.[84] These changes are not associated with the UL38 protein but may be important for a change in substrate specificities that accompany a localization of mTORC1 to the replication centers of CMV in infected cells.[85] These findings suggest that there are mechanisms of mTOR control in addition to UL38-mediated activation of mTOR.

### 5. Poxvirus Utilizes PI3K/Akt Activity to Drive Translation, and Viral Assembly

Poxviruses are some of the most genetically complex viruses known, expressing more than 200 genes from a DNA genome, including viral homologs of many cellular proteins. While a homolog of PI3K or Akt is not present, signaling through the PI3K/Akt pathway is critical for virus replication. For vaccinia, cowpox,[86] and myxoma viruses,[87] PI3K/Akt activity is rapidly induced following productive infection. This activation appears to play at least three distinct roles during poxvirus infection. One is the suppression of apoptosis, which has been suggested to be important to allow proper viral replication.[86] The second is the promotion of assembly of the eIF4F translation initiation factor complex through mTOR-mediated inhibition of 4E-BP1 function. This promotes virus mRNA translation.[88,89] PI3K/Akt activity appears to be especially critical for allowing the transcription of mRNAs that encode for proteins involved in the assembly and budding of poxviruses.[86] The third role for PI3K/Akt is at a late stage of viral infection, where it is important for the proper progression of morphogenesis.[89,90] Addition of PI3K inhibitors during infection or deletion of the p85 alpha and beta subunits of PI3K altered multiple aspects of vaccinia replication.[90] Of note were a decrease in late gene expression and the progression of the virus particle from an immature form to the intracellular mature virus.

At least one poxvirus, the rabbitpox virus (myxoma), produces a protein dedicated to harnessing and activating Akt. This protein, named M-T5, is an ankyrin-repeat protein. It was recently shown that this protein binds directly to Akt (see Fig. 3) and promotes its activation and signaling in a manner that is analogous to the cellular protein PikeA.[87,91] M-T5 not only activates Akt but also forces relocalization of nuclear Akt to the cytoplasm.[92] Removal of M-T5 from the myxomavirus genome results in a markedly attenuated virus that is incapable of causing lethal disease in rabbits.[93] This highlights the importance of Akt activity for poxvirus pathogenesis. How Akt activation supports the pathogenesis of virus replication in an animal model is not fully resolved. Also unknown at this point is whether other poxviruses such as smallpox or monkeypox have proteins analogous to M-T5 that function to activate PI3K/Akt signaling.

## 6. Influenza Benefits from Entry Assistance and Suppression of Apoptosis Through PI3K/Akt Pathway Activation

As one of the most significant endemic pathogens confronting humanity, influenza virus represents a virus that has repeatedly shown its ability to infect and adapt. As a virus that produces fewer than a dozen proteins, it is not surprising that host factors aid in influenza replication. What is striking about influenza is the extent to which the virus appears to utilize Akt-related signaling throughout the viral life cycle. Binding of influenza virus to the cell surface leads to the activation of Akt, potentially through clustering of RTKs such as the epidermal growth factor receptor (EGFR).[94] While a role for PI3K in flu virus entry seems clear, other signaling pathways (e.g., the ras signaling pathway) also appear to cooperate with the PI3K/Akt pathway to promote virus entry.[95]

The importance of PI3K/Akt signaling for influenza continues after infection. Following virus entry, influenza A virus directly activates Akt signaling[96,97] through the binding of the viral NSP1 to the P85 subunit of PI3K[96,98] (Fig. 3). NS1 binds to p85 at a coiled-coil region of p85,[99,100] relieving contacts between p85 and p110 subunit of PI3K that normally inhibit activity. The resulting activation of PI3K dramatically stimulates Akt activity in influenza-infected cells. For some strains of flu, it has been shown that NS1 will also bind the cellular adaptor Crk, an action which further potentiates PI3K/Akt signaling.[101] Functionally, this activation of Akt signaling has been proposed to lead to the inhibition of apoptosis and to allow the preservation of cell integrity during the viral replication cycle.[102,103] However, this may not be universally true. Viruses lacking NS1 function are stronger inducers of apoptosis during infection, but stimulation of Akt activity in viruses that lack the NS1 protein does not inhibit apoptosis. This suggests the true role for Akt activation during the replication of influenza is yet to be defined.

## D. The Other Side of the Coin: Akt in the Antiviral Response

The PI3K/Akt pathway is also a signaling pathway that is utilized in the cellular defense against invading pathogens. As described above, phosphorylation signals are the backbone of the antiviral response. Akt signaling is increasingly recognized to play a role in this response.[104] This is particularly true for the host response to interferon, where PI3K and Akt signaling potentiates the translation of interferon-stimulated genes through the actions of mTOR and p70S6K which stimulate interferon-responsive gene translation.[105,106] Akt also plays a role in stimulating the production of inflammatory cytokines by contributing to the activation of the NF-κB transcription factor.[57]

There are many examples of virus-mediated activation of Akt signaling leading to the activation of host innate immune genes such as inflammatory cytokines. Akt activation results in the increased production of proinflammatory cytokines IL-6 and IL-1β in avian reovirus infected cells.[68] In cells infected with RSV, the stimulation of Akt activity leads to the expression of inflammatory cytokines such as interleukin-8 (IL-8) during infection.[57,107,108] EMCV also stimulates inflammatory cytokine production through the activation of the PI3K/Akt pathway.[109]

Influenza-infected cells produce IL-8 and RANTES in a PI3K/Akt-dependent manner,[110] and virus activation of Akt potentiates the expression of interferon-β.[97] Influenza virus-infected macrophages can be kept alive by chemokine receptor-5-mediated activation of the Akt pathway, which allows the infected cell to clear other infected cells.[111] Thus activation of Akt signaling can act as a double-edged sword during infection, promoting viral infection and replication and driving the expression of host danger signals that can recruit immune cells to the site of infection. Thus, though many viruses appear to activate the PI3K/Akt pathway, this activation also seems to carry with it some negative consequences. One approach to avoiding these negative consequences is to inactivate this signaling pathway upon infection.

## E. Virus Suppression of Akt Signaling

With the importance of Akt signaling in the host antiviral response, it comes as little surprise that some viruses appear to avoid the use of the PI3K/Akt pathway altogether and instead appear to inhibit its function. This is a stark departure from the many examples of viruses that activate the signaling pathway, but underscores that while activation is common, it is not a requirement for virus replication.

### 1. Measles Virus Inhibition of PI3K/Akt Signaling

The first example of virus inhibition of PI3K/Akt signaling was described for measles virus, a pediatric pathogen whose hallmark of infection is the suppression of immune function. This immune suppression occurs through the decreased proliferation and responsiveness of lymphocytes.[112,113] While it is not fully understood how measles induces immune cell quiescence, recent studies point to a strong role of the PI3K/Akt pathway in this effect. Both *in vivo* and *in vitro*, the attachment of measles virus to the surface of T cells, can force Akt dephosphorylation in uninfected cells (Fig. 4). This inactivation of PI3K/Akt signaling is not a consequence of infection but merely requires virus binding to a cell surface receptor.[114] The receptor that is responsible for signaling this downregulation is currently unidentified, but following the binding of measles virus particles to the cell surface, there is both a decrease in PI3K and Akt

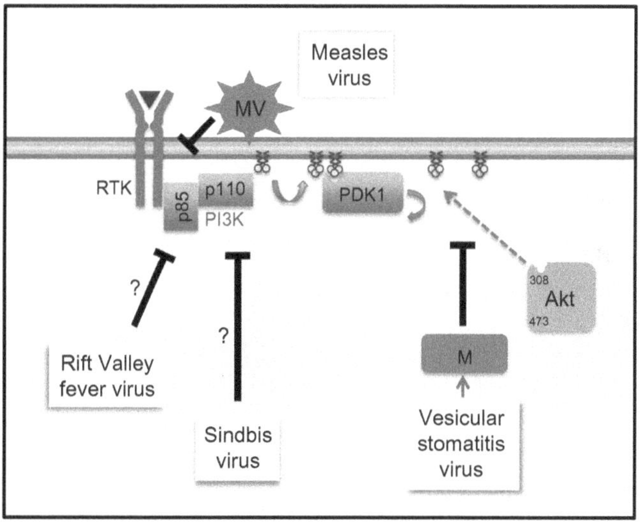


Fig. 4. Virus inactivation of Akt signaling. Four different viral approaches are shown. Measles virus suppresses Akt signaling from outside the cell through binding to a currently undefined receptor to block the accumulation of PIP3. The matrix protein of vesicular stomatitis virus blocks Akt activation through an alteration of Akt movement. PIP3 accumulation occurs in the presence of VSV infection, but phosphorylation of Akt is suppressed. Both the Rift Valley fever virus and Sindbis virus (an alphavirus) inhibit Akt activity during infection, but through a currently undefined mechanism (illustrated with a question mark). (See Color Insert.)

signaling and an upregulation of the inositol phosphatase SIP110[115] that lowers the overall level of PIP3 in the cell membrane. This promotes long-term suppression of PI3K/Akt signaling. The artificial stimulation of Akt activity overcomes measles-induced immunosuppression, suggesting that the inhibition of Akt is a central cause of measles virus-induced immune suppression.

In addition to the effect of virus particle binding on Akt signaling, active infection of cells by measles virus continues to block Akt activity even after entry. This suppression of a central cell-viability pathway has no demonstrable negative effect on measles replication.[116] Thus, unlike most viruses (even related viruses such as RSV and HPIV5), measles replication appears to be insensitive to the activity of the PI3K/Akt pathway and does not require the antiapoptotic protection that could be afforded by activating this pathway.

### 2. Rhabdovirus Inhibition of PI3K/Akt Signaling

A different mechanism of Akt suppression is seen upon infection by another RNA virus, vesicular stomatitis virus (VSV). VSV is a veterinary pathogen that is being developed both as an oncolytic agent[117] and as a platform for effective vaccines.[118] VSV replication is unaffected by Akt signaling.[119] As is seen in measles virus, infection of cells with VSV results in a rapid inhibition of Akt signaling in infected insect[120] and mammalian cells.[121] This inhibition, which extends to inhibit the downstream components mTOR,[122] 4E-BP1,[123] and rpS6,[124] makes cells insensitive to normal Akt-stimulating factors such as insulin and epidermal growth factor and is due in large part to the actions of the viral matrix protein (Fig. 4).[121] Unlike measles, VSV does not block Akt activity through inhibition of PI3K signaling. Instead, the virus blocks the activating phosphorylation of Akt by inhibiting the phosphorylation of Akt at the plasma membrane through the actions of the viral matrix protein.

### 3. Bunyavirus Inhibition of PI3K/Akt Signaling

The inhibition of PI3K/Akt signaling is also noted in cells infected with Rift Valley fever virus (RVFV). Recent phosphoproteome analysis of RVFV-infected cells showed that infection led to the dephosphorylation of the Akt-phosphorylated sites of FOXO1 and GSK3β.[125] The authors also observed a dephosphorylation of IRS1, an insulin-receptor adaptor that can facilitate PI3K/Akt activation. While the proteomics analysis did not identify Akt phosphorylation changes, the results are very much in line with RVFV inhibition of the Akt pathway. It is not clear how this virus evokes that inhibition of signaling at this point, but the analysis is consistent with the interpretation that it is downregulating RTK activity.

## 4. Alphavirus Inhibition of PI3K/Akt Signaling

Recent studies of alphaviruses show that these viruses also appear to downregulate the PI3K/Akt signaling pathway. Analysis of cells infected with the prototype alphavirus, Sindbis virus, showed that the virus downregulates PI3K/Akt activity. This downregulation of Akt phosphorylation had little effect on virus replication, and mTOR inhibitors did not significantly alter the replication of Sindbis virus.[126] This shows that alphaviruses are also capable of replicating without the contribution of these signaling pathways, which is directly analogous to what is seen in measles and VSV replication. The inhibition of PI3K/Akt signaling by Sindbis is likely relevant to the pathogenesis of alphaviruses, as these viruses are known to be neurotropic, and Akt activity is an important survival factor for these cells.[127]

## 5. Akt Antagonism as Antiviral Therapy?

The startling regularity with which viruses appear to utilize the Akt pathway has not gone unnoticed in the antiviral drug discovery field. There have been many suggestions that Akt inhibitors or inhibitors of downstream effectors such as mTOR might be used as either specific or general antiviral compounds, targeting steps of the virus lifestyle ranging from fusion and entry[128] to transcription inhibition[60] to triggering cell death of infected cells.[40] However, given the important role of Akt in promoting immune cell responsiveness[114,129] and the negative effect of PI3K/Akt inhibition on host immune cell function seen during measles virus infection, such an approach seems likely to dampen the host immune response. This latter conclusion is supported by reports showing that the PI3K/Akt pathway controls interferon production in dendritic cells[130,131] and that it is important for controlling an autophagy-based antiviral response.[120] This suggests that such an approach is unlikely to succeed as a broad-spectrum approach. However, for viruses that activate PI3K/Akt signaling to maintain a latent infection such as the EBV, inhibiting signaling through this pathway appear to be successful in limiting downstream effects of latency, such as cellular transformation.[132] This suggests that there may be limited situations in which this approach will be successful.

The PI3K/Akt signaling pathway itself may represent an evolutionary "tug-of-war" for viruses. If the PI3K/Akt signaling pathway is just as essential for host response and survival as it is for viral function, inactivation is not an effective response for the infected host. Similarly, for viruses that have adapted to activate the pathway and that rely on the downstream effect of this activation, they must also contend with "unwelcome" results of such pathway activation, including the stimulation of cytokine production. This suggests that PI3K/Akt signaling is part of the evolutionary "red queen race" between virus and host, where virus attempts at controlling the host cell are often thwarted by host evolution.

## IV. Conclusion and Outlook

While it has long been recognized that viruses are wholly dependent upon the use of cellular machinery for macromolecular events such as protein synthesis, it is now clear that viruses additionally rely upon cellular phosphoproteins and signaling cascades to mold the cellular environment to their advantage. The PI3K/Akt phosphorylation cascade is an emerging paradigm for how these interactions occur. Many viruses have incorporated methods to activate PI3K/Akt signaling in order to forestall apoptosis, drive the cell cycle, and thereby support their replication processes. Somewhat paradoxically, other viruses have evolved an opposite approach, short-circuiting the PI3K/Akt signaling axis. In this latter case, viruses appear to benefit from the repression of the host antiviral response that comes from blocking Akt signaling. Why do different viruses, some from the same families, take different approaches? This is a question that will need to be answered in future research.

It is likely that a better understanding of the importance of viral control of the PI3K/Akt signaling will be gained when the Akt substrates that are vital for viral replication are defined. Some significant progress has been made already. The phosphorylation of host factors such as the mTOR substrate 4E-BP is important for the robust translation of many viral messages, and it is likely that other Akt substrates that stimulate translation will also be sown to be important. It is likely that still more Akt-regulated phosphoproteins play a role in viral processes. Perhaps the most obvious but most poorly understood is the possibility that viral proteins themselves serve as substrates for Akt or effectors such as mTOR and GSK3. Recent work has suggested that Akt may phosphorylate the phosphoprotein of several negative-stranded RNA viruses[60] and that it may also control calicivirus polymerase function.[133] Future discoveries will likely greatly expand the number of viral substrates and more clearly point out the viral dependencies on this signaling axis.

### Acknowledgments

The authors would like to thank Rachel Fearns and Erin Hodges for helpful conversations, Rahm Gummuluru for insight regarding retroviruses and lentiviruses, and Rebecca Connor for significant help with editing and revision. We apologize that a number of references could not be incorporated because of space consideration.

## References


1. Neumann G, Noda T, Kawaoka Y. Emergence and pandemic potential of swine-origin H1N1 influenza virus. *Nature* 2009;**459**:931–9.
2. Frieman M, Baric R. Mechanisms of severe acute respiratory syndrome pathogenesis and innate immunomodulation. *Microbiol Mol Biol Rev* 2008;**72**:672–85. Table of contents.
3. Feldmann H. Truly emerging - a new disease caused by a novel virus. *N Engl J Med* 2011;**364** (16):1561–3.
4. Yu XJ, Liang MF, Zhang SY, Liu Y, Li JD, Sun YL, et al. Fever with thrombocytopenia associated with a novel bunyavirus in China. *N Engl J Med* 2011;**364**:1523–32.
5. Buchkovich NJ, Yu Y, Zampieri CA, Alwine JC. The TORrid affairs of viruses: effects of mammalian DNA viruses on the PI3K-Akt-mTOR signalling pathway. *Nat Rev Microbiol* 2008;**6**:266–75.
6. Munter S, Way M, Frischknecht F. Signaling during pathogen infection. *Sci STKE* 2006;**2006**: re5.
7. Hale BG, Randall RE, Ortin J, Jackson D. The multifunctional NS1 protein of influenza A viruses. *J Gen Virol* 2008;**89**:2359–76.
8. Langland JO, Jacobs BL. The role of the PKR-inhibitory genes, E3L and K3L, in determining vaccinia virus host range. *Virology* 2002;**299**:133–41.
9. Xiang Y, Condit RC, Vijaysri S, Jacobs B, Williams BR, Silverman RH. Blockade of interferon induction and action by the E3L double-stranded RNA binding proteins of vaccinia virus. *J Virol* 2002;**76**:5251–9.
10. Berro R, Pedati C, Kehn-Hall K, Wu WL, Klase Z, Even Y, et al. CDK13, a new potential human immunodeficiency virus type 1 inhibitory factor regulating viral mRNA splicing. *J Virol* 2008;**82**:7155–66.
11. Arias C, Walsh D, Harbell J, Wilson AC, Mohr I. Activation of host translational control pathways by a viral developmental switch. *PLoS Pathog* 2009;**5**:e1000334.
12. Vogt PK, Hart JR, Gymnopoulos M, Jiang H, Kang S, Bader AG, et al. Phosphatidylinositol 3-kinase: the oncoprotein. *Curr Top Microbiol Immunol* 2010;**347**:79–104.
13. Calleja V, Alcor D, Laguerre M, Park J, Vojnovic B, Hemmings BA, et al. Intramolecular and intermolecular interactions of protein kinase B define its activation in vivo. *PLoS Biol* 2007;**5**: e95.
14. Downward J. Mechanisms and consequences of activation of protein kinase B/Akt. *Curr Opin Cell Biol* 1998;**10**:262–7.
15. Facchinetti V, Ouyang W, Wei H, Soto N, Lazorchak A, Gould C, et al. The mammalian target of rapamycin complex 2 controls folding and stability of Akt and protein kinase C. *EMBO J* 2008;**27**:1932–43.
16. Andjelkovic M, Alessi DR, Meier R, Fernandez A, Lamb NJ, Frech M, et al. Role of translocation in the activation and function of protein kinase B. *J Biol Chem* 1997;**272**:31515–24.
17. Galetic I, Andjelkovic M, Meier R, Brodbeck D, Park J, Hemmings BA. Mechanism of protein kinase B activation by insulin/insulin-like growth factor-1 revealed by specific inhibitors of phosphoinositide 3-kinase—significance for diabetes and cancer. *Pharmacol Ther* 1999;**82**:409–25.
18. Biggs WH, 3rd, Meisenhelder J, Hunter T, Cavenee WK, Arden KC. Protein kinase B/Akt-mediated phosphorylation promotes nuclear exclusion of the winged helix transcription factor FKHR1. *Proc Natl Acad Sci USA* 1999;**96**:7421–6.
19. Hahn-Windgassen A, Nogueira V, Chen CC, Skeen JE, Sonenberg N, Hay N. Akt activates the mammalian target of rapamycin by regulating cellular ATP level and AMPK activity. *J Biol Chem* 2005;**280**:32081–9.

20. Robey RB, Hay N. Is Akt the "Warburg kinase"?—Akt-energy metabolism interactions and oncogenesis. *Semin Cancer Biol* 2009;**19**:25–31.
21. Wullschleger S, Loewith R, Hall MN. TOR signaling in growth and metabolism. *Cell* 2006;**124**:471–84.
22. Gao X, Zhang Y, Arrazola P, Hino O, Kobayashi T, Yeung RS, et al. Tsc tumour suppressor proteins antagonize amino-acid-TOR signalling. *Nat Cell Biol* 2002;**4**:699–704.
23. Ramaswamy S, Nakamura N, Vazquez F, Batt DB, Perera S, Roberts TM, et al. Regulation of G1 progression by the PTEN tumor suppressor protein is linked to inhibition of the phosphatidylinositol 3-kinase/Akt pathway. *Proc Natl Acad Sci USA* 1999;**96**:2110–5.
24. Kane LP, Shapiro VS, Stokoe D, Weiss A. Induction of NF-[kappa]B by the Akt/PKB kinase. *Curr Biol* 1999;**9**:601–4.
25. Cooray S. The pivotal role of phosphatidylinositol 3-kinase-Akt signal transduction in virus survival. *J Gen Virol* 2004;**85**:1065–76.
26. Norman KL, Sarnow P. Herpes Simplex Virus is Akt-ing in translational control. *Genes Dev* 2010;**24**:2583–6.
27. Bellacosa A, Testa JR, Staal SP, Tsichlis PN. A retroviral oncogene, akt, encoding a serine-threonine kinase containing an SH2-like region. *Science* 1991;**254**:274–7.
28. Bellacosa A, Franke TF, Gonzalez-Portal ME, Datta K, Taguchi T, Gardner J, et al. Structure, expression and chromosomal mapping of c-akt: relationship to v-akt and its implications. *Oncogene* 1993;**8**:745–54.
29. Coffer PJ, Woodgett JR. Molecular cloning and characterisation of a novel putative protein-serine kinase related to the cAMP-dependent and protein kinase C families. *Eur J Biochem* 1991;**201**:475–81.
30. Jones PF, Jakubowicz T, Pitossi FJ, Maurer F, Hemmings BA. Molecular cloning and identification of a serine/threonine protein kinase of the second-messenger subfamily. *Proc Natl Acad Sci USA* 1991;**88**:4171–5.
31. Chang HW, Aoki M, Fruman D, Auger KR, Bellacosa A, Tsichlis PN, et al. Transformation of chicken cells by the gene encoding the catalytic subunit of PI 3-kinase. *Science* 1997;**276**:1848–50.
32. Aoki M, Batista O, Bellacosa A, Tsichlis P, Vogt PK. The akt kinase: molecular determinants of oncogenicity. *Proc Natl Acad Sci USA* 1998;**95**:14950–5.
33. Aoki M, Jiang H, Vogt PK. Proteasomal degradation of the FoxO1 transcriptional regulator in cells transformed by the P3k and Akt oncoproteins. *Proc Natl Acad Sci USA* 2004;**101**:13613–7.
34. Aoki M, Blazek E, Vogt PK. A role of the kinase mTOR in cellular transformation induced by the oncoproteins P3k and Akt. *Proc Natl Acad Sci USA* 2001;**98**:136–41.
35. Reichman CT, Mayer BJ, Keshav S, Hanafusa H. The product of the cellular crk gene consists primarily of SH2 and SH3 regions. *Cell Growth Differ* 1992;**3**:451–60.
36. Tsuchie H, Chang CH, Yoshida M, Vogt PK. A newly isolated avian sarcoma virus, ASV-1, carries the crk oncogene. *Oncogene* 1989;**4**:1281–4.
37. Matsuda M, Mayer BJ, Hanafusa H. Identification of domains of the v-crk oncogene product sufficient for association with phosphotyrosine-containing proteins. *Mol Cell Biol* 1991;**11**:1607–13.
38. Akagi T, Murata K, Shishido T, Hanafusa H. v-Crk activates the phosphoinositide 3-kinase/AKT pathway by utilizing focal adhesion kinase and H-Ras. *Mol Cell Biol* 2002;**22**:7015–23.
39. Akagi T, Shishido T, Murata K, Hanafusa H. v-Crk activates the phosphoinositide 3-kinase/AKT pathway in transformation. *Proc Natl Acad Sci USA* 2000;**97**:7290–5.
40. Chugh P, Bradel-Tretheway B, Monteiro-Filho CM, Planelles V, Maggirwar SB, Dewhurst S, et al. Akt inhibitors as an HIV-1 infected macrophage-specific anti-viral therapy. *Retrovirology* 2008;**5**:11.

41. Gao T, Furnari F, Newton AC. PHLPP: a phosphatase that directly dephosphorylates Akt, promotes apoptosis, and suppresses tumor growth. *Mol Cell* 2005;**18**:13–24.
42. Arroyo JD, Hahn WC. Involvement of PP2A in viral and cellular transformation. *Oncogene* 2005;**24**:7746–55.
43. Yuan H, Veldman T, Rundell K, Schlegel R. Simian virus 40 small tumor antigen activates AKT and telomerase and induces anchorage-independent growth of human epithelial cells. *J Virol* 2002;**76**:10685–91.
44. Yang SI, Lickteig RL, Estes R, Rundell K, Walter G, Mumby MC. Control of protein phosphatase 2A by simian virus 40 small-t antigen. *Mol Cell Biol* 1991;**11**:1988–95.
45. Liacini A, Seamone ME, Muruve DA, Tibbles LA. Anti-BK virus mechanisms of sirolimus and leflunomide alone and in combination: toward a new therapy for BK virus infection. *Transplantation* 2010;**90**:1450–7.
46. Andrabi S, Gjoerup OV, Kean JA, Roberts TM, Schaffhausen B. Protein phosphatase 2A regulates life and death decisions via Akt in a context-dependent manner. *Proc Natl Acad Sci USA* 2007;**104**:19011–6.
47. Bollag B, Hofstetter CA, Reviriego-Mendoza MM, Frisque RJ. JC virus small T antigen binds phosphatase PP2A and Rb family proteins and is required for efficient viral DNA replication activity. *PLoS One* 2010;**5**:e10606.
48. Utermark T, Schaffhausen BS, Roberts TM, Zhao JJ. The p110alpha isoform of phosphatidylinositol 3-kinase is essential for polyomavirus middle T antigen-mediated transformation. *J Virol* 2007;**81**:7069–76.
49. Yu Y, Alwine JC. Interaction between simian virus 40 large T antigen and insulin receptor substrate 1 is disrupted by the K1 mutation, resulting in the loss of large T antigen-mediated phosphorylation of Akt. *J Virol* 2008;**82**:4521–6.
50. Yu Y, Alwine JC. Human cytomegalovirus major immediate-early proteins and simian virus 40 large T antigen can inhibit apoptosis through activation of the phosphatidylinositide 3′-OH kinase pathway and the cellular kinase Akt. *J Virol* 2002;**76**:3731–8.
51. Pim D, Massimi P, Dilworth SM, Banks L. Activation of the protein kinase B pathway by the HPV-16 E7 oncoprotein occurs through a mechanism involving interaction with PP2A. *Oncogene* 2005;**24**:7830–8.
52. Cheng S, Schmidt-Grimminger DC, Murant T, Broker TR, Chow LT. Differentiation-dependent up-regulation of the human papillomavirus E7 gene reactivates cellular DNA replication in suprabasal differentiated keratinocytes. *Genes Dev* 1995;**9**:2335–49.
53. McLaughlin-Drubin ME, Münger K. The human papillomavirus E7 oncoprotein. *Virology* 2009;**384**:335–44.
54. Lin GY, Lamb RA. The paramyxovirus simian virus 5 V protein slows progression of the cell cycle. *J Virol* 2000;**74**:9152–66.
55. Gustin KE. Inhibition of nucleo-cytoplasmic trafficking by RNA viruses: targeting the nuclear pore complex. *Virus Res* 2003;**95**:35–44.
56. Lyles DS. Cytopathogenesis and inhibition of host gene expression by RNA viruses. *Microbiol Mol Biol Rev* 2000;**64**:709–24.
57. Thomas KW, Monick MM, Staber JM, Yarovinsky T, Carter AB, Hunninghake GW. Respiratory syncytial virus inhibits apoptosis and induces NF-kappa B activity through a phosphatidylinositol 3-kinase-dependent pathway. *J Biol Chem* 2002;**277**:492–501.
58. Monick MM, Cameron K, Powers LS, Butler NS, McCoy D, Mallampalli RK, et al. Sphingosine kinase mediates activation of extracellular signal-related kinase and Akt by respiratory syncytial virus. *Am J Respir Cell Mol Biol* 2004;**30**:844–52.
59. Peters K, Chattopadhyay S, Sen GC. IRF-3 activation by Sendai virus infection is required for cellular apoptosis and avoidance of persistence. *J Virol* 2008;**82**:3500–8.

60. Sun M, Fuentes SM, Timani K, Sun D, Murphy C, Lin Y, et al. Akt plays a critical role in replication of nonsegmented negative-stranded RNA viruses. *J Virol* 2008;**82**:105–14.
61. Autret A, Martin-Latil S, Brisac C, Mousson L, Colbere-Garapin F, Blondel B. Early phosphatidylinositol 3-kinase/Akt pathway activation limits poliovirus-induced JNK-mediated cell death. *J Virol* 2008;**82**:3796–802.
62. Bentley JK, Newcomb DC, Goldsmith AM, Jia Y, Sajjan US, Hershenson MB. Rhinovirus activates interleukin-8 expression via a Src/p110beta phosphatidylinositol 3-kinase/Akt pathway in human airway epithelial cells. *J Virol* 2007;**81**:1186–94.
63. Lau C, Wang X, Song L, North M, Wiehler S, Proud D, et al. Syk associates with clathrin and mediates phosphatidylinositol 3-kinase activation during human rhinovirus internalization. *J Immunol* 2008;**180**:870–80.
64. Esfandiarei M, Luo H, Yanagawa B, Suarez A, Dabiri D, Zhang J, et al. Protein kinase B/Akt regulates coxsackievirus B3 replication through a mechanism which is not caspase dependent. *J Virol* 2004;**78**:4289–98.
65. Esfandiarei M, Boroomand S, Suarez A, Si X, Rahmani M, McManus B. Coxsackievirus B3 activates nuclear factor kappa B transcription factor via a phosphatidylinositol-3 kinase/protein kinase B-dependent pathway to improve host cell viability. *Cell Microbiol* 2007;**9**:2358–71.
66. Halasz P, Holloway G, Turner SJ, Coulson BS. Rotavirus replication in intestinal cells differentially regulates integrin expression by a phosphatidylinositol 3-kinase-dependent pathway, resulting in increased cell adhesion and virus yield. *J Virol* 2008;**82**:148–60.
67. Bagchi P, Dutta D, Chattopadhyay S, Mukherjee A, Halder UC, Sarkar S, et al. Rotavirus nonstructural protein 1 suppresses virus-induced cellular apoptosis to facilitate viral growth by activating the cell survival pathways during early stages of infection. *J Virol* 2010;**84**:6834–45.
68. Lin PY, Liu HJ, Liao MH, Chang CD, Chang CI, Cheng HL, et al. Activation of PI 3-kinase/Akt/NF-kappaB and Stat3 signaling by avian reovirus S1133 in the early stages of infection results in an inflammatory response and delayed apoptosis. *Virology* 2010;**400**:104–14.
69. Halasz P, Holloway G, Coulson BS. Death mechanisms in epithelial cells following rotavirus infection, exposure to inactivated rotavirus or genome transfection. *J Gen Virol* 2010;**91**:2007–18.
70. Lee C-J, Liao C-L, Lin Y-L. Flavivirus activates phosphatidylinositol 3-kinase signaling to block caspase-dependent apoptotic cell death at the early stage of virus infection. *J Virol* 2005;**79**:8388–99.
71. Peng L, Liang D, Tong W, Li J, Yuan Z. Hepatitis C virus NS5A activates the mammalian target of rapamycin (mTOR) pathway, contributing to cell survival by disrupting the interaction between FK506-binding protein 38 (FKBP38) and mTOR. *J Biol Chem* 2010;**285**:20870–81.
72. Bai X, Ma D, Liu A, Shen X, Wang QJ, Liu Y, et al. Rheb activates mTOR by antagonizing its endogenous inhibitor, FKBP38. *Science* 2007;**318**:977–80.
73. Moody CA, Scott RS, Amirghahari N, Nathan CO, Young LS, Dawson CW, et al. Modulation of the cell growth regulator mTOR by Epstein-Barr virus-encoded LMP2A. *J Virol* 2005;**79**:5499–506.
74. Bultema R, Longnecker R, Swanson-Mungerson M. Epstein-Barr virus LMP2A accelerates MYC-induced lymphomagenesis. *Oncogene* 2009;**28**:1471–6.
75. Chen YR, Liu MT, Chang YT, Wu CC, Hu CY, Chen JY. Epstein-Barr virus latent membrane protein 1 represses DNA repair through the PI3K/Akt/FOXO3a pathway in human epithelial cells. *J Virol* 2008;**82**:8124–37.
76. Darr CD, Mauser A, Kenney S. Epstein-Barr virus immediate-early protein BRLF1 induces the lytic form of viral replication through a mechanism involving phosphatidylinositol-3 kinase activation. *J Virol* 2001;**75**:6135–42.

77. Cen O, Longnecker R. Rapamycin reverses splenomegaly and inhibits tumor development in a transgenic model of Epstein-Barr virus-related Burkitt's lymphoma. *Mol Cancer Ther* 2011;**10**:679–86.
78. Benetti L, Roizman B. Protein kinase B/Akt is present in activated form throughout the entire replicative cycle of deltaU(S)3 mutant virus but only at early times after infection with wild-type herpes simplex virus 1. *J Virol* 2006;**80**:3341–8.
79. Wagner MJ, Smiley JR. Herpes simplex virus requires VP11/12 to activate Src family kinase-phosphoinositide 3-kinase-Akt signaling. *J Virol* 2011;**85**:2803–12.
80. MacLeod IJ, Minson T. Binding of herpes simplex virus type-1 virions leads to the induction of intracellular signalling in the absence of virus entry. *PLoS One* 2010;**5**:e9560.
81. Chuluunbaatar U, Roller R, Feldman ME, Brown S, Shokat KM, Mohr I. Constitutive mTORC1 activation by a herpesvirus Akt surrogate stimulates mRNA translation and viral replication. *Genes Dev* 2010;**24**:2627–39.
82. Moorman NJ, Cristea IM, Terhune SS, Rout MP, Chait BT, Shenk T. Human cytomegalovirus protein UL38 inhibits host cell stress responses by antagonizing the tuberous sclerosis protein complex. *Cell Host Microbe* 2008;**3**:253–62.
83. Kudchodkar SB, Yu Y, Maguire TG, Alwine JC. Human cytomegalovirus infection induces rapamycin-insensitive phosphorylation of downstream effectors of mTOR kinase. *J Virol* 2004;**78**:11030–9.
84. Kudchodkar SB, Yu Y, Maguire TG, Alwine JC. Human cytomegalovirus infection alters the substrate specificities and rapamycin sensitivities of raptor- and rictor-containing complexes. *Proc Natl Acad Sci USA* 2006;**103**:14182–7.
85. Clippinger AJ, Maguire TG, Alwine JC. Human cytomegalovirus infection maintains mTOR activity and its perinuclear localization during amino acid deprivation. *J Virol* 2011;**85**:9369–76.
86. Soares JA, Leite FG, Andrade LG, Torres AA, De Sousa LP, Barcelos LS, et al. Activation of the PI3K/Akt pathway early during vaccinia and cowpox virus infections is required for both host survival and viral replication. *J Virol* 2009;**83**:6883–99.
87. Wang G, Barrett JW, Stanford M, Werden SJ, Johnston JB, Gao X, et al. Infection of human cancer cells with myxoma virus requires Akt activation via interaction with a viral ankyrin-repeat host range factor. *Proc Natl Acad Sci USA* 2006;**103**:4640–5.
88. Zaborowska I, Walsh D. PI3K signaling regulates rapamycin-insensitive translation initiation complex formation in vaccinia virus-infected cells. *J Virol* 2009;**83**:3988–92.
89. Hu N, Yu R, Shikuma C, Shiramizu B, Ostrwoski MA, Yu Q. Role of cell signaling in poxvirus-mediated foreign gene expression in mammalian cells. *Vaccine* 2009;**27**:2994–3006.
90. McNulty S, Bornmann W, Schriewer J, Werner C, Smith SK, Olson VA, et al. Multiple phosphatidylinositol 3-kinases regulate vaccinia virus morphogenesis. *PLoS One* 2010;**5**:e10884.
91. Werden SJ, Barrett JW, Wang G, Stanford MM, McFadden G. M-T5, the ankyrin repeat, host range protein of myxoma virus, activates Akt and can be functionally replaced by cellular PIKE-A. *J Virol* 2007;**81**:2340–8.
92. Werden SJ, Lanchbury J, Shattuck D, Neff C, Dufford M, McFadden G. The myxoma virus m-t5 ankyrin repeat host range protein is a novel adaptor that coordinately links the cellular signaling pathways mediated by Akt and Skp1 in virus-infected cells. *J Virol* 2009;**83**:12068–83.
93. Mossman K, Lee S, Barry M, Boshkov L, McFadden G. Disruption of M-T5, a novel myxoma virus gene member of poxvirus host range superfamily, results in dramatic attenuation of myxomatosis in infected European rabbits. *J Virol* 1996;**70**:4394–410.
94. Eierhoff T, Hrincius ER, Rescher U, Ludwig S, Ehrhardt C. The epidermal growth factor receptor (EGFR) promotes uptake of influenza A viruses (IAV) into host cells. *PLoS Pathog* 2010;**6**:12585–93.

95. Fujioka Y, Tsuda M, Hattori T, Sasaki J, Sasaki T, Miyazaki T, et al. The Ras-PI3K signaling pathway is involved in clathrin-independent endocytosis and the internalization of influenza viruses. *PLoS One* 2011;**6**:e16324.
96. Hale BG, Jackson D, Chen YH, Lamb RA, Randall RE. Influenza A virus NS1 protein binds p85beta and activates phosphatidylinositol-3-kinase signaling. *Proc Natl Acad Sci USA* 2006;**103**:14194–9.
97. Ehrhardt C, Marjuki H, Wolff T, Nurnberg B, Planz O, Pleschka S, et al. Bivalent role of the phosphatidylinositol-3-kinase (PI3K) during influenza virus infection and host cell defence. *Cell Microbiol* 2006;**8**:1336–48.
98. Shin YK, Liu Q, Tikoo SK, Babiuk LA, Zhou Y. Influenza A virus NS1 protein activates the phosphatidylinositol 3-kinase (PI3K)/Akt pathway by direct interaction with the p85 subunit of PI3K. *J Gen Virol* 2007;**88**:13–8.
99. Hale BG, Batty IH, Downes CP, Randall RE. Binding of influenza A virus NS1 protein to the inter-SH2 domain of p85 suggests a novel mechanism for phosphoinositide 3-kinase activation. *J Biol Chem* 2008;**283**:1372–80.
100. Hale BG, Kerry PS, Jackson D, Precious BL, Gray A, Killip MJ, et al. Structural insights into phosphoinositide 3-kinase activation by the influenza A virus NS1 protein. *Proc Natl Acad Sci USA* 2010;**107**:1954–9.
101. Heikkinen LS, Kazlauskas A, Melen K, Wagner R, Ziegler T, Julkunen I, et al. Avian and 1918 Spanish influenza a virus NS1 proteins bind to Crk/CrkL Src homology 3 domains to activate host cell signaling. *J Biol Chem* 2008;**283**:5719–27.
102. Ehrhardt C, Wolff T, Pleschka S, Planz O, Beermann W, Bode JG, et al. Influenza A virus NS1 protein activates the PI3K/Akt pathway to mediate antiapoptotic signaling responses. *J Virol* 2007;**81**:3058–67.
103. Zhirnov OP, Klenk HD. Control of apoptosis in influenza virus-infected cells by up-regulation of Akt and p53 signaling. *Apoptosis* 2007;**12**:1419–32.
104. Kaur S, Katsoulidis E, Platanias LC. Akt and mRNA translation by interferons. *Cell Cycle* 2008;**7**:2112–6.
105. Kaur S, Sassano A, Joseph AM, Majchrzak-Kita B, Eklund EA, Verma A, et al. Dual regulatory roles of phosphatidylinositol 3-kinase in IFN signaling. *J Immunol* 2008;**181**:7316–23.
106. Kaur S, Sassano A, Dolniak B, Joshi S, Majchrzak-Kita B, Baker DP, et al. Role of the Akt pathway in mRNA translation of interferon-stimulated genes. *Proc Natl Acad Sci USA* 2008;**105**:4808–13.
107. Mastronarde JG, He B, Monick MM, Mukaida N, Matsushima K, Hunninghake GW. Induction of interleukin (IL)-8 gene expression by respiratory syncytial virus involves activation of nuclear factor (NF)-kappa B and NF-IL-6. *J Infect Dis* 1996;**174**:262–7.
108. Mastronarde JG, Monick MM, Hunninghake GW. Oxidant tone regulates IL-8 production in epithelium infected with respiratory syncytial virus. *Am J Respir Cell Mol Biol* 1995;**13**:237–44.
109. Freudenburg W, Moran JM, Lents NH, Baldassare JJ, Buller RM, Corbett JA. Phosphatidylinositol 3-kinase regulates macrophage responses to double-stranded RNA and encephalomyocarditis virus. *J Innate Immun* 2009;**2**:77–86.
110. Guillot L, Le Goffic R, Bloch S, Escriou N, Akira S, Chignard M, et al. Involvement of toll-like receptor 3 in the immune response of lung epithelial cells to double-stranded RNA and influenza A virus. *J Biol Chem* 2005;**280**:5571–80.
111. Tyner JW, Uchida O, Kajiwara N, Kim EY, Patel AC, O'Sullivan MP, et al. CCL5-CCR5 interaction provides antiapoptotic signals for macrophage survival during viral infection. *Nat Med* 2005;**11**:1180–7.
112. Griffin DE. Measles virus-induced suppression of immune responses. *Immunol Rev* 2010;**236**:176–89.

113. Avota E, Gassert E, Schneider-Schaulies S. Measles virus-induced immunosuppression: from effectors to mechanisms. *Med Microbiol Immunol* 2010;**199**:227–37.
114. Avota E, Avots A, Niewiesk S, Kane LP, Bommhardt U, ter Meulen V, et al. Disruption of Akt kinase activation is important for immunosuppression induced by measles virus. *Nat Med* 2001;**7**:725–31.
115. Avota E, Harms H, Schneider-Schaulies S. Measles virus induces expression of SIP110, a constitutively membrane clustered lipid phosphatase, which inhibits T cell proliferation. *Cell Microbiol* 2006;**8**:1826–39.
116. Carsillo M, Kim D, Niewiesk S. Role of AKT kinase in measles virus replication. *J Virol* 2010;**84**:2180–3.
117. Stojdl DF, Lichty B, Knowles S, Marius R, Atkins H, Sonenberg N, et al. Exploiting tumor-specific defects in the interferon pathway with a previously unknown oncolytic virus. *Nat Med* 2000;**6**:821–5.
118. Geisbert TW, Bausch DG, Feldmann H. Prospects for immunisation against Marburg and Ebola viruses. *Rev Med Virol* 2010;**20**:344–57.
119. Dunn EF, Fearns R, Connor JH. Akt inhibitor Akt-IV blocks virus replication through an Akt-independent mechanism. *J Virol* 2009;**83**:11665–72.
120. Shelly S, Lukinova N, Bambina S, Berman A, Cherry S. Autophagy is an essential component of Drosophila immunity against vesicular stomatitis virus. *Immunity* 2009;**30**:588–98.
121. Dunn EF, Connor JH. Dominant inhibition of Akt/protein kinase B signaling by the matrix protein of a negative-strand RNA virus. *J Virol* 2011;**85**:422–31.
122. Alain T, Lun X, Martineau Y, Sean P, Pulendran B, Petroulakis E, et al. Vesicular stomatitis virus oncolysis is potentiated by impairing mTORC1-dependent type I IFN production. *Proc Natl Acad Sci USA* 2010;**107**:1576–81.
123. Connor JH, Lyles DS. Vesicular stomatitis virus infection alters the eIF4F translation initiation complex and causes dephosphorylation of the eIF4E binding protein 4E-BP1. *J Virol* 2002;**76**:10177–87.
124. Minami K, Tambe Y, Watanabe R, Isono T, Haneda M, Isobe K, et al. Suppression of viral replication by stress-inducible GADD34 protein via the mammalian serine/threonine protein kinase mTOR pathway. *J Virol* 2007;**81**:11106–15.
125. Popova TG, Turell MJ, Espina V, Kehn-Hall K, Kidd J, Narayanan A, et al. Reverse-phase phosphoproteome analysis of signaling pathways induced by Rift valley fever virus in human small airway epithelial cells. *PLoS One* 2010;**5**:e13805.
126. Mohankumar V, Dhanushkodi NR, Raju R. Sindbis virus replication, is insensitive to rapamycin and torin1, and suppresses Akt/mTOR pathway late during infection in HEK cells. *Biochem Biophys Res Commun* 2011;**406**:262–7.
127. Yuan J, Lipinski M, Degterev A. Diversity in the mechanisms of neuronal cell death. *Neuron* 2003;**40**:401–13.
128. Sharma NR, Mani P, Nandwani N, Mishra R, Rana A, Sarkar DP. Reciprocal regulation of AKT and MAP kinase dictates virus-host cell fusion. *J Virol* 2010;**84**:4366–82.
129. Hand TW, Cui W, Jung YW, Sefik E, Joshi NS, Chandele A, et al. Differential effects of STAT5 and PI3K/AKT signaling on effector and memory CD8 T-cell survival. *Proc Natl Acad Sci USA* 2010;**107**:16601–6.
130. Guiducci C, Ghirelli C, Marloie-Provost MA, Matray T, Coffman RL, Liu YJ, et al. PI3K is critical for the nuclear translocation of IRF-7 and type I IFN production by human plasmacytoid predendritic cells in response to TLR activation. *J Exp Med* 2008;**205**:315–22.
131. Cao W, Manicassamy S, Tang H, Kasturi SP, Pirani A, Murthy N, et al. Toll-like receptor-mediated induction of type I interferon in plasmacytoid dendritic cells requires the rapamycin-sensitive PI(3)K-mTOR-p70S6K pathway. *Nat Immunol* 2008;**9**:1157–64.

132. Nepomuceno RR, Balatoni CE, Natkunam Y, Snow AL, Krams SM, Martinez OM. Rapamycin inhibits the interleukin 10 signal transduction pathway and the growth of Epstein Barr virus B-cell lymphomas. *Cancer Res* 2003;**63**:4472–80.
133. Eden JS, Sharpe LJ, White PA, Brown AJ. Norovirus RNA-dependent RNA-polymerase is phosphorylated by an important survival kinase, Akt. *J Virol* 2011;**85**:10894–8.

# Section III

# Protein Phosphatases

# Protein Tyrosine Phosphatases in Cancer: Friends and Foes!

David P. Labbé,[*,†] Serge Hardy,[*] and Michel L. Tremblay[*,†,‡]

[*]*Goodman Cancer Research Centre, McGill University, Montréal, Québec, Canada*
[†]*Department of Medicine, Division of Experimental Medicine, McGill University, Montréal, Québec, Canada*
[‡]*Department of Biochemistry and Oncology, McGill University, Montréal, Québec, Canada*



Tyrosine phosphorylation of proteins serves as an exquisite switch in controlling several key oncogenic signaling pathways involved in cell proliferation, apoptosis, migration, and invasion. Since protein tyrosine phosphatases (PTPs) counteract protein kinases by removing phosphate moieties on target proteins, one may intuitively think that PTPs would act as tumor suppressors. Indeed, one of the most described PTPs, namely, the phosphatase and tensin homolog (PTEN), is a tumor suppressor. However, a

Progress in Molecular Biology
and Translational Science, Vol. 106
DOI: 10.1016/B978-0-12-396456-4.00009-2

growing body of evidence suggests that PTPs can also function as potent oncoproteins. In this chapter, we provide a broad historical overview of the PTPs, their mechanism of action, and posttranslational modifications. Then, we focus on the dual properties of classical PTPs (receptor and nonreceptor) and dual-specificity phosphatases in cancer and summarize the current knowledge of the signaling pathways regulated by key PTPs in human cancer. In conclusion, we present our perspective on the potential of these PTPs to serve as therapeutic targets in cancer.

## I. Introduction

Regulated protein phosphorylation is an evolutionarily conserved means of intra- and intercellular communication. While regulated phosphorylation of serine (pSer) and threonine (pThr) residues arose among the first single-celled eukaryotes, phosphotyrosine (pTyr) base signal transduction emerged later in evolution, coincident with the evolution of multicellular animals. In fact, the tyrosine signaling system has been proposed to be a critical event in evolution that enabled the expansion of multicellular species. Indeed, tyrosine phosphorylation is required for the activation of signaling pathways that regulate a plethora of important cellular activities, such as cell growth, hormone response, immune defense, and many others.[1]

Transduction of pTyr signaling requires two classes of enzymes: tyrosine kinases (TyrKs), which phosphorylate tyrosine residues; and protein tyrosine phosphatases (PTPs), which removes phosphate moieties. Although there are a handful of PTPs present in *S. cerevisiae*, the human "PTPome" contain 107 PTPs, which are grouped into four families based on the amino acid sequence of their catalytic domains. Almost all PTPs are part of the Class I cysteine-based family (99 genes), which includes the 38 classical PTPs (receptor-like PTPs (RPTPs, 21 genes)), nonreceptor-like PTPs (NRPTPs, 17 genes), and the dual-specificity phosphatases (DSPs, 61 genes).[2] They share a common fold and the same $HC(X)_5R$ catalytic motif. Class II and Class III families are also cysteine-based PTPs and they include the low-molecular-weight PTP (LMPTP) and the three Cdc25 proteins, respectively. Finally, Asp-based PTPs represent the last family in which there are four *EYA*.[2]

Any changes in the expression or activities of these enzymes might tip the balance of cellular homeostasis and contribute positively or negatively to various diseases such as cancer. However, as the PTP field is expanding rapidly, this chapter cannot cover all aspects of PTP biology in cancer. For complementary information on the genetics and/or epigenetic alterations of PTP genes, refer to Julien *et al.*[3] More detailed information on the expression of Class I

PTP members in various cancer tissues[4] and on the DSPs in human cancers[5] was described elsewhere. Herein, the oncogenic activity of the 38 classical pTyr-specific PTPs is discussed in the context of human cancer cell lines and animal models. We also cover the DSP subgroup that can dephosphorylate pTyr-, pSer-, and pThr-containing substrates.

## II. PTPs and Their Mechanism of Action

Tyrosine phosphorylation was first reported in 1978 in conjunction with the identification of the first TyrK, namely, Src.[6,7] The demonstration of an opposing regulator, however, required an additional decade of research. Initial studies involved the purification, characterization, and cloning of the prototypic PTP family member PTP1B by Nick Tonks and colleagues[8–10] and a year later by the laboratory of Jack Dixon (Fig. 1).[11] In 1994, the crystal structure of PTP1B was reported, resolving for the first time the three-dimensional conformation of a PTP.[12] The structure of PTP1B provided a basis for the understanding of the mechanism of action of PTPs.

Classical PTPs and DSPs share the same mechanism of action, requiring an embedded conserved $HC(X)_5R$ catalytic motif. Central to this "PTP-signature motif" is the catalytic cysteine residue, located at the bottom of the catalytic cleft. In addition, an aspartate residue that is part of a highly conserved WPD motif located on the side of the catalytic cleft is essential for the two-step mechanism of action of PTPs (reviewed in Ref. 24). In classical PTPs, the depth of the cleft dictates the specificity towards the longer pTyr-containing

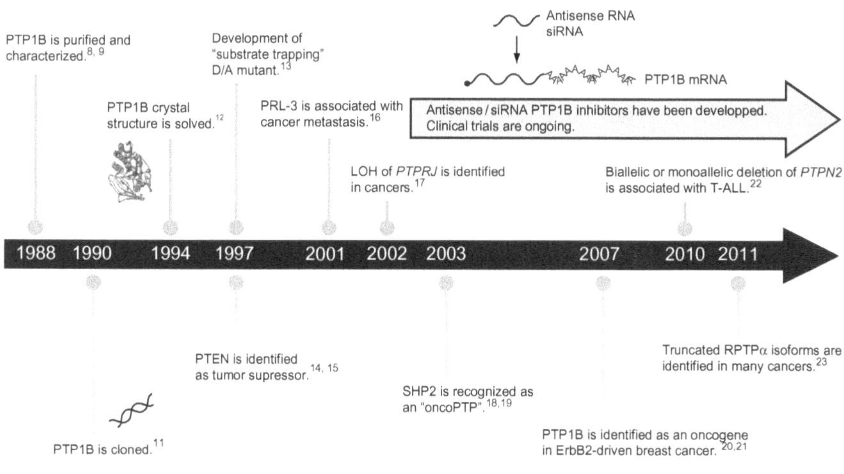


FIG. 1. Key events in the cancer-related PTP field.

substrates over the shorter pSer and pThr residues.[24] As the PTP signature motif of DSPs also needs to accommodate the shorter pSer and pThr protein substrates, their catalytic domain cleft is 3 Å shallower compared to classical PTPs.[25] Of interest, while mutation of the critical catalytic cysteine to serine, or replacement of other critical residues such as aspartate to alanine in the WPD loop, abolishes catalytic activity, it still allows recognition of phosphorylated substrates.[13,26]. The identification of physiological substrates using these "substrate trapping mutants" showed an intrinsic specificity of the PTP catalytic domains in their recognition of substrates.[26]

Among PTPs, the RPTPs are a unique subgroup that possess an N-terminal extracellular region and a single transmembrane region linked to one or two typical PTP catalytic domains. Like their counterparts, receptor tyrosine kinases (RTKs), RPTPs enable the transmission of extracellular environmental cues to the intracellular cell signaling circuitry, impacting cell fate. For example, binding of pleiotrophin, a PDGF-inducible heparin-binding cytokine, to RPTPβ/ζ disrupts the receptor's function and downregulates its catalytic activity.[27] Although some results suggest that the activities of RPTPs are regulated through dimerization-induced inactivation,[28,29] this mechanism remains controversial.[30,31] As another example, binding of chondroitin sulfate proteoglycans, a component of the extracellular matrix upregulated following neural injury, to RPTPσ results in the reduction of axonal growth following spinal cord injury.[32,33] However, after more than 20 years of research, only a relatively small number of RPTP ligands have been identified and their mechanism of regulation is still largely undefined.[34]

## III. Posttranslational Modifications of PTPs

Classical PTPs and DSPs share the same mechanism of action to achieve dephosphorylation of pTyr residues, yet they all demonstrate substrate specificity through various mechanisms. PTPs present different expression levels in organs, tissues, and cell types.[34] Subcellular localization also regulates PTP specificity. While RPTPs are localized at the plasma membrane and NRPTPs and DSPs are primarily in the cytoplasm, PTPs can also reside in different subcellular compartments including the membrane of secretory vesicles,[35] in the nucleus,[36] in the mitochondria,[37] or even in the endoplasmic reticulum (ER).[36,38] In addition to the cell-lineage-restricted expression and subcellular localization, posttranslational modifications also regulate PTP function and activity.

### A. Phosphorylation

A mechanism of PTP regulation is through serine/threonine-specific phosphorylation. This covalent posttranslational modification can regulate PTPs in a negative manner, as in the case of PTP-PEST when phosphorylated on Ser39/

Ser434 by the protein kinase A or protein kinase C.[39] Alternatively, phosphorylation of CD45 on multiple serine residues by casein kinase 2 promotes activity.[40] Serine phosphorylation of SHP1, as well as the structurally-close relative SHP2, was shown to differentially affect phosphatase activity. In fact, SHP2 activity is not modulated by serine phosphorylation,[41] while Ser591 phosphorylation on SHP1 does inhibit its catalytic activity.[42] The prototypic PTP1B was also shown to be phosphorylated on several serine residues.[43,44] Interestingly, phosphorylation on PTP1B Ser50 by Akt proved to be important for the dephosphorylation of the insulin receptor, which leads to a negative feedback mechanism since the insulin receptor is a substrate of PTP1B.[45] Phosphorylation on two serine residues (Ser180 and Ser204) located in the juxtamembrane domain of RPTPα stimulates its catalytic activity[46] and were found to be essential for the ability of RPTPα to activate Src during mitosis.[47]

Phosphorylation on tyrosine residues can also modulate PTP activity. Phosphorylation of SHP1 on Tyr538 in response to insulin[48] or on Tyr564 by the Lyn TyrK[49] results in an increase in phosphatase activity. Tyrosine phosphorylation following different stimuli can also have opposite effects. For example, the injection of insulin in mice induces PTP1B tyrosine phosphorylation and a concomitant decrease in its activity.[50] On the other hand, EGF-induced tyrosine phosphorylation of PTP1B on Tyr66 enhances its activity.[51] Interestingly, tyrosine phosphorylation on SHP1[52] and SHP2[53] also generates a binding site for the SH2 domain of the adaptor protein Grb2. Another example is the tyrosine phosphorylation of a C-terminal residue on RPTPα, which also provides a Grb2 binding site.[54,55] Such association with Grb2 serves as a platform from which to direct activation of the Ras/ERK pathway, central to cell growth, differentiation, and survival processes, all of which are deregulated during carcinogenesis.

## B. Proteolytic Cleavage

Proteolysis is another common posttranslational modification that regulates PTP activity. Calpains are calcium-regulated cysteine proteases that are activated in response to a temporal and localized calcium elevation following numerous stimuli and are implicated in cytoskeletal reorganization, cell proliferation, apoptosis, cell motility, and hemostasis.[56] Increased calcium in platelets leads to PTP1B,[57] SHP1,[58] or PTP-MEG[59] calpain-mediated cleavage of negative regulatory domains and causes enzyme activation. Ultimately, platelet aggregation causes PTP1B degradation and inactivation, allowing thrombus formation *in vivo*.[60] Cytoplasmic forms of RPTPα (RPTPα66) and RPTPε (RPTPε65) are also induced following calpain-mediated cleavage in the intracellular juxtamembrane domain.[61] Changes in the subcellular localization of those generated fragments result in a decreased capacity to dephosphorylate and activate Src or the voltage-gated potassium channel Kv2.1.

Caspases are cysteine proteases central to the apoptotic process. Since deregulation of apoptotic signals can contribute to the development of various pathologies such as neurodegenerative and cardiovascular disorders, acquired immune deficiency syndrome (AIDS), or cancer,[62] caspases are regarded as potential therapeutic targets.[63] Activation of the executioner caspase-3 cleaves PTP-PEST on its $^{549}$DSPD motif, producing an N-terminal fragment with increased catalytic activity. Furthermore, the remodeling of PTP-PEST protein complexes with adaptor molecules such as paxillin, following caspase-3 mediated cleavage, facilitates cellular detachment during the execution phase of apoptosis.[64] Interestingly, neither PTP1B, TCPTP, nor SHP2 were found to be cleaved during apoptosis.[64] However, many of the classical PTPs might be regulated by caspases since many of them harbor putative caspase cleavage sites within their amino acid sequences, as highlighted by *in silico* analysis.[65]

Alternatively, numerous other proteases can cleave phosphatases and affect various cellular events. An interesting example is the infection of humans by *Leishmania*, which are vectored by sandflies. This parasitism results in leishmaniasis, a disease threatening more than 350 million people worldwide.[66] This tropical infectious disease typically invades macrophages, causing symptoms ranging from skin lesions to death. The surface protease GP63 is a key *Leishmania* virulence factor and mediates PTP-PEST,[67] SHP1,[68] PTP1B, and TCPTP but not the phosphatase and tensin homolog (PTEN) or SHP2 cleavage.[69] These posttranslational modifications of SHP1, PTP1B, and TCPTP were found to stimulate PTP enzymatic activation. Excitingly, PTP1B was also identified as a key player in the *in vivo* progression of *Leishmania* infection.[69] Recently, several RPTPs were shown to be cleaved by proteolysis during tumorigenesis. This process, mediated by proteases such as matrix metalloproteinases, ADAM-like metalloproteases, or furin-like proteases, influences RPTP cellular function and generates fragments that may have oncogenic functions (reviewed in Ref. 70).

## C. Reversible Oxidation

Reactive oxygen species (ROS) are fundamental to certain cell types such as macrophages that generate them during the innate immune response following an infection.[71] Moreover, high levels of ROS are present in cancer cells, a consequence of cross-talk with infiltrating immune cells or intrinsically produced by the tumor cells through one or more of these following events: increased metabolic activity, mitochondrial dysfunction, peroxisome activity, increased cellular receptor signaling, oncogene activity, increased activity of oxidases, cyclooxygenases, lipoxigenases, or thymidine phosphorylase.[72] The transient and localized modulation of ROS levels within cancer cells regulates a number of cancer hallmarks[73] (i.e., cell cycle progression/proliferation, cell survival/apoptosis and others).[72] Additionally, ROS such as hydrogen peroxide ($H_2O_2$) are now recognized as second messengers and regulators of PTKs,[74] transcription factors,[75] and PTPs[76–78] through reversible oxidation.

The low $pK_a$ at the bottom of the PTP catalytic cleft maintains the conserved cysteine residue in a negatively charged thiolate form, allowing nucleophilic attack on the substrate phosphate. On the other hand, being negatively charged makes the cysteine particularly sensitive to oxidative stress. Cysteine thiolate is oxidized to sulfenic acid (S-OH) by $H_2O_2$, which in turn forms cyclic sulphenamide, a structure created between the cysteine sulfur atom and the adjacent serine residue following oxygen elimination. This reaction is readily reversible under physiological conditions and prevents the classical PTPs to be further oxidized to sulfinic acid ($S{-}O_2H$) or sulfonic acid ($S{-}O_3H$), which are irreversible events. Although the cyclic sulfenamide state is transient, it does inactivate the enzyme activity by inducing profound changes in the catalytic site architecture, making some buried residues in solvent-exposed positions thereby available for reduction and reversion to an active state.[76,79] Because DSPs contain a second cysteine residue within their active site, a disulfide bond is formed between these two cysteines following oxidation instead of a cyclic sulfenamide. Similar to classical PTPs, this bond is reversible through reduction and has the purpose of preventing irreversible inactivation following higher oxidation.[78] PTPs can also be protected from oxidation-induced irreversible inactivation through reversible S-nitrosylation mediated by cellular nitric oxide (NO), expressed in response to many extracellular stimuli.[80] Therefore PTP reversible oxidation is a mechanism that controls the steady-state level of tyrosine phosphorylation in the cell for the time being.

Additionally, reversible oxidation was shown to regulate calpain-mediated degradation of PTPs. In an original paper, Gulati *et al.* observed that PTPs were inactivated following UVA irradiation in the range of physiological exposure using human keratinocyte. Although UVA irradiation causes an increase in intracellular calcium concentration, treatment with ionomycin alone induces a rise in the intracellular level of calcium, which leads to a three- to fourfold elevation of calpain activity that did not induce significant PTP degradation. However, UVA irradiation also induces ROS production. Cell treatment with $H_2O_2$, was not sufficient to induce PTP degradation or calpain activation, but when immunoprecipitated, the oxidized form of PTP1B was readily degraded when exposed to calpain *in vitro*. This particular recognition of the oxidized form of PTP1B over the reduced one by calpain was also observed with LAR. This exciting mechanism of PTP regulation implying both reversible oxidation and proteolytic cleavage adds another layer of complexity that requires additional investigation.[81] In line with this, oxidation was also shown to induce conformational changes in the cytoplasmic domain of RPTP$\alpha$, LAR, RPTP$\mu$, and CD45, altering rotational coupling within RPTP dimers. Whether these changes regulate the activity or the function of RPTP dimers is still unknown, but it is tempting to speculate that the role of the second and inactive PTP domain in RPTP may be to regulate the oxidative milieu in order to regulate the active first catalytic domain and as well modify ligand binding properties.[82]

## D. Sumoylation

Small-uniquitin-related modifier or SUMO, was initially identified in *Saccharomyces cerevisiae* in the mid-1990s.[83] The human genome encodes four distinct SUMO proteins that act as posttranslational protein modifiers.[84,85] Sumoylation is a reversible process that refers to the covalent attachment of SUMO to the lysine side that falls within the consensus sequence ΨKXE/D, where Ψ is a hydrophobic residue. The mechanism of sumoylation is a three-step process that requires nascent SUMO to be cleaved by SUMO-specific isopeptidases (sentrin-specific proteases; SENPs) in order to reveal the Gly-Gly motif of its C-terminal tail. Thereafter, mature SUMO is (1) activated by the E1 heterodimer AOS1-UBA2 in an ATP-dependent reaction, (2) transferred to the catalytic Cys residue of the E2 enzyme UBC9, and (3) bound to a Lys residue in the substrate by its C-terminal Gly residue with the help of an E3 ligase. This whole process can therefore be reverted, since sumoylated substrates can be targeted by SENPs, removing the SUMO.[86] Although target proteins are typically conjugated to single SUMO moieties,[87] poly-SUMO chains are also observed under certain circumstances.[88] Consequences of this posttranslational modification are many, since sumoylation can altered the interaction between the target and its partner by providing a binding site for an interacting partner or resulting in a conformational change of the sumoylated protein. Therefore, sumoylation can alter localization, activity, or stability of the protein.[86]

Sumoylation controls the activity of numerous targets, including crucial transcription factors in carcinogenesis such as p53, the androgen receptor, or the estrogen receptor through direct modification or indirectly as observed with NF-κB *via* IκBα sumoylation.[89] Although most SUMO-targeted proteins are located in the nucleus, other targets were found in the mitochondria, at the plasma membrane, or at the ER membrane.[86] PTP1B is localized at the ER membrane and nuclear envelope on the cytoplasmic side through a C-terminal targeting motif. PTP1B was the first-identified ER-associated SUMO target protein and its ER localization was found to be essential for its optimal sumoylation.[90] After point mutation of four lysine residues to arginine (K/R) at consensus sumoylation sites (73, 335, 347, 389), $PTP1B^{K/R}$ failed to undergo sumoylation. Co-expression of wild-type PTP1B with SUMO-1 induced its sumoylation and a 60% decrease in catalytic activity, a decrease that was not observed with the $PTP1B^{K/R}$ mutant. In a physiological context, up to 20% of total PTP1B is sumoylated in mouse embryonic fibroblasts (MEFs) following insulin stimulation, a proportion that declined after 5min. Because this kinetic was comparable to the insulin-receptor activity after insulin stimulation, the insulin-dependent sumoylation of PTP1B might be necessary to control its activity and to insure maximal insulin signaling.[90,91] Furthermore, in an oncogenic context, the overexpression of a catalytically active PTP1B was shown to suppress transformation mediated by v-*crk*.[92] Interestingly, PTP1B$^{K/R}$ was more effective at suppressing v-*crk*-induced anchorage-independent growth of rat 3Y1 cells compared to the wild-type and nonsumoylated PTP1B.[90]

Although there is no other reported PTP regulated through direct sumoylation, indirect regulation of PTEN through the Akt–EGR1–ARF–PTEN axis has been demonstrated. The early growth response gene 1 (EGR1) is a transcription factor that gets phosphorylated on Ser350 and Thr309 by Akt following IGF1-R activation. EGR1 phosphorylation enables its migration to the nucleus. In the nucleus, phosphorylated EGR1 undergoes sumoylation on Lys272 in an ARF-dependent mechanism. Only then, EGR1 is able to directly transactivate the PTEN promoter and induce mRNA and protein expression, suppressing cell growth and proliferation.[93] Interestingly, other PTPs might also be regulated through transcription factors affected by sumoylation. PTP1B expression is repressed by EGR1, which prevents Sp-3 mediated *PTPN1* (PTP1B gene) promoter activation.[94] Furthermore, PTP1B is also positively regulated by inflammation through NF-κB transcriptional activation,[95] which is regulated via IκBα sumoylation.[96] Finally, we recently demonstrated that PTP1B is positively regulated by the androgen receptor in prostate cancer (L. Lessard, DP. Labbé, and ML. Tremblay, unpublished data), a transcription factor also regulated through sumoylation.[97] Therefore, regulation of these and probably other transcription factor activities by sumoylation may be important in PTP1B regulation and other PTPs.

## E. Prenylation

Prenylation is a covalent posttranslational modification that tightly controls the signaling activities of several cellular proteins.[98] Examples of prenylated proteins include the small GTPases (including Ras, Rac1, Rab)[99] and the phosphatase of regenerating liver (PRL) family of PTPs.[100] Protein prenylation involves the addition of a 15-carbon (farnesyl) or a 20-carbon (geranylgeranyl) isoprenoid moiety via a thioether linkage to one or more cysteines located at or near the C terminus of a protein.[101] This type of reaction can be catalyzed by three different protein prenyl transferases: protein farnesyltransferase (FTase), protein geranylgeranyl transferase-I (GGTase-I), and Rab geranylgeranyl transferase (RabGGTase or GGTase-II).[101] In general, the consensus prenylation sequence contains the CAAX motif (referred to as the CAAX box; C is cysteine, A is usually an aliphatic amino acid, and X can be a variety of amino acids). The X residue of this motif largely determines the choice of the isoprenoid.[102]

Interest in protein prenylation (farnesylation) was stimulated by the important discovery that inhibitors of FTases (FTIs) can reverse the phenotypes of cancer cells, suggesting their potential as anticancer drugs.[103] At first, this effect was attributed to farnesylation inhibition of the oncoprotein Ras, which is often mutated in human tumors.[104] Although, a large number of FTIs have been tested in various stages of preclinical and clinical trials,[103] it was subsequently realized that the task of inhibiting Ras prenylation is more complex than was initially hoped. One of the reasons is that the mechanism of FTI antitumor activity is still poorly understood, in part because there exist about 50 known potential substrates for farnesyltransferase, including the PRLs.[105]

The PRL-1, -2, -3 phosphatases are prenylated PTPs with oncogenic activity that are proposed to drive tumor metastasis (see Section V.B). The PRLs are the only PTPs that are farnesylated by the FTase and this posttranslational modification is important for their localization, structure, and function.[106] Interestingly, human PRL-1 and -2 were first discovered using an *in vitro* prenylation screen of a human breast carcinoma cDNA expression library for cDNA-encoded FTase substrates.[107] Later, Zeng *et al.* showed that the association of PRLs with the membrane of the cell surface and that the early endosome is dependent on their prenylation.[100] Interestingly, cells treated with FTI-277, a selective farnesyltransferase inhibitor, shifted PRLs into the nucleus. Furthermore, mutant forms of PRLs lacking the C-terminal prenylation signal are associated with the nucleus due to a polybasic region that has been proposed to act as a nuclear localization sequence.[108] On the other hand, this conserved polybasic region located in front of the CAAX domain was shown to play a crucial role in phospholipid binding (presumably via electrostatic interactions), but was not sufficient for membrane targeting.[109] This suggests that both prenyl modification and this polybasic region are essential for membrane targeting and attachment of the PRLs. Interestingly, this phenomenon is also observed with K-ras and Rac-1, in which the polybasic domain near the prenylation site appears to act as a strong targeting signal for the plasma membrane.[18]

One of the first reports indicating that prenylation of PRLs was important for their biological activities arose from mutating the critical cysteine in the CAAX domain of PRL-1 and -3 to show that these nonfarnesylated mutants were unable to promote migration and invasion of SW480 colon adenocarcinoma cells.[19] Other independent studies have confirmed these observations for these two PRLs in other cell types,[109–111] suggesting that proper localization for enzyme–substrate interaction is essential for their physiological action. Recently, ectopic expression of a C-terminal CAAX deleted PRL-2 mutant also revealed a requirement of this prenylation site for migration and invasion of A549 lung cancer cells.[112]

PRLs have been shown to form oligomers both *in vitro* and *in vivo*, suggesting that this association could regulate their activity.[109,113–115] Importantly, trimerization of PRL-1 requires its C-terminal prenylation, but disruption of trimer formation had no effect on PRL-1 association with the membrane.[109] Nevertheless, disruption of PRL-1 trimer formation abolished the ability of this phosphatase to promote cell growth and migration, indicating that PRL-1 trimer formation is essential for its function.[109] Similarly, prenylation of PRL-3 also promoted oligomer formation,[115] but the biological consequences of this association have not yet been reported.

The presence of the CAAX domain in PRLs is unique among PTPs, suggesting that they may have distinct functions compared to other PTPs. Lastly, since prenylation is essential for their biological activities, specifically targeting this posttranslational modification of these oncoPTPs could be a new avenue for cancer therapy.

## IV. PTPs as Tumor Suppressors

TyrKs were discovered a decade before PTPs and many of them were found to be oncoproteins.[116] Early on, many attempts were made to better understand the role of TyrKs in carcinogenesis[117] and develop to drug treatments.[118] Therefore, when PTPs were revealed later on as being the natural counterpart of TyrKs, it was assumed that they would act as tumor suppressors. This speculation took nearly a decade to be proven in human cancers. Identified in 1997, PTEN was classified as a putative tumor suppressor following identification of numerous deletions of the *PTEN* gene in brain, breast, and prostate cancers.[14,15] PTEN dampens the downstream signaling of RTKs through 3-phosphoinositide dephosphorylation (a product of phosphatidylinositol 3-kinase), which is thereafter unable to activate key survival kinases such as the phosphoinositide-dependant kinase 1 and Akt.[119] Monoallelic mutations at the *PTEN* locus occur at high frequencies (50–80%) in sporadic tumors such as endometrial carcinoma, glioblastoma, and prostate cancer, and at lower rates (30–50%) in breast, colon, and lung tumors. Complete loss of *PTEN* is also observed at higher incidence in endometrial cancer and glioblastoma and correlates with advanced cancers and metastases.[120] Furthermore, *PTEN* hypermorphic mice expressing 80% of normal PTEN levels were still found to develop various tumor types, breast cancer being the most sensitive to subtle modulations of PTEN expression.[121] There are mutiple examples of PTEN tumor-suppressing activities throughout the literature that would be difficult to cover in this chapter therefore we refer the reader to other reviews.[122,123]

In addition to PTEN, other PTPs have been identified as tumor suppressors. Along this line, we have reviewed the tumor suppressor activity of RPTPs, NRPTPs, and DSPs (Table I–III) in human cancer cell lines and animal models. Furthermore, signaling associated with these PTPs was also described when characterized (Table IV). Within these groups of PTPs, we highlight and more thoroughly discuss below DEP 1 and TCPTP as examples.

## A. DEP1

Cloning of DEP1 (also known as CD148, encoded by *PTPRJ*) was achieved in 1994 by the Tonks research group.[246] Initially obtained from an HeLa cDNA librairy, DEP1 was also found to be widely expressed in many cell types such a fibroblasts,[246] endothelial cells,[247] hematopoietic cells,[248] and human tissues.[249] Its expression was reported as being greatly increased in near confluent cells, and therefore it was named "high cell *d*ensity-*e*nhanced *PTP 1*" or DEP1. The extracellular segment of this RPTP comprises 8 fibronectin repeats and 34 potential N-linked glycosylation sites.[246] Three different approaches were taken in the attempt to understand the role of DEP1 *in vivo*. The first one, the replacement of the DEP1 catalytic domain with green fluorescent protein leads to embryonic lethality. Analysis of the embryos demonstrates important vasculature defects,[250] confirming its role in VEGFR-2[251] and PDGFR[252] signaling, which are key RTKs in tumor angiogenesis. Another model disrupted DEP1 expression through deletion of *PTPRJ* exons 3, 4, and 5. Surprisingly, genetic ablation of *PTPRJ* in mice results in normal growth and development.[253] Finally, Zhu *et al*. generated a mutant protein by deletion of the transmembrane domain, which causes the production of a soluble truncated protein representing the extracellular portion of DEP1. While no alteration in thymic development or peripheral T-cell function was observed, they demonstrated a positive regulatory function for DEP1 in B-cell and macrophage development.[254]

Interestingly, although none of these animal models demonstrate spontaneous tumor apparition, there is a large body of evidence linking DEP1 to tumor-suppressing functions. In mice, Ruivenkamp *et al*. identified *PTPRJ* as a candidate for the mouse-colon-cancer susceptibility locus Scc1. They confirm that loss of heterozigocity (LOH) in human colon cancer was quite common, occurring in 49% of their samples. Furthermore, they identified LOH in lung (50%) and breast (78%) cancer. Interestingly, five single nucleotide polymorphisms (SNPs) coding for amino acid substitutions were sequenced in human colorectal cancer samples, all of which mapped to the extracellular fibronectin domains of DEP1. These substitutions are probably functional since secondary structure prediction and homology through modeling predicted that amino acid substitution would occur in exposed regions available for interactions.[17] To further support this idea, three of these SNPs (coding for a Gln276Pro, Arg326Gln, and Asp873Glu) were observed in about 25% of thyroid

TABLE I

RPTPs Involved in Human Cancer Cell Lines and Animal Models

| Gene[a] | Protein | Cell line | Origin | Overexpression/ downregulation | Onco/ TS | Proliferation/ apoptosis | Soft agar/3D | Colony formation | Invasion | Migration | EMT | Mouse model | Human samples | Refs | Signaling[b] |
|---|---|---|---|---|---|---|---|---|---|---|---|---|---|---|---|
| *PTPRA* | RPTPα | MCF-7 | Breast | Over | Onco | ✓ | | | | | | ✓ | ✓ | 124 | |
| | | MDA-MB-231, MDA-435S, MDA-468, SKBR3 | Breast | siRNA | Onco | ✓ | | ✓ | ✓ | | | | ✓ | 125 | ✓ |
| | | SW480 | Colon | Over/siRNA | Onco | | | | ✓ | | | | ✓ | 126 | |
| *PTPRB/Z* | RPTPβ/ζ | C6 | Glioblastoma | mAb | | | | | | ✓ | | | ✓ | 127 | |
| | | D566, G122 | Glioblastoma | siRNA | Onco | | | | | ✓ | | | | 128 | |
| | | U87 | Glioblastoma | Over | Onco | | | | | ✓ | | | ✓ | 129 | |
| | | U251MG | Glioblastoma | shRNA | Onco | ✓ | | | | ✓ | | ✓ | | 130 | |
| | | U87 | Glioblastoma | mAb | Onco | ✓ | | ✓ | | | | ✓ | | 131 | |
| | | LNCaP | Prostate | siRNA | Onco | | | | | ✓ | | | | 132 | |
| | | A549 | Lung | siRNA | Onco | | | | | ✓ | | | | 133 | ✓ |
| *PTPRE* | RPTPε | | Breast | Transgenic | Onco | | | | | | | ✓ | | 134 | |
| *PTPRG* | RPTPγ | MCF-7 | Breast | Over | TS | ✓ | ✓ | | | | | | | 135 | |
| | | HONE1 | Nasopharyngeal | over | TS | | ✓ | ✓ | | | | ✓ | | 136 | ✓ |
| | | MCF-7 | Breast | Over | TS | | | | | | | ✓ | | 137 | ✓ |
| *PTPRH* | SAP1 | HLE, HLF | Liver | Over | TS | ✓ | | | | ✓ | | | ✓ | 138 | |
| | | | | KO mouse | Onco | | | | | | | ✓ | | 139 | |
| *PTPRJ* | DEP1 | ZR75-1, SKBR-3, MCF-7 | Breast | Over | TS | | | ✓ | | | | | | 140 | |
| | | AsPC1, PSN1 | Pancreas | Over | TS | ✓ | | ✓ | | | | ✓ | | 141 | |
| | | U373 | Glioblastoma | Over | TS | ✓ | ✓ | | | | | | ✓ | 142 | |
| | | TPC1, FB2, | Thyroid carcinoma | Over | TS | | | ✓ | | | | | | 143 | ✓ |
| | | SW480 | Colon | Over | TS | ✓ | | | | ✓ | | | | 144 | |
| | | Ln229 | Glioblastoma | KO mouse | TS | ✓ | | | | | | | | 145 | |

*(Continues)*

TABLE I (*Continued*)

| Gene[a] | Protein | Cell line | Origin | Overexpression/ downregulation | Onco/ TS | Proliferation/ apoptosis | Soft agar/3D | Colony formation | Invasion | Migration | EMT | Mouse model | Human samples | Refs | Signaling[b] |
|---|---|---|---|---|---|---|---|---|---|---|---|---|---|---|---|
| | | KT21-MG1, SF3061 | Meningioma | KO mouse | TS | ✓ | | | ✓ | ✓ | | ✓ | ✓ | 146 | ✓ |
| *PTPRK* | RPTPκ | Me1 | Melanoma | Over | TS | ✓ | | | | ✓ | | | | 147 | ✓ |
| | | KM-H2 | H lymphoma | Over/siRNA | TS | ✓ | | | | | | | | 148 | |
| *PTPRM* | RPTPμ | U-87 Mg | Glioblastoma | shRNA | TS | | | | | ✓ | | | | 149 | ✓ |
| | | T98G | Glioblastoma | shRNA | TS | | | | | ✓ | | | | 150 | |
| *PTPRO* | GLEPP1 | A549 | Lung | Over | TS | ✓ | ✓ | | | | | | ✓ | 151 | |
| | | DHL-4, DHL-10 | Lymphoma | Over | TS | ✓ | | | | | | | | 152 | |
| | | WaC3CD5 | CLL | Over | TS | ✓ | | | | | | | ✓ | 153 | |
| | | MCF-7 | Breast | Over | TS | ✓ | | | | | | | ✓ | 154 | |
| | | K562 | CML | Over | TS | ✓ | ✓ | | | | | ✓ | | 155 | ✓ |
| | | Raji cells | Lymphoma | Over | TS | ✓ | | | | | | | | 156 | ✓ |
| *PTPRS* | RPTPσ | A431 | Skin | Over/antisens | TS | | ✓ | | | | | | | 157 | ✓ |
| *PTPRT* | RPTPρ | HCT116 | Colon | Over | TS | | | ✓ | | | | | ✓ | 158 | |
| *PTPRV* | OST-PTP | | Epidermal papillomas | KO mouse | TS | | | | | | | ✓ | | 159 | |

[a] Includes all RPTPs except *PTPRC* (CD45).
[b] Associated signaling is detailed in Table IV and V.

TABLE II

NRPTPs Involved in Human Cancer Cell Lines and Animal Models

| Gene | Protein | Cell line | Origin | Overexpression/ downregulation | Onco/ TS | Proliferation/ apoptosis | Soft agar/3D | Colony formation | Invasion | Migration | EMT | Mouse model | Human samples | Refs | Signaling[a] |
|---|---|---|---|---|---|---|---|---|---|---|---|---|---|---|---|
| *PTPN1* | PTP1B | K562 | CML | Over | TS | ✓ | | | | | | ✓ | | 201 | ✓ |
| | | | B-cell leukemia | KO mouse | TS | | | | | | | ✓ | | 202 | |
| | | NE (LNCaP) | Prostate | Over | Onco | ✓ | | | | | | | ✓ | 203 | |
| | | | Breast | KO mouse | Onco | | | | | | | ✓ | | 204 | |
| | | | Breast | Transgenic/KO mouse | Onco | | | | | | | ✓ | | 176 | ✓ |
| | | SW48, DLD-1 | Colon | Over/siRNA | Onco | | ✓ | | | | | ✓ | | 181 | ✓ |
| | | MCF-7 | Breast | siRNA | Onco | ✓ | | | | | | | | 21 | ✓ |
| | | MCF10A/ErbB2 | Breast | Over/siRNA | Onco | | ✓ | | | | | | | 177 | ✓ |
| | | MCF-7 | Breast | siRNA | Onco | | | | | ✓ | | | | 178 | ✓ |
| | | MKN45 | Gastric | Over | Onco | ✓ | ✓ | | | | | ✓ | ✓ | 186 | ✓ |
| | | MCF10A-NeuNT, MDA-MB-231 | Breast | siRNA/KO mouse | Onco | | | | | | | ✓ | | 205 | |
| *PTPN2* | TCPTP | U87MG/EGFR | Glioblastoma | Over | TS | ✓ | ✓ | ✓ | | | | ✓ | | 206 | |
| | | | B-cell leukemia | KO mouse | TS | | | | | | | ✓ | | 207 | |
| | | MCF-7 | Breast | siRNA | Onco | ✓ | | | | | | | | 21 | ✓ |
| | | MCF-7 | Breast | siRNA | Onco | | | | | ✓ | | | | 178 | ✓ |
| *PTPN3* | PTPH1 | HCT116 | Colon | shRNA | Onco | | | | | | | ✓ | | 185 | ✓ |
| | | MDA-MB-231, MCF-7, T47D | Breast | Over/shRNA | Onco | | | ✓ | | | | | ✓ | 208 | |
| *PTPN6* | SHP1 | MCF-7 | Breast | Over | TS | ✓ | | | | | | | | 209 | |
| | | K562 | CML | Over | TS | ✓ | | | | | | | | 210 | |
| | | PC3 | Prostate | Over | TS | ✓ | | | | | | | ✓ | 211 | |
| | | K-562 | Lymphoma/leuke | Over | TS | ✓ | | | | | | | ✓ | 212 | |
| | | Romas, H9, Jurkat | Lymphoma/leuke | Over | TS | ✓ | | | | | | | | 167 | ✓ |
| | | HTB26 | Breast | Over | TS | ✓ | | | | | | | | 167 | |
| | | TF-1 | Leukemia | siRNA | TS | ✓ | | | | | | | | 213 | |
| | | SU-DHL-1 | Lymphoma | Over | TS | ✓ | | | | | | | | 168 | ✓ |
| | | SUDHL-1, Karpas 299 | Lymphoma | Over | TS | ✓ | | | | | | | ✓ | 214 | |
| | | PC3 | Prostate | siRNA | Onco | ✓ | | | | | | | | 199 | ✓ |

(*Continues*)

TABLE II (*Continued*)

| Gene | Protein | Cell line | Origin | Overexpression/ downregulation | Onco/ TS | Proliferation/ apoptosis | Soft agar/3D | Colony formation | Invasion | Migration | EMT | Mouse model | Human samples | Refs | Signaling[a] |
|---|---|---|---|---|---|---|---|---|---|---|---|---|---|---|---|
| *PTPN9* | PTP-MEG2 | MDA-MB-231, SKBr3 | Breast | Over | TS | | ✓ | | ✓ | | | | | 162 | |
| *PTPN11* | SHP2 | | Noonan (leukemia) | KI mouse | Onco | | | | | | | ✓ | | 187 | ✓ |
| | | | Noonan (leukemia) | Over | Onco | | | | | | | ✓ | | 188 | ✓ |
| | | SUDHL-1, Karpas 299 | Lymphoma | shRNA | Onco | ✓ | | | | ✓ | | | | 189 | ✓ |
| | | BT474, BT20, T47D, MCF7 | Breast | Over | Onco | | ✓ | | | | | | ✓ | 215 | |
| | | BT474, BT20, sum225, MDA468 | Breast | shRNA | Onco | | ✓ | | | ✓ | | | | 216 | |
| | | | Noonan (leukemia) | KI mouse | Onco | | | | | | | ✓ | | 217 | |
| | | U87MG/EGFR, LN229/EGFR | Glioblastoma | Over/siRNA | Onco | ✓ | ✓ | | | | | ✓ | | 218 | |
| | | DU145 | Prostate | shRNA | Onco | ✓ | | | | | | ✓ | | 194 | |
| | | H292 | Lung | shRNA | Onco | ✓ | | | | | | ✓ | | 194 | ✓ |
| | | MDA-MB-231 | Breast | Over | Onco | | | | ✓ | | | | | 219 | |
| | | FaDu | Hypopharyngeal | siRNA | Onco | ✓ | | | | | | | ✓ | 220 | |
| | | | Leukemia | KI mouse | Onco | | | | | | | ✓ | | 190 | ✓ |
| | | | Liver | KO mouse | TS | | | | | | | ✓ | ✓ | 169 | ✓ |
| *PTPN12* | PTP-PEST | PC3 | Prostate | Over | TS | | | | | ✓ | | | | 221 | |
| | | KM12C | Colon | Over/siRNA | TS | | | | | ✓ | ✓ | | | 165 | ✓ |
| | | MCF10A, MDA-MB-231, HCC38, HCC1937 | Breast | Over/shRNA | TS | | ✓ | ✓ | | | | ✓ | ✓ | 163 | ✓ |
| | | SKOV-3 | Ovary | shRNA | TS | | | | | ✓ | | | | 222 | |
| *PTPN13* | PTP-BAS | MCF-7, T47D | Breast | siRNA/shRNA | TS | ✓ | | ✓ | ✓ | | | ✓ | ✓ | 161 | ✓ |
| *PTPN21* | PTPD1 | J82 | Bladder | siRNA | Onco | ✓ | | | | ✓ | | | ✓ | 223 | |
| *PTPN23* | HD-PTP | T24 | Bladder | siRNA | TS | | | | | ✓ | | | | 224 | |
| | | ACHN, 786-0 | Renal cell carcinoma | Over | TS | | | ✓ | | | | | | 225 | |
| | | MCF10A | Breast | shRNA | TS | | | | ✓ | ✓ | ✓ | | | 164 | |

[a]Associated signaling is detailed in Table IV and V.

TABLE III

DSPs Role in Human Cancer Cell Lines and Animal Models

| Gene[a] | Protein | Cell line | Origin | Overexpression/ downregulation | Onco/TS | Proliferation/ apoptosis | Soft agar/3D | Colony formation | Invasion | Migration | EMT | Mouse model | Human samples | Refs | Signaling[b] |
|---|---|---|---|---|---|---|---|---|---|---|---|---|---|---|---|
| *DUSP1* | MKP-1 | A2780, UCI101 | Ovarian | Over | TS | ✓ | ✓ | | | | | ✓ | ✓ | 173 | ✓ |
| | | PANC-1, T3M4 | Pancreas | Antisens | Onco | ✓ | ✓ | | | | | ✓ | ✓ | 196 | ✓ |
| | | H441GL | Lung | Over | TS | ✓ | | | ✓ | ✓ | | ✓ | | 226 | |
| | | MOLT-3, DND41 | T-ALL | shRNA | Onco | ✓ | | | | | | ✓ | | 191 | ✓ |
| | | H460 | Lung | shRNA | Onco | | | | ✓ | | | ✓ | ✓ | 227 | |
| *DUSP2* | PAC-1 | HeLa | Cervical | Over/shRNA | TS | ✓ | | | | | | ✓ | | 228 | ✓ |
| *DUSP3* | VHR | HeLa | Cervical | siRNA | Onco | ✓ | | | | | | | | 180 | ✓ |
| | | LNCaP, CWR22 | Prostate | siRNA | Onco | ✓ | | | | | | | | 198 | ✓ |
| | | H1299 | Lung | Over/shRNA | TS | ✓ | ✓ | | | | | | ✓ | 172 | ✓ |
| *DUSP4* | MKP-2 | U373M, A172 | Glioma | Over | TS | ✓ | | | | | | | ✓ | 229 | |
| *DUSP5* | hVH3? | HCT116, H1299 | Colon and Lung | Over | TS | ✓ | | | | | | | | 170 | ✓ |
| *DUSP6* | MKP-3 | PCI-35, PK-8 | Pancreas | Over | TS | ✓ | | | | | | | | 175 | ✓ |
| | | A2780 | Ovary | Over/shRNA | TS | ✓ | ✓ | | | | | ✓ | ✓ | 174 | ✓ |
| | | A549 | Lung | Over | TS | ✓ | | | | | | | ✓ | 230 | |
| | | H1975 | Lung | Over/shRNA | TS | ✓ | | | | | | | ✓ | 171 | ✓ |
| | | SLMT-1, HONE1 | Esophageal and nasopharyngeal | Over | TS | ✓ | ✓ | | | | ✓ | ✓ | ✓ | 231 | |
| *DUSP23/25* | VHZ | MCF-7 | Breast | Over/shRNA | Onco | ✓ | | | | | | | ✓ | 232 | |
| *DUSP24/26* | (MKP-8) | KTA1, KTA3, TTA1, 8305C, 8505C, HTC/C3 | Thyroid | Over/siRNA | Onco | ✓ | | ✓ | | | | | ✓ | 200 | ✓ |
| *DUSP26* | VHP | MCF10A, HOSE17.1 | Breast and ovary | Over/siRNA | TS | ✓ | ✓ | | | | | | ✓ | 233 | |
| *PTP4A1* | PRL-1 | D27 | Pancreas | Over | Onco | ✓ | | | | | | | | 234 | |
| | | SW480 | Colon | Over | Onco | | | | ✓ | ✓ | | | | 19 | ✓ |
| | | A549 | Lung | shRNA | Onco | ✓ | | | ✓ | | | | | 192 | ✓ |
| | | MIA PaCa-2, PANC-1 | Pancreas | siRNA | Onco | ✓ | ✓ | | | ✓ | | | ✓ | 197 | ✓ |
| | | A549 | Lung | Over | Onco | | | | ✓ | ✓ | | | | 195 | ✓ |
| *PTP4A2* | PRL-2 | D27 | Pancreas | Over | Onco | ✓ | | | | | | | | 234 | |

(*Continues*)

TABLE III (*Continued*)

| Gene[a] | Protein | Cell line | Origin | Overexpression/ downregulation | Onco/TS | Proliferation/ apoptosis | Soft agar/3D | Colony formation | Invasion | Migration | EMT | Mouse model | Human samples | Refs | Signaling[b] |
|---|---|---|---|---|---|---|---|---|---|---|---|---|---|---|---|
| | | MIA PaCa-2, PANC-1 | Pancreas | siRNA | Onco | ✓ | ✓ | | | ✓ | | | ✓ | 197 | ✓ |
| | | MDA-MB-231 | Breast | shRNA/siRNA | Onco | | ✓ | | | ✓ | | ✓ | ✓ | 179 | ✓ |
| | | A549 | Lung | shRNA | Onco | | | | ✓ | ✓ | | | | 112 | ✓ |
| *PTP4A3* | PRL-3 | A2780, SKOV-3, Igrov-1 | Ovarian | siRNA | Onco | ✓ | | ✓ | | | | | ✓ | 235 | |
| | | SW480 | Colon | Over | Onco | | | | ✓ | ✓ | | | | 19 | ✓ |
| | | DLD-1 | Colon | Over | Onco | | | | ✓ | | | | | 236 | |
| | | MCF-7 | Breast | siRNA | Onco | | | | ✓ | | | | | 236 | |
| | | RKO | Colon | siRNA | Onco | | ✓ | | | | | | | 182 | ✓ |
| | | SGC7901 | Gastric | shRNA | Onco | ✓ | | | | | | ✓ | | 237 | |
| | | SGC7901 | Gastric | shRNA | Onco | ✓ | | | ✓ | ✓ | | ✓ | ✓ | 238 | |
| | | INA-6 | Myeloma (plasma) | siRNA | Onco | ✓ | | | | ✓ | | | ✓ | 239 | |
| | | SGC7901 | Gastric | shRNA | Onco | ✓ | | | ✓ | | | ✓ | ✓ | 240 | |
| | | A549 | Lung | Anti-PRL3 | Onco | ✓ | | | ✓ | ✓ | | | ✓ | 193 | ✓ |
| | | 5-8F, HONE1 | Nasopharyngeal | siRNA | Onco | | | | ✓ | ✓ | | | ✓ | 241 | |
| | | SW480 | Colon | Over/shRNA | Onco | | | | | | ✓ | ✓ | | 183 | ✓ |
| | | LoVo | Colon | Over/shRNA | Onco | | | | ✓ | ✓ | | ✓ | | 184 | ✓ |
| | | SW480 | Colon | over | Onco | ✓ | | | | | | | | 242 | |
| | | SH101-P4 | Gastric | shRNA | Onco | ✓ | | | ✓ | | | ✓ | | 243 | |
| *KAP* | CDKN3 | LNCaP | Prostate | Antisens | Onco | | ✓ | | | | | | | 244 | |
| | | U87 | Glioblastoma | Over/shRNA | TS | ✓ | | | | ✓ | | | ✓ | 200 | ✓ |
| *TENC1* | Tensin 2 | BEL7402, Hep3B | Liver | Over | Onco | ✓ | | | ✓ | ✓ | | ✓ | ✓ | 245 | |

[a]Includes all DSPs except PTEN.
[b]Associated signaling is detailed in Table IV and V.

TABLE IV
SIGNALING EVENTS ASSOCIATED WITH PTP TUMOR SUPPRESSOR FUNCTION

| Origin of cancer | Gene | Signaling | Refs |
|---|---|---|---|
| Astrocytes (glioblastoma) | *KAP* | Overexpression decreases p-cdk2; siRNA increases p-cdk2 | 160 |
| | *PTPRM* | *In vitro* substrate trapping of PLCγ1 | 149 |
| Breast | *PTPRG* | Overexpression decreases p-ERK1/2 | 137 |
| | *PTPN13* | siRNA increases pTyr418 Src | 161 |
| | *PTPN9* | *In vitro* substrate trapping of EGFR and ErbB2, siRNA increases p-EGFR and p-ErbB2 and overexpression decreases it | 162 |
| | *PTPN12* | Overexpression decreases p-ErbB2 and p-PDGFR | 163 |
| | *PTPN23* | shRNA increases MMP9 activity and activates Src and β-catenin | 164 |
| Colon | *PTPN12* | siRNA enhances Rac activity and suppress RhoA activity | 165 |
| Leukemia | | | |
| Chronic myelogenous Leukemia | *PTPRO* | *In vitro* substrate trapping of BCR-Abl; overexpression decreases p-BCR-Abl | 155 |
| Megakaryoblastic leukemia | *PTPN1* | *In vitro* substrate trapping of BCR-Abl | 166 |
| T-ALL | *PTPN6* | Overexpression decreases p-JAK1 and p-TYK2 | 167 |
| Lymphoma | | | |
| Anaplastic large-cell and T-cell lymphoma | *PTPN6* | Overexpression decreases p-STAT3 and p-JAK3; siRNA increases p-STAT3 and p-JAK3 | 168 |
| Burkitt's lymphoma | *PTPRO* | Overexpression decreases pTyr418 Src | 156 |
| Liver | *PTPN11* | Specific deletion in the liver increases p-STAT3 | 169 |
| Lung | *DUSP5* | Overexpression decreases p-ERK1/2 | 170 |
| | *DUSP6* | Overexpression decreases p-ERK1/2; shRNA increases p- ERK1/2 | 171 |
| | *DUSP3* | Overxpression decreases p-Tyr992 EGFR | 172 |
| Melanocytes | *PTPRK* | Overexpression decreases β-catenin transcription | 147 |
| Meningioma | *PTPRJ* | siRNA increases p-PDGFR and p-PLCγ | 146 |
| Ovaries | *DUSP1* | Overexpression DUSP1 decreases kinase activities of ERK-2 and JNK1 | 173 |
| | *DUSP6* | shRNA increases p-ERK1/2; overexpression decreases p-ERK1/2 | 174 |
| Pancreas | *DUSP6* | Overexpression decreases p-ERK1/2 | 175 |
| Skin | *PTPRS* | Antisense increases p-EGFR | 157 |
| Thyroid, head and neck | | | |
| Nasopharyngeal | *PTPRG* | Overexpression downregulates cyclin D1 | 136 |
| Thyroid | *PTPRJ* | Overexpression decreases p-RET | 143 |

carcinomas[255] and the association between Arg326Gln and/or Gln276Pro and the risk of developing colorectal cancer (40%), head and neck squamous cell carcinoma (29%), or oesophageal squamous cell carcinoma (40%) was uncovered.[256] Finally, the Asp873Glu amino acid substitution was equally associated with an increased risk to develop papillary thyroid carcinoma.[257] All of these associations combine to make a strong case for a DEP1-associated tumor suppressor function.

Similarly, overexpression of DEP1 in a variety of cancer cell lines was found to impede important cancer-related processes such as cell proliferation, colony formation, or cell migration/invasion (Table I). Its role in tumor suppression is mediated through the dephosphorylation of many key substrates at different levels of the cellular circuitry. DEP1 can dephosphorylate different RTKs central to carcinogenesis such as VEGFR-2,[251] PDGFR,[252] or Met.[258] Using an acute monocytic leukemia human cell line, Arora *et al.* recently demonstrated that TyrK FLT3, a positive contributor of acute myeloid leukemia development, is also a putative DEP1 substrate. Furthermore, DEP1 depletion was found to stimulate proliferation and clonal growth[259] and has been identified as a regulator of vesicular trafficking. In human glioblastoma cells, DEP1 was found to suppress EGFR-mediated MAPK activation by limiting endocytosis of activated EGFR.[145] Since endosomal EGFR is thought to support sustained RTK signaling,[260] prevention of EGFR translocation to endosomes by DEP1 would dampen its signal. Downstream of RTKs, DEP1 can also negatively modulate the EGF-activated RAS–MAPK pathway. In order to be fully activated, the key MAPK members ERK1/2 needs to be phosphorylated both on threonine and tyrosine regulatory residues. DEP1 dephosphorylates Tyr204 directly in the activation loop of the ERK1/2 kinases, which is sufficient to inactivate ERK1/2.[261] Therefore, although its deletion is not sufficient to induce tumor growth in mice, the role of DEP1 as a tumor suppressor is supported by a wealth of human data, cell line assays, and mechanistic evidence.

## B. TCPTP

The discovery of TCPTP was made in 1989, a year after the identification of the prototypic PTP1B, from a screen of a human peripheral T-cell cDNA library. Therefore, T-cell phosphatase (TCPTP) was named after this screen, even though it is a ubiquitously expressed enzyme.[262] Interestingly, the catalytic domain of TCPTP shares over 70% of its amino acid sequence identity with PTP1B. Both NRPTPs share a role in the hematopoiesis process[202,207] and a few substrates such as the RTK CSF1R[263,264]and the transcription factors Stat5a and Stat5b.[265,266] TCPTP and PTP1B, however, drive many different phenotypes in cells and animals. In fact, both phosphatases do not have the same access to substrates since PTP1B is located at the ER membrane on the cytosolic side while TCPTP is mainly nuclear. Because of alternative splicing, two forms of TCPTP are present in humans. The 45-kDa protein is the principal form expressed in humans and the only one found in mice. It contains a RKRKR nuclear signature signal that targets TCPTP to the nucleus.[36] Interestingly, it was shown that, under certain circumstances such as cellular stress, the 45 kDa form can be redistributed to the cytoplasm.[267] The other TCPTP form (48 kDa) contains a C-terminal hydrophobic segment and seems to be localized at the ER membrane.[36]

Analysis of TCPTP-null mice suggests a role for TCPTP in carcinogenesis. TCPTP knockout mice develop severe systemic inflammation,[268] a process central to carcinogenesis that was not observed in the PTP1B-null mice.[269] In fact, TCPTP knockout mice die between 3 and 5 weeks of age mainly because of anemia[268] and severe inflammation.[270] A genome-wide scan by the Wellcome Trust Case Control Consortium reported, in 2007, a potential role for TCPTP in inflammatory bowel disease (IBD),[271] which was also described in an Italian cohort of patients.[272] Ulcerative colitis and Crohn's disease are the two main types of IBD and they confer a high risk of developing colorectal cancer.[273] Using dextran sulfate sodium (DSS) to induce colitis in mice, loss of only one *PTPN2* (encoding TCPTP) allele sensitized mice to the DSS treatment and resulted in a more severe inflammatory response.[274] Recently, TCPTP protein levels were also found to be higher in patients with Crohn's disease compared to control patients.[275] These studies have provided a functional link between TCPTP and IBD, the latter being an important risk factor in colorectal cancer.

The direct link supporting tumor-suppressing functions for TCPTP, however, came in 2010 with work on human T-cell acute lymphoblastic leukemia (T-ALL). Kleppe *et al.* identified that focal biallelic or a monoallelic deletion of *PTPN2* associated with lower mRNA expression occurs in 6% of patients with T-ALL. Functionally, depletion of TCPTP in T-ALL cells increased cell proliferation and their sensitivity to cytokines, corroborating a tumor-suppressing function for this PTP. Furthermore, 33% of those individuals had aberrant expression of the TLX1 transcription factor oncogene, with some patients co-expressing the NUP214-ABL1 fusion protein TyrK.[22] They revealed that NUP214-ABL1 TyrK, found to be amplified in T-ALL,[276] is a putative TCPTP substrate that is negatively regulated when dephosphorylated. Excitingly, the downregulation of TCPTP in NUP214-ABL1 Tyr expressing ALL-SIL cells results in a significant increase in cellular proliferation, providing functional evidence for TCPTP's role in tumor suppression.[22] Furthermore, activation of JAK1 kinase through the activation of *JAK1* mutations is documented in around 10% of human T-ALL cases. TCPTP is a negative regulator of the JAK/STAT pathway and, interestingly, *JAK1* activating mutations are found in patients where *PTPN2* is deleted, further potentiating the JAK/STAT pathway.[277] Also, *PTPN2* inactivation by nonsense mutations was observed in as much as 5% of the analyzed patients with peripheral T-cell lymphoma, not otherwise specified.[278]

In solid tumors, *PTPN2* was also identified as a potent negative regulator of Akt by a screen in Ras-activated A459 lung adenocarcinoma cancer cells,[279] and its mRNA expression was found to be downregulated in human hepatocellular carcinoma lymphatic metastasis.[280] In glioblastomas, TCPTP was found to inhibit the activity of ΔEGFR, an EGFR truncated form that results from the most glioblastoma-related EGFR mutation. In this case, TCPTP expression suppressed the anchorage-independent growth, *in vitro* proliferation, and *in vivo* tumorigenicity of ΔEGFR-U87MG.[206]

As is the case for many PTPs that can act in some situations as a tumor suppressor and in others as an oncogene, TCPTP demonstrated tumor-promoting functions in B-cell lymphomas. TCPTP was shown to be overexpressed in some activated B-cell-like diffuse large B-cell lymphoma cell lines, giving some indication of a potential tumor-promoting function.[281] Such an overexpression was confirmed for other human and murine B-cell lymphomas.[282] In murine *MYC*-driven B-cell lymphomas, TCPTP expression was found to correlate with overexpression of the MYC transcription factor. Additionally, TCPTP knockdown seriously impeded murine B-cell lymphoma cell proliferation and abrogated tumor maintenance *in vivo*.[282] Although to date there are no mechanistic explanations for the oncogenic role of TCPTP, this clearly demonstrates that, depending on the cellular context, the same PTP might have opposing roles in tumor development.

## V. Role of "oncoPTPs" in Cancer

Tyrosine phosphorylation does have a variety of effects that are not limited to activation of target proteins. In fact, tyrosine phosphorylation might do the complete opposite. An example is the cytosolic TyrK Src that is regulated through phosphorylation on Tyr529 at its C-terminal tail by the C-terminal Src kinase (Csk) or the Csk homologous kinase. When phosphorylated on this tyrosine residue, the Src oncogene is kept in a basal inactive state only to be activated following dephosphorylation. Thus, since 90–95% of Src is phosphorylated on Tyr529 under basal conditions, its activation relies on tyrosine dephosphorylation by PTPs.[283] This is an interesting paradigm, since PTPs were considered by many, mainly because of historical reasons, as simple "erasers" acting in reaction to TyrKs and not as main actors capable of initiating signaling themselves.[1]

Along this line, PTPs were found to positively regulate carcinogenesis; however, it was only in 2003, 6 years after the first description of PTEN as a tumor suppressor, that the breakthrough was made. Somatic mutations in *PTPN11* (which encodes SHP2) were identified in about 35% of the patients harboring juvenile myelomonocytic leukemia.[284–286] Composed of two SH2 domains in its N-terminal segment, SHP2 is kept in a low-activity state under basal conditions, its N-terminal SH2 domain occluding the active site through an intramolecular interaction. Only when engaged with tyrosine-phosphorylated substrates via its SH2 domains, phosphotyrosyl peptide binding to the N-terminal SH2 domain disrupts the autoinhibitory interaction, exposing the PTP domain and activating its phosphatase activity.[287] Most of the somatic mutations in *PTPN11* were found to affect this particular auto-inhibition mechanism, creating "activated mutants" with increased basal activity. Interestingly, the outcome of these mutations varies depending on the level of SHP2 activation. High degrees of SHP2 activity are more likely to cause various neoplastic diseases, whereas low levels of activation causes Noonan syndrome, an autosomal-dominant disorder characterized by facial dysmorphia, short stature, and cardiac defects.[288] Actually, there are still many hypotheses concerning the mechanisms through which activated SHP2 causes leukemia. One of the original hypotheses suggested that SHP2 might activate Src by directly dephosphorylating Tyr529.[289] Although this idea has never been confirmed, it has been demonstrated that SHP2 activity is indeed required for the dephosphorylation of Tyr529, but more likely by controlling Csk recruitment.[290,291] Tumor-promoting pathways implicated with SHP2 activity are many and extend beyond the scope of this chapter (see Refs. 288,292,293 for reviews).

Although SHP2 was the first "oncoPTP" characterized, it is not the last that demonstrated tumor-promoting properties. In this section, we have reviewed the oncogenic activities of RPTPs, NRPTPs, and DSPs (Table I–III) both in human cancer cell lines and in animal models. When available, the pro-oncogenic signaling associated with these PTPs was also described (Table V). Interestingly, as of today, some of them such as the RPTPα or the PRL family have been recognized only as tumor-promoting enzymes, while others such as the prototypic PTP1B have demonstrated both pro-oncogenic and tumor-suppressing properties. These three PTPs serve as interesting examples and we have selected them for more thorough discussion.

## A. RPTPα

The oncogenic potential of RPTPα was discovered at the very beginning of the PTP field. Zheng *et al.* demonstrated in 1992 that overexpression of RPTPα in Fisher rat embryo fibroblasts results in anchorage-independent growth in soft agar and enhances colony formation and the ability to form tumors in nude mice. Already authors reported a persistent Src activation mediated by the dephosphorylation on Tyr529 by RPTPα, a dephosphorylation observed *in vitro* in RPTPα-transfected cells.[294] This observation proved to be correct since RPTPα knockout mice have a dramatic decrease (50–70%) in Src activity in the brain, an organ expressing particularly high PTPRα levels.[295] Fibroblasts derived from such mice had also impaired TyrK activity of Src together with an

TABLE V
SIGNALING EVENTS ASSOCIATED WITH PTP ONCOGENIC FUNCTION

| Origin of cancer | Gene | Signaling | Refs |
|---|---|---|---|
| Breast | *PTPN1* | Deletion in the ErbB2 transgenic mouse decreases the Ras-MAPK signaling and p-Akt | 176 |
| | *PTPRA* | siRNA decreases Src activity | 125 |
| | *PTPN1* | siRNA dephosphorylates pTyr529 Src | 177 |
| | *PTPN2* | siRNA decreases p-ERK1/2 | 21,178 |
| | *PTPN1* | siRNA decreases p-ERK1/2 | 21,178 |
| | *PTP4A2* | Overexpression increases p-ERK1/2 | 179 |
| Cervix | *DUSP3* | siRNA increases p-ERK1/2 and p-JNK | 180 |
| Colon | *PTP4A1* | Overexpression activates Rho | 19 |
| | *PTP4A3* | Overexpression activates Rho | 19 |
| | *PTPN1* | Overexpression decreases pTyr529 Src | 181 |
| | *PTP4A3* | siRNA decreases p-p130cas | 182 |
| | *PTP4A3* | siRNA leads to reexpression E-cadherin | 183 |
| | *PTP4A3* | PRL-3 interacts with integrin β1 and decreases its tyrosine phosphorylation; overexpression increases p-ERK1/2 | 184 |
| | *PTPN3* | shRNA increases p-p38MAPK | 185 |
| Gastric cancer | *PTPN1* | siRNA decreases p-Akt, p-ERK, p-FAK and increases pTyr529 Src; Overexpression increases p-Akt et p-ERK1/2 | 186 |
| Leukemia | | | |
| Leukemia | *PTPN11* | KI of the mutant (D61G) increases Ras pathway | 187 |
| | *PTPN11* | Expression of SHP-2 mutants lead to hyperactivation of the ERK, Akt, and STAT5 pathways | 188 |
| | *PTPN11* | shRNA decreases p-ERK1/2 and pTyr418 Src | 189 |
| | *PTPN11* | Gab2 mediates SHP-2 (D61G) proliferative effect | 190 |
| T-ALL | *DUSP1* | shRNA increases p-p38MAPK | 191 |
| Lung | *PTP4A1* | shRNA increases p-FAK | 192 |
| | *PTP4A3* | PTP4A3 increases P-ERK1/2 and activates Rho | 193 |
| | *PTPN11* | shRNA decreases p-ERK1/2 and pTyr416 Src | 194 |
| | *PTPRB/Z* | shRNA decreases p-ERK1/2 and p-FAK | 133 |
| | *PTP4A1* | Overexpression decreases Cdc42 and Rac activities | 195 |
| | *PTP4A2* | shRNA decreases p-ERK1/2 and increases phospho Ezrin | 112 |
| Pancreas | *DUSP1* | Antisens leads to sustained p-ERK1/2 in response to mitogens | 196 |
| | *PTP4A1* | siRNA decreases p-Akt and p-ERK1/2 | 197 |
| | *PTP4A2* | siRNA decreases p-Akt | 197 |
| Prostate | *DUSP3* | Overespression decreases p-JNK | 198 |
| | *PTPN6* | siRNA increases p27 levels and decreases p-Akt | 199 |
| Thyroid, head and neck | | | |
| Thyroid | *DUSP24/26* | Overexpression decreases p-p38MAPK; siRNA increases p-p38MAPK | 200 |

increase in Tyr529 phosphorylation. A consequence of a defect in RPTPα was a reduction in the rate of spreading on fibronectin substrates and deficiencies in integrin-mediated signaling responses. Similar results were obtained for the Src-family kinase Fyn.[296] Mechanistically, constitutive phosphorylation on RPTPα Tyr789 residue provides a high-affinity binding site to the Grb2 SH2 domain.[297] Further phosphorylation on Ser180 and Ser204 located in RPTPα intracellular juxtamembrane region reduces affinity of Grb2 to the tyrosyl-phosphorylated SH2 domain of RPTPα through conformational changes.[47] The newly liberated pTyr789 site then binds to the Src SH2 domain, displacing the autoinhibitory Src pTyr529 and activating Src.[298] Src activation by RPTPα, however, might be independent of its phosphorylation on Tyr789, but the mechanism is not yet completely understood.[299]

Therefore, RPTPα was thought to act as a tumor promoter and, accordingly, it was shown as being overexpressed in human cancers such as colorectal,[126,300] head and neck,[301] and gastric cancer.[302] Interestingly, in a study by Ardini *et al.*, RPTPα protein levels were found to be significantly overexpressed in 29% of cases in human breast tumors. This subgroup also correlates with positive estrogen receptor (ER) status and, unexpectedly, with low tumor grade. All tumors analyzed had already reached the stage of clinically diagnosable disease; maybe RPTPα was more related to the tumor initiation stage rather than the promotion/progression stages. Furthermore, RPTPα overexpression in the ER-positive MCF-7 breast cancer cells inhibits *in vitro* growth rate.[124] Was this a result of a particular cellular context? This is possible because Zheng *et al.* demonstrated that siRNA-mediated suppression of RPTPα induced apoptosis in ER-negative breast cancer or colon cancer cells, but not in ER-positive breast cancer (including MCF-7) or other cancer cells tested; yet Src activity was effectively downregulated in all cell lines.[125] Recently, RPTPα was found to be expressed in more than 70% of human colon cancers, whereas it was undetectable in normal colonocytes by immunohistochemistry. Using SW480, a human colon cancer cell line, Krndija *et al.* demonstrated an RPTPα-mediated increase in contractility. Along this line, RPTPα knockdown in these cells decreased their ability to grow chicken chorioallantoic membrane and drastically diminished their capability to invade the membrane.[126]

Finally, different isoforms of RPTPα are expressed in humans. Two of them differ only in their extracellular region. The shorter form, expressed in most tissues, has 793 amino acids (RPTPα793) of which 123 are extracellular. The longer form, RPTPα802, has nine extra amino acids located just before the transmembrane region and is expressed only in few tissues. Interestingly, while both forms are able to dephosphorylate and activate Src in a similar manner, only the shorter form was able to induce focus formation and anchorage-independent growth of NIH3T3. Considering that both isoforms only differ in their extracellular domain, this sheds light on a regulatory function of RPTPα extracellular domain in cell transformation, which is probably independent of Src activation.[303] In a recent study, Huang *et al.* sequenced RPTPα cDNAs from five types of human tumors and paired normal samples. They found three sequences encoding truncated proteins lacking the D1 domain or both the D1 and D2 domains. One mutant, RPTPα245, lacking the D1 and D2 domain, was expressed in colon, breast, and liver tumors. What they found was surprising; although RPTPα245 lacks a catalytic domain, RPTPα245 retains its ability to dephosphorylate Src529. This sheds light on a novel and exciting mechanism of regulation, which involves the displacement of Grb2 binding on the full-length RPTPα by RPTPα245. This results in an increased dephosphorylation of Src by RPTPα, granting the ability to transform Fisher rat embryo fibroblasts that were tumorigenic in nude mice.[23] Therefore, expression of cancer specific RPTPα isoforms provides new and unique potential targets as therapeutic tools in cancer treatment.

## B. PRLs

The PRL phosphatases comprise three members, PRL-1, -2, and -3. Based on the conserved amino acid sequences of their catalytic domain, PRLs have been classified as a unique subgroup of VH1-like PTPs with dual specificity.[2] PRL-1 was first identified as an immediate early gene owing to its induction in mitogen-stimulated cells and in regenerating liver after partial hepatectomy.[304] Using the PRL-1 sequence to search the mouse EST database, Zeng identified PRL-2 and -3 phosphatases.[305] All three family members have a prenylation sequence at the C-terminus that is critical for their localization in the endosomal compartment and their function (see Section III.E).

In the past 10 years, PRL phosphatases have gained much attention since they have been constantly associated with cell proliferation, cell invasion/migration, and metastasis. In 2001, the Vogelstein research group was one of the first groups to report an important role for a member of the PRL subfamily in cancer metastasis.[16] Using gene expression profiling, they showed that among 144 upregulated genes detected in metastatic colorectal liver samples, PRL-3 was the only gene consistently overexpressed in all 18 of the cancer metastases examined.[16] Interestingly, this PTP was undetectable in normal colorectal epithelia, but was expressed at low levels in primary colonic adenocarcinomas and strongly was expressed in metastatic lesions derived from colorectal cancer. Since this pioneer study, high expressions of all PRL members have been shown to correlate with disease progression in many tumor types (reviewed in Ref. 4).

Of the PRL phosphatases, only PRL-3 has been associated with gene amplification. Using genomic FISH analysis, important differences in PRL-3 gene copy number on the chromosome 8q arm were observed in liver metastases derived from colorectal cancers.[16,306] On the other hand, gene amplification alone seems unlikely to account for the overexpression of PRLs seen at high frequency in multiple advanced tumors. For example, high levels of *PRL-3* were identified as a strong predictor of the occurrence of metastasis in melanoma, but this overexpression was not a consequence of 8q chromosome over-representation.[307] Therefore, deregulated gene expression of the PRLs could be responsible for their upregulation in cancer cells; however, the regulation of PRLs at the transcriptional levels by specific transcription factors is still poorly understood. Recently, PRL-1 and -3 have been identified as p53-inducible genes involved in cell-cycle regulation.[308–310] Interestingly, this increase of PRL-1 and -3 was followed by the downregulation of p53 via its increased ubiquitination and proteosomal degradation, suggesting that the PRLs may contribute to tumor development by the downregulation of p53 via a negative feedback mechanism. Additionally, it was shown that *PRL-3* is a direct regulatory target of TGFβ in colon cancer metastasis and that inhibiting the expression of this PTP might be an important mechanism through which TGFβ suppresses metastasis in colon cancer.[311] Posttranscriptional alterations have also been suggested as a consequence of the overexpression of PRLs. The binding of the poly(C)-binding protein 1 (PCBP1) on the triple GCCCAG motifs in the 5′-UTR of PRL-3 mRNAs suppresses the translation of the PTP, showing that the overexpression of this protein is not directly associated with its transcript levels, further indicating the existence of an underlying posttranscriptional regulation.[312] Interestingly, an inverse correlation between protein levels of this phosphatase and PCBP1 in different human primary cancers supported the clinical relevance of this observation. More recently, the peptidyl prolyl cis/trans isomerase FK506-binding protein 38 (FKBP38) was identified as a new PRL-3 binding partner that promotes degradation of endogenous PRL-3 protein via the protein–proteasome pathway, suggesting that alteration in the stability of PRL-3 can also have a dramatic impact on its ability to promote tumorigenesis.[313]

Although the exact biological functions of the PRL phosphatases remain unclear, mechanistically the PRLs have been linked to several pathways involved in cancer-related phenotypes in many human cancer cell lines. The major signaling pathways implicated in proliferation/survival of cancer cells are mainly related to the stimulation of ERK1/2 and AKT pathways. In MIA PaCa-2 and PANC-1 pancreatic cancer cell lines, small interfering RNA-mediated knockdown of PRL-1 and PRL-2 in combination resulted in a reduction of cellular growth and migration.[197] Consistent with this, D27 pancreatic ductal cells stably transfected with PRL-1 or PRL-2 increased proliferation and exhibited enhanced cyclin-dependent kinase-2 activity and lower p21 protein levels.[234] In breast cancer, co-expression of PRL-2 and ErbB2 in transgenic mouse resulted in faster mammary tumor formation due to an increase in ERK1/2 activity.[179] More recently, in A549 lung adenocarcinoma cells, the overexpression of wild-type PRL-2, but not the catalytically inactive and truncated C-terminal CAAX mutants, caused ERK phosphorylation followed by its nuclear translocation.[112] In lung cancer cells, PRL-3 also positively regulates ERK1/2 phosphorylation and Rho activation, which facilitates VEGF expression and accelerates angiogenesis followed by distant metastasis.[193] The same observation was noticed in colorectal cancer cells, where PRL-1 and -3 activated Rho GTPase, thus promoting invasion/motility,[19] increasing AKT phosphorylation,[314] and controlling the integrin beta1-ERK1/2-matrix metalloproteinases-2 (MMP-2) signaling.[184] It was also noticed that PRL-3 behaves as an oncogene by negatively regulating E-cadherin levels,[183,314] *PTEN* expression,[314] and p130cas phosphorylation in colon cancer cells.[182] Of interest, MMPs have been strongly implicated in multiple stages of cancer progression and PRL-3 may contribute to this event in hepatocarcinogenesis by acting through cell adhesion molecules, including matrix MMP-2, MMP-9, and E-cadherin.[315]

## C. PTP1B

A literature review of PTP1B in carcinogenesis demonstrates opposite roles for PTP1B.[316] Since PTP1B was the first tyrosine phosphatase isolated, it was initially hypothesized to be a tumor suppressor. Very early on, different groups attempted to validate this hypothesis. Using transformed NIH 3T3 MEF cells transformed with either v-src[317] or *neu* oncogene,[318] PTP1B *overexpression* diminished tumorigenicity, particularly in the *neu*-transformed fibroblast, whereas xenograft growth was severely impeded. Moreover, p210-BCR-Abl-transformed Rat-1 fibroblast cells expressing PTP1B or the substrate trapping mutant (PTP1B-D/A) greatly diminished the capacity of those cells to grow in soft agar, reduced serum, and formed tumors subcutaneously in nude mice. PTP1B was not able to inhibit v-Abl-induced transformation but its transient overexpression in K562 cells, a chronic myelogenous leukemia cell line expressing p210-BRC-Abl, induced erythroid differentiation.[201] A striking example, however, of PTP1B's role as a tumor suppressor came from its genetic ablation in p53-deficient mice. Mice lacking p53 develop a range of spontaneous tumors, 30% of solid tumors and soft sarcomas tumors, with the remainder predominantly lymphomas to an extent of 70%. Three-quarters of these lymphomas are of thymic origin (70–75%), and the remaining lymphoid tumor cells are characterized as B-cell lymphomas. Surprisingly, complete PTP1B genetic ablation together with p53 makes mice more susceptible to develop lymphomas composing almost 85% of the developing tumors, which appear earlier and with an equal proportion of B-cell and T-cell lymphomas. An explanation for this seemingly tumor-suppressive effect of PTP1B lies in its regulation of B-cell development since $PTP1B^{-/-}$ have a slight increased in pre-B population. This higher number of pre-B cells could then explain the shift in ratio in B-cell lymphomas versus T-cell lymphomas when p53 expression is mutated. This suggests that PTP1B acts indirectly as a tumor suppressor by modifying the B-cell differentiation process and not by acting on the lymphomagenesis process *per se.*[202] Therefore, the effect of PTP1B on tumorigenesis can be linked to a particular context and changing of cell lineage in conjonction with the presence of a mutation in a strong tumor suppressor like p53.

Although PTP1B mRNA in oesophageal cancer was significantly decreased in cancer lesions compared to surrounding normal mucosa, which supports a tumor suppressor role for PTP1B,[319] all other human studies reported a divergent finding. Amplification in gene copy number on chromosome 20q13, where *PTPN1* is located, was observed in multiple tumors including pancreatic cancer[320] and in a group with recurrent gastric cancer.[321] The latter also correlates with poor survival. Interestingly, this region was also shown to be amplified in prostate cancer,[322,323] supporting a potential role for PTP1B in this disease.[203] Indeed, we recently demonstrated that PTP1B is an AR-regulated PTP overexpressed in primary prostate cancer tissues (L. Lessard, DP. Labbé, and ML. Tremblay, unpublished data), although in a recent study *PTPN1* was not associated with breast cancer susceptibility, clinical outcome, or survival.[324] Its expression and protein levels were found to increase in breast carcinomas[325,326] and the 20q13 region amplified in a breast cancer subgroup,[20] advocating a potential role for PTP1B in this important cancer type.

Using siRNA, the downregulation of PTP1B was shown to impede MCF-7 cell proliferation and to be essential for resistance to tamoxifen.[21] Noteworthy, STAT5 phosphorylation is frequently lost during breast cancer progression and PTP1B was identified as a negative regulator of prolactin-induced STAT5 phosphorylation in invasive breast cancer.[327] Expression of Src and members of the EGFR family such as ErbB2 are associated with specific breast tumor subsets.[328,329] ErbB2-transgenic mice develop spontaneous mammary tumors, and the complete genetic deletion of PTP1B results in an increase in mammary tumor latency and resistance to lung metastasis.[176,204] Furthermore, specific PTP1B overexpression in the mammary gland was sufficient to drive spontaneous tumor formation.[176] Together with the fact that PTP1B-null mice do not develop tumors, even at old age,[269] this finding advocates an oncogenic role for PTP1B in cancer. Although PTP1B has been suggested to dephosphorylate Src on Tyr529 in metastatic renal cell carcinomas that retain wild-type von Hippel-Lindau protein expression[330] or during MCF-10A ErbB2-mediated transformation,[177] Src activity was unaltered in these *in vivo* models. The Ras-MAPK-induced ErbB2 axis, however, was impaired in PTP1B null mice, in part because of the increased phosphorylation of p62Dok, a PTP1B substrate, which when phosphorylated attenuates Ras activity and thus MAPK signaling.[331] More recently, the specific deletion of PTP1B in the mammary epithelium was equally shown to delay the onset of mammary tumors in ErbB2-transgenic mice, establishing a cell

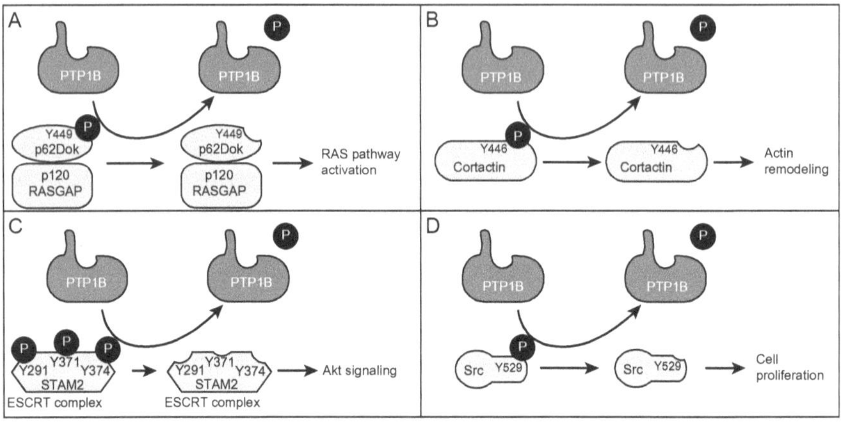


FIG. 2. PTP1B substrates in oncogenic signaling. (A) Phosphorylated p62Dok is a negative regulator of Ras/MAPK pathway and its dephosphorylation by PTP1B promotes Ras activation.[331] (B) PTP1B targets cortactin,[332] leading to actin remodeling during RTK internalization. (C) PTP1B-mediated dephosphorylation of STAM2[333] increases Akt signaling following RTK internalization and recycling.[334] (D) Src autoinhibitory phosphotyrosine residue is targeted by PTP1B leading to increased cell proliferation.[335]

autonomous role for PTP1B in this disease. This finding was reproduced using MCF-10A-NeuNT or MDA-MB-231 breast cancer cells implanted into the fat pad of immunodeficient mice, where PTP1B was knocked down at the stage of palpable tumors.[205] These findings are crucial, since they suggest that PTP1B is not essential for breast tumor maintenance but its inhibition may be of utmost importance in breast cancer prevention. Additionally, since dephosphorylation of many PTP1B substrates results in tumor-promoting functions (Fig. 2), PTP1B might also be a potential therapeutic target in other cancer types.

## VI. Therapeutic Tools Targeting the Tyrosine Phosphatases

As we write this chapter, it is important to note that the development of small inhibitors of selected PTPs has proven to be a very difficult process for many individual medicinal teams across the world. Aside from the relative success made using antisense oligonucleotides against PTP1B mRNA in phase two clinical trials (Earlier Phase 2 studies of ISIS 113715, and a novel generation of PTP1B antisense (ISIS-PTP1BRx) that is just starting phase 1) (Fig. 3A), small molecular inhibitors have not advanced beyond phase 1. The considerable financial and developmental efforts, which likely equates to well over half a billion dollars in investments, have dampened the enthusiasm of private sector management and pharmaceutical researchers alike.

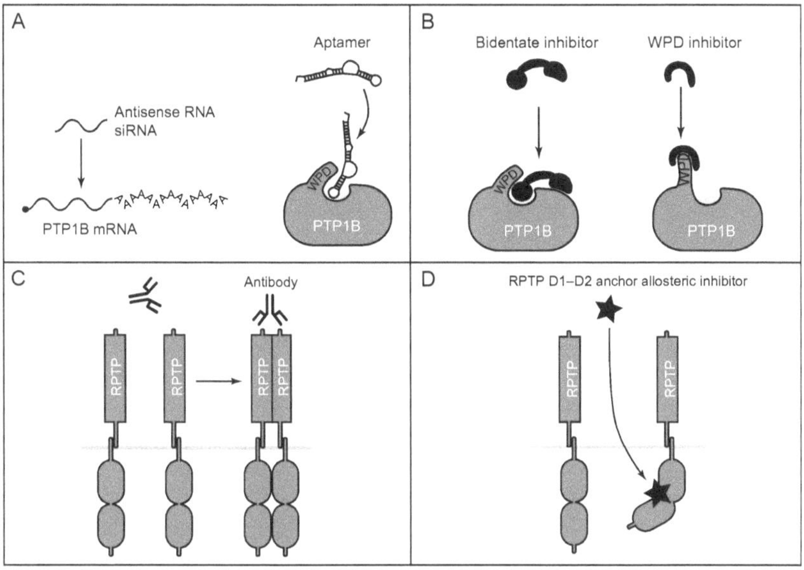


FIG. 3. Examples of PTP inhibitor design approaches. In addition to standard drug design by molecular modeling and natural compound screens, several approaches in inhibitor development are particularly interesting for protein tyrosine phosphatases. (A) Nucleic acid base inhibitors. PTP1B antisense RNAs are in clinical trial. Anti-PTP1B RNA aptamers have shown great specificity in low nanomolar range. (B) Bidentate inhibitors have been generated with a portion of the compounds having high affinity to the catalytic pocket and the other portion recognizing closely located, yet specific, features of each one of the PTPs targeted. WPD loop targeted compounds act as allosteric inhibitors and prevent the closing of this important motif on the catalytic pocket, thus leading to an inactive PTP enzyme. (C) Therapeutic bivalent antibodies are designed to induce the dimerization of RPTP monomers and thus promote enzyme inactivation. (D) RPTP D1–D2 anchor allosteric inhibitors are based on the concept that small molecules can recognize both D1 and D2 and anchor them in a conformation that sterically prevents the function of the catalytically active D1 domains.

Interestingly, the difficulty in developing small-molecule PTP inhibitors appears to be a common theme for all PTP inhibitor programs. It mainly resides in the use of the catalytic domain of the phosphatase as the molecular landscape for the compound design. Cytoplasmic enzymes such as PTP1B (diabetes, obesity, and cancer), SHP2 (cancer), and PRLs (cancer) as well as RPTPs such as RPTP$\gamma$ (schizophrenia), RPTP$\sigma$ (neuronal regeneration), and CD45 (inflammation) were found to be excellent biological targets; yet the difficulty in finding ideal competitive inhibitor compounds with high specificity, oral availability, and sufficient potency remains an insurmountable hurdle to date. This was thoroughly reviewed by Scott *et al.* for PTP targets in cancer.[336]

Using PTP1B as an example, one clearly comprehends the scope of the challenge. First, like all PTPs, PTP1B is an enzyme that in its inner catalytic pocket uses multiple hydrophilic interactions to stabilize its substrates. Very little space is left to develop compounds that will not be intrinsically composed of hydrophilic features. Second, the cysteine-based active donor for the nucleophilic attack is extremely sensitive to oxidative agents, often leading to nonspecific compounds that would not be suitable for further advances. Indeed, large screens of compound libraries often identify such small active molecules. Third, although the surrounding amino acid of the PTP pocket are in part used for substrate recognition, there are few major features in the three-dimensional conformation and hydrophobicity, that can be an easy additional docking site for tyrosine phosphate mimetics. The whole structure of the PTP1B catalytic domain has been humoristically compared by one medicinal chemist to "a golf ball with little less than wet pimples to grasp on". Hence, with a broad consensus that the PTP gene family represents fantastic biological targets, a major task remains for both academics and private sector researchers to identify novel directions to bypass these major medicinal chemistry hurdles.

In addition to many large screens of inhibitory activity of the catalytic domain, several interesting approaches are now being explored in order to develop inhibitors of PTPs. Potential alternative directions have been explored in order to open new doors for inhibiting the PTP family members. The approaches that are the most diverse, promising, and/or innovative include the use of bidentate inhibitors, WPD-loop modulators, and intracellular antibodies. Moreover, some nucleotide-based methods such as the use of antisense oligonucleotides and aptamers (Fig. 3A) have also been examined. Finally, for RPTPs one could explore additional approaches such as the use of extracellular antibody inhibitors that dimerize and block the catalytic activity or the use of new classes of allosteric inhibitors, which may act by anchoring the two PTP domains and blocking their function.

## A. The Bidentate Inhibitors Approach

The recognition that the majority of the PTPs possess a relatively hydrophilic, shallow, and conserved catalytic pocket supports the need for a different mechanism in providing both specificity and potency. A feature of PTP1B that elicited some excitement is the discovery of a secondary binding pocket near the active cysteine.[337,338] Moreover, in a survey of a large number of PTP structures, the existence of this second substrate binding was also recognized on a great number of PTPs.[29,339] The concept is to develop highly specific PTP1B inhibitors that can span both the active site and the adjacent noncatalytic site (Fig. 3B). Hence, such an inhibitor would be made of small compounds that are linked together but that recognize, simultaneously, the active and the proximal noncatalytic sites.[336,337] The application of such bidentate molecules is, however, not limited to the second aryl binding domain. As Liu *et al.* has demonstrated several, other amino acid features on PTP1B could also be potential anchor sites for the noncatalytic portion of the molecules.[340] New combinations of small molecules based on this theme are ongoing in several laboratories and their relative successes to date justifies an interest in this approach.

## B. The WPD-Loop Inhibitors

A structural feature of most PTPs that is important for the design of competitive inhibitors is the deep pocket that accommodates the pTyr residue being dephosphorylated. The small WPD loop is a unique motif of PTPs that closes over the phosphor substrate. It is the specific aspartate residue within the loop that allows proper substrate recognition.[25] Importantly, it is clear that mutations affecting this loop and preventing the covering of the substrate on the active pocket lead to deficient reaction.[341] Hence, the idea of attempting to design inhibitors that will either target the loop directly or by steric hindrance maintain the loop in its open conformation in order to block catalytic activity is appealing (Fig. 3B). Interestingly, such small molecules would result in a noncompetitive allosteric inhibitor.

Several laboratories have used this feature successfully to derive novel sets of PTP inhibitors. A series of compounds reported by Sunesis Inc. were acting in a low micromole range and seemingly interacted with the WPD loop.[342] The laboratory of Zhang at Indiana University also generated several compounds, one of which was more thoroughly characterized by crystallizing it with the PTP1B protein.[343] Moreover, Zhang and Bishop reported a very interesting twist to this approach by creating an inducible biarsenical fluorescein derivative (FlAsH) that was capable of recognizing a specifically engineered form of the WPD loop only recognized by the activatable compound.[344] Naturally, this creates interesting ways to modulate PTP in cells; however, it remains unusable in clinical settings. Nevertheless, the paradigm of targeting the "essential movable feature" of PTP exposes new avenues in the screening and modeling of various allosteric compounds.

Although the mechanisms of inhibition described above can also be applied to RPTPs, other unique approaches can be employed to this segment of the PTP gene family. RPTPs in some ways resemble the features of RTKs. The presence of an external domain, a transmembrane segment, and the intracellular portion of the RPTP proteins with two catalytic PTP domains provide other opportunities to target those PTPs.

## C. Inhibiting PTPs Using Anti-RPTP Receptor Antibody

The paradigm of using specific anti-RTK antibodies to block downstream signaling has been validated in cancer patients and is well known. The best example is likely trastuzumab (Herceptin), which is widely used in patients with ErbB2-overexpressing metastatic breast cancer. There is plenty of evidence that similar approaches could potentially work for the PTPs, with several important caveats. To date, very few of the RPTPs have been found to have specific ligands. Similarly, the downstream signaling events controlled by RPTPs remain sketchy at best. Both of these gaps make the development of specific anti-RPTP antibodies difficult: first by having no specific motif to target in the receptor extracellular domain, and second by having incomplete readouts for the activities of the enzymes.

A third feature also brings an additional difficulty. The RPTP literature still presents some controversy in the mechanism of action of RPTPs. It is clear that some subtypes such as the LAR PTP are found not only as homodimers but also as heterodimers. Moreover, the general model proposed for RPTPs in contrast to RTKs would be that dimerization inhibits their enzymatic activities. The recent work by Barr *et al.*, which presented an extensive comparison of several RPTP structures, led to the suggestion that, indeed, dimerization of RPTPs causes inhibition of activity.[29] Therefore, developing a bivalent antibody that would recognize both partners and through dimerization block their activity could be a potentially interesting approach (Fig. 3C). Oncogenic PTPs such as RPTPα[303] and modulators of the nervous system such as GLEPP1 (encode by *PTPRO*)[345] and RPTPσ[32] are already excellent candidates for this paradigm.

It is worth noticing that an enduring controversy exists in the RPTP field on the mechanism of dimerization leading to inhibition. The Barr *et al.* study presents a general mechanism of inhibition in which the carboxyl termini of the each monomer sterically block the active catalytic domain I of the partner protein. Other groups have presented a model of inhibition that includes a wedge domain close to the transmembrane motif sterically inhibiting each other active catalytic domains of the dimers. This controversy is described in a short preview of the Barr *et al.* report.[339]

To complicate things even further, DEP1 was activated following dimerization.[346] Part of the answer to this dichotomy may lie in the fact that several conformations of both intra- and extracellular segments of the RPTPs have been detected during dimerization, as well as a complex redox regulation of the D1–D2 domain interaction.[347] These findings clearly indicate that much remains to be understood on their mechanism of regulation. The difficulties and the lack of detailed knowledge about their mechanism of action explain why, as yet, no such therapeutic antibodies have been developed. With the rapid increase in knowledge on RPTP function and downstream signaling effects, however, one would expect that this approach could soon be used to modulate RPTP in a disease state context.

### D. RPTP Domains Anchor Allosteric Inhibitors

A last example of a new potential direction to inhibit RPTPs is the development of short molecules that interact with the linker domain separating the two PTP domains (D1–D2) and thus block any activity towards potential substrates (Fig. 3D). These molecules would then be quite specific, as they could be designed to recognize a portion of the linker domain that is known to be not as well-conserved as the PTP catalytic segments of the proteins. They would also be noncompetitive inhibitors, as they are not targeting the catalytic pocket of the RPTPs. Such an approach was successfully employed in the laboratory of U. Saragovi (JGH, McGill University) by Perron *et al*,.[348] who identified a small-molecule inhibitor that was able to recognize CD45 and completely block its activity. CD45 is a primary component of the inflammatory response by the immune system and appears to have significant anti-inflammatory activity.

## VII. Conclusions and Perspectives

Much has been revealed in the last decade concerning the role of PTPs in cancer (Fig. 1). Though at the beginning as simple caretakers in response to TyrK activity, PTPs arise as key players in the carcinogenesis process. Their mutation, deletion, or overexpression can have great consequenses on cellular homeostasis and can drive tumorigenesis. Although many PTPs are potential therapeutical targets in cancer, the conserved motif among the PTP superfamily is a significant hurdle in the development of specific drugs with no "off-target" effects. Nucleotide based inhibitors, including the antisense oligonucleotides, and also other avenues such as aptamers,[349] may become more appealing approaches for tackling medicinal applications of this difficult gene family. The development of increasingly powerful computer modeling software and the vast efforts that are being made in a broad array of natural products and synthetic chemistry will no doubt bring fruit to the development of novel PTP inhibitor-based medicine. Finally, the rapid knowledge gain from studying these enzymes is also bringing us closer to our objective of tapping into this extensive gene family for new therapeutic use.

## Acknowledgments

The authors regret that, owing to space limitation, the work of many investigators who have contributed to define the role of PTPs in cancer had to be omitted. We thank Joseph J. Bowden and Kelly Pike for a critical review of the manuscript and Noriko Uetani for technical assistance with figure design and drawing. DPL is a recipient of a CIHR Frederick Banting and Charles Best Doctoral Research Award. MLT is the holder of the Jeanne and Jean-Louis Levesque Chair in Cancer Research.

## References

1. Lim WA, Pawson T. Phosphotyrosine signaling: evolving a new cellular communication system. *Cell* 2010;**142**:661–7.
2. Alonso A, Sasin J, Bottini N, Friedberg I, Friedberg I, Osterman A, et al. Protein tyrosine phosphatases in the human genome. *Cell* 2004;**117**:699–711.
3. Julien SG, Dube N, Hardy S, Tremblay ML. Inside the human cancer tyrosine phosphatome. *Nat Rev Cancer* 2011;**11**:35–49.
4. Hardy S, Julien SG, Tremblay ML. Impact of oncogenic protein tyrosine phosphatases in cancer. *Anticancer Agents Med Chem* 2011; Epub ahead of print.
5. Nunes-Xavier C, Roma-Mateo C, Rios P, Tarrega C, Cejudo-Marin R, Tabernero L, et al. Dual-specificity MAP kinase phosphatases as targets of cancer treatment. *Anticancer Agents Med Chem* 2011;**11**:109–32.
6. Collett MS, Erikson RL. Protein kinase activity associated with the avian sarcoma virus src gene product. *Proc Natl Acad Sci USA* 1978;**75**:2021–4.
7. Levinson AD, Oppermann H, Levintow L, Varmus HE, Bishop JM. Evidence that the transforming gene of avian sarcoma virus encodes a protein kinase associated with a phosphoprotein. *Cell* 1978;**15**:561–72.
8. Tonks NK, Diltz CD, Fischer EH. Characterization of the major protein-tyrosine-phosphatases of human placenta. *J Biol Chem* 1988;**263**:6731–7.
9. Tonks NK, Diltz CD, Fischer EH. Purification of the major protein-tyrosine-phosphatases of human placenta. *J Biol Chem* 1988;**263**:6722–30.
10. Charbonneau H, Tonks NK, Kumar S, Diltz CD, Harrylock M, Cool DE, et al. Human placenta protein-tyrosine-phosphatase: amino acid sequence and relationship to a family of receptor-like proteins. *Proc Natl Acad Sci USA* 1989;**86**:5252–6.
11. Guan KL, Haun RS, Watson SJ, Geahlen RL, Dixon JE. Cloning and expression of a protein-tyrosine-phosphatase. *Proc Natl Acad Sci USA* 1990;**87**:1501–5.
12. Barford D, Flint AJ, Tonks NK. Crystal structure of human protein tyrosine phosphatase 1B. *Science* 1994;**263**:1397–404.
13. Flint AJ, Tiganis T, Barford D, Tonks NK. Development of "substrate-trapping" mutants to identify physiological substrates of protein tyrosine phosphatases. *Proc Natl Acad Sci USA* 1997;**94**:1680–5.
14. Li J, Yen C, Liaw D, Podsypanina K, Bose S, Wang SI. PTEN, a putative protein tyrosine phosphatase gene mutated in human brain, breast, and prostate cancer. *Science* 1997;**275**:1943–7.
15. Steck PA, Pershouse MA, Jasser SA, Yung WK, Lin H, Ligon AH. Identification of a candidate tumour suppressor gene, MMAC1, at chromosome 10q23.3 that is mutated in multiple advanced cancers. *Nat Genet* 1997;**15**:356–62.
16. Saha S, Bardelli A, Buckhaults P, Velculescu VE, Rago C, St Croix B. A phosphatase associated with metastasis of colorectal cancer. *Science* 2001;**294**:1343–6.

17. Ruivenkamp CA, van Wezel T, Zanon C, Stassen AP, Vlcek C, Csikos T. Ptprj is a candidate for the mouse colon-cancer susceptibility locus Scc1 and is frequently deleted in human cancers. *Nat Genet* 2002;**31**:295–300.
18. Williams CL. The polybasic region of Ras and Rho family small GTPases: a regulator of protein interactions and membrane association and a site of nuclear localization signal sequences. *Cell Signal* 2003;**15**:1071–80.
19. Fiordalisi JJ, Keller PJ, Cox AD. PRL tyrosine phosphatases regulate rho family GTPases to promote invasion and motility. *Cancer Res* 2006;**66**:3153–61.
20. Tanner MM, Tirkkonen M, Kallioniemi A, Isola J, Kuukasjarvi T, Collins C. Independent amplification and frequent co-amplification of three nonsyntenic regions on the long arm of chromosome 20 in human breast cancer. *Cancer Res* 1996;**56**:3441–5.
21. Blanquart C, Karouri SE, Issad T. Implication of protein tyrosine phosphatase 1B in MCF-7 cell proliferation and resistance to 4-OH tamoxifen. *Biochem Biophys Res Commun* 2009;**387**:748–53.
22. Kleppe M, Lahortiga I, El Chaar T, De Keersmaecker K, Mentens N, Graux C. Deletion of the protein tyrosine phosphatase gene PTPN2 in T-cell acute lymphoblastic leukemia. *Nat Genet* 2010;**42**:530–5.
23. Huang J, Yao L, Xu R, Wu H, Wang M, White BS. Activation of Src and transformation by an RPTPalpha splice mutant found in human tumours. *EMBO J* 2011;**30**:3200–11.
24. Tonks NK. PTP1B: from the sidelines to the front lines!. *FEBS Lett* 2003;**546**:140–8.
25. Jia Z, Barford D, Flint AJ, Tonks NK. Structural basis for phosphotyrosine peptide recognition by protein tyrosine phosphatase 1B. *Science* 1995;**268**:1754–8.
26. Blanchetot C, Chagnon M, Dube N, Halle M, Tremblay ML. Substrate-trapping techniques in the identification of cellular PTP targets. *Methods* 2005;**35**:44–53.
27. Meng K, Rodriguez-Pena A, Dimitrov T, Chen W, Yamin M, Noda M, et al. Pleiotrophin signals increased tyrosine phosphorylation of beta beta-catenin through inactivation of the intrinsic catalytic activity of the receptor-type protein tyrosine phosphatase beta/zeta. *Proc Natl Acad Sci USA* 2000;**97**:2603–8.
28. Tonks NK. Protein tyrosine phosphatases: from genes, to function, to disease. *Nat Rev Mol Cell Biol* 2006;**7**:833–46.
29. Barr AJ, Ugochukwu E, Lee WH, King ON, Filippakopoulos P, Alfano I, et al. Large-scale structural analysis of the classical human protein tyrosine phosphatome. *Cell* 2009;**136**:352–63.
30. Nam HJ, Poy F, Krueger NX, Saito H, Frederick CA. Crystal structure of the tandem phosphatase domains of RPTP LAR. *Cell* 1999;**97**:449–57.
31. Nam HJ, Poy F, Saito H, Frederick CA. Structural basis for the function and regulation of the receptor protein tyrosine phosphatase CD45. *J Exp Med* 2005;**201**:441–52.
32. Fry EJ, Chagnon MJ, Lopez-Vales R, Tremblay ML, David S. Corticospinal tract regeneration after spinal cord injury in receptor protein tyrosine phosphatase sigma deficient mice. *Glia* 2010;**58**:423–33.
33. Shen Y, Tenney AP, Busch SA, Horn KP, Cuascut FX, Liu K, et al. PTPsigma is a receptor for chondroitin sulfate proteoglycan, an inhibitor of neural regeneration. *Science* 2009;**326**:592–6.
34. den Hertog J, Ostman A, Bohmer FD. Protein tyrosine phosphatases: regulatory mechanisms. *FEBS J* 2008;**275**:831–47.
35. Huynh H, Bottini N, Williams S, Cherepanov V, Musumeci L, Saito K, et al. Control of vesicle fusion by a tyrosine phosphatase. *Nat Cell Biol* 2004;**6**:831–9.
36. Lorenzen JA, Dadabay CY, Fischer EH. COOH-terminal sequence motifs target the T cell protein tyrosine phosphatase to the ER and nucleus. *J Cell Biol* 1995;**131**:631–43.
37. Zhang J, Guan Z, Murphy AN, Wiley SE, Perkins GA, Worby CA, et al. Mitochondrial phosphatase PTPMT1 is essential for cardiolipin biosynthesis. *Cell Metab* 2011;**13**:690–700.

38. Frangioni JV, Beahm PH, Shifrin V, Jost CA, Neel BG. The nontransmembrane tyrosine phosphatase PTP-1B localizes to the endoplasmic reticulum via its 35 amino acid C-terminal sequence. *Cell* 1992;**68**:545–60.
39. Garton AJ, Tonks NK. PTP-PEST: a protein tyrosine phosphatase regulated by serine phosphorylation. *EMBO J* 1994;**13**:3763–71.
40. Wang Y, Guo W, Liang L, Esselman WJ. Phosphorylation of CD45 by casein kinase 2. Modulation of activity and mutational analysis. *J Biol Chem* 1999;**274**:7454–61.
41. Strack V, Krutzfeldt J, Kellerer M, Ullrich A, Lammers R, Haring HU. The Protein-tyrosine-phosphatase SHP2 is phosphorylated on serine residues 576 and 591 by protein kinase C isoforms alpha, beta 1, beta 2, and eta. *Biochemistry* 2002;**41**:603–8.
42. Liu Y, Kruhlak MJ, Hao JJ, Shaw S. Rapid T cell receptor-mediated SHP-1 S591 phosphorylation regulates SHP-1 cellular localization and phosphatase activity. *J Leukoc Biol* 2007;**82**:742–51.
43. Brautigan DL, Pinault FM. Serine phosphorylation of protein tyrosine phosphatase (PTP1B) in HeLa cells in response to analogues of cAMP or diacylglycerol plus okadaic acid. *Mol Cell Biochem* 1993;**127–128**:121–9.
44. Flint AJ, Gebbink MF, Franza BR, Jr., Hill DE, Tonks NK. Multi-site phosphorylation of the protein tyrosine phosphatase, PTP1B: identification of cell cycle regulated and phorbol ester stimulated sites of phosphorylation. *EMBO J* 1993;**12**:1937–46.
45. Ravichandran LV, Chen H, Li Y, Quon MJ. Phosphorylation of PTP1B at Ser(50) by Akt impairs its ability to dephosphorylate the insulin receptor. *Mol Endocrinol* 2001;**15**:1768–80.
46. den Hertog J, Sap J, Pals CE, Schlessinger J, Kruijer W. Stimulation of receptor protein-tyrosine phosphatase alpha activity and phosphorylation by phorbol ester. *Cell Growth Differ* 1995;**6**:303–7.
47. Zheng XM, Resnick RJ, Shalloway D. Mitotic activation of protein-tyrosine phosphatase alpha and regulation of its Src-mediated transforming activity by its sites of protein kinase C phosphorylation. *J Biol Chem* 2002;**277**:21922–9.
48. Uchida T, Matozaki T, Noguchi T, Yamao T, Horita K, Suzuki T, et al. Insulin stimulates the phosphorylation of Tyr538 and the catalytic activity of PTP1C, a protein tyrosine phosphatase with Src homology-2 domains. *J Biol Chem* 1994;**269**:12220–8.
49. Yoshida K, Kharbanda S, Kufe D. Functional interaction between SHPTP1 and the Lyn tyrosine kinase in the apoptotic response to DNA damage. *J Biol Chem* 1999;**274**:34663–8.
50. Tao J, Malbon CC, Wang HY. Insulin stimulates tyrosine phosphorylation and inactivation of protein-tyrosine phosphatase 1B *in vivo*. *J Biol Chem* 2001;**276**:29520–5.
51. Liu F, Chernoff J. Protein tyrosine phosphatase 1B interacts with and is tyrosine phosphorylated by the epidermal growth factor receptor. *Biochem J* 1997;**327**(Pt 1):139–45.
52. Kon-Kozlowski M, Pani G, Pawson T, Siminovitch KA. The tyrosine phosphatase PTP1C associates with Vav, Grb2, and mSos1 in hematopoietic cells. *J Biol Chem* 1996;**271**:3856–62.
53. Bennett AM, Tang TL, Sugimoto S, Walsh CT, Neel BG. Protein-tyrosine-phosphatase SHPTP2 couples platelet-derived growth factor receptor beta to Ras. *Proc Natl Acad Sci USA* 1994;**91**:7335–9.
54. den Hertog J, Tracy S, Hunter T. Phosphorylation of receptor protein-tyrosine phosphatase alpha on Tyr789, a binding site for the SH3-SH2-SH3 adaptor protein GRB-2 *in vivo*. *EMBO J* 1994;**13**:3020–32.
55. Su J, Batzer A, Sap J. Receptor tyrosine phosphatase R-PTP-alpha is tyrosine-phosphorylated and associated with the adaptor protein Grb2. *J Biol Chem* 1994;**269**:18731–4.
56. Berridge MJ, Lipp P, Bootman MD. The versatility and universality of calcium signalling. *Nat Rev Mol Cell Biol* 2000;**1**:11–21.
57. Frangioni JV, Oda A, Smith M, Salzman EW, Neel BG. Calpain-catalyzed cleavage and subcellular relocation of protein phosphotyrosine phosphatase 1B (PTP-1B) in human platelets. *EMBO J* 1993;**12**:4843–56.

58. Falet H, Pain S, Rendu F. Tyrosine unphosphorylated platelet SHP-1 is a substrate for calpain. *Biochem Biophys Res Commun* 1998;**252**:51–5.
59. Gu M, Majerus PW. The properties of the protein tyrosine phosphatase PTPMEG. *J Biol Chem* 1996;**271**:27751–9.
60. Kuchay SM, Kim N, Grunz EA, Fay WP, Chishti AH. Double knockouts reveal that protein tyrosine phosphatase 1B is a physiological target of calpain-1 in platelets. *Mol Cell Biol* 2007;**27**:6038–52.
61. Gil-Henn H, Volohonsky G, Elson A. Regulation of protein-tyrosine phosphatases alpha and epsilon by calpain-mediated proteolytic cleavage. *J Biol Chem* 2001;**276**:31772–9.
62. Rupinder SK, Gurpreet AK, Manjeet S. Cell suicide and caspases. *Vascul Pharmacol* 2007;**46**:383–93.
63. Howley B, Fearnhead HO. Caspases as therapeutic targets. *J Cell Mol Med* 2008;**12**:1502–16.
64. Halle M, Liu YC, Hardy S, Theberge JF, Blanchetot C, Bourdeau A, et al. Caspase-3 regulates catalytic activity and scaffolding functions of the protein tyrosine phosphatase PEST, a novel modulator of the apoptotic response. *Mol Cell Biol* 2007;**27**:1172–90.
65. Halle M, Tremblay ML, Meng TC. Protein tyrosine phosphatases: emerging regulators of apoptosis. *Cell Cycle* 2007;**6**:2773–81.
66. Kamhawi S. Phlebotomine sand flies and Leishmania parasites: friends or foes? *Trends Parasitol* 2006;**22**:439–45.
67. Halle M, Gomez MA, Stuible M, Shimizu H, McMaster WR, Olivier M, et al. The Leishmania surface protease GP63 cleaves multiple intracellular proteins and actively participates in p38 mitogen-activated protein kinase inactivation. *J Biol Chem* 2009;**284**:6893–908.
68. Forget G, Gregory DJ, Whitcombe LA, Olivier M. Role of host protein tyrosine phosphatase SHP-1 in Leishmania donovani-induced inhibition of nitric oxide production. *Infect Immun* 2006;**74**:6272–9.
69. Gomez MA, Contreras I, Halle M, Tremblay ML, McMaster RW, Olivier M. Leishmania GP63 alters host signaling through cleavage-activated protein tyrosine phosphatases. *Sci Signal* 2009;**2**:ra58.
70. Craig SE, Brady-Kalnay SM. Tumor-derived extracellular fragments of receptor protein tyrosine phosphatases (RPTPs) as cancer molecular diagnostic tools. *Anticancer Agents Med Chem* 2011;**11**:133–40.
71. Hussain SP, Hofseth LJ, Harris CC. Radical causes of cancer. *Nat Rev Cancer* 2003;**3**:276–85.
72. Storz P, Liou GY. Reactive oxygen species in cancer. *Free Radic Res* 2010;**44**:479–96.
73. Hanahan D, Weinberg RA. Hallmarks of cancer: the next generation. *Cell* 2011;**144**:646–74.
74. Chiarugi P, Giannoni E, Taddei ML. Src redox regulation: again in the front line. *Free Radic Biol Med* 2010;**49**:516–27.
75. Li N, Karin M. Is NF-kappaB the sensor of oxidative stress? *FASEB J* 1999;**13**:1137–43.
76. Salmeen A, Andersen JN, Myers MP, Meng TC, Hinks JA, Tonks NK, et al. Redox regulation of protein tyrosine phosphatase 1B involves a sulphenyl-amide intermediate. *Nature* 2003;**423**:769–73.
77. Yang J, Groen A, Lemeer S, Jans A, Slijper M, Roe SM, et al. Reversible oxidation of the membrane distal domain of receptor PTPalpha is mediated by a cyclic sulfenamide. *Biochemistry* 2007;**46**:709–19.
78. Salmeen A, Barford D. Functions and mechanisms of redox regulation of cysteine-based phosphatases. *Antioxid Redox Signal* 2005;**7**:560–77.
79. van Montfort RL, Congreve M, Tisi D, Carr R, Jhoti H. Oxidation state of the active-site cysteine in protein tyrosine phosphatase 1B. *Nature* 2003;**423**:773–7.
80. Chen YY, Chu HM, Pan KT, Teng CH, Wang DL, Wang AH, et al. Cysteine S-nitrosylation protects protein-tyrosine phosphatase 1B against oxidation-induced permanent inactivation. *J Biol Chem* 2008;**283**:35265–72.

81. Gulati P, Markova B, Gottlicher M, Bohmer FD, Herrlich PA. UVA inactivates protein tyrosine phosphatases by calpain-mediated degradation. *EMBO Rep* 2004;**5**:812–7.
82. Groen A, Overvoorde J, van der Wijk T, den Hertog J. Redox regulation of dimerization of the receptor protein-tyrosine phosphatases RPTPalpha, LAR, RPTPmu and CD45. *FEBS J* 2008;**275**:2597–604.
83. Meluh PB, Koshland D. Evidence that the MIF2 gene of Saccharomyces cerevisiae encodes a centromere protein with homology to the mammalian centromere protein CENP-C. *Mol Biol Cell* 1995;**6**:793–807.
84. Guo D, Li M, Zhang Y, Yang P, Eckenrode S, Hopkins D, et al. A functional variant of SUMO4, a new I kappa B alpha modifier, is associated with type 1 diabetes. *Nat Genet* 2004;**36**:837–41.
85. Melchior F. SUMO—nonclassical ubiquitin. *Annu Rev Cell Dev Biol* 2000;**16**:591–626.
86. Geiss-Friedlander R, Melchior F. Concepts in sumoylation: a decade on. *Nat Rev Mol Cell Biol* 2007;**8**:947–56.
87. Johnson ES. Protein modification by SUMO. *Annu Rev Biochem* 2004;**73**:355–82.
88. Mullen JR, Das M, Brill SJ. Genetic evidence that polysumoylation bypasses the need for a SUMO-targeted Ub ligase. *Genetics* 2011;**187**:73–87.
89. Kim KI, Baek SH. SUMOylation code in cancer development and metastasis. *Mol Cells* 2006;**22**:247–53.
90. Dadke S, Cotteret S, Yip SC, Jaffer ZM, Haj F, Ivanov A, et al. Regulation of protein tyrosine phosphatase 1B by sumoylation. *Nat Cell Biol* 2007;**9**:80–5.
91. Matsuo K, Bettaieb A, Nagata N, Matsuo I, Keilhack H, Haj FG. Regulation of brown fat adipogenesis by protein tyrosine phosphatase 1B. *PLoS One* 2011;**6**:e16446.
92. Liu F, Sells MA, Chernoff J. Transformation suppression by protein tyrosine phosphatase 1B requires a functional SH3 ligand. *Mol Cell Biol* 1998;**18**:250–9.
93. Yu J, Zhang SS, Saito K, Williams S, Arimura Y, Ma Y, et al. PTEN regulation by Akt-EGR1-ARF-PTEN axis. *EMBO J* 2009;**28**:21–33.
94. Fukada T, Tonks NK. The reciprocal role of Egr-1 and Sp family proteins in regulation of the PTP1B promoter in response to the p210 Bcr-Abl oncoprotein-tyrosine kinase. *J Biol Chem* 2001;**276**:25512–9.
95. Zabolotny JM, Kim YB, Welsh LA, Kershaw EE, Neel BG, Kahn BB. Protein-tyrosine phosphatase 1B expression is induced by inflammation *in vivo*. *J Biol Chem* 2008;**283**:14230–41.
96. Desterro JM, Rodriguez MS, Hay RT. SUMO-1 modification of IkappaBalpha inhibits NF-kappaB activation. *Mol Cell* 1998;**2**:233–9.
97. Callewaert L, Verrijdt G, Haelens A, Claessens F. Differential effect of small ubiquitin-like modifier (SUMO)-ylation of the androgen receptor in the control of cooperativity on selective versus canonical response elements. *Mol Endocrinol* 2004;**18**:1438–49.
98. Magee T, Seabra MC. Fatty acylation and prenylation of proteins: what's hot in fat. *Curr Opin Cell Biol* 2005;**17**:190–6.
99. Pechlivanis M, Kuhlmann J. Hydrophobic modifications of Ras proteins by isoprenoid groups and fatty acids—more than just membrane anchoring. *Biochim Biophys Acta* 2006;**1764**:1914–31.
100. Zeng Q, Si X, Horstmann H, Xu Y, Hong W, Pallen CJ. Prenylation-dependent association of protein-tyrosine phosphatases PRL-1, -2, and -3 with the plasma membrane and the early endosome. *J Biol Chem* 2000;**275**:21444–52.
101. Hannoush RN, Sun J. The chemical toolbox for monitoring protein fatty acylation and prenylation. *Nat Chem Biol* 2010;**6**:498–506.
102. Maurer-Stroh S, Eisenhaber F. Refinement and prediction of protein prenylation motifs. *Genome Biol* 2005;**6**:R55.

103. Agrawal AG, Somani RR. Farnesyltransferase inhibitor as anticancer agent. *Mini Rev Med Chem* 2009;**9**:638–52.
104. Rowinsky EK, Windle JJ, Von Hoff DD. Ras protein farnesyltransferase: a strategic target for anticancer therapeutic development. *J Clin Oncol* 1999;**17**:3631–52.
105. Konstantinopoulos PA, Karamouzis MV, Papavassiliou AG. Post-translational modifications and regulation of the RAS superfamily of GTPases as anticancer targets. *Nat Rev Drug Discov* 2007;**6**:541–55.
106. Al-Aidaroos AQ, Zeng Q. PRL-3 phosphatase and cancer metastasis. *J Cell Biochem* 2010;**111**:1087–98.
107. Cates CA, Michael RL, Stayrook KR, Harvey KA, Burke YD, Randall SK, et al. Prenylation of oncogenic human PTP(CAAX) protein tyrosine phosphatases. *Cancer Lett* 1996;**110**:49–55.
108. Wang J, Kirby CE, Herbst R. The tyrosine phosphatase PRL-1 localizes to the endoplasmic reticulum and the mitotic spindle and is required for normal mitosis. *J Biol Chem* 2002;**277**:46659–68.
109. Sun JP, Luo Y, Yu X, Wang WQ, Zhou B, Liang F, et al. Phosphatase activity, trimerization, and the C-terminal polybasic region are all required for PRL1-mediated cell growth and migration. *J Biol Chem* 2007;**282**:29043–51.
110. Song R, Qian F, Li YP, Sheng X, Cao SX, Xu Q. Phosphatase of regenerating liver-3 localizes to cyto-membrane and is required for B16F1 melanoma cell metastasis *in vitro* and *in vivo*. *PLoS One* 2009;**4**:e4450.
111. Skinner AL, Vartia AA, Williams TD, Laurence JS. Enzyme activity of phosphatase of regenerating liver is controlled by the redox environment and its C-terminal residues. *Biochemistry* 2009;**48**:4262–72.
112. Wang Y, Lazo JS. Metastasis-associated phosphatase PRL-2 regulates tumor cell migration and invasion. *Oncogene* 2011; Epub ahead of print.
113. Jeong DG, Kim SJ, Kim JH, Son JH, Park MR, Lim SM, et al. Trimeric structure of PRL-1 phosphatase reveals an active enzyme conformation and regulation mechanisms. *J Mol Biol* 2005;**345**:401–13.
114. Sun JP, Wang WQ, Yang H, Liu S, Liang F, Fedorov AA, et al. Structure and biochemical properties of PRL-1, a phosphatase implicated in cell growth, differentiation, and tumor invasion. *Biochemistry* 2005;**44**:12009–21.
115. Pascaru M, Tanase C, Vacaru AM, Boeti P, Neagu E, Popescu I, et al. Analysis of molecular determinants of PRL-3. *J Cell Mol Med* 2009;**13**:3141–50.
116. Lemmon MA, Schlessinger J. Cell signaling by receptor tyrosine kinases. *Cell* 2010;**141**:1117–34.
117. Kolch W, Pitt A. Functional proteomics to dissect tyrosine kinase signalling pathways in cancer. *Nat Rev Cancer* 2010;**10**:618–29.
118. Zhang J, Yang PL, Gray NS. Targeting cancer with small molecule kinase inhibitors. *Nat Rev Cancer* 2009;**9**:28–39.
119. Cully M, You H, Levine AJ, Mak TW. Beyond PTEN mutations: the PI3K pathway as an integrator of multiple inputs during tumorigenesis. *Nat Rev Cancer* 2006;**6**:184–92.
120. Salmena L, Carracedo A, Pandolfi PP. Tenets of PTEN tumor suppression. *Cell* 2008;**133**:403–14.
121. Alimonti A, Carracedo A, Clohessy JG, Trotman LC, Nardella C, Egia A, et al. Subtle variations in Pten dose determine cancer susceptibility. *Nat Genet* 2010;**42**:454–8.
122. Hollander MC, Blumenthal GM, Dennis PA. PTEN loss in the continuum of common cancers, rare syndromes and mouse models. *Nat Rev Cancer* 2011;**11**:289–301.
123. Zbuk KM, Eng C. Cancer phenomics: RET and PTEN as illustrative models. *Nat Rev Cancer* 2007;**7**:35–45.

124. Ardini E, Agresti R, Tagliabue E, Greco M, Aiello P, Yang LT. Expression of protein tyrosine phosphatase alpha (RPTPalpha) in human breast cancer correlates with low tumor grade, and inhibits tumor cell growth *in vitro* and *in vivo*. *Oncogene* 2000;**19**:4979–87.
125. Zheng X, Resnick RJ, Shalloway D. Apoptosis of estrogen-receptor negative breast cancer and colon cancer cell lines by PTP alpha and src RNAi. *Int J Cancer* 2008;**122**:1999–2007.
126. Krndija D, Schmid H, Eismann JL, Lother U, Adler G, Oswald F. Substrate stiffness and the receptor-type tyrosine-protein phosphatase alpha regulate spreading of colon cancer cells through cytoskeletal contractility. *Oncogene* 2010;**29**:2724–38.
127. Ulbricht U, Brockmann MA, Aigner A, Eckerich C, Muller S, Fillbrandt R. Expression and function of the receptor protein tyrosine phosphatase zeta and its ligand pleiotrophin in human astrocytomas. *J Neuropathol Exp Neurol* 2003;**62**:1265–75.
128. Muller S, Kunkel P, Lamszus K, Ulbricht U, Lorente GA, Nelson AM. A role for receptor tyrosine phosphatase zeta in glioma cell migration. *Oncogene* 2003;**22**:6661–8.
129. Lorente G, Nelson A, Mueller S, Kuo J, Urfer R, Nikolich K. Functional comparison of long and short splice forms of RPTPbeta: implications for glioblastoma treatment. *Neuro Oncol* 2005;**7**:154–63.
130. Ulbricht U, Eckerich C, Fillbrandt R, Westphal M, Lamszus K. RNA interference targeting protein tyrosine phosphatase zeta/receptor-type protein tyrosine phosphatase beta suppresses glioblastoma growth *in vitro* and *in vivo*. *J Neurochem* 2006;**98**:1497–506.
131. Foehr ED, Lorente G, Kuo J, Ram R, Nikolich K, Urfer R. Targeting of the receptor protein tyrosine phosphatase beta with a monoclonal antibody delays tumor growth in a glioblastoma model. *Cancer Res* 2006;**66**:2271–8.
132. Polytarchou C, Hatziapostolou M, Poimenidi E, Mikelis C, Papadopoulou A, Parthymou A. Nitric oxide stimulates migration of human endothelial and prostate cancer cells through up-regulation of pleiotrophin expression and its receptor protein tyrosine phosphatase beta/zeta. *Int J Cancer* 2009;**124**:1785–93.
133. Feng ZJ, Gao SB, Wu Y, Xu XF, Hua X, Jin GH. Lung cancer cell migration is regulated via repressing growth factor PTN/RPTP beta/zeta signaling by menin. *Oncogene* 2010;**29**:5416–26.
134. Elson A. Protein tyrosine phosphatase epsilon increases the risk of mammary hyperplasia and mammary tumors in transgenic mice. *Oncogene* 1999;**18**:7535–42.
135. Liu S, Sugimoto Y, Sorio C, Tecchio C, Lin YC. Function analysis of estrogenically regulated protein tyrosine phosphatase gamma (PTPgamma) in human breast cancer cell line MCF-7. *Oncogene* 2004;**23**:1256–62.
136. Cheung AK, Lung HL, Hung SC, Law EW, Cheng Y, Yau WL. Functional analysis of a cell cycle-associated, tumor-suppressive gene, protein tyrosine phosphatase receptor type G, in nasopharyngeal carcinoma. *Cancer Res* 2008;**68**:8137–45.
137. Shu ST, Sugimoto Y, Liu S, Chang HL, Ye W, Wang LS. Function and regulatory mechanisms of the candidate tumor suppressor receptor protein tyrosine phosphatase gamma (PTPRG) in breast cancer cells. *Anticancer Res* 2010;**30**:1937–46.
138. Nagano H, Noguchi T, Inagaki K, Yoon S, Matozaki T, Itoh H. Downregulation of stomach cancer-associated protein tyrosine phosphatase-1 (SAP-1) in advanced human hepatocellular carcinoma. *Oncogene* 2003;**22**:4656–63.
139. Sadakata H, Okazawa H, Sato T, Supriatna Y, Ohnishi H, Kusakari S. SAP-1 is a microvillus-specific protein tyrosine phosphatase that modulates intestinal tumorigenesis. *Genes Cells* 2009;**14**:295–308.
140. Keane MM, Lowrey GA, Ettenberg SA, Dayton MA, Lipkowitz S. The protein tyrosine phosphatase DEP-1 is induced during differentiation and inhibits growth of breast cancer cells. *Cancer Res* 1996;**56**:4236–43.

141. Trapasso F, Yendamuri S, Dumon KR, Iuliano R, Cesari R, Feig B. Restoration of receptor-type protein tyrosine phosphatase eta function inhibits human pancreatic carcinoma cell growth *in vitro* and *in vivo*. *Carcinogenesis* 2004;**25**:2107–14.
142. Massa A, Barbieri F, Aiello C, Arena S, Pattarozzi A, Pirani P. The expression of the phosphotyrosine phosphatase DEP-1/PTPeta dictates the responsivity of glioma cells to somatostatin inhibition of cell proliferation. *J Biol Chem* 2004;**279**:29004–12.
143. Iervolino A, Iuliano R, Trapasso F, Viglietto G, Melillo RM, Carlomagno F. The receptor-type protein tyrosine phosphatase J antagonizes the biochemical and biological effects of RET-derived oncoproteins. *Cancer Res* 2006;**66**:6280–7.
144. Balavenkatraman KK, Jandt E, Friedrich K, Kautenburger T, Pool-Zobel BL, Ostman A. DEP-1 protein tyrosine phosphatase inhibits proliferation and migration of colon carcinoma cells and is upregulated by protective nutrients. *Oncogene* 2006;**25**:6319–24.
145. Tarcic G, Boguslavsky SK, Wakim J, Kiuchi T, Liu A, Reinitz F. An unbiased screen identifies DEP-1 tumor suppressor as a phosphatase controlling EGFR endocytosis. *Curr Biol* 2009;**19**:1788–98.
146. Petermann A, Haase D, Wetzel A, Balavenkatraman KK, Tenev T, Guhrs KH. Loss of the protein-tyrosine phosphatase DEP-1/PTPRJ drives meningioma cell motility. *Brain Pathol* 2011;**21**:405–18.
147. Novellino L, De Filippo A, Deho P, Perrone F, Pilotti S, Parmiani G. PTPRK negatively regulates transcriptional activity of wild type and mutated oncogenic beta-catenin and affects membrane distribution of beta-catenin/E-cadherin complexes in cancer cells. *Cell Signal* 2008;**20**:872–83.
148. Flavell JR, Baumforth KR, Wood VH, Davies GL, Wei W, Reynolds GM. Down-regulation of the TGF-beta target gene, PTPRK, by the Epstein-Barr virus encoded EBNA1 contributes to the growth and survival of Hodgkin lymphoma cells. *Blood* 2008;**111**:292–301.
149. Phillips-Mason PJ, Kaur H, Burden-Gulley SM, Craig SE, Brady-Kalnay SM. Identification of phospholipase C gamma1 as a protein tyrosine phosphatase mu substrate that regulates cell migration. *J Cell Biochem* 2011;**112**:39–48.
150. Quintavalle C, Garofalo M, Zanca C, Romano G, Iaboni M, Del Basso De Caro M. miR-221/222 overexpession in human glioblastoma increases invasiveness by targeting the protein phosphate PTPmu. *Oncogene* 2011; Epub ahead of print.
151. Motiwala T, Kutay H, Ghoshal K, Bai S, Seimiya H, Tsuruo T. Protein tyrosine phosphatase receptor-type O (PTPRO) exhibits characteristics of a candidate tumor suppressor in human lung cancer. *Proc Natl Acad Sci USA* 2004;**101**:13844–9.
152. Chen L, Juszczynski P, Takeyama K, Aguiar RC, Shipp MA. Protein tyrosine phosphatase receptor-type O truncated (PTPROt) regulates SYK phosphorylation, proximal B-cell-receptor signaling, and cellular proliferation. *Blood* 2006;**108**:3428–33.
153. Motiwala T, Majumder S, Kutay H, Smith DS, Neuberg DS, Lucas DM. Methylation and silencing of protein tyrosine phosphatase receptor type O in chronic lymphocytic leukemia. *Clin Cancer Res* 2007;**13**:3174–81.
154. Ramaswamy B, Majumder S, Roy S, Ghoshal K, Kutay H, Datta J. Estrogen-mediated suppression of the gene encoding protein tyrosine phosphatase PTPRO in human breast cancer: mechanism and role in tamoxifen sensitivity. *Mol Endocrinol* 2009;**23**:176–87.
155. Motiwala T, Majumder S, Ghoshal K, Kutay H, Datta J, Roy S. PTPROt inactivates the oncogenic fusion protein BCR/ABL and suppresses transformation of K562 cells. *J Biol Chem* 2009;**284**:455–64.
156. Motiwala T, Datta J, Kutay H, Roy S, Jacob ST. Lyn kinase and ZAP70 are substrates of PTPROt in B-cells: lyn inactivation by PTPROt sensitizes leukemia cells to VEGF-R inhibitor pazopanib. *J Cell Biochem* 2010;**110**:846–56.

157. Suarez Pestana E, Tenev T, Gross S, Stoyanov B, Ogata M, Bohmer FD. The transmembrane protein tyrosine phosphatase RPTPsigma modulates signaling of the epidermal growth factor receptor in A431 cells. *Oncogene* 1999;**18**:4069–79.
158. Wang Z, Shen D, Parsons DW, Bardelli A, Sager J, Szabo S. Mutational analysis of the tyrosine phosphatome in colorectal cancers. *Science* 2004;**304**:1164–6.
159. Doumont G, Martoriati A, Beekman C, Bogaerts S, Mee PJ, Bureau F. G1 checkpoint failure and increased tumor susceptibility in mice lacking the novel p53 target Ptprv. *EMBO J* 2005;**24**:3093–103.
160. Yu Y, Jiang X, Schoch BS, Carroll RS, Black PM, Johnson MD. Aberrant splicing of cyclin-dependent kinase-associated protein phosphatase KAP increases proliferation and migration in glioblastoma. *Cancer Res* 2007;**67**:130–8.
161. Glondu-Lassis M, Dromard M, Lacroix-Triki M, Nirde P, Puech C, Knani D. PTPL1/PTPN13 regulates breast cancer cell aggressiveness through direct inactivation of Src kinase. *Cancer Res* 2010;**70**:5116–26.
162. Yuan T, Wang Y, Zhao ZJ, Gu H. Protein-tyrosine phosphatase PTPN9 negatively regulates ErbB2 and epidermal growth factor receptor signaling in breast cancer cells. *J Biol Chem* 2010;**285**:14861–70.
163. Sun T, Aceto N, Meerbrey KL, Kessler JD, Zhou C, Migliaccio I. Activation of multiple proto-oncogenic tyrosine kinases in breast cancer via loss of the PTPN12 phosphatase. *Cell* 2011;**144**:703–18.
164. Lin G, Aranda V, Muthuswamy SK, Tonks NK. Identification of PTPN23 as a novel regulator of cell invasion in mammary epithelial cells from a loss-of-function screen of the 'PTP-ome'. *Genes Dev* 2011;**25**:1412–25.
165. Espejo R, Rengifo-Cam W, Schaller MD, Evers BM, Sastry SK. PTP-PEST controls motility, adherens junction assembly, and Rho GTPase activity in colon cancer cells. *Am J Physiol Cell Physiol* 2010;**299**:C454–63.
166. LaMontagne KR, Jr., Flint AJ, Franza BR, Jr., Pandergast AM, Tonks NK. Protein tyrosine phosphatase 1B antagonizes signalling by oncoprotein tyrosine kinase p210 bcr-abl *in vivo*. *Mol Cell Biol* 1998;**18**:2965–75.
167. Wu C, Guan Q, Wang Y, Zhao ZJ, Zhou GW. SHP-1 suppresses cancer cell growth by promoting degradation of JAK kinases. *J Cell Biochem* 2003;**90**:1026–37.
168. Han Y, Amin HM, Franko B, Frantz C, Shi X, Lai R. Loss of SHP1 enhances JAK3/STAT3 signaling and decreases proteosome degradation of JAK3 and NPM-ALK in ALK+ anaplastic large-cell lymphoma. *Blood* 2006;**108**:2796–803.
169. Bard-Chapeau EA, Li S, Ding J, Zhang SS, Zhu HH, Princen F. Ptpn11/Shp2 acts as a tumor suppressor in hepatocellular carcinogenesis. *Cancer Cell* 2011;**19**:629–39.
170. Ueda K, Arakawa H, Nakamura Y. Dual-specificity phosphatase 5 (DUSP5) as a direct transcriptional target of tumor suppressor p53. *Oncogene* 2003;**22**:5586–91.
171. Zhang Z, Kobayashi S, Borczuk AC, Leidner RS, Laframboise T, Levine AD. Dual specificity phosphatase 6 (DUSP6) is an ETS-regulated negative feedback mediator of oncogenic ERK signaling in lung cancer cells. *Carcinogenesis* 2010;**31**:577–86.
172. Wang JY, Yeh CL, Chou HC, Yang CH, Fu YN, Chen YT. Vaccinia H1-related phosphatase is a phosphatase of ErbB receptors and is down-regulated in non-small cell lung cancer. *J Biol Chem* 2011;**286**:10177–84.
173. Manzano RG, Montuenga LM, Dayton M, Dent P, Kinoshita I, Vicent S. CL100 expression is down-regulated in advanced epithelial ovarian cancer and its re-expression decreases its malignant potential. *Oncogene* 2002;**21**:4435–47.
174. Chan DW, Liu VW, Tsao GS, Yao KM, Furukawa T, Chan KK. Loss of MKP3 mediated by oxidative stress enhances tumorigenicity and chemoresistance of ovarian cancer cells. *Carcinogenesis* 2008;**29**:1742–50.

175. Furukawa T, Sunamura M, Motoi F, Matsuno S, Horii A. Potential tumor suppressive pathway involving DUSP6/MKP-3 in pancreatic cancer. *Am J Pathol* 2003;**162**:1807–15.
176. Julien SG, Dube N, Read M, Penney J, Paquet M, Han Y. Protein tyrosine phosphatase 1B deficiency or inhibition delays ErbB2-induced mammary tumorigenesis and protects from lung metastasis. *Nat Genet* 2007;**39**:338–46.
177. Arias-Romero LE, Saha S, Villamar-Cruz O, Yip SC, Ethier SP, Zhang ZY. Activation of Src by protein tyrosine phosphatase 1B Is required for ErbB2 transformation of human breast epithelial cells. *Cancer Res* 2009;**69**:4582–8.
178. Blanquart C, Karouri SE, Issad T. Protein tyrosine phosphatase-1B and T-cell protein tyrosine phosphatase regulate IGF-2-induced MCF-7 cell migration. *Biochem Biophys Res Commun* 2010;**392**:83–8.
179. Hardy S, Wong NN, Muller WJ, Park M, Tremblay ML. Overexpression of the protein tyrosine phosphatase PRL-2 correlates with breast tumor formation and progression. *Cancer Res* 2010;**70**:8959–67.
180. Rahmouni S, Cerignoli F, Alonso A, Tsutji T, Henkens R, Zhu C. Loss of the VHR dual-specific phosphatase causes cell-cycle arrest and senescence. *Nat Cell Biol* 2006; **8**:524–31.
181. Zhu S, Bjorge JD, Fujita DJ. PTP1B contributes to the oncogenic properties of colon cancer cells through Src activation. *Cancer Res* 2007;**67**:10129–37.
182. Daouti S, Li WH, Qian H, Huang KS, Holmgren J, Levin W. A selective phosphatase of regenerating liver phosphatase inhibitor suppresses tumor cell anchorage-independent growth by a novel mechanism involving p130Cas cleavage. *Cancer Res* 2008;**68**:1162–9.
183. Liu Y, Zhou J, Chen J, Gao W, Le Y, Ding Y. PRL-3 promotes epithelial mesenchymal transition by regulating cadherin directly. *Cancer Biol Ther* 2009;**8**:1352–9.
184. Peng L, Xing X, Li W, Qu L, Meng L, Lian S. PRL-3 promotes the motility, invasion, and metastasis of LoVo colon cancer cells through PRL-3-integrin beta1-ERK1/2 and-MMP2 signaling. *Mol Cancer* 2009;**8**:110.
185. Hou SW, Zhi HY, Pohl N, Loesch M, Qi XM, Li RS. PTPH1 dephosphorylates and cooperates with p38gamma MAPK to increase ras oncogenesis through PDZ-mediated interaction. *Cancer Res* 2010;**70**:2901–10.
186. Wang J, Liu B, Chen X, Su L, Wu P, Wu J. PTP1B expression contributes to gastric cancer progression. *Med Oncol* 2011; Epub ahead of print.
187. Araki T, Mohi MG, Ismat FA, Bronson RT, Williams IR, Kutok JL. Mouse model of Noonan syndrome reveals cell type- and gene dosage-dependent effects of Ptpn11 mutation. *Nat Med* 2004;**10**:849–57.
188. Mohi MG, Williams IR, Dearolf CR, Chan G, Kutok JL, Cohen S. Prognostic, therapeutic, and mechanistic implications of a mouse model of leukemia evoked by Shp2 (PTPN11) mutations. *Cancer Cell* 2005;**7**:179–91.
189. Voena C, Conte C, Ambrogio C, Boeri Erba E, Boccalatte F, Mohammed S. The tyrosine phosphatase Shp2 interacts with NPM-ALK and regulates anaplastic lymphoma cell growth and migration. *Cancer Res* 2007;**67**:4278–86.
190. Xu D, Wang S, Yu WM, Chan G, Araki T, Bunting KD. A germline gain-of-function mutation in Ptpn11 (Shp-2) phosphatase induces myeloproliferative disease by aberrant activation of hematopoietic stem cells. *Blood* 2010;**116**:3611–21.
191. Masiero M, Minuzzo S, Pusceddu I, Moserle L, Persano L, Agnusdei V. Notch3-mediated regulation of MKP-1 levels promotes survival of T acute lymphoblastic leukemia cells. *Leukemia* 2011;**25**:588–98.
192. Achiwa H, Lazo JS. PRL-1 tyrosine phosphatase regulates c-Src levels, adherence, and invasion in human lung cancer cells. *Cancer Res* 2007;**67**:643–50.

193. Ming J, Liu N, Gu Y, Qiu X, Wang EH. PRL-3 facilitates angiogenesis and metastasis by increasing ERK phosphorylation and up-regulating the levels and activities of Rho-A/C in lung cancer. *Pathology* 2009;**41**:118–26.
194. Ren Y, Chen Z, Chen L, Fang B, Win-Piazza H, Haura E. Critical role of Shp2 in tumor growth involving regulation of c-Myc. *Genes Cancer* 2010;**1**:994–1007.
195. Nakashima M, Lazo JS. Phosphatase of regenerating liver-1 promotes cell migration and invasion and regulates filamentous actin dynamics. *J Pharmacol Exp Ther* 2010;**334**:627–33.
196. Liao Q, Guo J, Kleeff J, Zimmermann A, Buchler MW, Korc M. Down-regulation of the dual-specificity phosphatase MKP-1 suppresses tumorigenicity of pancreatic cancer cells. *Gastroenterology* 2003;**124**:1830–45.
197. Stephens B, Han H, Hostetter G, Demeure MJ, Von Hoff DD. Small interfering RNA-mediated knockdown of PRL phosphatases results in altered Akt phosphorylation and reduced clonogenicity of pancreatic cancer cells. *Mol Cancer Ther* 2008;**7**:202–10.
198. Arnoldussen YJ, Lorenzo PI, Pretorius ME, Waehre H, Risberg B, Maelandsmo GM. The mitogen-activated protein kinase phosphatase vaccinia H1-related protein inhibits apoptosis in prostate cancer cells and is overexpressed in prostate cancer. *Cancer Res* 2008;**68**:9255–64.
199. Rodriguez-Ubreva FJ, Cariaga-Martinez AE, Cortes MA, Romero-De Pablos M, Ropero S, Lopez-Ruiz P. Knockdown of protein tyrosine phosphatase SHP-1 inhibits G1/S progression in prostate cancer cells through the regulation of components of the cell-cycle machinery. *Oncogene* 2010;**29**:345–55.
200. Yu W, Imoto I, Inoue J, Onda M, Emi M, Inazawa J. A novel amplification target, DUSP26, promotes anaplastic thyroid cancer cell growth by inhibiting p38 MAPK activity. *Oncogene* 2007;**26**:1178–87.
201. LaMontagne KR, Jr., Hannon G, Tonks NK. Protein tyrosine phosphatase PTP1B suppresses p210 bcr-abl-induced transformation of rat-1 fibroblasts and promotes differentiation of K562 cells. *Proc Natl Acad Sci USA* 1998;**95**:14094–9.
202. Dube N, Bourdeau A, Heinonen KM, Cheng A, Loy AL, Tremblay ML. Genetic ablation of protein tyrosine phosphatase 1B accelerates lymphomagenesis of p53-null mice through the regulation of B-cell development. *Cancer Res* 2005;**65**:10088–95.
203. Wu C, Zhang L, Bourne PA, Reeder JE, di Sant'Agnese PA, Yao JL. Protein tyrosine phosphatase PTP1B is involved in neuroendocrine differentiation of prostate cancer. *Prostate* 2006;**66**:1125–35.
204. Bentires-Alj M, Neel BG. Protein-tyrosine phosphatase 1B is required for HER2/Neu-induced breast cancer. *Cancer Res* 2007;**67**:2420–4.
205. Balavenkatraman KK, Aceto N, Britschgi A, Mueller U, Bence KK, Neel B. Epithelial protein-tyrosine phosphatase 1B (PTP1B) contributes to the induction of mammary tumors by HER2/Neu but is not essential for tumor maintenance. *Mol Cancer Res* 2011;**9** (10):1377–84.
206. Klingler-Hoffmann M, Fodero-Tavoletti MT, Mishima K, Narita Y, Cavenee WK, Furnari FB. The protein tyrosine phosphatase TCPTP suppresses the tumorigenicity of glioblastoma cells expressing a mutant epidermal growth factor receptor. *J Biol Chem* 2001;**276**:46313–8.
207. Bourdeau A, Dube N, Heinonen KM, Theberge JF, Doody KM, Tremblay ML. TC-PTP-deficient bone marrow stromal cells fail to support normal B lymphopoiesis due to abnormal secretion of interferon-{gamma}. *Blood* 2007;**109**:4220–8.
208. Zhi HY, Hou SW, Li RS, Basir Z, Xiang Q, Szabo A. PTPH1 cooperates with vitamin D receptor to stimulate breast cancer growth through their mutual stabilization. *Oncogene* 2011;**30**:1706–15.
209. Thangaraju M, Sharma K, Leber B, Andrews DW, Shen SH, Srikant CB. Regulation of acidification and apoptosis by SHP-1 and Bcl-2. *J Biol Chem* 1999;**274**:29549–57.

210. Bruecher-Encke B, Griffin JD, Neel BG, Lorenz U. Role of the tyrosine phosphatase SHP-1 in K562 cell differentiation. *Leukemia* 2001;**15**:1424–32.
211. Zapata PD, Ropero RM, Valencia AM, Buscail L, Lopez JI, Martin-Orozco RM. Autocrine regulation of human prostate carcinoma cell proliferation by somatostatin through the modulation of the SH2 domain containing protein tyrosine phosphatase (SHP)-1. *J Clin Endocrinol Metab* 2002;**87**:915–26.
212. Koyama M, Oka T, Ouchida M, Nakatani Y, Nishiuchi R, Yoshino T. Activated proliferation of B-cell lymphomas/leukemias with the SHP1 gene silencing by aberrant CpG methylation. *Lab Invest* 2003;**83**:1849–58.
213. Chen P, Levis M, Brown P, Kim KT, Allebach J, Small D. FLT3/ITD mutation signaling includes suppression of SHP-1. *J Biol Chem* 2005;**280**:5361–9.
214. Honorat JF, Ragab A, Lamant L, Delsol G, Ragab-Thomas J. SHP1 tyrosine phosphatase negatively regulates NPM-ALK tyrosine kinase signaling. *Blood* 2006;**107**:4130–8.
215. Zhou X, Coad J, Ducatman B, Agazie YM. SHP2 is up-regulated in breast cancer cells and in infiltrating ductal carcinoma of the breast, implying its involvement in breast oncogenesis. *Histopathology* 2008;**53**:389–402.
216. Zhou XD, Agazie YM. Inhibition of SHP2 leads to mesenchymal to epithelial transition in breast cancer cells. *Cell Death Differ* 2008;**15**:988–96.
217. Chan G, Kalaitzidis D, Usenko T, Kutok JL, Yang W, Mohi MG. Leukemogenic Ptpn11 causes fatal myeloproliferative disorder via cell-autonomous effects on multiple stages of hematopoiesis. *Blood* 2009;**113**:4414–24.
218. Zhan Y, Counelis GJ, O'Rourke DM. The protein tyrosine phosphatase SHP-2 is required for EGFRvIII oncogenic transformation in human glioblastoma cells. *Exp Cell Res* 2009; **315**:2343–57.
219. Yang X, Dutta U, Shaw LM. SHP2 mediates the localized activation of Fyn downstream of the alpha6beta4 integrin to promote carcinoma invasion. *Mol Cell Biol* 2010;**30**:5306–17.
220. Kikkawa N, Hanazawa T, Fujimura L, Nohata N, Suzuki H, Chazono H. miR-489 is a tumour-suppressive miRNA target PTPN11 in hypopharyngeal squamous cell carcinoma (HSCC). *Br J Cancer* 2010;**103**:877–84.
221. Sahu SN, Nunez S, Bai G, Gupta A. Interaction of Pyk2 and PTP-PEST with leupaxin in prostate cancer cells. *Am J Physiol Cell Physiol* 2007;**292**:C2288–96.
222. Villa-Moruzzi E. Tyrosine phosphatases in the HER2-directed motility of ovarian cancer cells: involvement of PTPN12, ERK5 and FAK. *Anal Cell Pathol* 2011;**34**(3):101–12.
223. Carlucci A, Porpora M, Garbi C, Galgani M, Santoriello M, Mascolo M. PTPD1 supports receptor stability and mitogenic signaling in bladder cancer cells. *J Biol Chem* 2010;**285**: 39260–70.
224. Mariotti M, Castiglioni S, Maier JA. Inhibition of T24 human bladder carcinoma cell migration by RNA interference suppressing the expression of HD-PTP. *Cancer Lett* 2009;**273**:155–63.
225. Gingras MC, Zhang YL, Kharitidi D, Barr AJ, Knapp S, Tremblay ML. HD-PTP is a catalytically inactive tyrosine phosphatase due to a conserved divergence in its phosphatase domain. *PLoS One* 2009;**4**:e5105.
226. Tai CJ, Wu AT, Chiou JF, Jan HJ, Wei HJ, Hsu CH. The investigation of mitogen-activated protein kinase phosphatase-1 as a potential pharmacological target in non-small cell lung carcinomas, assisted by non-invasive molecular imaging. *BMC Cancer* 2010;**10**:95.
227. Moncho-Amor V, Ibanez de Caceres I, Bandres E, Martinez-Poveda B, Orgaz JL, Sanchez-Perez I. DUSP1/MKP1 promotes angiogenesis, invasion and metastasis in non-small-cell lung cancer. *Oncogene* 2011;**30**:668–78.
228. Lin SC, Chien CW, Lee JC, Yeh YC, Hsu KF, Lai YY. Suppression of dual-specificity phosphatase-2 by hypoxia increases chemoresistance and malignancy in human cancer cells. *J Clin Invest* 2011;**121**:1905–16.

229. Waha A, Felsberg J, Hartmann W, von dem Knesebeck A, Mikeska T, Joos S. Epigenetic downregulation of mitogen-activated protein kinase phosphatase MKP-2 relieves its growth suppressive activity in glioma cells. *Cancer Res* 2010;**70**:1689–99.
230. Okudela K, Yazawa T, Woo T, Sakaeda M, Ishii J, Mitsui H. Down-regulation of DUSP6 expression in lung cancer: its mechanism and potential role in carcinogenesis. *Am J Pathol* 2009;**175**:867–81.
231. Wong VC, Chen H, Ko JM, Chan KW, Chan YP, Law S. Tumor suppressor dual-specificity phosphatase 6 (DUSP6) impairs cell invasion and epithelial-mesenchymal transition (EMT)-associated phenotype. *Int J Cancer* 2012;**130**(1):83–95.
232. Tang JP, Tan CP, Li J, Siddique MM, Guo K, Chan SW. VHZ is a novel centrosomal phosphatase associated with cell growth and human primary cancers. *Mol Cancer* 2010;**9**:128.
233. Patterson KI, Brummer T, Daly RJ, O'Brien PM. DUSP26 negatively affects the proliferation of epithelial cells, an effect not mediated by dephosphorylation of MAPKs. *Biochim Biophys Acta* 2010;**1803**:1003–12.
234. Werner SR, Lee PA, DeCamp MW, Crowell DN, Randall SK, Crowell PL. Enhanced cell cycle progression and down regulation of p21(Cip1/Waf1) by PRL tyrosine phosphatases. *Cancer Lett* 2003;**202**:201–11.
235. Polato F, Codegoni A, Fruscio R, Perego P, Mangioni C, Saha S. PRL-3 phosphatase is implicated in ovarian cancer growth. *Clin Cancer Res* 2005;**11**:6835–9.
236. Rouleau C, Roy A, St Martin T, Dufault MR, Boutin P, Liu D. Protein tyrosine phosphatase PRL-3 in malignant cells and endothelial cells: expression and function. *Mol Cancer Ther* 2006;**5**:219–29.
237. Cai SR, Wang Z, Chen CQ, Wu WH, He YL, Zhan WH. Role of silencing phosphatase of regenerationg liver-3 expression by microRNA interference in the growth of gastric cancer. *Chin Med J (Engl)* 2008;**121**:2534–8.
238. Wang Z, He YL, Cai SR, Zhan WH, Li ZR, Zhu BH. Expression and prognostic impact of PRL-3 in lymph node metastasis of gastric cancer: its molecular mechanism was investigated using artificial microRNA interference. *Int J Cancer* 2008;**123**:1439–47.
239. Fagerli UM, Holt RU, Holien T, Vaatsveen TK, Zhan F, Egeberg KW. Overexpression and involvement in migration by the metastasis-associated phosphatase PRL-3 in human myeloma cells. *Blood* 2008;**111**:806–15.
240. Wang Z, Cai SR, He YL, Zhan WH, Chen CQ, Cui J. High expression of PRL-3 can promote growth of gastric cancer and exhibits a poor prognostic impact on patients. *Ann Surg Oncol* 2009;**16**:208–19.
241. Zhou J, Wang S, Lu J, Li J, Ding Y. Over-expression of phosphatase of regenerating liver-3 correlates with tumor progression and poor prognosis in nasopharyngeal carcinoma. *Int J Cancer* 2009;**124**:1879–86.
242. Semba S, Mizuuchi E, Yokozaki H. Requirement of phosphatase of regenerating liver-3 for the nucleolar localization of nucleolin during the progression of colorectal carcinoma. *Cancer Sci* 2010;**101**:2254–61.
243. Matsukawa Y, Semba S, Kato H, Koma Y, Yanagihara K, Yokozaki H. Constitutive suppression of PRL-3 inhibits invasion and proliferation of gastric cancer cell *in vitro* and *in vivo*. *Pathobiology* 2010;**77**:155–62.
244. Lee SW, Reimer CL, Fang L, Iruela-Arispe ML, Aaronson SA. Overexpression of kinase-associated phosphatase (KAP) in breast and prostate cancer and inhibition of the transformed phenotype by antisense KAP expression. *Mol Cell Biol* 2000;**20**:1723–32.
245. Yam JW, Ko FC, Chan CY, Yau TO, Tung EK, Leung TH. Tensin2 variant 3 is associated with aggressive tumor behavior in human hepatocellular carcinoma. *Hepatology* 2006;**44**:881–90.
246. Ostman A, Yang Q, Tonks NK. Expression of DEP-1, a receptor-like protein-tyrosine-phosphatase, is enhanced with increasing cell density. *Proc Natl Acad Sci USA* 1994;**91**: 9680–4.

247. Borges LG, Seifert RA, Grant FJ, Hart CE, Disteche CM, Edelhoff S, et al. Cloning and characterization of rat density-enhanced phosphatase-1, a protein tyrosine phosphatase expressed by vascular cells. *Circ Res* 1996;**79**:570–80.
248. Lin J, Zhu JW, Baker JE, Weiss A. Regulated expression of the receptor-like tyrosine phosphatase CD148 on hemopoietic cells. *J Immunol* 2004;**173**:2324–30.
249. Autschbach F, Palou E, Mechtersheimer G, Rohr C, Pirotto F, Gassler N, et al. Expression of the membrane protein tyrosine phosphatase CD148 in human tissues. *Tissue Antigens* 1999;**54**:485–98.
250. Takahashi T, Takahashi K, St John PL, Fleming PA, Tomemori T, Watanabe T, et al. A mutant receptor tyrosine phosphatase, CD148, causes defects in vascular development. *Mol Cell Biol* 2003;**23**:1817–31.
251. Grazia Lampugnani M, Zanetti A, Corada M, Takahashi T, Balconi G, Breviario F, et al. Contact inhibition of VEGF-induced proliferation requires vascular endothelial cadherin, beta-catenin, and the phosphatase DEP-1/CD148. *J Cell Biol* 2003;**161**:793–804.
252. Kovalenko M, Denner K, Sandstrom J, Persson C, Gross S, Jandt E, et al. Site-selective dephosphorylation of the platelet-derived growth factor beta-receptor by the receptor-like protein-tyrosine phosphatase DEP-1. *J Biol Chem* 2000;**275**:16219–26.
253. Trapasso F, Drusco A, Costinean S, Alder H, Aqeilan RI, Iuliano R, et al. Genetic ablation of Ptprj, a mouse cancer susceptibility gene, results in normal growth and development and does not predispose to spontaneous tumorigenesis. *DNA Cell Biol* 2006;**25**:376–82.
254. Zhu JW, Brdicka T, Katsumoto TR, Lin J, Weiss A. Structurally distinct phosphatases CD45 and CD148 both regulate B cell and macrophage immunoreceptor signaling. *Immunity* 2008;**28**:183–96.
255. Iuliano R, Le Pera I, Cristofaro C, Baudi F, Arturi F, Pallante P, et al. The tyrosine phosphatase PTPRJ/DEP-1 genotype affects thyroid carcinogenesis. *Oncogene* 2004;**23**:8432–8.
256. Mita Y, Yasuda Y, Sakai A, Yamamoto H, Toyooka S, Gunduz M, et al. Missense polymorphisms of PTPRJ and PTPN13 genes affect susceptibility to a variety of human cancers. *J Cancer Res Clin Oncol* 2010;**136**:249–59.
257. Iuliano R, Palmieri D, He H, Iervolino A, Borbone E, Pallante P, et al. Role of PTPRJ genotype in papillary thyroid carcinoma risk. *Endocr Relat Cancer* 2010;**17**:1001–6.
258. Palka HL, Park M, Tonks NK. Hepatocyte growth factor receptor tyrosine kinase met is a substrate of the receptor protein-tyrosine phosphatase DEP-1. *J Biol Chem* 2003; **278**:5728–35.
259. Arora D, Stopp S, Bohmer SA, Schons J, Godfrey R, Masson K, et al. Protein-tyrosine phosphatase DEP-1 controls receptor tyrosine kinase FLT3 signaling. *J Biol Chem* 2011; **286**:10918–29.
260. Polo S, Di Fiore PP. Endocytosis conducts the cell signaling orchestra. *Cell* 2006;**124**: 897–900.
261. Sacco F, Tinti M, Palma A, Ferrari E, Nardozza AP, Hooft van Huijsduijnen R, et al. Tumor suppressor density-enhanced phosphatase-1 (DEP-1) inhibits the RAS pathway by direct dephosphorylation of ERK1/2 kinases. *J Biol Chem* 2009;**284**:22048–58.
262. Cool DE, Tonks NK, Charbonneau H, Walsh KA, Fischer EH, Krebs EG. cDNA isolated from a human T-cell library encodes a member of the protein-tyrosine-phosphatase family. *Proc Natl Acad Sci USA* 1989;**86**:5257–61.
263. Lammers R, Bossenmaier B, Cool DE, Tonks NK, Schlessinger J, Fischer EH, et al. Differential activities of protein tyrosine phosphatases in intact cells. *J Biol Chem* 1993;**268**:22456–62.
264. Simoncic PD, Bourdeau A, Lee-Loy A, Rohrschneider LR, Tremblay ML, Stanley ER, et al. T-cell protein tyrosine phosphatase (Tcptp) is a negative regulator of colony-stimulating factor 1 signaling and macrophage differentiation. *Mol Cell Biol* 2006;**26**:4149–60.

265. Aoki N, Matsuda T. A cytosolic protein-tyrosine phosphatase PTP1B specifically dephosphorylates and deactivates prolactin-activated STAT5a and STAT5b. *J Biol Chem* 2000;**275**:39718–26.
266. Aoki N, Matsuda T. A nuclear protein tyrosine phosphatase TC-PTP is a potential negative regulator of the PRL-mediated signaling pathway: dephosphorylation and deactivation of signal transducer and activator of transcription 5a and 5b by TC-PTP in nucleus. *Mol Endocrinol* 2002;**16**:58–69.
267. Lam MH, Michell BJ, Fodero-Tavoletti MT, Kemp BE, Tonks NK, Tiganis T. Cellular stress regulates the nucleocytoplasmic distribution of the protein-tyrosine phosphatase TCPTP. *J Biol Chem* 2001;**276**:37700–7.
268. You-Ten KE, Muise ES, Itie A, Michaliszyn E, Wagner J, Jothy S, et al. Impaired bone marrow microenvironment and immune function in T cell protein tyrosine phosphatase-deficient mice. *J Exp Med* 1997;**186**:683–93.
269. Elchebly M, Payette P, Michaliszyn E, Cromlish W, Collins S, Loy AL, et al. Increased insulin sensitivity and obesity resistance in mice lacking the protein tyrosine phosphatase-1B gene. *Science* 1999;**283**:1544–8.
270. Heinonen KM, Nestel FP, Newell EW, Charette G, Seemayer TA, Tremblay ML, et al. T-cell protein tyrosine phosphatase deletion results in progressive systemic inflammatory disease. *Blood* 2004;**103**:3457–64.
271. Wellcome Trust Case Control Consortium. Genome-wide association study of 14,000 cases of seven common diseases and 3,000 shared controls. *Nature* 2007;**447**:661–78.
272. Latiano A, Palmieri O, Latiano T, Corritore G, Bossa F, Martino G, et al. Investigation of multiple susceptibility Loci for inflammatory bowel disease in an Italian cohort of patients. *PLoS One* 2011;**6**:e22688.
273. Jawad N, Direkze N, Leedham SJ. Inflammatory bowel disease and colon cancer. *Recent Results Cancer Res* 2011;**185**:99–115. (Fortschritte der Krebsforschung. Progres dans les recherches sur le cancer).
274. Hassan SW, Doody KM, Hardy S, Uetani N, Cournoyer D, Tremblay ML. Increased susceptibility to dextran sulfate sodium induced colitis in the T cell protein tyrosine phosphatase heterozygous mouse. *PLoS One* 2010;**5**:e8868.
275. Scharl M, McCole DF, Weber A, Vavricka SR, Frei P, Kellermeier S, et al. Protein tyrosine phosphatase N2 regulates TNFalpha-induced signalling and cytokine secretion in human intestinal epithelial cells. *Gut* 2011;**60**:189–97.
276. Graux C, Cools J, Melotte C, Quentmeier H, Ferrando A, Levine R, et al. Fusion of NUP214 to ABL1 on amplified episomes in T-cell acute lymphoblastic leukemia. *Nat Genet* 2004;**36**:1084–9.
277. Kleppe M, Soulier J, Asnafi V, Mentens N, Hornakova T, Knoops L, et al. PTPN2 negatively regulates oncogenic JAK1 in T-cell acute lymphoblastic leukemia. *Blood* 2011;**117**:7090–8.
278. Kleppe M, Tousseyn T, Geissinger E, Kalender Atak Z, Aerts S, Rosenwald A, et al. Mutation analysis of the tyrosine phosphatase PTPN2 in Hodgkin lymphoma and T-cell non-Hodgkin lymphoma. *Haematologica* 2011.
279. Omerovic J, Clague MJ, Prior IA. Phosphatome profiling reveals PTPN2, PTPRJ and PTEN as potent negative regulators of PKB/Akt activation in Ras-mutated cancer cells. *Biochem J* 2010;**426**:65–72.
280. Lee CF, Ling ZQ, Zhao T, Fang SH, Chang WC, Lee SC, et al. Genomic-wide analysis of lymphatic metastasis-associated genes in human hepatocellular carcinoma. *World J Gastroenterol* 2009;**15**:356–65.
281. Lu X, Chen J, Sasmono RT, Hsi ED, Sarosiek KA, Tiganis T, et al. T-cell protein tyrosine phosphatase, distinctively expressed in activated-B-cell-like diffuse large B-cell lymphomas, is the nuclear phosphatase of STAT6. *Mol Cell Biol* 2007;**27**:2166–79.

282. Young RM, Polsky A, Refaeli Y. TC-PTP is required for the maintenance of MYC-driven B-cell lymphomas. *Blood* 2009;**114**:5016–23.
283. Roskoski R. Jr., Src kinase regulation by phosphorylation and dephosphorylation. *Biochem Biophys Res Commun* 2005;**331**:1–14.
284. Loh ML, Vattikuti S, Schubbert S, Reynolds MG, Carlson E, Lieuw KH, et al. Mutations in PTPN11 implicate the SHP-2 phosphatase in leukemogenesis. *Blood* 2004;**103**:2325–31.
285. Tartaglia M, Niemeyer CM, Fragale A, Song X, Buechner J, Jung A, et al. Somatic mutations in PTPN11 in juvenile myelomonocytic leukemia, myelodysplastic syndromes and acute myeloid leukemia. *Nat Genet* 2003;**34**:148–50.
286. Tartaglia M, Martinelli S, Iavarone I, Cazzaniga G, Spinelli M, Giarin E, et al. Somatic PTPN11 mutations in childhood acute myeloid leukaemia. *Br J Haematol* 2005;**129**:333–9.
287. Neel BG, Gu H, Pao L. The 'Shp'ing news: SH2 domain-containing tyrosine phosphatases in cell signaling. *Trends Biochem Sci* 2003;**28**:284–93.
288. Mohi MG, Neel BG. The role of Shp2 (PTPN11) in cancer. *Curr Opin Genet Dev* 2007;**17**:23–30.
289. Peng ZY, Cartwright CA. Regulation of the Src tyrosine kinase and Syp tyrosine phosphatase by their cellular association. *Oncogene* 1995;**11**:1955–62.
290. Zhang SQ, Yang W, Kontaridis MI, Bivona TG, Wen G, Araki T, et al. Shp2 regulates SRC family kinase activity and Ras/Erk activation by controlling Csk recruitment. *Mol Cell* 2004;**13**:341–55.
291. Ren Y, Meng S, Mei L, Zhao ZJ, Jove R, Wu J. Roles of Gab1 and SHP2 in paxillin tyrosine dephosphorylation and Src activation in response to epidermal growth factor. *J Biol Chem* 2004;**279**:8497–505.
292. Chan G, Kalaitzidis D, Neel BG. The tyrosine phosphatase Shp2 (PTPN11) in cancer. *Cancer Metastasis Rev* 2008;**27**:179–92.
293. Dance M, Montagner A, Salles JP, Yart A, Raynal P. The molecular functions of Shp2 in the Ras/Mitogen-activated protein kinase (ERK1/2) pathway. *Cell Signal* 2008;**20**:453–9.
294. Zheng XM, Wang Y, Pallen CJ. Cell transformation and activation of pp 60c-src by overexpression of a protein tyrosine phosphatase. *Nature* 1992;**359**:336–9.
295. Ponniah S, Wang DZ, Lim KL, Pallen CJ. Targeted disruption of the tyrosine phosphatase PTPalpha leads to constitutive downregulation of the kinases Src and Fyn. *Curr Biol* 1999;**9**:535–8.
296. Su J, Muranjan M, Sap J. Receptor protein tyrosine phosphatase alpha activates Src-family kinases and controls integrin-mediated responses in fibroblasts. *Curr Biol* 1999;**9**:505–11.
297. Zheng XM, Resnick RJ, Shalloway D. A phosphotyrosine displacement mechanism for activation of Src by PTPalpha. *EMBO J* 2000;**19**:964–78.
298. Zheng XM, Shalloway D. Two mechanisms activate PTPalpha during mitosis. *EMBO J* 2001;**20**:6037–49.
299. Pallen CJ. Protein tyrosine phosphatase alpha (PTPalpha): a Src family kinase activator and mediator of multiple biological effects. *Curr Top Med Chem* 2003;**3**:821–35.
300. Tabiti K, Smith DR, Goh HS, Pallen CJ. Increased mRNA expression of the receptor-like protein tyrosine phosphatase alpha in late stage colon carcinomas. *Cancer Lett* 1995;**93**:239–48.
301. Berndt A, Luo X, Bohmer FD, Kosmehl H. Expression of the transmembrane protein tyrosine phosphatase RPTPalpha in human oral squamous cell carcinoma. *Histochem Cell Biol* 1999;**111**:399–403.
302. Wu CW, Kao HL, Li AF, Chi CW, Lin WC. Protein tyrosine-phosphatase expression profiling in gastric cancer tissues. *Cancer Lett* 2006;**242**:95–103.
303. Tremper-Wells B, Resnick RJ, Zheng X, Holsinger LJ, Shalloway D. Extracellular domain dependence of PTPalpha transforming activity. *Genes Cells* 2010;**15**:711–24.

304. Diamond RH, Peters C, Jung SP, Greenbaum LE, Haber BA, Silberg DG, et al. Expression of PRL-1 nuclear PTPase is associated with proliferation in liver but with differentiation in intestine. *Am J Physiol* 1996;**271**:G121–9.
305. Zeng Q, Hong W, Tan YH. Mouse PRL-2 and PRL-3, two potentially prenylated protein tyrosine phosphatases homologous to PRL-1. *Biochem Biophys Res Commun* 1998;**244**:421–7.
306. Bardelli A, Saha S, Sager JA, Romans KE, Xin B, Markowitz SD, et al. PRL-3 expression in metastatic cancers. *Clin Cancer Res* 2003;**9**:5607–15.
307. Laurent C, Valet F, Planque N, Silveri L, Maacha S, Anezo O, et al. High PTP4A3 phosphatase expression correlates with metastatic risk in uveal melanoma patients. *Cancer Res* 2011;**71**(3):666–74.
308. Min SH, Kim DM, Heo YS, Kim YI, Kim HM, Kim J, et al. New p53 target, phosphatase of regenerating liver 1 (PRL-1) downregulates p53. *Oncogene* 2009;**28**:545–54.
309. Basak S, Jacobs SB, Krieg AJ, Pathak N, Zeng Q, Kaldis P, et al. The metastasis-associated gene Prl-3 is a p53 target involved in cell-cycle regulation. *Mol Cell* 2008;**30**:303–14.
310. Min SH, Kim DM, Heo YS, Kim HM, Kim IC, Yoo OJ. Downregulation of p53 by phosphatase of regenerating liver 3 is mediated by MDM2 and PIRH2. *Life Sci* 2010;**86**:66–72.
311. Jiang Y, Liu XQ, Rajput A, Geng L, Ongchin M, Zeng Q, et al. Phosphatase PRL-3 Is a direct regulatory target of TGF{beta} in colon cancer metastasis. *Cancer Res* 2010;**71**:234–44.
312. Wang H, Vardy LA, Tan CP, Loo JM, Guo K, Li J, et al. PCBP1 suppresses the translation of metastasis-associated PRL-3 phosphatase. *Cancer Cell* 2010;**18**:52–62.
313. Choi MS, Min SH, Jung H, Lee JD, Lee TH, Lee HK, et al. The essential role of FKBP38 in regulating phosphatase of regenerating liver 3 (PRL-3) protein stability. *Biochem Biophys Res Commun* 2011;**406**:305–9.
314. Wang H, Quah SY, Dong JM, Manser E, Tang JP, Zeng Q. PRL-3 down-regulates PTEN expression and signals through PI3K to promote epithelial-mesenchymal transition. *Cancer Res* 2007;**67**:2922–6.
315. Zhao WB, Li Y, Liu X, Zhang LY, Wang X. Evaluation of PRL-3 expression, and its correlation with angiogenesis and invasion in hepatocellular carcinoma. *Int J Mol Med* 2008;**22**:187–92.
316. Lessard L, Stuible M, Tremblay ML. The two faces of PTP1B in cancer. *Biochim Biophys Acta* 2010;**1804**:613–9.
317. Woodford-Thomas TA, Rhodes JD, Dixon JE. Expression of a protein tyrosine phosphatase in normal and v-src-transformed mouse 3T3 fibroblasts. *J Cell Biol* 1992;**117**:401–14.
318. Brown-Shimer S, Johnson KA, Hill DE, Bruskin AM. Effect of protein tyrosine phosphatase 1B expression on transformation by the human neu oncogene. *Cancer Res* 1992;**52**: 478–82.
319. Warabi M, Nemoto T, Ohashi K, Kitagawa M, Hirokawa K. Expression of protein tyrosine phosphatases and its significance in esophageal cancer. *Exp Mol Pathol* 2000;**68**:187–95.
320. Mahlamaki EH, Barlund M, Tanner M, Gorunova L, Hoglund M, Karhu R, et al. Frequent amplification of 8q24, 11q, 17q, and 20q-specific genes in pancreatic cancer. *Genes Chromosomes Cancer* 2002;**35**:353–8.
321. Yang SH, Seo MY, Jeong HJ, Jeung HC, Shin J, Kim SC, et al. Gene copy number change events at chromosome 20 and their association with recurrence in gastric cancer patients. *Clin Cancer Res* 2005;**11**:612–20.
322. Bar-Shira A, Pinthus JH, Rozovsky U, Goldstein M, Sellers WR, Yaron Y, et al. Multiple genes in human 20q13 chromosomal region are involved in an advanced prostate cancer xenograft. *Cancer Res* 2002;**62**:6803–7.
323. Tabach Y, Kogan-Sakin I, Buganim Y, Solomon H, Goldfinger N, Hovland R, et al. Amplification of the 20q chromosomal arm occurs early in tumorigenic transformation and may initiate cancer. *PLoS One* 2011;**6**:e14632.

324. Shi H, Bevier M, Johansson R, Grzybowska E, Chen B, Eyfjord JE, et al. Single nucleotide polymorphisms in the 20q13 amplicon genes in relation to breast cancer risk and clinical outcome. *Breast Cancer Res Treat* 2011;**130**(3):905–16.
325. Wiener JR, Hurteau JA, Kerns BJ, Whitaker RS, Conaway MR, Berchuck A, et al. Overexpression of the tyrosine phosphatase PTP1B is associated with human ovarian carcinomas. *Am J Obstet Gynecol* 1994;**170**:1177–83.
326. Wiener JR, Kerns BJ, Harvey EL, Conaway MR, Iglehart JD, Berchuck A, et al. Overexpression of the protein tyrosine phosphatase PTP1B in human breast cancer: association with p185c-erbB-2 protein expression. *J Natl Cancer Inst* 1994;**86**:372–8.
327. Johnson KJ, Peck AR, Liu C, Tran TH, Utama FE, Sjolund AB, et al. PTP1B suppresses prolactin activation of Stat5 in breast cancer cells. *Am J Pathol* 2010;**177**:2971–83.
328. Finn RS. Targeting Src in breast cancer. *Ann Oncol* 2008;**19**:1379–86.
329. Hynes NE, MacDonald G. ErbB receptors and signaling pathways in cancer. *Curr Opin Cell Biol* 2009;**21**:177–84.
330. Suwaki N, Vanhecke E, Atkins KM, Graf M, Swabey K, Huang P, et al. A HIF-regulated VHL-PTP1B-Src signaling axis identifies a therapeutic target in renal cell carcinoma. *Sci Transl Med* 2011;**3**:85ra47.
331. Dube N, Cheng A, Tremblay ML. The role of protein tyrosine phosphatase 1B in Ras signaling. *Proc Natl Acad Sci USA* 2004;**101**:1834–9.
332. Stuible M, Dube N, Tremblay ML. PTP1B regulates cortactin tyrosine phosphorylation by targeting Tyr446. *J Biol Chem* 2008;**283**:15740–6.
333. Stuible M, Abella JV, Feldhammer M, Nossov M, Sangwan V, Blagoev B. PTP1B targets the endosomal sorting machinery: dephosphorylation of regulatory sites on the endosomal sorting complex required for transport component STAM2. *J Biol Chem* 2010;**285**:23899–907.
334. Stuible M, Tremblay ML. In control at the ER: PTP1B and the down-regulation of RTKs by dephosphorylation and endocytosis. *Trends Cell Biol* 2010;**20**:672–9.
335. Bjorge JD, Pang A, Fujita DJ. Identification of protein-tyrosine phosphatase 1B as the major tyrosine phosphatase activity capable of dephosphorylating and activating c-Src in several human breast cancer cell lines. *J Biol Chem* 2000;**275**:41439–46.
336. Scott LM, Lawrence HR, Sebti SM, Lawrence NJ, Wu J. Targeting protein tyrosine phosphatases for anticancer drug discovery. *Curr Pharm Des* 2010;**16**:1843–62.
337. Puius YA, Zhao Y, Sullivan M, Lawrence DS, Almo SC, Zhang ZY. Identification of a second aryl phosphate-binding site in protein-tyrosine phosphatase 1B: a paradigm for inhibitor design. *Proc Natl Acad Sci USA* 1997;**94**:13420–5.
338. Salmeen A, Andersen JN, Myers MP, Tonks NK, Barford D. Molecular basis for the dephosphorylation of the activation segment of the insulin receptor by protein tyrosine phosphatase 1B. *Mol Cell* 2000;**6**:1401–12.
339. Tremblay ML. The PTP family photo album. *Cell* 2009;**136**:213–4.
340. Liu GX, Tan JZ, Niu CY, Shen JH, Luo XM, Shen X, et al. Molecular dynamics simulations of interaction between protein-tyrosine phosphatase 1B and a bidentate inhibitor. *Acta Pharmacol Sin* 2006;**27**:100–10.
341. Zhang ZY, Wang Y, Dixon JE. Dissecting the catalytic mechanism of protein-tyrosine phosphatases. *Proc Natl Acad Sci USA* 1994;**91**:1624–7.
342. Wiesmann C, Barr KJ, Kung J, Zhu J, Erlanson DA, Shen W, et al. Allosteric inhibition of protein tyrosine phosphatase 1B. *Nat Struct Mol Biol* 2004;**11**:730–7.
343. Liu S, Zeng LF, Wu L, Yu X, Xue T, Gunawan AM, et al. Targeting inactive enzyme conformation: aryl diketoacid derivatives as a new class of PTP1B inhibitors. *J Am Chem Soc* 2008;**130**:17075–84.
344. Zhang XY, Bishop AC. Engineered inhibitor sensitivity in the WPD loop of a protein tyrosine phosphatase. *Biochemistry* 2008;**47**:4491–500.

345. Hower AE, Beltran PJ, Bixby JL. Dimerization of tyrosine phosphatase PTPRO decreases its activity and ability to inactivate TrkC. *J Neurochem* 2009;**110**:1635–47.
346. Takahashi T, Takahashi K, Mernaugh RL, Tsuboi N, Liu H, Daniel TO. A monoclonal antibody against CD148, a receptor-like tyrosine phosphatase, inhibits endothelial-cell growth and angiogenesis. *Blood* 2006;**108**:1234–42.
347. van der Wijk T, Blanchetot C, Overvoorde J, den Hertog J. Redox-regulated rotational coupling of receptor protein-tyrosine phosphatase alpha dimers. *J Biol Chem* 2003;**278**:13968–74.
348. Perron MSC, Aubry I, Tremblay ML, Uri Saragovi H. *Selective CD45 phosphatase inhibitors are immunosuppressive in vivo, through a non-competitive allosteric mechanism.* 2011 (Submitted).
349. Townshend B, Aubry I, Marcellus RC, Gehring K, Tremblay ML. An RNA aptamer that selectively inhibits the enzymatic activity of protein tyrosine phosphatase 1B *in vitro*. *Chembiochem* 2010;**11**:1583–93.

# Wip1-Dependent Signaling Pathways in Health and Diseases

YUN-HUA ZHU AND DMITRY V. BULAVIN

*Cell Cycle Control and Tumorigenesis Group, Institute of Molecular and Cell Biology, Proteos, Singapore*



Spatial and temporal regulation of protein phosphorylation is key to the control of different molecular networks. This regulation is achieved in part through dephosphorylation of numerous signaling molecules, and emerging evidence highlights the importance of a new member of the PP2C family of phosphatase, Wild-type p53 induced phosphatase 1 (Wip1), in regulating stress-induced and DNA damage-induced networks. In recent years, analysis of Wip1 has focused primarily on its role in tumorigenesis because of its overexpression in human tumors and a profound tumor-resistant phenotype of Wip1-deficient mice. Recently, Wip1 has also been shown to play an important role in several physiological processes including adult neurogenesis and organismal aging. This review addresses how Wip1 phosphatase regulates different signaling networks in a spatial and temporal manner and how these differences contribute to various biological outcomes in the context of physiological and pathological conditions.

## I. Introduction

Signal transduction pathways play a crucial role in a variety of cellular processes including self renewal and differentiation, proliferation, and cell death, and ultimately regulate the balance between regenerative potential and the risk of neoplastic transformation. The major regulatory mechanism controlling signal transduction is reversible phosphorylation, which involves a

Progress in Molecular Biology
and Translational Science, Vol. 106
DOI: 10.1016/B978-0-12-396456-4.00001-8

delicate balance between multiple kinases and phosphatases. Kinases have a classic role in the regulation of numerous signaling pathways and are considered potent therapeutic targets. Recent findings that phosphatases are just as important in the same pathways involving kinases have brought phosphatases to the forefront of biomedical research.

The type 2C family of protein phosphatases (PP2C) is one of the eight major protein serine/threonine phosphatase families that also include PP1, PP2A, PP2B, PP4, PP5, PP6, and PP7 in eukaryotes. PP2C is a structurally and functionally distinct family of phosphatases which currently contains at least 19 different members.[1,2] In contrast to other serine/threonine-specific phosphatases, PP2Cs function as monomers with N- and C-terminal extensions that localize enzymes to specific domains and substrates. In addition, PP2C phosphatases depend on divalent cations for their activity and are insensitive to the broad-spectrum phosphatase inhibitor okadaic acid.[3]

In this review, we discuss the role of a newly emerging member of the PP2C family of phosphatases, Wip1, in regulating several signaling networks in the context of different conditions. Wip1 phosphatase has been implicated in oncogenic transformation at multiple levels; however, newly emerging functions of Wip1 in a wider range of conditions support its broad role as a regulator of homeostasis, especially related to the aging organism.[4–6] Importantly, depending on the condition, cell type, and potentially the age of the organism, Wip1 phosphatase regulates different molecular networks, creating a spatial and temporal way for controlling protein phosphorylation.

The protein phosphatase Wip1 was identified using a differential display methodology as a gene with expression induced in response to $\gamma$ or UV radiation in a p53-dependent manner.[3] Wip1 mRNA was found to be rapidly induced by ionizing radiation in p53-positive but not in p53-negative cells, confirming it as a direct p53 target, and giving it its name wild-type p53-induced phosphatase 1 (Wip1).[3] The Wip1 sequence shows significant homology to PP2C phosphatases, and *in vitro* recombinant Wip1 displays characteristics of the PP2Cs, one of which is relative insensitivity to okadaic acid.[3] Wip1 is also known as protein phosphatase magnesium-dependent 1 delta (Ppm1d) because it requires magnesium for its phosphatase activity.

The gene encoding Wip1, namely, *PPM1D*, is located on human chromosome 17q23 (mouse chromosome 11), and its mRNA expression was found to be ubiquitous in many embryonic and adult mouse tissues, with the highest level of expression in the testis.[7] The murine *Ppm1d* gene is spread over six exons spanning over 36 Kb of DNA. The gene encodes a protein of 598 amino acids, while the human version encodes a protein of 605 amino acids with a molecular mass of approximately 66 kDa.[7] Recently, an alternative splice form was described that contains the first 420 amino acids of the full-length Wip1 protein and an additional 10 amino acids.[8] In the same study, mRNA analysis of human tissues revealed that the shorter form is expressed exclusively in testes and leukocytes. Wip1 protein is shown to be localized mostly to the nucleus. Wip1 is induced in response to many environmental factors; however, in recent years, it has become clear that induction of Wip1 expression does not always depend on p53 as was originally suggested.[3] For example, accumulation of Wip1 after UV radiation relies on p38MAPK-p53 signaling pathway, whereas, in mouse intestinal stem cells, the p53-independent mechanism of Wip1 accumulation is currently unknown.[9]

Molecular studies of Wip1 have focused mainly on its interactions with DNA damage-response molecules. It has become increasingly clear that a primary role of Wip1 is to shut down several signaling pathways within the DNA damage-response network in which various molecules have been shown to be direct targets of Wip1 *in vitro* (Fig. 1). In addition to inhibiting p38MAPK after UV radiation,[9] Wip1 has been shown to dephosphorylate numerous molecules within the ataxia telangiectasia mutated kinase (ATM)-dependent

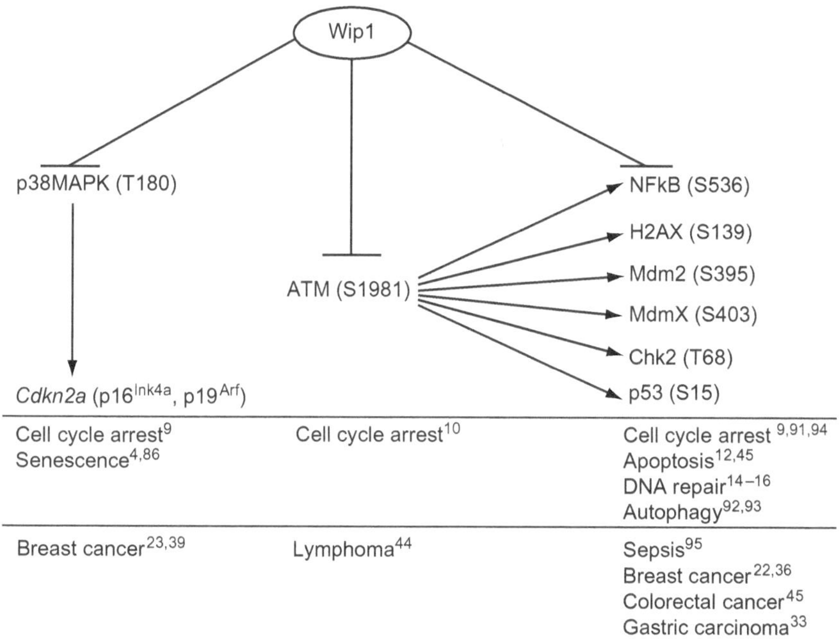


FIG. 1. Wip1 substrates that function to control tumorigenesis. Two major signaling pathways have been found to be regulated by Wip1 phosphatase, p38MAPK and ATM, both *in vitro* and *in vivo*. The large group of ATM substrates also has been identified to be direct targets of Wip1 *in vitro* (right side of the figure). Note that these proteins are also targets of ATM; thus Wip1 could affect their phosphorylation either directly or through ATM. Note also that ATM does not directly phosphorylate NF-kB but activates a cascade that ultimately results in increase in S536 phosphorylation. Phosphorylation sites that have been shown to be regulated by Wip1 are indicated for each protein. The role of these proteins at the cellular level and in specific types of cancer is highlighted.

signaling network. Wip1 dephosphorylates the critical autophosphorylation sites on ATM[10,11] and cell cycle Checkpoint Kinase 2 (Chk2), downstream of ATM.[12] Wip1 has also been reported to play a role in promoting stabilization of the oncogenic proteins murine double minute (MDM2) and MdmX,[13,14] both of which are ATM substrates. Finally, several groups have reported the requirement for Wip1 phosphatase in controlling H2A histone family, member X (H2AX) phosphorylation in some but not all[4] cell types after DNA damage.[15–17] While many of these targets downstream of ATM are regulated by Wip1 *in vitro*, it remains highly debatable and controversial whether this regulation results from a direct interaction with these proteins or through an ATM kinase. To understand the role of Wip1 *in vivo*, a knockout (KO) mouse strain has been generated by replacing exons 4 and 5 with a neomycin cassette. These mice have been used to answer questions about pathway specificity *in vivo*.[18] The mouse genetic data, as highlighted below, indicate that in many instances the effect of Wip1 deletion under normal conditions and in diseases occurs either through regulation of a key DNA damage-response mediator ATM, or a stress-response kinase p38MAPK.

## II. Wip1 Phosphatase in Tumorigenesis

An elegant and intricate network of protective mechanisms has evolved to prevent malignant transformation and tumor development. Perturbation of these mechanisms, in turn, is important in cancer development, considering the number of "hits," including myriad mutagenic and epigenetic events required for malignant transformation.[19] The DNA damage checkpoints in this respect serve to preserve the genomic fidelity of proliferating cells, preventing the mutation accumulation required for cellular transformation.[20] Wip1 provides negative feedback to DNA damage checkpoints (Fig. 1); thus, theoretically, overexpression of Wip1 should promote oncogenesis by shortening the cell cycle, which otherwise would result in continuous proliferation without completing DNA repair. Alternatively, knocking out Wip1 should impose more stringent conditions for cell cycle re-entry after DNA damage, providing more time for DNA repair and thus potentially protecting against DNA damage-induced oncogenesis. Based on these initial postulations, functional studies of Wip1 have been almost exclusively focused on either the oncogenic effects of overexpression of Wip1 or the tumor-resistance phenotypes of knocking it out.

In 2002, two groups independently reported that Wip1 plays a role in breast cancer.[21,22] Wip1 is localized within the epicenter of the region of chromosome 17q23 which is amplified in some breast cancers.[21] Overexpression of Wip1 abrogates apoptosis induced by serum starvation and cooperates with several oncogenes in the transformation of primary mouse embryo fibroblasts, further confirming its role as a potential oncogene. *In vivo*, however,

Wip1 overexpression either in the mammary gland epithelium or ubiquitously is not sufficient to induce cancer because these transgenic lines are not prone to increased spontaneous tumorigenesis.[23] On the other hand, Wip1 overexpression promotes mammary gland tumorigenesis in cooperation with other oncogenes such as ErbB2.[23] As such, it is possible that Wip1 provides a selective advantage for expanding a clonal variant of cancer cells without being a cancer-initiating oncogene on its own.

The gene that encodes Wip1, namely, *PPM1D*, was found to be amplified in numerous primary human tumor types including breast cancer,[21,24–26] neuroblastoma,[27] medulloblastoma,[28–30] ovarian clear cell tumor,[31,32] gastric carcinoma,[33] pancreatic adenocarcinoma,[34] and chronic lymphocytic leukemia.[35] These reports further support the potential role of Wip1 in the regulation of tumorigenesis. Interestingly, the p53 tumor suppressor, which is commonly mutated in human cancers, is almost exclusively wild type in breast tumors with Wip1 overexpression.[22,36,37] One explanation is that Wip1 promotes tumor formation through its ability to inhibit p53 tumor suppressor functions directly or indirectly, thus reducing selective pressure for p53 mutations during the progression of breast cancer.

Wip1 amplification may promote tumor formation; therefore, inhibiting Wip1 in cancer could be an attractive strategy for therapeutic intervention. We have found that Wip1-deficient mice are resistant to mammary gland tumorigenesis driven by multiple types of oncogenes, acting through a p38MAPK-dependent upregulation of the inhibitor of cyclin-dependent kinase 4A/alternative reading frame ($p16^{Ink4a}/p19^{Arf}$) proteins.[38,39] Follow-up studies on that work showed that the tumor-promoting potential of Wip1 overexpression in mice was eliminated by Mkk6–p38MAPK activation in the mammary gland,[23] while overexpression of Wip1 correlated with inactivation of p38MAPK and silencing of $p16^{Ink4a}$ expression in primary human breast tumors.[36] These results further highlight the significance of the Wip1-p38 phosphorylation network in regulating mammary gland tumorigenesis in mice in the presence of at least certain oncogenes such as ErbB2. In turn, p38MAPK is emerging as an important negative regulator of tumorigenesis in several types of cancer.[38,40,41] Loss of p38 in the lung, for example, using a conditional KO of p38α results in a dramatic increase in lung tumors driven by oncogenic k-Ras.[42] Similar results have been obtained with carcinogen-induced liver cancer, indicating that p38 signaling has an important role(s) in protecting from deregulated growth.[43]

The value of inhibiting Wip1 in preventing cancer is not only limited to mammary gland tumors because this effect has also been demonstrated in lymphomagenesis,[44] colorectal cancer models,[45] and spontaneous tumors.[40] Further, inactivation of Wip1 does not always rely on p38MAPK pathways to suppress tumorigenesis. Wip1 deletion enhanced ATM/p53-mediated apoptosis and thus delayed onset of tumor formation in a model of Myc-driven lymphomagenesis.[44]

Likewise, Wip1 deficiency inhibited polyp formation and increased mouse survival by increasing apoptosis of cancer-initiating intestinal stem cells in an adenomatous polyposis coli-driven colorectal cancer model.[45] Based on this evidence, we proposed that the Wip1 deletion in intestinal stem cells lowers the threshold for committing cells to apoptosis.[45] In terms of molecular pathways, this process appears to depend on Chk2-mediated phosphorylation and activation of the tumor suppressor p53. In contrast to mammary gland tumorigenesis, which depends on p38MAPK, Wip1 uses a different molecular network to suppress cancer initiation in mouse intestinal epithelium. These differences have yet to be explored but could arise from various expression levels of downstream targets of Wip1 such as p38MAPK and Atm/Chk2. The cell type specificity and overall role of different signal transduction pathways in decision making upon activation of a certain oncogene could further play a role in discriminating substrates downstream of Wip1. For example, ionizing-radiation-induced apoptosis of intestinal stem cells is fully dependent on the presence of Chk2 but not on p38MAPK, which is consistent with the finding that Wip1 uses the Chk2-dependent signaling pathway to activate apoptosis of intestinal stem cells in response to oncogenes.[12,46]

The ATM-dependent regulatory network and its proper coordination through a series of phosphorylation and dephosphorylation events are important for controlling tumorigenesis as well as for directing an appropriate DNA damage response. Several groups have highlighted the significance of the ATM– p53–Wip1 loop in the generation of p53 pulses and subsequent activation of a cell cycle arrest or apoptosis depending on the strength of the signal and potentially on cell type.[11,47] DNA damage results in phosphorylation and activation of ATM, which, in turn, activates p53 followed by an increase in Wip1 levels. Wip1 then inhibits ATM activity as part of a negative feedback loop, and after a delay the p53 level falls while the Wip1 level decreases at a slower rate. If residual damage is present, ATM will be reactivated and p53 is subsequently activated again. In this model, p53 activity must reach a certain threshold level during the pulse to activate apoptosis and this could only be achieved if DNA damage is either present for longer than one pulse or if a negative regulator, Wip1, is deleted. In the latter case, Wip1-deficient cells become ultrasensitive to irradiation, which is consistent with the role of Wip1 as an oncogenic regulator of stress-induced apoptosis.[6,45] In contrast, in Wip1-proficient cells, p53 requires more time to reach a level sufficient to induce apoptosis. Further, when Wip1 is overexpressed, apoptosis induction is significantly postponed, and the cell becomes resistant to irradiation.[48] This result may partially explain how Wip1 acts as an oncogene and why it is frequently overexpressed in several cancers.

Some groups have shown recently that oncogene-induced activation of the ATM/ATR-dependent signaling network occurs early in the course of tumorigenesis and protects against cancer progression.[49–53] Consistent with these data, preneoplastic lesions from a variety of human cancers have been shown to express markers of an activated DNA damage response. Interestingly, late-stage tumors often show loss of these DNA damage-response markers, suggesting that disabling the DNA damage pathways is an important prerequisite for cancer progression. The Wip1–ATM-dependent regulation of tumorigenesis has been further highlighted in the delayed onset of B-cell lymphoma in Wip1-deficient mice overexpressing the c-myc oncogene.[44] Wip1-deficient mice carrying the Eμ-Myc transgene have a higher level of activated Atm in splenocytes and show a delay in the onset of lymphomas that depends on Atm and p53, but not on Arf. This finding contrasts with the tumor-resistant phenotype of Mdm2 heterozygous mice in a similar cancer model, which appeared to depend on Arf.[54] This work provides further evidence that Mdm2 and Wip1 engage different molecular pathways to modulate p53 responses in the Eμ-Myc mouse model of cancer, further highlighting the importance of Wip1 in modulating ATM activity in the course of tumorigenesis.

## III. Wip1 Phosphatase in Adult Neurogenesis

During the past two decades, multipotent primary neural stem cells/progenitors (NPCs) have been identified in the adult mammalian brain.[55] These adult NPCs continuously proliferate and differentiate into neuroblasts and functionally integrate into existing neural circuits as mature neurons. The process of evolving from NPCs into mature neurons is collectively termed "adult neurogenesis" and represents an important form of long-term brain plasticity and regeneration.[56,57] It also provides a promising resource for autonomous cell replacement therapy in neurodegenerative diseases.[58–60] In the mouse under physiological conditions, the proper activity of these NPCs is important for odor discrimination,[61] learning, and memory.[56,62] In neurodegenerative diseases and injury, NPCs can further respond to the environment by adjusting their proliferation and migration, which makes them a potential target for therapeutic interventions.[63,64]

In spite of the physiological importance and potential therapeutic value of NPCs, the molecular regulatory networks within NPCs remain poorly understood. In recent years, some niche factors, including growth factors, extracellular matrix molecules, and hormones, have been recognized as important regulators of the NPC pool size.[56] Comparatively, intrinsic pathways that regulate the NPC pool size are less well studied.

Wip1 phosphatase has been recently identified as a critical regulator of adult neurogenesis.[5] In the forebrain, Wip1 deficiency results in a 90% decrease in new cell formation in the adult olfactory bulb (OB), accompanied by an aberrant reduction in NPC amplification, stem cell frequency, and self-renewal. Primary stem cells in the subventricular zone (SVZ) are relatively quiescent, with an estimated cell cycle of about 15 days.[65] These stem cells produce transient amplifying progenitors which proliferate and further differentiate into neuroblasts.[56] The number of all cell types is reduced in the Wip1-deficient SVZ, suggesting that Wip1 possibly regulates primary stem cells, in turn, affecting the transient amplifying cells and neurons. At the cellular level, Wip1-deficient NPCs exhibit a prolonged cell cycle, an accumulation of cells in $G_2$ phase, and elevated expression of cell cycle inhibitors including p21 and Reprimo.[5]

Multiple cell cycle proteins, including the DNA damage-response molecules p38MAPK, ATM, and p53, have been implicated in adult neurogenesis and shown to be regulated by Wip1 phosphatase directly or indirectly (Figs. 1 and 2). In addition, multiple cell cycle regulators downstream of these Wip1 substrates, including p21, which is downstream of p53, have been reported to function in adult neurogenesis. Consistent with these data, p53

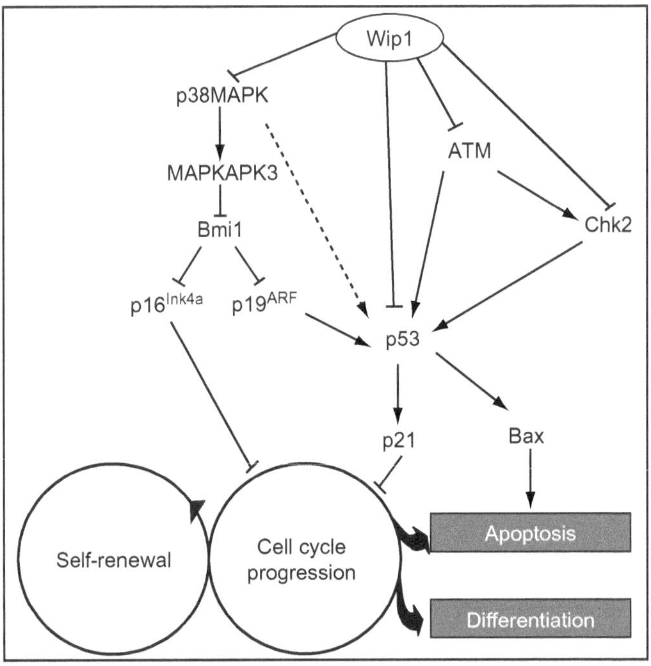


FIG. 2. Wip1-dependent regulation of neural stem cell/progenitor proliferation. Wip1 is implicated in the regulation of proliferation of neural stem cells and progenitors through several signaling pathways as summarized in the figure. Dashed line between p38 and p53 represents a connection that has been established in other cell lines but has not been shown yet for neural stem cells or progenitors. For more details, see text.

phosphorylation is upregulated in Wip1-deficient NPCs both *in vivo* and *in vitro*, resulting in elevated p53-dependent transcription of p21 and Reprimo in Wip1-deficient NPCs.[5]

The p53 protein plays an important functional role in NPCs. Early studies showed that p53 is expressed in the SVZ area of the forebrain,[66,67] while knocking out p53 promotes the self-renewal of NPCs *in vitro*.[66] Another group used electron microscopy to characterize changes in the cellular composition of the ventricular zone and the SVZ in p53-deficient mice.[67] They found that SVZ consists of three different cell types,[56] each with unique structural characteristics,[67] and that p53-deficient mice showed an increase in all of them. Consistent with the structural analysis, a functional study showed that the p53 deletion increased cell proliferation, as demonstrated by the number of both short- and long-term BrdU-labeled cells in the SVZ. Accordingly, neuroblast formation was also increased with p53 deletion both *in vivo* and *in vitro*.[67]

Further support for the role of p53 in regulating adult neurogenesis emerged with the generation of a mouse model with elevated p53 expression.[68] These mice exhibited a phenotype largely opposite to that of p53-deficient mice. More specifically, increased expression of p53 resulted in reduced proliferation of stem cells and progenitors, as shown by BrdU label-retaining experiments in 1-year-old mice. Extra copies of p53 also induced expression of the CDK inhibitor p21, and the NPCs exhibited a longer $G_0$ phase that correlated with a reduced OB size and reduced olfactory discrimination.[68]

The involvement of p53 in Wip1-dependent regulation of NPCs was tested in Wip1 KO mice crossed onto a p53-deficient background. The Wip1/p53 double KO mice fully reversed the impairment in cell cycle progression observed in the Wip1-deficient NPCs[5] and showed behavior comparable to the p53 KO mice. This functional rescue observed in the Wip1/p53 double KO mice was further confirmed *in vivo*, with the observation of an increased number of NPCs and neuroblasts.[5] Therefore, Wip1 regulated the generation of new neural cells in the adult OB, specifically through p53-dependent regulation of cell cycle progression in NPCs.

The significance of proper regulation of stem cell proliferation and self-renewal is emerging in several pathological conditions such as cancer.[69,70] Several recent reports have shown that conditional deletions of tumor suppressors, including p53, in adult NPCs lead to the formation of brain tumors, in particular, glioblastoma and medulloblastoma.[71–74] For instance, conditional deletion of p53 in mice heterozygous for a phosphatase and tensin homolog (another important factor in suppressing NPC self-renewal and proliferation by repressing the $G_0$–$G_1$ transition[75]) triggered high-grade glioblastoma formation.[71] Inducible deletion of p53 specifically in NPCs induced glioblastoma with a 100% penetrance.[73] These studies suggest that p53 is an important tumor suppressor that inhibits the conversion of NPCs to glioblastoma cancer stem cells.

Eliminating p53 enhances NPC transformation into tumor-initiating cells, while elevation of p53 activity upon Wip1 deletion diminishes the regeneration capacity of NPCs.[67] Thus, the Wip1–p53 pathway plays a critical role in the balance between maintaining NPC plasticity and potentially minimizing the risk of neoplastic transformation.

## IV. NPC Aging and Wip1 Phosphatase

NPCs are susceptible to malfunctions. The regenerative capacity of NPCs is progressively restricted by aging, while deregulation of the NPC pool size by genomic mutations that alter proliferation and self-renewal can lead to brain tumors.[71,73,76,77] Both NPC aging and transformation are intrinsically linked to physiological proliferation and self-renewal. In fact, both age-related tissue degeneration and cancer have been proposed to arise as a result of deregulation of stem cells.[69,70,78] Study of the mechanisms underlying proliferation and self-renewal conducted under physiological conditions will ultimately provide insights into the pathways involved in NPC aging and how deregulation of these mechanisms could lead to transformation of NPCs and brain tumors.[69] A question remains, however, as to how to control the regenerative capacity of stem cells to use them for tissue regeneration and repair. The fact that neurodegenerative diseases predominate in older individuals is supported by the limited regenerative capacity of aging tissue stem cells,[69] further emphasizing the need to understand how NPCs age. In fact, the regenerative capacity of NPCs starts to decline during early development and does so more profoundly with advanced age.[79–81] *In vitro* studies with multilineage neurosphere formation assays have indicated that stem cell frequency decreases gradually with aging.[79] Accordingly, there is also an aging-induced decrease in the number of long-term BrdU-retaining stem cells in the SVZ and a dramatic reduction in new neuron formation in the OB. Different studies have partially characterized the underlying mechanisms of NPC aging; however, the full picture remains unclear.

Several molecules have been implicated recently in the regulation of a decline in proliferation of NPCs with aging. The expression of a cyclin-dependent kinase inhibitor, namely, $p16^{Ink4a}$, in the adult SVZ region has been shown to increase gradually from 2 months to 2 years of age.[79] In turn, deletion of $p16^{Ink4a}$ partially rescued a decline in proliferation and self-renewal of NPCs in 2-year-old mice, which was confirmed both *in vitro* and *in vivo*. Further, that work showed that deletion of $p16^{Ink4a}$ also partially rescued the decline in new neuron formation in the aged OB with an increase from 22% to 38% compared to that of young mice.[79] The same group suggested a mechanism upstream of $p16^{Ink4a}$ that regulates its expression during development and aging of NPCs.[79] The study showed that the microRNA let-7b represses the high-mobility group AT-hook

2 (Hmga2), which suppresses p16$^{Ink4a}$ expression in NPCs. Throughout development and aging, let-7b is gradually upregulated by an unknown mechanism. The increase of let-7b expression incrementally inhibits Hmga2 at the RNA level, which, in turn, releases the Hmga2-dependent repression of p16$^{Ink4a}$ transcription. While the increase in let-7b expression upregulates p16$^{Ink4a}$ expression during NPC aging, deficiency of p16$^{Ink4a}$ only partially rescues the aging-induced phenotypes, suggesting that additional mechanisms are involved in mediating the age-related decline in proliferation and self-renewal of NPCs in SVZ and new neuron formation in the OB.

Similar to p16$^{Ink4a}$, another gene in the same Cdkn2a genomic locus, namely, p19$^{Arf}$, has also been shown to be upregulated during aging.[79,82,83] In fact, deletion of Hmga2 in embryonic NPCs also elevates p19$^{Arf}$ expression, a potent regulator of the cell cycle inhibitor p53. Studies have shown that the p19$^{Arf}$–p53 pathway functions during embryonic development but most likely not in advanced aging.[80] During the embryo-to-adult transition, p53 in cooperation with myc protein coordinates self-renewal and differentiation. In turn, expression of myc is increased in embryonic NPCs, mediating their proliferation and ultimately neurogenesis. In contrast, p19$^{Arf}$ becomes abundant upon induction with growth factors and gliogenic stimuli later in life and, through induction of the p53-dependent pathway, restricts NPC proliferation and neurogenesis while promoting gliogenesis.

To date, only a limited number of reports have highlighted the role of the p19$^{Arf}$–p53 pathway in advanced aging. One group has suggested that p53 might play a role in NPC aging, showing that transgenic overexpression of a short isoform of p53 led to an aging-like phenotype in NPCs.[68] Interestingly, this phenotype was more apparent when mice were getting older, suggesting that simply overexpressing p53 is not sufficient to mimic the effects of aging on NPCs but rather that an additional signaling pathway leading to activation of p53 must be engaged.[68] Consistent with these data, we also found that the level of p53 phosphorylation at the activatory sites, serine 15 and 20, is increased in the SVZ in old mice (unpublished data). The mechanism of this increase in p53 phosphorylation remains unclear.

As the level of Wip1 drops with aging at least in NPCs and pancreatic islets,[4] this decrease may trigger subsequent activation of ATM and phosphorylation of p53 at serines 15 and 20 directly or through Chk2. This observation would be consistent with an aging-induced increase in p53 phosphorylation at specific sites and the fact that Wip1-deficient NPCs behave as though their aging program has been accelerated both *in vitro* and *in vivo*.[5] This behavior includes reduction in proliferation, stem cell frequency, self-renewal of NPCs, and new neuron formation in the OB.[79] A recent study showed that several age-related phenotypes could depend on Wip1 in different models of aging. An interesting mechanism of Wip1 regulation has been proposed through the analysis of Zmptse24-deficient mice, a mouse model of Hutchinson–Gilford progeria. This study showed that deletion of Zmptse24 results in upregulation of the microRNA miR-29 which targets Wip1 for degradation and results in subsequent activation of p53.[84] The miR-29 upregulation, in turn, appeared to result from a very specific genotoxic stress caused by nuclear envelope dysfunction but was not initiated in response to intrinsic genotoxic stress driven by repair machinery impairment.[84] Altogether, accumulating evidence suggests that Wip1 in fact may play a role in NPC aging (Fig. 3). The question is: if the Wip1 decrease does play a role in aging-induced changes in the brain, which pathway downstream of Wip1 is engaged?

One of the Wip1 substrates and a potent regulator of p53 and Cdkn2a, namely, p38MAPK, has been recently implicated in the induction of the permanent cell cycle arrest (senescence), which is commonly seen in aged organisms, as well as in a decline in proliferation of certain self-renewing cell types, including β-cells of pancreatic islets, with aging.[4,85] Wip1 functions as a negative feedback regulator that dephosphorylates and inactivates p38MAPK, and studies in various cell types and conditions have reported that Wip1 deletion results in an elevation of p38MAPK phosphorylation and activity.[39,86,87] Further, our group

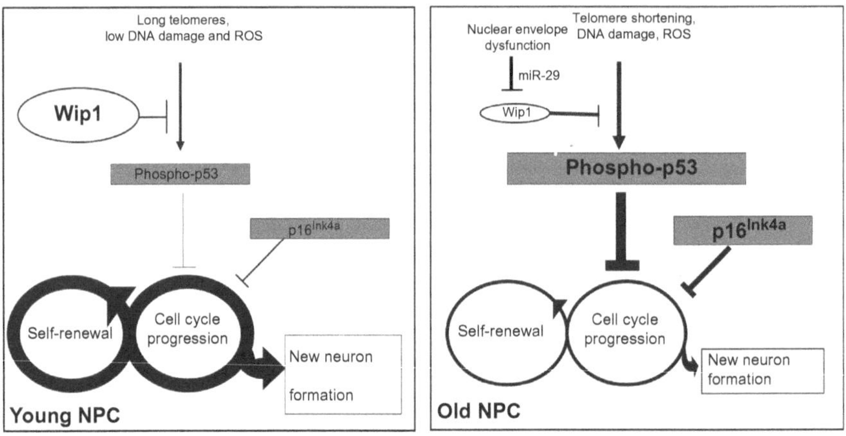


Fig. 3. A role for Wip1 in NPC homeostasis. p53 acts as a potent suppressor of NPC proliferation and consequently represses their self-renewal as well as neuronal differentiation (formation of new neurons). The activity of p53 is cumulatively determined by upstream factors including DNA damage, telomere shortening, the level of reactive oxygen species (ROS), and the nuclear envelop integrity. Many of these factors control p53 through signaling pathways regulated by Wip1 as shown in Fig. 2. In young animals, Wip1 is abundant and efficiently represses signals that come from p53 activating factors. In contrast, in old animals, expression of Wip1 is reduced, while the intensity of p53 activating factors (as shown in the figure) rises sharply. This results in the activation of p53, restricting NPC proliferation and self-renewal and suppressing neuronal differentiation. The arrow thickness defines the intensity of the signal.

recently found that p38MAPK activity and expression of p16$^{Ink4a}$ are upregulated in multiple tissues, including pancreatic islets, during aging. In turn, Wip1 deletion promoted whereas Wip1 overexpression diminished p38MAPK activity and p16$^{Ink4a}$ expression with aging. Functionally, transgenic expression of Wip1 increased β-cell proliferation and enhanced resistance to drug-induced diabetes in aged mice.[87] This study highlights the role of Wip1 in regulating the aging process in the pancreas through the regulation of the p38MAPK–p16$^{Ink4a}$ pathway. Similarly, the Wip1–p38MAPK–p16$^{Ink4a}$ pathway has been implicated in human cells in an *in vitro* aging system. Proliferative senescence of human mesenchymal stem cells during *in vitro* passaging correlated with an increase in p38MAPK activity and p16$^{Ink4a}$ expression. Inhibition of p38MAPK by overexpressing Wip1 abolished the increase of p16$^{Ink4a}$ and rescued the proliferative capacity of mesenchymal stem cells.[86]

p38MAPK, a potent regulator of aging-induced gene expression from the *Cdkn2a* locus in different tissues,[87] has been implicated in positive regulation of p53[88] and in regulation of the cell cycle inhibitor p27 *in vitro* in NPCs.[89] Therefore, the level of phospho-p38MAPK and expression of p27, p16$^{Ink4a}$/p19$^{Arf}$, and p53 have been examined in Wip1-deficient NPCs to delineate the pathway involved in the regulation of NPC proliferation downstream of Wip1. Surprisingly, with the exception of p53, none of the analyzed molecules was elevated in Wip1-deficient when compared to wild- type NPCs (Ref. 5 and unpublished data). Further, the expression of a dominate-negative double-mutant p38MAPK also failed to normalize the effect of Wip1 deletion on NPC proliferation and self-renewal (Ref. 5 and unpublished data). These results clearly suggest that the signaling pathway(s) regulated by Wip1 phosphatase in NPCs with aging could differ from the one(s) operational in other aging tissues (such as pancreatic islets) and emphasize that p38MAPK is not the major Wip1 downstream target in regulating aging-induced induction of p53 in the brain. Therefore, a more detailed knowledge of how Wip1 modulates NPCs will provide further insights into the physiological process of NPC regulation and may offer a unique opportunity for developing interventions against brain tumors and potentially against aging-related decline in brain functions.

## V. Conclusion

Wip1 phosphatase is emerging as a potent regulator of tumorigenesis and recently has been implicated in controlling neurogenesis and certain aspects of organismal aging. Like many, if not all, phosphatases, Wip1 does not seem to have a consensus dephosphorylation motif and, as such, a wide range of substrates have been identified (Fig. 1). In turn, mouse genetic studies continue to provide an important tool for understanding molecular pathways downstream of Wip1 that are engaged in regulating different physiological conditions. This engagement of downstream pathways depends on several factors, including the level of expression of a full-length and potentially a splice variant of Wip1,[8] the ability to be localized to a particular cellular compartment, and the abundance of downstream targets. In some instances, the abundance and overall biological significance of Wip1's targets define the signaling pathway downstream of Wip1. The cell type/tissue-specific expression of Atm/Chk2 and p38MAPK could also partially explain why Wip1 controls one signaling pathway to regulate NPCs while engaging another in other cell types such as pancreatic β cells. p38 MAPK has been implicated in the regulation of cell differentiation in several tissues and could be more abundant and thus physiologically relevant in differentiated cells such as pancreatic β cells.[4,90] In contrast, p53 has an important role in adult stem cells, and in intestinal stem cells, a Chk2-dependent activation of p53 is the predominant program to activate apoptosis after different forms of stress.[46]

Deletion or an aging-induced decline in expression of Wip1 could engage different downstream pathways potentially in a cell-type-specific manner, further emphasizing the molecular diversity of aging mechanisms. In turn, overexpression or activation of Wip1, as commonly seen in primary human tumors, could have a more general effect on multiple signaling pathways (Fig. 1), contributing to deregulation of several molecular networks. Thus, finding ways to control Wip1 activity or levels could represent an interesting therapeutic approach for the treatment of cancer and of certain aging-induced pathological conditions.

## References

1. Lammers T, Lavi S. Role of type 2C protein phosphatases in growth regulation and in cellular stress signaling. *Crit Rev Biochem Mol Biol* 2007;**42**:437–61.
2. Sugiura T, Noguchi Y, Sakurai K, Hattori C. Protein phosphatase 1H, overexpressed in colon adenocarcinoma, is associated with CSE1L. *Cancer Biol Ther* 2008;**7**:285–92.
3. Fiscella M, Zhang H, Fan S, Sakaguchi K, Shen S, Mercer ME, et al. Wip1, a novel human protein phosphatase that is induced in response to ionizing radiation in a p53-dependent manner. *Proc Natl Acad Sci USA* 1997;**94**:6048–53.
4. Wong ES, Le Guezennec X, Demidov ON, Marshall NT, Wang ST, Krishnamurthy J, et al. p38MAPK controls expression of multiple cell cycle inhibitors and islet proliferation with advancing age. *Dev Cell* 2009;**17**:142–9.
5. Zhu YH, Zhang CW, Lu L, Demidov ON, Sun L, Yang L, et al. Wip1 regulates the generation of new neural cells in the adult olfactory bulb through p53-dependent cell cycle control. *Stem Cells* 2009;**27**:1433–42.
6. Le Guezennec X, Bulavin DV. WIP1 phosphatase at the crossroads of cancer and aging. *Trends Biochem Sci* 2010;**35**:109–14.
7. Choi J, Appella E, Donehower LA. The structure and expression of the murine wildtype p53-induced phosphatase 1 (Wip1) gene. *Genomics* 2000;**64**:298–306.

8. Chuman Y, Kurihashi W, Mizukami Y, Nashimoto T, Yagi H, Sakaguchi K. PPM1D430, a novel alternative splicing variant of the human PPM1D, can dephosphorylate p53 and exhibits specific tissue expression. *J Biochem* 2009;**145**:1–12.
9. Takekawa M, Adachi M, Nakahata A, Nakayama I, Itoh F, Tsukuda H, et al. p53-inducible wip1 phosphatase mediates a negative feedback regulation of p38 MAPK-p53 signaling in response to UV radiation. *EMBO J* 2000;**19**:6517–26.
10. Shreeram S, Demidov ON, Hee WK, Yamaguchi H, Onishi N, Kek C, et al. Wip1 phosphatase modulates ATM-dependent signaling pathways. *Mol Cell* 2006;**23**:757–64.
11. Batchelor E, Mock CS, Bhan I, Loewer A, Lahav G. Recurrent initiation: a mechanism for triggering p53 pulses in response to DNA damage. *Mol Cell* 2008;**30**:277–89.
12. Fujimoto H, Onishi N, Kato N, Takekawa M, Xu XZ, Kosugi A, et al. Regulation of the antioncogenic Chk2 kinase by the oncogenic Wip1 phosphatase. *Cell Death Differ* 2006;**13**:1170–80.
13. Lu X, Ma O, Nguyen TA, Jones SN, Oren M, Donehower LA. The Wip1 phosphatase acts as a gatekeeper in the p53-Mdm2 autoregulatory loop. *Cancer Cell* 2007;**12**:342–54.
14. Zhang X, Lin L, Guo H, Yang J, Jones SN, Jochemsen A, et al. Phosphorylation and degradation of MdmX is inhibited by Wip1 phosphatase in the DNA damage response. *Cancer Res* 2009;**69**:7960–8.
15. Cha H, Lowe JM, Li H, Lee JS, Belova GI, Bulavin DV, et al. Wip1 directly dephosphorylates gamma-H2AX and attenuates the DNA damage response. *Cancer Res* 2010;**70**:4112–22.
16. Macurek L, Lindqvist A, Voets O, Kool J, Vos HR, Medema RH. Wip1 phosphatase is associated with chromatin and dephosphorylates gammaH2AX to promote checkpoint inhibition. *Oncogene* 2010;**29**:2281–91.
17. Moon SH, Nguyen TA, Darlington Y, Lu X, Donehower LA. Dephosphorylation of gamma-H2AX by WIP1: an important homeostatic regulatory event in DNA repair and cell cycle control. *Cell Cycle* 2010;**9**:2092–6.
18. Choi J, Nannenga B, Demidov ON, Bulavin DV, Cooney A, Brayton C, et al. Mice deficient for the wild-type p53-induced phosphatase gene (Wip1) exhibit defects in reproductive organs, immune function, and cell cycle control. *Mol Cell Biol* 2002;**22**:1094–105.
19. Hanahan D, Weinberg RA. Hallmarks of cancer: the next generation. *Cell* 2011;**144**:646–74.
20. Sancar A, Lindsey-Boltz LA, Unsal-Kacmaz K, Linn S. Molecular mechanisms of mammalian DNA repair and the DNA damage checkpoints. *Annu Rev Biochem* 2004;**73**:39–85.
21. Li J, Yang Y, Peng Y, Austin RJ, van Eyndhoven WG, Nguyen KC, et al. Oncogenic properties of PPM1D located within a breast cancer amplification epicenter at 17q23. *Nat Genet* 2002;**31**:133–4.
22. Bulavin DV, Demidov ON, Saito S, Kauraniemi P, Phillips C, Amundson SA, et al. Amplification of PPM1D in human tumors abrogates p53 tumor-suppressor activity. *Nat Genet* 2002;**31**:210–5.
23. Demidov ON, Kek C, Shreeram S, Timofeev O, Fornace AJ, Appella E, et al. The role of the MKK6/p38 MAPK pathway in Wip1-dependent regulation of ErbB2-driven mammary gland tumorigenesis. *Oncogene* 2007;**26**:2502–6.
24. Sinclair CS, Rowley M, Naderi A, Couch FJ. The 17q23 amplicon and breast cancer. *Breast Cancer Res Treat* 2003;**78**:313–22.
25. Barlund M, Kuukasjarvi T, Syrjakoski K, Auvinen A, Kallioniemi A. Frequent amplification and overexpression of CCND1 in male breast cancer. *Int J Cancer* 2004;**111**:968–71.
26. Natrajan R, Lambros MB, Rodriguez-Pinilla SM, Moreno-Bueno G, Tan DS, Marchio C, et al. Tiling path genomic profiling of grade 3 invasive ductal breast cancers. *Clin Cancer Res* 2009;**15**:2711–22.
27. Saito-Ohara F, Imoto I, Inoue J, Hosoi H, Nakagawara A, Sugimoto T, et al. PPM1D is a potential target for 17q gain in neuroblastoma. *Cancer Res* 2003;**63**:1876–83.

28. Mendrzyk F, Radlwimmer B, Joos S, Kokocinski F, Benner A, Stange DE, et al. Genomic and protein expression profiling identifies CDK6 as novel independent prognostic marker in medulloblastoma. *J Clin Oncol* 2005;**23**:8853–62.
29. Castellino RC, De Bortoli M, Lu X, Moon SH, Nguyen TA, Shepard MA, et al. Medulloblastomas overexpress the p53-inactivating oncogene WIP1/PPM1D. *J Neurooncol* 2008;**86**:245–56.
30. Ehrbrecht A, Muller U, Wolter M, Hoischen A, Koch A, Radlwimmer B, et al. Comprehensive genomic analysis of desmoplastic medulloblastomas: identification of novel amplified genes and separate evaluation of the different histological components. *J Pathol* 2006;**208**:554–63.
31. Tan DS, Lambros MB, Rayter S, Natrajan R, Vatcheva R, Gao Q, et al. PPM1D is a potential therapeutic target in ovarian clear cell carcinomas. *Clin Cancer Res* 2009;**15**:2269–80.
32. Hirasawa A, Saito-Ohara F, Inoue J, Aoki D, Susumu N, Yokoyama T, et al. Association of 17q21-q24 gain in ovarian clear cell adenocarcinomas with poor prognosis and identification of PPM1D and APPBP2 as likely amplification targets. *Clin Cancer Res* 2003;**9**:1995–2004.
33. Fuku T, Semba S, Yutori H, Yokozaki H. Increased wild-type p53-induced phosphatase 1 (Wip1 or PPM1D) expression correlated with downregulation of checkpoint kinase 2 in human gastric carcinoma. *Pathol Int* 2007;**57**:566–71.
34. Loukopoulos P, Shibata T, Katoh H, Kokubu A, Sakamoto M, Yamazaki K, et al. Genome-wide array-based comparative genomic hybridization analysis of pancreatic adenocarcinoma: identification of genetic indicators that predict patient outcome. *Cancer Sci* 2007;**98**:392–400.
35. Lopez-Guerra M, Trigueros-Motos L, Molina-Arcas M, Villamor N, Casado FJ, Montserrat E, et al. Identification of TIGAR in the equilibrative nucleoside transporter 2-mediated response to fludarabine in chronic lymphocytic leukemia cells. *Haematologica* 2008;**93**:1843–51.
36. Yu E, Ahn YS, Jang SJ, Kim MJ, Yoon HS, Gong G, et al. Overexpression of the wip1 gene abrogates the p38 MAPK/p53/Wip1 pathway and silences p16 expression in human breast cancers. *Breast Cancer Res Treat* 2007;**101**:269–78.
37. Rauta J, Alarmo EL, Kauraniemi P, Karhu R, Kuukasjarvi T, Kallioniemi A. The serine-threonine protein phosphatase PPM1D is frequently activated through amplification in aggressive primary breast tumours. *Breast Cancer Res Treat* 2006;**95**:257–63.
38. Harrison M, Li J, Degenhardt Y, Hoey T, Powers S. Wip1-deficient mice are resistant to common cancer genes. *Trends Mol Med* 2004;**10**:359–61.
39. Bulavin DV, Phillips C, Nannenga B, Timofeev O, Donehower LA, Anderson CW, et al. Inactivation of the Wip1 phosphatase inhibits mammary tumorigenesis through p38 MAPK-mediated activation of the p16(Ink4a)-p19(Arf) pathway. *Nat Genet* 2004;**36**:343–50.
40. Nannenga B, Lu X, Dumble M, Van Maanen M, Nguyen TA, Sutton R, et al. Augmented cancer resistance and DNA damage response phenotypes in PPM1D null mice. *Mol Carcinog* 2006;**45**:594–604.
41. Bulavin DV, Fornace Jr. AJ. p38 MAP kinase's emerging role as a tumor suppressor. *Adv Cancer Res* 2004;**92**:95–118.
42. Ventura JJ, Tenbaum S, Perdiguero E, Huth M, Guerra C, Barbacid M, et al. p38alpha MAP kinase is essential in lung stem and progenitor cell proliferation and differentiation. *Nat Genet* 2007;**39**:750–8.
43. Hui L, Bakiri L, Mairhorfer A, Schweifer N, Haslinger C, Kenner L, et al. p38alpha suppresses normal and cancer cell proliferation by antagonizing the JNK-c-Jun pathway. *Nat Genet* 2007;**39**:741–9.
44. Shreeram S, Hee WK, Demidov ON, Kek C, Yamaguchi H, Fornace Jr. AJ, et al. Regulation of ATM/p53-dependent suppression of myc-induced lymphomas by Wip1 phosphatase. *J Exp Med* 2006;**203**:2793–9.
45. Demidov ON, Timofeev O, Lwin HN, Kek C, Appella E, Bulavin DV. Wip1 phosphatase regulates p53-dependent apoptosis of stem cells and tumorigenesis in the mouse intestine. *Cell Stem Cell* 2007;**1**:180–90.

46. Takai H, Naka K, Okada Y, Watanabe M, Harada N, Saito S, et al. Chk2-deficient mice exhibit radioresistance and defective p53-mediated transcription. *EMBO J* 2002;**21**:5195–205.
47. Zhang XP, Liu F, Wang W. Two-phase dynamics of p53 in the DNA damage response. *Proc Natl Acad Sci USA* 2011;**108**:8990–5.
48. Lu X, Nguyen TA, Donehower LA. Reversal of the ATM/ATR-mediated DNA damage response by the oncogenic phosphatase PPM1D. *Cell Cycle* 2005;**4**:1060–4.
49. Reddy JP, Peddibhotla S, Bu W, Zhao J, Haricharan S, Du YC, et al. Defining the ATM-mediated barrier to tumorigenesis in somatic mammary cells following ErbB2 activation. *Proc Natl Acad Sci USA* 2010;**107**:3728–33.
50. Bartkova J, Horejsi Z, Koed K, Kramer A, Tort F, Zieger K, et al. DNA damage response as a candidate anti-cancer barrier in early human tumorigenesis. *Nature* 2005;**434**:864–70.
51. Bartek J, Bartkova J, Lukas J. DNA damage signalling guards against activated oncogenes and tumour progression. *Oncogene* 2007;**26**:7773–9.
52. Oka K, Tanaka T, Enoki T, Yoshimura K, Ohshima M, Kubo M, et al. DNA damage signaling is activated during cancer progression in human colorectal carcinoma. *Cancer Biol Ther* 2010;**9**:246–52.
53. Cipriano R, Kan CE, Graham J, Danielpour D, Stampfer M, Jackson MW. TGF-{beta} signaling engages an ATM-CHK2-p53-independent RAS-induced senescence and prevents malignant transformation in human mammary epithelial cells. *Proc Natl Acad Sci USA* 2011;**108**:8668–73.
54. Wang P, Greiner TC, Lushnikova T, Eischen CM. Decreased Mdm2 expression inhibits tumor development induced by loss of ARF. *Oncogene* 2006;**25**:3708–18.
55. Ma DK, Bonaguidi MA, Ming GL, Song H. Adult neural stem cells in the mammalian central nervous system. *Cell Res* 2009;**19**:672–82.
56. Ming GL, Song H. Adult neurogenesis in the mammalian central nervous system. *Annu Rev Neurosci* 2005;**28**:223–50.
57. Taupin P. Adult neurogenesis in the mammalian central nervous system: functionality and potential clinical interest. *Med Sci Monit* 2005;**11**:RA247–52.
58. Arias-Carrion O, Yuan TF. Autologous neural stem cell transplantation: a new treatment option for Parkinson's disease? *Med Hypotheses* 2009;**73**(5):757–9.
59. Muraoka K, Shingo T, Yasuhara T, Kameda M, Yuan W, Hayase H, et al. The high integration and differentiation potential of autologous neural stem cell transplantation compared with allogeneic transplantation in adult rat hippocampus. *Exp Neurol* 2006;**199**:311–27.
60. Lie DC, Song H, Colamarino SA, Ming GL, Gage FH. Neurogenesis in the adult brain: new strategies for central nervous system diseases. *Annu Rev Pharmacol Toxicol* 2004;**44**:399–421.
61. Gheusi G, Cremer H, McLean H, Chazal G, Vincent JD, Lledo PM. Importance of newly generated neurons in the adult olfactory bulb for odor discrimination. *Proc Natl Acad Sci USA* 2000;**97**:1823–8.
62. Bruel-Jungerman E, Rampon C, Laroche S. Adult hippocampal neurogenesis, synaptic plasticity and memory: facts and hypotheses. *Rev Neurosci* 2007;**18**:93–114.
63. Curtis MA, Eriksson PS, Faull RL. Progenitor cells and adult neurogenesis in neurodegenerative diseases and injuries of the basal ganglia. *Clin Exp Pharmacol Physiol* 2007;**34**:528–32.
64. Curtis MA, Faull RL, Eriksson PS. The effect of neurodegenerative diseases on the subventricular zone. *Nat Rev Neurosci* 2007;**8**:712–23.
65. Morshead CM, Craig CG, van der Kooy D. In vivo clonal analyses reveal the properties of endogenous neural stem cell proliferation in the adult mammalian forebrain. *Development* 1998;**125**:2251–61.
66. Meletis K, Wirta V, Hede SM, Nister M, Lundeberg J, Frisen J. p53 suppresses the self-renewal of adult neural stem cells. *Development* 2006;**133**:363–9.

67. Gil-Perotin S, Marin-Husstege M, Li J, Soriano-Navarro M, Zindy F, Roussel MF, et al. Loss of p53 induces changes in the behavior of subventricular zone cells: implication for the genesis of glial tumors. *J Neurosci* 2006;**26**:1107–16.
68. Medrano S, Burns-Cusato M, Atienza MB, Rahimi D, Scrable H. Regenerative capacity of neural precursors in the adult mammalian brain is under the control of p53. *Neurobiol Aging* 2009;**30**:483–97.
69. Rossi DJ, Jamieson CH, Weissman IL. Stems cells and the pathways to aging and cancer. *Cell* 2008;**132**:681–96.
70. Beachy PA, Karhadkar SS, Berman DM. Tissue repair and stem cell renewal in carcinogenesis. *Nature* 2004;**432**:324–31.
71. Zheng H, Ying H, Yan H, Kimmelman AC, Hiller DJ, Chen AJ, et al. p53 and Pten control neural and glioma stem/progenitor cell renewal and differentiation. *Nature* 2008;**455**:1129–33.
72. Zheng H, Ying H, Yan H, Kimmelman AC, Hiller DJ, Chen AJ, et al. Pten and p53 converge on c-Myc to control differentiation, self-renewal, and transformation of normal and neoplastic stem cells in glioblastoma. *Cold Spring Harb Symp Quant Biol* 2008;**73**:427–37.
73. Alcantara Llaguno S, Chen J, Kwon CH, Jackson EL, Li Y, Burns DK, et al. Malignant astrocytomas originate from neural stem/progenitor cells in a somatic tumor suppressor mouse model. *Cancer Cell* 2009;**15**:45–56.
74. Kwon CH, Zhao D, Chen J, Alcantara S, Li Y, Burns DK, et al. Pten haploinsufficiency accelerates formation of high-grade astrocytomas. *Cancer Res* 2008;**68**:3286–94.
75. Groszer M, Erickson R, Scripture-Adams DD, Dougherty JD, Le Belle J, Zack JA, et al. PTEN negatively regulates neural stem cell self-renewal by modulating G0-G1 cell cycle entry. *Proc Natl Acad Sci USA* 2006;**103**:111–6.
76. Bachoo RM, Maher EA, Ligon KL, Sharpless NE, Chan SS, You MJ, et al. Epidermal growth factor receptor and Ink4a/Arf: convergent mechanisms governing terminal differentiation and transformation along the neural stem cell to astrocyte axis. *Cancer Cell* 2002;**1**:269–77.
77. Sanai N, Alvarez-Buylla A, Berger MS. Neural stem cells and the origin of gliomas. *N Engl J Med* 2005;**353**:811–22.
78. Dontu G, Liu S, Wicha MS. Stem cells in mammary development and carcinogenesis: implications for prevention and treatment. *Stem Cell Rev* 2005;**1**:207–13.
79. Molofsky AV, Slutsky SG, Joseph NM, He S, Pardal R, Krishnamurthy J, et al. Increasing p16INK4a expression decreases forebrain progenitors and neurogenesis during ageing. *Nature* 2006;**443**:448–52.
80. Nagao M, Campbell K, Burns K, Kuan CY, Trumpp A, Nakafuku M. Coordinated control of self-renewal and differentiation of neural stem cells by Myc and the p19ARF-p53 pathway. *J Cell Biol* 2008;**183**:1243–57.
81. Nishino J, Kim I, Chada K, Morrison SJ. Hmga2 promotes neural stem cell self-renewal in young but not old mice by reducing p16Ink4a and p19Arf Expression. *Cell* 2008;**135**:227–39.
82. Janzen V, Forkert R, Fleming HE, Saito Y, Waring MT, Dombkowski DM, et al. Stem-cell ageing modified by the cyclin-dependent kinase inhibitor p16INK4a. *Nature* 2006;**443**:421–6.
83. Krishnamurthy J, Ramsey MR, Ligon KL, Torrice C, Koh A, Bonner-Weir S, et al. p16INK4a induces an age-dependent decline in islet regenerative potential. *Nature* 2006;**443**:453–7.
84. Ugalde AP, Ramsay AJ, de la Rosa J, Varela I, Marino G, Cadinanos J, et al. Aging and chronic DNA damage response activate a regulatory pathway involving miR-29 and p53. *EMBO J* 2011;**30**:2219–32.
85. Freund A, Patil CK, Campisi J. p38MAPK is a novel DNA damage response-independent regulator of the senescence-associated secretory phenotype. *EMBO J* 2011;**30**:1536–48.
86. Lee JS, Lee MO, Moon BH, Shim SH, Fornace Jr., AJ, Cha HJ. Senescent growth arrest in mesenchymal stem cells is bypassed by Wip1-mediated downregulation of intrinsic stress signaling pathways. *Stem Cells* 2009;**27**(8):1963–75.

87. Wong E, Le Guezennec X, Wang ST, Krishnamurthy J, Dunn NR, Sharpless NE, et al. p38MAPK signalling induces Ink4a/Arf expression with aging. *Dev Cell* 2009;**17**:142–9.
88. Bulavin DV, Saito S, Hollander MC, Sakaguchi K, Anderson CW, Appella E, et al. Phosphorylation of human p53 by p38 kinase coordinates N-terminal phosphorylation and apoptosis in response to UV radiation. *EMBO J* 1999;**18**:6845–54.
89. Kim J, Wong PKY. Loss of ATM impairs proliferation of neural stem cells through oxidative stress-mediated p38 MAPK signaling. *Stem Cells* 2009;**27**(8):1987–98.
90. Keren A, Tamir Y, Bengal E. The p38 MAPK signaling pathway: a major regulator of skeletal muscle development. *Mol Cell Endocrinol* 2006;**252**:224–30.
91. Lindqvist A, de Bruijn M, Macurek L, Bras A, Mensinga A, Bruinsma W, et al. Wip1 confers G2 checkpoint recovery competence by counteracting p53-dependent transcriptional repression. *EMBO J* 2009;**28**:3196–206.
92. Maiuri MC, Galluzzi L, Morselli E, Kepp O, Malik SA, Kroemer G. Autophagy regulation by p53. *Curr Opin Cell Biol* 2010;**22**:181–5.
93. Balaburski GM, Hontz RD, Murphy ME. p53 and ARF: unexpected players in autophagy. *Trends Cell Biol* 2010;**20**:363–9.
94. Oliva-Trastoy M, Berthonaud V, Chevalier A, Ducrot C, Marsolier-Kergoat MC, Mann C, et al. The Wip1 phosphatase (PPM1D) antagonizes activation of the Chk2 tumour suppressor kinase. *Oncogene* 2007;**26**:1449–58.
95. Chew J, Biswas S, Shreeram S, Humaidi M, Wong ET, Dhillion MK, et al. WIP1 phosphatase is a negative regulator of NF-kappaB signalling. *Nat Cell Biol* 2009;**11**:659–66.

# Phosphatases Driving Mitosis: Pushing the Gas and Lifting the Brakes

Erika Segear Johnson and
Sally Kornbluth

*Department of Pharmacology and Cancer Biology, Box 3813, Duke University School of Medicine, Durham, North Carolina, USA*



Entry into and progression through mitosis depends critically on the establishment and maintenance of protein phosphorylation. For this reason, studies on mitotic progression have focused heavily on the activation of MPF (M phase promoting factor), a cyclin-dependent kinase responsible for phosphorylating proteins that execute the dynamic events of mitosis. Recent work, however, has significantly expanded our understanding of mechanisms that allow accumulation of phosphoproteins at M phase, suggesting that mitotic entry relies not only on MPF activation but also on the inhibition of antimitotic phosphatases. It is now clear that there exists a separate, albeit equally important, signaling pathway for the inactivation of protein phosphatases at the G2/M transition. This pathway, which is governed by the kinase Greatwall is essential for both entry into and maintenance of M phase. This chapter will outline the molecular events regulating entry into mitosis, specifically highlighting the role that protein phosphorylation plays in triggering both MPF activation and the inhibition of phosphatase activity that would otherwise prevent accumulation of mitotic phosphoproteins. These intricate regulatory pathways are essential for maintaining normal cell division and preventing inappropriate cell proliferation, a central hallmark of cancer cells.

Progress in Molecular Biology
and Translational Science, Vol. 106
DOI: 10.1016/B978-0-12-396456-4.00008-0

## I. Introduction

Cell division is a tightly regulated process through which cells duplicate their DNA and divide, forming two distinct daughter cells. In order to properly segregate the duplicated DNA, cells must undergo elaborate structural changes, including chromosomal condensation, nuclear envelope breakdown, and spindle formation in preparation for the separation of sister chromatids. The dramatic physical changes associated with mitosis were long recognized by histologists, but it was not until the late twentieth century that it was fully recognized that the coordinated phosphorylation of multiple substrates was responsible for the finely orchestrated structural changes observed in dividing cells. Since that time, remarkable progress has been made in understanding the events that govern the process of cell division, including the identification of significant factors and pathways that cooperate in intricate ways to promote entry into mitosis.

An early cell cycle study utilizing oocytes from the African clawed frog *Xenopus laevis* revealed that M phase is characterized by the appearance of many newly phosphorylated proteins.[1] These proteins were shown to be phosphorylated in response to an activity known as M-phase promoting factor (MPF), leading researchers to hypothesize that MPF-driven protein phosphorylation was controlling the key events of mitosis. It appeared from this work that MPF might contain a kinase activity, but whether MPF was directly phosphorylating mitotic phosphoproteins or might be activating secondary kinases was not fully settled until MPF was characterized molecularly; purified, active MPF was shown to contain a protein kinase catalytic subunit, now known as Cdk1, and an activating subunit, Cyclin B.[2–9] It was determined that active MPF could both directly and indirectly (through the activation of mitotic kinases) phosphorylate proteins involved in mitotic progression, and it became clear that the activation of Cdk1 kinase activity was a central event in the initiation and maintenance of mitosis. Work over the next decade aimed to understand the precise mechanisms regulating MPF activation and focused primarily on kinases and phosphatases acting upstream of the Cdk1/Cyclin B complex.

Given the role that MPF-driven protein phosphorylation plays in promoting mitotic entry, it was broadly acknowledged that exit from mitosis must involve reversal of the mitotic state through dephosphorylation of MPF targets. However, the notion that such phosphatase activity(ies) must be suppressed in order to allow accumulation of mitotic phosphoproteins beyond the threshold required for mitotic entry was recognized only recently. Experiments performed first in *Xenopus* egg extracts demonstrated that a kinase called Greatwall is essential for the restraint of such dephosphorylation. When activated, Greatwall can inhibit phosphatase activity(ies) directed against MPF phosphosites.

This chapter will examine the regulation of entry into mitosis, discussing how the cell activates MPF to push mitosis forward, while at the same time releasing the brake on mitotic entry through the inhibition of antimitotic phosphatases. The role of protein phosphorylation in regulating each of these pathways is specifically highlighted, while the cooperation of these pathways in promoting the burst of mitotic phosphorylation required for proper cell cycle progression is discussed. (Note that this review will use *Xenopus laevis* protein naming and numbering, as many of the discussed signaling pathways were originally defined using this model and may or may not have been validated in human cells to date. Reference will be made to human proteins where applicable.)

# II. Part I: Powering into Mitosis: Activation of MPF

## A. The Cell Cycle

Cyclin-dependent kinases drive each of the major cell cycle transition points (G1, S, G2, M), inducing downstream processes by phosphorylating select proteins. While CDK protein levels remain stable throughout the cell cycle, cyclin levels fluctuate and, in this way, periodically activate CDKs. Progression through each phase of the cell cycle requires different CDK/cyclin pairs and, thus, phase-specific cyclin expression allows ordered activation of CDKs. In addition to cyclin binding, CDK activity is further controlled through regulatory phosphorylation and dephosphorylation, ensuring well-delineated transitions between cell cycle phases.

## B. Entry into Mitosis and MPF Activation

During the G2 phase of the cell cycle, Cdk1/Cyclin B (also referred to as Cdc2/Cyclin B in *Xenopus*) complexes accumulate and are held inactive by phosphorylations on Thr14 and Tyr15 of Cdk1. These inhibitory phosphorylations are catalyzed by the Wee1 family kinases, namely, Wee1 and Myt1.[10,11] At the G2/M transition, Wee1 and Myt1 are inactivated, while the dual specificity phosphatase Cdc25 is activated. Cdc25 dephosphorylates Cdc2 at Thr14 and Tyr15, converting Cdc2/Cyclin B into active MPF.[12–14] This active complex then drives entry into mitosis through the direct and indirect phosphorylation of target proteins. We will focus here on the regulation and activation of Cdc25, which is now understood in considerable detail (Fig. 1).

## C. Cdc25: Posttranslational Regulation and Activation

Mammalian cells contain three isoforms of Cdc25 (A, B, and C), all of which can dephosphorylate Cdc2/Cyclin B. For simplicity, the generic term "Cdc25" will be used here to describe the isoforms functioning to promote

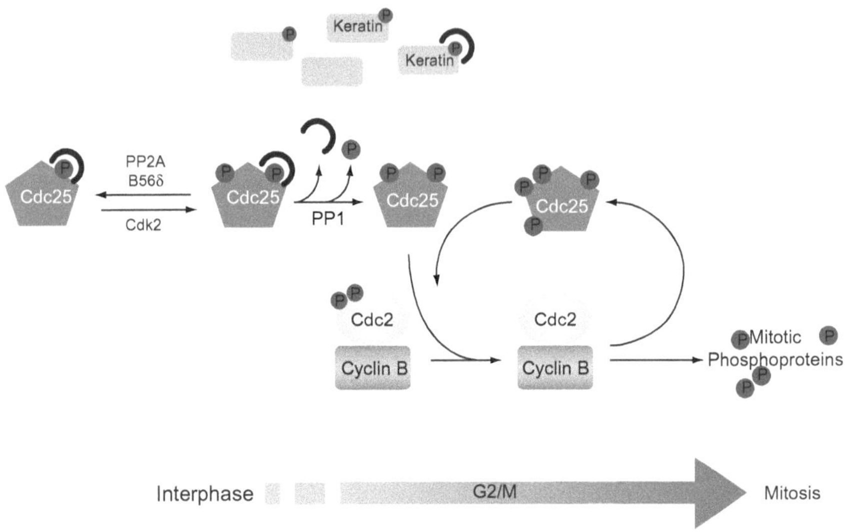


FIG. 1. During interphase, Cdc25 is held inactive by an inhibitory phosphorylation on Ser287 and binding of the small acidic protein 14-3-3. The phosphatase PP2A/B56δ keeps Thr138 dephosphorylated until the G2/M transition when Cdc25 is activated. Thr138 phosphorylation accumulates through the constitutive action of Cdk2 and this triggers the release of 14-3-3, which is then sequestered from Cdc25 by phosphorylated keratins. Ser287 is then dephosphorylated by PP1, Cdc25 is activated, and a small pool of Cdc2/Cyclin B is converted into active MPF. Active MPF then phosphorylates many downstream targets, including Cdc25, as part of a positive feedback loop. (See Color Insert.)

mitotic entry. During interphase, Cdc25 is held inactive by an inhibitory phosphorylation at Ser287 (*Xenopus* numbering; Ser216 in human Cdc25C). This phosphorylation, which can be catalyzed by a number of kinases including Chk1, Chk2, C-Tak, CaMKII, and PKA,[15–18] confers binding to 14-3-3 proteins and ultimately dampens Cdc25's catalytic activity, while sequestering it in the cytoplasm.[19–23] Originally, the checkpoint kinases Chk1 and Chk2 were found to phosphorylate Ser287, suggesting that DNA-responsive checkpoints were primarily responsible for inactivating Cdc25.[16,18] However, work by Dunphy and colleagues demonstrated that, in unperturbed cell cycles, interphase Cdc25 was phosphorylated at Ser287 and bound to 14-3-3, indicating that this mechanism is responsible for suppressing Cdc25 even in the absence of DNA damage.[20] Additional experiments in which Ser287 was mutated to a nonphosphorylatable alanine confirmed the importance of this residue in maintaining Cdc25 in an inactive state during interphase, as this mutation resulted in unrestrained nuclear accumulation of Cdc25, rapid dephosphorylation of Cdc2, and premature entry into mitosis.[24,25]

Once it was understood how Cdc25 activity was suppressed, the mechanism of its activation at the G2/M transition could be investigated. Given the importance of Ser287 phosphorylation in suppressing Cdc25 during interphase, it was assumed that activation would require the dephosphorylation of this residue. It was originally thought that release of 14-3-3 at the time of mitotic entry was due to the loss of a phosphodocking site following loss of Ser287-directed kinase activity (and presumably activation of S287-directed phosphatases). However, data emerged to contradict this notion. First, in *Xenopus* egg extracts able to reconstitute *in vitro* the transition from interphase to mitosis, 14-3-3 was released from Cdc25 prior to Ser287 dephosphorylation, and inhibition of dephosphorylation did not inhibit the release.[26] Indeed, it appeared that loss of 14-3-3 binding could actually accelerate Ser287 dephosphorylation in that a Cdc25 protein unable to bind 14-3-3 (constructed by mutation of downstream residues forming part of the 14-3-3 binding consensus) was dephosphorylated more rapidly than wild-type protein. Additional work revealed that Ser287 was dephosphorylated by protein phosphatase 1 (PP1), which bound directly to Cdc25 without the aid of a targeting subunit.[26] Moreover, a PP1-binding mutant failed to promote entry into mitosis, suggesting that dephosphorylation of Ser287 is, indeed, required for activation of Cdc25's phosphatase activity.[26] However, despite its clear role in catalyzing Ser287 dephosphorylation, PP1 activity did not appear to be activated at the G2/M transition. Together, these findings raised the interesting possibility that release of 14-3-3 from Cdc25 was the locus of Cdc25 regulation by cell cycle and DNA-responsive checkpoint cues and that PP1-mediated Cdc25 dephosphorylation simply followed passively after the release of 14-3-3. How then was 14-3-3 release controlled?

It had been known for some time that multiple sites on Cdc25 are phosphorylated at the time of mitotic entry.[27] Interestingly, phosphorylation of one of these residues, Thr138, was shown to be important for weakening the interaction between 14-3-3 and Cdc25, but, while this phosphorylation was required for 14-3-3 release, it was not sufficient. Efficient 14-3-3 removal, instead, seemed to involve the creation of alternative docking sites for 14-3-3 on other proteins at G2/M. In *Xenopus* egg extracts, these critical docking proteins were identified as keratins 8/18, which were shown to act as a "sink," binding to and reducing the availability of free 14-3-3 in the cytosol.[28] This, in combination with Thr138 phosphorylation, was enough to sequester 14-3-3 from Cdc25. These data suggest that two events are required for the release of 14-3-3: (1) a decrease in binding affinity between 14-3-3 and Cdc25 brought about by Thr138 phosphorylation; and (2) the formation of alternative binding sites for 14-3-3 on other proteins.

Although the scheme above was appealing, it neglected the key question of how the entire 14-3-3 release process was regulated. Biochemical analysis revealed that (at least in *Xenopus* egg extracts) another cyclin-dependent kinase, Cdk2, was responsible for phosphorylating T138 on Cdc25;[28] however, the activity of Cdk2 is not regulated at mitotic entry, suggesting that Thr138 phosphorylation must be controlled by an alternative mechanism.[29–31] Indeed, it was demonstrated that the phosphorylation state of Thr138 is determined by T138-directed phosphatase activity, with the relevant phosphatase being identified as protein phosphatase 2A (PP2A).[28]

The catalytic core of PP2A exists as an obligate dimer, which consists of a catalytic C subunit and a structural A subunit (A/C) thought to be targeted to its substrates by a variable third subunit, the B subunit (B). To date, four families of B subunits have been identified, B (B55), B′ (B56), B″, B‴, each consisting of multiple isoforms. In the case of T138-directed phosphatase activity, it was found that the PP2A catalytic core associated with the B56δ regulatory subunit.[28] During interphase, when DNA replication is ongoing (or under conditions of DNA damage/replication stalling when DNA-responsive checkpoints are activated), the kinase Chk1 phosphorylates B56δ to enhance its interaction with the catalytic core, forming the active PP2A holoenzyme. This active phosphatase dephosphorylates Thr138 and holds Cdc25 in an inactive state until replication has been successfully completed and/or any DNA damage has been repaired. At this point, Chk1 activity wanes and dephosphorylation of Thr138 by PP2A/B56δ diminishes, allowing constitutive Cdk2 activity to predominate and phospho-T138 to accumulate.[28]

The transition from the inactive to the active state of Cdc25 is further regulated by a positive feedback loop in which Cdc25 activation is enhanced through MPF-mediated phosphorylation. While Cdc25 is activated following Ser287 dephosphorylation, the activity of this phosphatase is minimal and only a small pool of MPF can be dephosphorylated and activated. Since mitotic entry relies on a quick and irreversible activation of MPF activity, MPF acts to control its regulators, participating in a positive feedback loop to maximally activate its activator, Cdc25. This type of regulation creates a bistable system, such that MPF activity is either off or on and does not exist in an intermediate state of activation.[32] Studies performed in both *Xenopus* and human systems have demonstrated that Cdc25 must be phosphorylated by MPF for complete activation of its phosphatase activity.[33–36]

Several of the residues on Cdc25 that are phosphorylated at the time of mitotic entry have been identified as MPF target sites, including Thr48, Thr67, and Ser285. While phosphorylation at each of these sites is required for maximal activation of Cdc25 at mitotic entry, phosphorylation of Ser285 has been shown to play a unique role, helping to maintain Ser287 in a dephosphorylated state.[34,35,37] As mentioned above, PP1 associates with Cdc25 following the release of 14-3-3. However, the binding between PP1 and Cdc25 at this point is weak, resulting in inefficient dephosphorylation of Ser287 and activation of only a small fraction of Cdc25. This activated Cdc25 dephosphorylates Thr14 and Tyr15 on Cdc2, converting a small pool of inactive Cdc2/Cyclin B into active MPF. This small pool of active MPF then phosphorylates Cdc25 on Ser285, which increases the affinity of PP1 for Cdc25 and results in more efficient dephosphorylation of Ser287. At the same time, MPF can phosphorylate Thr48 and Thr67 in the N-terminus of Cdc25, enhancing its activity even further to promote the maintenance and activation of MPF kinase activity. Accordingly, mutation of Ser285 short-circuits the feedback loop, severely impairing the ability of Cdc25 to fully activate MPF and promote mitotic entry.[34,35,37]

Given the explosive activation of MPF at the G2/M transition, it is easy to speculate that a sudden increase in Cdc2/Cyclin B kinase activity is all that is needed for a cell to enter mitosis. For example, if the kinase activity of active MPF was potent enough to overcome a basal level of phosphatase activity within the cell, then activation of MPF would result in the accumulation of mitotic phosphorylations and entry into M phase. The cell could then progress through mitosis and, upon exiting M phase, inactivate MPF, allowing mitotic phosphoproteins to be dephosphorylated by unopposed intracellular phosphatase activity. This model, which suggests that the regulation of protein phosphatase activity would not be required for entry into mitosis, was for many years a widely accepted one for understanding mitotic progression.

However, recent work has challenged this idea, suggesting that mitotic phosphorylations depend not only on the activation of MPF but also on the inhibition of phosphatases directed against MPF phosphosites. The second part of this chapter will focus on molecular regulation of the recently discovered signaling pathway responsible for inactivating protein phosphatases at mitotic entry.

# III. Part II: Releasing the Brake on Mitosis: Inactivation of Phosphatases by Greatwall Kinase

## A. Phosphatases at Mitotic Entry

As movement through the cell cycle relies predominantly on signaling through reversible phosphorylation, protein phosphatases have significant effects on cell cycle progression. For example, early experiments demonstrated that treatment of *Xenopus* oocytes and extracts with the phosphatase inhibitor okadaic acid resulted in an accelerated entry into M phase, resulting at least in part from the accumulation of activating phosphorylations on Cdc25.[36,38–40] These data suggested that there must be Cdc25-directed phosphatases acting to otherwise inhibit entry into mitosis. Since phosphorylation is essential for promoting the robust Cdc25 phosphatase activity required for mitotic entry, this raised the question of whether these phosphatases opposing mitotic entry were inactivated at the G2/M transition or were simply overpowered by the abundance of MPF kinase activity. With no concrete evidence to support the former conclusion, investigators began to speculate that M phase relies predominantly on MPF activation and that the regulation of protein phosphatase activity would be nonessential for both entering and maintaining the mitotic state. However, as described below, this viewpoint has been challenged by recent findings.

## B. Greatwall Kinase: Negative Regulator of Antimitotic Phosphatases

In the past several years, studies investigating a newly discovered kinase, namely, Greatwall (Gwl), have modified our ideas on control of mitotic progression, suggesting that the regulation of protein phosphatase activity is essential for both entry into and maintenance of M phase. Using *Xenopus* egg extracts, Goldberg and colleagues have identified a functional role for Gwl, demonstrating that the kinase is responsible for inactivating antimitotic phosphatases which function to inhibit Cdc25 at mitotic entry.[41,42] In the absence of Gwl, MPF activity is too weak to promote entry into or maintenance of M-phase, suggesting that Gwl is essential for the positive feedback loop that results in hyperphosphorylation of Cdc25, generating and maintaining high MPF activity.[41]

Originally identified in *Drosophila*, Gwl kinase appeared to function as a regulator of mitotic entry, with mutation of the *Drosophila Gwl* gene resulting in improper chromosomal condensation and a delay in the G2/M phase transition.[43] Gwl belongs to the AGC family of Ser/Thr kinases and is well conserved across evolution. The human ortholog of Gwl has been identified as microtubule-associated serine/threonine kinase-like (MASTL), which has also been demonstrated to play a role in chromosomal condensation and entry into mitosis.[44,45] The conservation of such a dramatic cell cycle phenotype strongly supports the idea that Gwl is functioning as a regulator of mitotic entry.

Following the biochemical identification of a *Xenopus* Gwl homolog, Goldberg and colleagues demonstrated a functional role for the kinase as a participant in the MPF-mediated positive feedback loop. Specifically, Gwl was shown to be activated via MPF phosphorylation and it appeared to be required for both generating and maintaining MPF kinase activity.[41] In the absence of Gwl, extracts that cycle between interphase and mitosis fail to enter M phase, whereas extracts that are already mitotic exhibit decreased MPF kinase activity, resulting from an accumulation of inhibitory phosphorylations on Cdk1.[41] Gwl appears to be able to regulate Cdk1 phosphorylation via the phosphatase Cdc25, as extracts lacking Gwl are unable to promote Cdc25 phosphorylation and activation. Interestingly, these findings place Gwl both upstream and downstream of MPF and suggest that this kinase is important for both activating and maintaining the positive feedback loop which enhances Cdc25 activity to ensure substantial MPF activation.[41]

Shortly after this initial publication, a follow-up study revealed that Gwl's function extends beyond simply activating MPF as part of a positive feedback loop. It was demonstrated that, in the absence of any MPF kinase activity, Gwl could induce activating phosphorylations on Cdc25.[42] These phosphorylations were not directly catalyzed by Gwl and therefore suggested that Gwl might be acting to inhibit a Cdc25-directed phosphatase. In agreement with this hypothesis, Gwl appeared to be able to induce changes similar to those observed with the phosphatase inhibitor okadaic acid, including an accumulation of activating phosphorylations on Cdc25 and accelerated entry into mitosis.[42] Moreover, the inability of Gwl-depleted extracts to enter mitosis could be rescued by cotreating extracts with okadaic acid.[42] These findings strongly suggest that the inactivation of antimitotic phosphatases by Gwl is essential for both entry into and maintenance of M phase.

## C. PP2A: The Major Phosphatase Acting on MPF Phosphosites During Interphase

Around the same time that Gwl was shown to be involved in suppressing antimitotic phosphatases at mitotic entry, a separate and unrelated series of investigations revealed that phosphatases targeting MPF phosphosites were also inactivated during M phase.[46,47] Using model MPF substrates, Hunt and colleagues demonstrated an abundance of phosphatase activity directed against MPF phosphosites during interphase but were unable to detect any dephosphorylation of MPF substrates during mitosis.[46] The interphase dephosphorylation of MPF substrates was inhibited by okadaic acid and, therefore, to determine the identity of the phosphatase, each of the okadaic acid-sensitive catalytic subunits were removed by immunodepletion. PP2A was identified as the phosphatase targeting MPF phosphosites, as the removal of its catalytic subunit blocked the interphase dephosphorylation of mitotic phosphoproteins.[47]

To determine which of the associated B subunits, if any, were involved in the PP2A-mediated dephosphorylation of MPF substrates, each of the regulatory B subunits was removed by immunodepletion and the MPF substrate phosphorylation was assessed.[47] Removal of a single B subunit, namely, B55δ, resulted in increased phosphorylation of mitotic phosphoproteins and accelerated entry into mitosis. Moreover, overexpression of this subunit could significantly delay the accumulation of mitotic phosphorylations, despite sufficient accumulation of MPF (Cdc2/Cyclin B).[47] Based on these findings, it was concluded that PP2A, in association with the B55δ subunit, was the phosphatase acting on mitotic phosphoproteins during interphase and that inhibition of this phosphatase would be required for mitotic phosphorylations to accumulate.[47]

Given the ability of Gwl to inhibit an antimitotic phosphatase at mitotic entry and the identification of PP2A/B55δ as a phosphatase targeting MPF phosphosites during interphase, it was possible that these pathways were directly linked.[47] Indeed, Goldberg and colleagues were able to demonstrate that the loss of Gwl resulted in an inability to inactivate PP2A/B55δ in interphase *Xenopus* egg extracts.[48] These extracts, which were unable to enter mitosis, could be rescued through the removal of PP2A/B55δ, making it clear that Gwl functions upstream of PP2A/B55δ to inhibit its ability to dephosphorylate mitotic phosphosites.[48] A second independent study further validated this conclusion, showing that depletion of Gwl from M-phase *Xenopus* egg extracts induces the activation of PP2A and results in the dephosphorylation of MPF phosphosites.[49] Most recently, these effects have been confirmed in human cells, where the functional ortholog of Gwl, MASTL, has been shown to regulate PP2A to control the level of mitotic phosphorylations.[50] In summary, these studies have expanded our understanding of the functional role of Gwl, demonstrating that at the G2/M transition this kinase acts to inhibit PP2A/B55δ-mediated dephosphorylation of MPF substrates, including Cdc25.

## D. The Inhibitory Role of ARPP-19/α-Endosulfine

Despite functional evidence linking Gwl to PP2A/B55δ inactivation, Gwl does not directly phosphorylate any of the PP2A subunits; thus Gwl is not a direct inhibitor of PP2A.[48,49] Instead, Gwl appeared to be acting indirectly, potentially by activating an inhibitor of PP2A. To identify potential intermediates between Gwl and PP2A, Gwl-phosphorylated substrates were identified in *Xenopus* egg extracts and then examined for their ability to inhibit PP2A/B55δ.[51,52]

Two-independent studies identified the cAMP-regulated phosphoprotein-19 (ARPP-19) as the predominant Gwl substrate in extracts.[51,52] ARPP-19 is a small heat-stable protein that is closely related to α-endosulfine (ENSA) and is most commonly known for being a substrate of PKA. It was demonstrated that both ARPP-19 and ENSA could acts as substrates of Gwl *in vitro*, and that Gwl-mediated phosphorylation could convert both ARPP-19 and ENSA into potent inhibitors of PP2A/B55δ.[52] Interestingly, phosphorylated ARPP-19/ENSA can only inactivate the B55δ-associated PP2A trimer and has no apparent inhibitory effect on the A/C dimer or monomeric catalytic subunit.[52] Unphosphorylated forms of ARPP-19/ENSA also show no effect on PP2A activity, suggesting that Gwl phosphorylation is required to enable the inhibitory activity of ARPP-19/ENSA. Moreover, phosphorylation of ARPP-19/ENSA can promote binding to PP2A/B55δ, making it likely that the inhibition of PP2A results from the physical interaction between these two proteins.[51,52]

In agreement with the biochemical data on Gwl and ARPP-19/ENSA, functional analyses indicate that recombinant ARPP-19/ENSA added to *Xenopus* egg extracts can only accelerate entry into mitosis when it is pretreated with Gwl kinase.[51] Moreover, ARPP-19 depletion from *Xenopus* extracts prevents entry into mitosis. Mitotic entry can be rescued by codepletion of PP2A/B55δ, indicating, again, that ARPP-19 is responsible for inhibiting this antimitotic phosphatase.[52]

Taken together, these studies indicate that PP2A/B55δ must be inhibited at the G2/M transition to allow progression into M phase. As summarized in Fig. 2, PP2A/B55δ acts as a brake during interphase, preventing the accumulation of mitotic phosphorylations through the constitutive dephosphorylation of

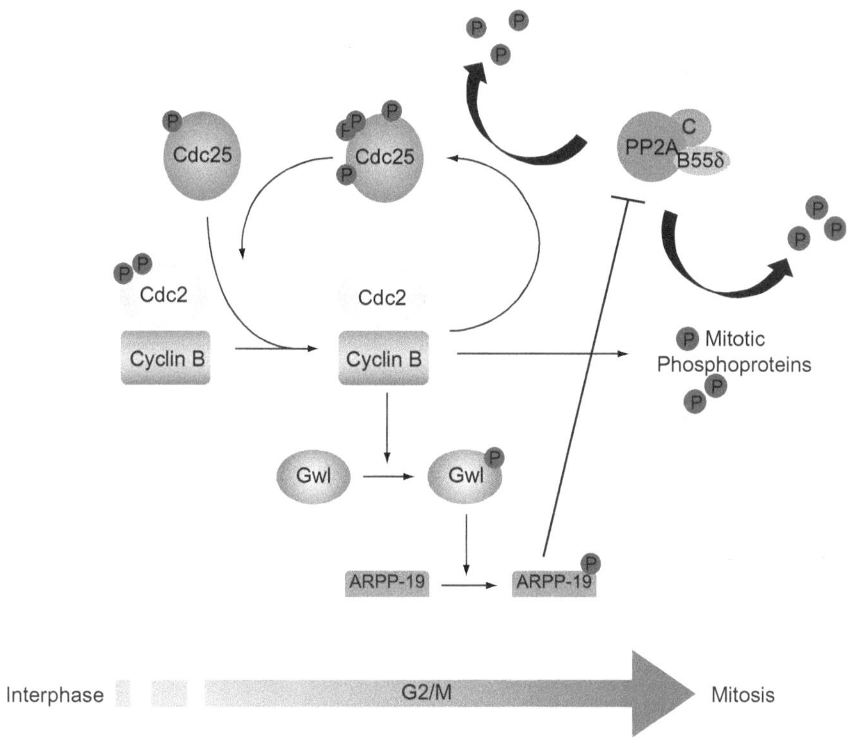


FIG. 2. Prior to mitotic entry, PP2A/B55δ acts as a brake and prevents the early accumulation of mitotic phosphorylations through the constitutive dephosphorylation of MPF substrates. Following the conversion of a small pool of Cdc2/Cyclin B into active MPF, Gwl is activated via phosphorylation. Active Gwl is able to inactivate the interphase dephosphorylation of mitotic phosphoproteins by phosphorylating ARPP-19/ENSA, an inhibitor that binds to and inactivates the antimitotic phosphatase PP2A/B55δ. (See Color Insert.)

MPF substrates. Following the conversion of a small pool of Cdc2/Cyclin B into active MPF, Gwl gets activated and phosphorylates ARPP-19/ENSA at a conserved serine residue. Phosphorylated ARPP-19/ENSA acts as a potent and specific inhibitor of the PP2A/B55δ holoenzyme through direct association with the phosphatase. This inhibits PP2A-mediated dephosphorylation of MPF substrates and releases the brake on mitotic entry, allowing accumulation of mitotic phosphorylations to accumulate and consequent entry into mitosis.

## E. Conclusion and Perspectives

It has been appreciated for some time that kinases participate in intricate regulatory networks to govern the precise timing of cell cycle transitions. In principle, it was possible that all of the regulation resided with kinase control and that phosphatases were passive partners, constitutively opposing kinase action. Under such circumstances, the dynamic changes in protein phosphorylation would be driven solely by changes in kinase levels or activities. However, this scenario is clearly far from the truth; both in normal cell cycle control and under conditions where DNA is damaged, the Cdc25 phosphatase plays a pivotal role in determining the timing of MPF activation (and mitosis) and, as the work described above makes clear, Cdc25 is itself controlled in a precise way through the combined actions of PP1 and PP2A. On a broader level, MPF activity is opposed by PP2A/B55δ through dephosphorylation of MPF substrates and this pathway is under tight control by Gwl and its substrates, which are themselves targets of MPF-mediated positive feedback. As a regulator of PP2A/B55δ, Gwl is able to govern the phosphorylation of MPF substrates, including Cdc25 (as part of the autoregulatory loop), thereby influencing both activation of MPF and accumulation of its phosphorylated targets. Therefore, the cell must coordinate the activation of MPF with the inhibition of PP2A/B55δ, both pushing the gas pedal and releasing the brake to allow smooth mitotic entry.

## F. Phosphatases as Targets in Cancer Therapy

Many laboratories have attempted to develop therapeutic strategies for impeding cancer cell proliferation through the targeting of cell cycle regulators. In this regard, broadly functional phosphatases (e.g., PP1 and PP2A) might not be viewed as promising targets because of their involvement in a myriad of cellular processes. However, the work described here may open a new chapter in this regard, as highly specific features of phosphatase regulation have been revealed that could be exploited. Indeed, small-molecule targeting of kinases has been more successful than similar approaches for inhibiting phosphatases;

however, with the discovery of Gwl kinase as a key regulator of PP2A antimitotic phosphatase activity, it may be that Gwl inhibitors could effectively prevent mitotic progression. Similarly, although the Cdc25 phosphatase has been explored as a therapeutic target, elucidation of the complex control of Cdc25 might offer additional points of vulnerability in addition to direct inhibition of its catalytic activity.

Although beyond the realm of current capabilities, it is also conceivable that selective activation of phosphatase activity to drive cell cycle progression in otherwise postmitotic tissues could be harnessed for the purposes of tissue repair or replacement of tissues that have degenerated pathologically.

## References


1. Maller JL, Smith DS. Two-dimensional polyacrylamide gel analysis of changes in protein phosphorylation during maturation of Xenopus oocytes. *Dev Biol* 1985;**109**:150–6.
2. Dunphy WG, Brizuela L, Beach D, Newport J. The Xenopus cdc2 protein is a component of MPF, a cytoplasmic regulator of mitosis. *Cell* 1988;**54**:423–31.
3. Gautier J, Minshull J, Lohka M, Glotzer M, Hunt T, Maller JL. Cyclin is a component of maturation-promoting factor from Xenopus. *Cell* 1990;**80**:487–94.
4. Gautier J, Norbury C, Lohka M, Nurse P, Maller JL. Purified maturation-promoting factor contains the product of a Xenopus homolog of the fission yeast cell cycle control gene cdc2. *Cell* 1988;**54**:433–9.
5. Gerhart J, Wu M, Kirschner M. Cell cycle dynamics of an M-phase specific cytoplasmic factor in *Xenopus laevis* oocytes and eggs. *J Cell Biol* 1984;**98**:1247–55.
6. Labbe JC, Capony JP, Caput D, Cavadore JC, Derancourt J, Kaghad M, et al. MPF from starfish oocytes at first meiotic metaphase is a heterodimer containing one molecule of cdc2 and one molecule of cyclin B. *EMBO J* 1989;**8**(10):3053–8.
7. Maller JL, Wu M, Gerhart JC. Changes in protein phosphorylation accompanying maturation of Xenopus laevis oocytes. *Dev Biol* 1977;**58**:295–312.
8. Nielson PJ, Thomas G, Maller JL. Increased phosphorylation of ribosomal protein S6 during meiotic maturation of Xenopus oocytes. *Proc Natl Acad Sci USA* 1982;**79**:2937–41.
9. Wu M, Gerhart JC. Partial purification and characterization of the maturation-promoting factor from eggs of Xenopus laevis. *Dev Biol* 1980;**79**:465–77.
10. Meuller PR, Coleman TR, Kumagai A, Dunphy WG. Myt1: a membrane-associated inhibitory kinase that phosphorylates Cdc2 on both threonine-14 and tyrosine-15. *Science* 1995;**270**:86–90.
11. Russell P, Nurse P. Negative regulation of mitosis by wee1, a gene encoding a protein kinase homolog. *Cell* 1987;**49**:559–67.
12. Dunphy WG, Kumagai A. The cdc25 protein contains an intrinsic phosphatase activity. *Cell* 1991;**67**:189–96.
13. Gautier J, Solomon MJ, Booher RN, Bazan JF, Kirschner MW. Cdc25 is a specific tyrosine phosphatase that directly activates p34cdc2. *Cell* 1991;**67**:197–211.
14. Kumagai A, Dunphy WG. The cdc25 protein controls tyrosine dephosphorylation of the cdc2 protein in a cell-free system. *Cell* 1991;**64**:903–14.
15. Duckworth BC, Weaver JS, Ruderman JV. G2 arrest in Xenopus oocytes depends on phosphorylation of cdc25 by protein kinase A. *Proc Natl Acad Sci USA* 2002;**99**:16794–9.

16. Furnari B, Rhind N, Russell P. Cdc25 mitotic inducer targeted by chk1 DNA damage checkpoint kinase. *Science* 1997;**277**:1495–7.
17. Hutchins JR, Dikovskaya D, Clarke PR. Regulation of Cdc2/cyclin B activation in Xenopus egg extracts via inhibitory phosphorylation of Cdc25C phosphatase by Ca(2+)/calmodulin-dependent protein kinase II. *Mol Biol Cell* 2003;**14**:4003–14.
18. Kumagai A, Guo Z, Emami KH, Wang SX, Dunphy WG. The Xenopus Chk1 protein kinase mediates a caffeine-sensitive pathway of checkpoint control in cell-free extracts. *J Cell Biol* 1998;**142**:1559–69.
19. Graves PR, Lovly CM, Uy GL, Piwnica-Worms H. Localization of human Cdc25C is regulated both by nuclear export and 14-3-3 protein binding. *Oncogene* 2001;**20**:1839–51.
20. Kumagai A, Yakowec PS, Dunphy WG. 14-3-3 proteins act as negative regulators of the mitotic inducer Cdc25 in Xenopus egg extracts. *Mol Biol Cell* 1998;**9**(2):345–54.
21. Kumagai A, Dunphy WG. Binding of 14-3-3 proteins and nuclear export control the intracellular localization of the mitotic inducer Cdc25. *Genes Dev* 1999;**13**:1067–72.
22. Peng CY, Graves PR, Ogg S, Thoma RS, Byrnes MJ, Wu Z, et al. C-TAK1 protein kinase phosphorylates human Cdc25C on serine 216 and promotes 14-3-3 protein binding. *Cell Growth Differ* 1998;**9**:197–208.
23. Peng CY, Graves PR, Thoma RS, Wu Z, Shaw AS, Piwnica-Worms H. Mitotic and G2 checkpoint control: regulation of 14-3-3 protein binding by phosphorylation of Cdc25C on serine-216. *Science* 1997;**277**(5331):1501–5.
24. Lopez-Girona A, Furnari B, Mondesert O, Russell P. Nuclear localization of Cdc25 is regulated by DNA damage and a 14-3-3 protein. *Nature* 1999;**397**:172–5.
25. Yang J, Winkler K, Yoshida M, Kornbluth S. Maintenance of G2 arrest in the Xenopus oocyte: a role for 14-3-3-mediated inhibition of Cdc25 nuclear import. *EMBO J* 1999;**18**:2174–83.
26. Margolis SS, Walsh S, Weiser DC, Yoshida M, Shenolikar S, Kornbluth S. PP1 control of M phase entry exerted through 14-3-3-regulated Cdc25 dephosphorylation. *EMBO J* 2003;**22**:5734–45.
27. Izumi T, Walker DH, Maller JL. Periodic changes in phosphorylation of the Xenopus cdc25 phosphatase regulate its activity. *Mol Biol Cell* 1992;**3**(8):927–39.
28. Margolis SS, Perry JA, Forester CM, Nutt LK, Guo Y, Jardim MJ, et al. Role for the PP2A/B56delta phosphatase in regulating 14-3-3 release from Cdc25 to control mitosis. *Cell* 2006;**127**:759–73.
29. Guadagno TM, Newport JW. Cdk2 kinase is required for entry into mitosis as a positive regulator of Cdc2-cyclin B kinase activity. *Cell* 1996;**84**:73–82.
30. Rempel RE, Sleight SB, Maller JL. Maternal Xenopus Cdk2-cyclin E complexes function during meiotic and early embryonic cell cycles that lack a G1 phase. *J Biol Chem* 1995;**270**:6843–55.
31. Rosenblatt J, Gu Y, Morgan DO. Human cyclin-dependent kinase 2 is activated during the S and G2 phases of the cell cycle and associates with cyclin A. *Proc Natl Acad Sci* 1992;**89**:2824–8.
32. Pomerening JR, Sontag ED, Ferrell Jr. JE. Building a cell cycle oscillator: hysteresis and bistability in the activation of Cdc2. *Nat Cell Biol* 2003;**5**:346–51.
33. Bulavin DV, Demidenko ZN, Phillips C, Moody SA, Fornace Jr. AJ. Phosphorylation of Xenopus Cdc25C at Ser285 interferes with ability to activate a DNA damage replication checkpoint in pre-midblastula embryos. *Cell Cycle* 2003;**2**:263–6.
34. Bulavin DV, Higashimoto Y, Demidenko ZN, Meek S, Graves P, Phillips C, et al. Dual phosphorylation controls Cdc25 phosphatases and mitotic entry. *Nat Cell Biol* 2003;**5**:545–51.
35. Izumi T, Maller JL. Elimination of cdc2 phosphorylation sites in the cdc25 phosphatase blocks initiation of M-phase. *Mol Biol Cell* 1993;**4**:1337–50.
36. Kumagai A, Dunphy WG. Regulation of the cdc25 protein during the cell cycle in Xenopus extracts. *Cell* 1992;**70**:139–51.

37. Margolis SS, Perry JA, Weitzel DH, Freel CD, Yoshida M, Haystead TA, et al. A role for PP1 in the Cdc2/Cyclin B-mediated positive feedback activation of Cdc25. *Mol Biol Cell* 2006;**17**:1779–89.
38. Felix MA, Cohen P, Karsenti E. Cdc2 Hl kinase is negatively regulated by a type 2A phosphatase in the Xenopus early embryonic cell cycle: evidence from the effects of okadaic acid. *EMBO J* 1990;**9**:675–83.
39. Field J, Nikawa J, Broek JessusC, Rime H, Haccard O, Van Lint J, et al. Tyrosine phosphorylation of p34cdc2 and p42 during meiotic maturation of *Xenopus oocyte*. Antagonistic action of okadaic acid and 6-DMAP. *Development* 1991;**111**:813–20.
40. Lorca T, Fesquet D, Zindy F, Le Bouffant F, Cerruti M, Brechot C, et al. An okadaic acid-sensitive phosphatase negatively controls the cyclin degradation pathway in amphibian eggs. *Mol Cell Biol* 1991;**11**:1171–5.
41. Yu J, Zhao Y, Li Z, Galas S, Goldberg ML. Greatwall kinase participates in the Cdc2 autoregulatory loop in Xenopus egg extracts. *Mol Cell* 2006;**22**:83–91.
42. Zhao Y, Haccard O, Wang R, Yu J, Kuang J, Jessus C, et al. Roles of greatwall kinase in the regulation of Cdc25 phosphatase. *Mol Biol Cell* 2008;**19**:1317–27.
43. Yu J, Fleming SL, Williams B, Williams EV, Li Z, Somma P, et al. Greatwall kinase: a nuclear protein required for proper chromosome condensation and mitotic progression in Drosophila. *J Cell Biol* 2004;**164**:487–92.
44. Hanks SK, Hunter T. Protein kinases 6. The eukaryotic protein kinase superfamily: kinase (catalytic) domain structure and classification. *FASEB J* 1995;**9**:576–96.
45. Voets E, Wolthius MF. MASTL is the human ortholog of Greatwall kinase that facilitates mitotic entry, anaphase, and cytokinesis. *Cell Cycle* 2010;**9**(17):3591–601.
46. Mochida S, Hunt T. Calcineurin is required to release Xenopus egg extracts from meiotic M phase. *Nature* 2007;**449**:336–40.
47. Mochida S, Ikeo S, Gannon J, Hunt T. Regulated activity of PP2A–B55δ is crucial for controlling entry into and exit from mitosis in Xenopus egg extracts. *EMBO J* 2009;**28**:2777–85.
48. Castilho PV, Williams BC, Mochida S, Zhao Y, Goldberg ML. The M phase kinase greatwall (gwl) promotes inactivation of PP2A/B55δ, a phosphatase directed against CDK phosphosites. *Mol Biol Cell* 2009;**20**:4777–89.
49. Vigneron S, Brioudes E, Burgess A, Labbe JC, Lorca T, Castro A. Greatwall maintains mitosis through regulation of PP2A. *EMBO J* 2009;**28**:2786–93.
50. Burgess A, Vigneron S, Brioudes E, Labbe JC, Lorca T, Castro A. Loss of human greatwall results in G2 arrest and multiple mitotic defects due to deregulation of the cyclin B-Cdc2/PP2A balance. *Proc Natl Acad Sci USA* 2010;**107**:12564–9.
51. Gharbi-Ayachi A, Labbe JC, Burgess A, Vigneron S, Strub JM, Brioudes E, et al. The substrate of greatwall kinase, ARPP19, controls mitosis by inhibiting protein phosphatase 2A. *Science* 2010;**300**:1673–7.
52. Mochida S, Maslen SL, Skehel M, Hunt T. Greatwall phosphorylates an inhibitor of protein phosphatase 2A that is essential for mitosis. *Science* 2010;**300**:1670–3.

# Protein Phosphatases and Alzheimer's Disease

Steven P. Braithwaite,[*] Jeffry B. Stock,[†] Paul J. Lombroso,[‡] and Angus C. Nairn[§]

[*]*Signum Biosciences, Monmouth Junction, New Jersey, USA*

[†]*Department of Molecular Biology, Princeton University, Princeton, New Jersey, USA*

[‡]*Child Study Center, Yale University School of Medicine, New Haven, Connecticut, USA*

[§]*Department of Psychiatry, Yale University School of Medicine, New Haven, Connecticut, USA*



Progress in Molecular Biology
and Translational Science, Vol. 106
DOI: 10.1016/B978-0-12-396456-4.00012-2

Alzheimer's Disease (AD) is characterized by progressive loss of cognitive function, linked to marked neuronal loss. Pathological hallmarks of the disease are the accumulation of the amyloid-β (Aβ) peptide in the form of amyloid plaques and the intracellular formation of neurofibrillary tangles (NFTs). Accumulating evidence supports a key role for protein phosphorylation in both the normal and pathological actions of Aβ as well as the formation of NFTs. NFTs contain hyperphosphorylated forms of the microtubule-binding protein tau, and phosphorylation of tau by several different kinases leads to its aggregation. The protein kinases involved in the generation and/or actions of tau or Aβ are viable drug targets to prevent or alleviate AD pathology. However, it has also been recognized that the protein phosphatases that reverse the actions of these protein kinases are equally important. Here, we review recent advances in our understanding of serine/threonine and tyrosine protein phosphatases in the pathology of AD.

## I. Introduction

Alzheimer's Disease (AD), the most common neurodegenerative disorder, is a major and growing public health concern because of increases in both median age and life expectancy.[1] With the enormous economic cost of AD patient care and loss of productivity, its impact as a major clinical, social, and economic issue has been widely acknowledged. AD is characterized by progressive loss of cognitive function, starting with mild cognitive impairment that eventually evolves to include more severe cognitive deficiencies followed by death from secondary complications. At a cellular level, AD is characterized by marked neuronal and neuritic loss.[2]

The major pathological hallmarks of AD are the aberrant accumulation of the amyloid-β peptide (Aβ) in the form of amyloid plaques and the intracellular formation of hyperphosphorylated tau protein inclusions (neurofibrillary tangles, or NFTs).[1] Oligomeric assemblies of Aβ and tau are increasingly recognized as the most pathogenic forms, probably more so than amyloid plaques or NFTs.[3] Aβ peptides are generated by the processing of the amyloid precursor protein (APP). The central role of APP in AD is underscored by the fact that mutations in APP [familial AD (FAD)] cosegregate with an early onset AD pathology.[4,5] Notably, various independent lines of transgenic mice expressing APP with FAD mutations display pathological and cognitive deficits that correlate with those found in human AD.[6]

Formation of Aβ is catalyzed by β- and γ-secretase (γ-secretase/presenilin).[7] The formation and subsequent aggregation of Aβ initiate a complex cascade of molecular and cellular changes that gradually leads to the clinical features of AD. Recent studies have indicated that soluble Aβ oligomers act initially to disrupt synaptic function.[8,9] Thus, Aβ-mediated synaptopathology represents a critical component in the cognitive decline associated with the disease. Aβ oligomeric preparations inhibit long-term potentiation (LTP) likely through its effects on AMPA and NMDA glutamate receptor trafficking. However, while much has been learned about the role of Aβ in AD, particularly in synaptopathology, it is still unknown what mechanisms are involved in the transition from impaired synaptic function to loss of synapses and to eventual cell death.

Accumulating evidence supports a key role for protein phosphorylation in both the normal and pathological actions of APP, Aβ, and tau. APP itself is phosphorylated at multiple sites by several protein kinases,[10–15] and this may modulate its processing and influence the production of Aβ.[16,17] For example, phosphorylation of APP may influence the ability of β-secretase to cleave APP to produce the most amyloidogenic and toxic peptide $A\beta_{1-42}$.[16,18] Both presenilin (the γ-secretase) and BACE (the β-secretase) responsible for generation of Aβ are regulated by phosphorylation.[19–23]

AD is a member of the family of tauopathies that are characterized by the presence of hyperphosphorylated tau.[24] Up to 50 different sites of phosphorylation are present in tau (out of a total of 85 serine, threonine, and tyrosine residues)[25–27] (Table I), and many of these are found in tau isolated from AD brain. Phosphorylation of tau in AD brain is likely mediated by various protein kinases, including Cdk5, GSK3β, PKA, and MARK, which target serine and threonine residues in proline-rich domains of the protein.[30] Phosphorylation of tau interferes with its ability to interact with microtubules, leading to its aggregation and, ultimately, to the generation of NFTs.

Studies of the protein kinases that phosphorylate APP, tau, and the proteases involved in generation of Aβ are clearly important, as protein kinases are viable drug targets, and inhibitors may prevent or alleviate AD pathology. However, it has also been recognized that the protein phosphatases that reverse the actions of these protein kinases are equally important and warrant detailed analysis. For example, it appears that a major contributor to the

TABLE I

ASSESSMENT OF DEPHOSPHORYLATION OF HYPERPHOSPHORYLATED TAU BY SER/THR PPASES

| Site | | | | | Site | | | | |
|---|---|---|---|---|---|---|---|---|---|
| Y18 | | | | | S235 | | | | |
| T39 | | | | | S237 | | | | |
| S46 | | | | | S238 | | | | |
| S68 | | | | | S241 | | | | |
| T69 | | | | | S258 | | | | |
| T71 | | | | | S262 | PP1 | PP2A+++ | PP2B +++[2,3] | PP5 |
| S113 | | | | | S285 | | | | |
| T123 | | | | | S289 | | | | |
| S137 | | | | | S305 | | | | |
| T153 | | | | | S324 | | | | |
| T175 | | | | | S352 | | | | |
| T181 | PP1 | PP2A | PP2B[1] | | S356 | | PP2A | | |
| S184 | | | | | Y294 | | | | |
| S185 | | | | | S396 | PP1 | PP2A* | PP2B +++[2,3] | PP5 |
| S191 | | | | | S400 | | | | |
| Y197 | | | | | T403 | | PP2A | | |
| S198 | | | PP2B[2,3] | | S404 | PP1 | PP2A | PP2B[2,3] | PP5 |
| S199 | PP1 | PP2A | PP2B[2,3] | PP5 | S409 | PP1+++ | PP2A+++ | PP2B+++ | PP5+++ |
| S202 | PP1 | PP2A | PP2B[2,3] | PP5 | S412 | | | | |
| T205 | PP1+++ | PP2A+++* | PP2B | PP5+++ | S413 | | | | |
| S208 | | | | | T414 | | | | |
| S210 | | | | | S416 | | | | |
| T212 | PP1+++ | PP2A+++* | PP2B | PP5+++ | S422 | PP1 | PP2A | PP2B | |
| S214 | PP1+++ | PP2A | PP2B | PP5 | T427 | | | | |
| T217 | PP1 | PP2A* | PP2B | | S433 | | | | |
| T231 | | PP2A | PP2B[1] | | S435 | | | | |

More than 50 sites of phosphorylation have been identified in tau.[26] The dephosphorylation of a limited number has been analyzed using available phospho-specific antibodies, and in *in vitro* assays various preparations of PP1,[28,29] PP2A,[28,30,31] PP2B,[28,32] and PP5[28,33,34] exhibit selectivity for different sites (+++ indicates highest activity in *in vitro* assays). *Note that studies of PP2A came to a different assessment of the dephosphorylation of some sites. Studies of the role of PP2B were also assessed *in vivo*: [1]PP2B knockdown in rat brain[35]; [2]cyclosporin infusion into mouse left lateral ventricle[36]; [3]FK506 infusion in mouse[37] hyperphosphorylation of tau at so many sites is that there is a significant decrease in the levels or activity of the protein phosphatase(s) that dephosphorylate tau in AD brain. Protein phosphatases are classified on the basis of their ability to dephosphorylate either serine and threonine residues (PPPs and PPMs) or tyrosine residues (protein tyrosine phosphatase, PTPs or dual-specificity DSPs). The PPP family includes PP1, PP2A, PP2B (also known as calcineurin), PP4, PP5, and PP6, while the PPM family includes various forms of PP2C and mitochondrial PPases. High levels of the serine/threonine PPases are found throughout the brain, with PP1, PP2A, PP2B, and PP5 being abundant and implicated in AD. Recent studies have also begun to reveal roles for PTPases, especially striatal-enriched tyrosine phosphatase (STEP), in AD. Here, we review recent advances in our understanding of the role of these serine/threonine and tyrosine protein phosphatases in the pathology of AD.

## II. Protein Phosphatase 2A

### A. General Properties

Protein phosphatase 2A (PP2A) is widely expressed throughout the body and plays a predominant role in the dephosphorylation of thousands of different phosphoproteins in vertebrate tissues.[38] PP2A is expressed at high levels in the brain, where it acts on diverse substrates.[39–42] It has been estimated that perhaps more than 50% of all proteins are regulated by one or more of several hundred different protein kinases whose activities are, in turn, regulated by a wide range of signal transduction pathways.[43,44] This regulatory network is highly interconnected, as kinases are invariably subject to cross talk regulation wherein one or more kinase phosphorylate and thereby activate or inhibit one or more other kinase. In this respect, in addition to its general role in the direct dephosphorylation of many cellular phosphoproteins, PP2A is responsible for the dephosphorylation of many protein kinases.[45] The broad spectrum of PP2A activity means that any dysfunction in PP2A can have profound consequences on diverse cellular processes. In neurodegenerative disorders, exemplified by AD, any PP2A dysfunction would lead to an imbalance in kinase-mediated signaling pathways that contribute to multiple pathologies at the core of the disease process.

The majority of PP2A exists as a functional heterotrimer consisting of a catalytic C subunit, a scaffold-like A subunit, and one of a series of alternative regulatory B subunits.[38] Four structurally unrelated families of B subunits have been identified, termed B, B', B", and B"'. Splice variants lead to at least 24 alternative forms. Holoenzymes with different B subunits direct PP2A to different spectra of substrates and different subcellular compartments.

Additionally, the specificity and subcellular localization of the PP2A catalytic subunit are posttranslationally regulated by a chemistry that appears to be completely specific to PP2A involving the formation and hydrolysis of a leucine carboxyl methyl ester at the carboxyl terminal leucine L309.[46–49] PP2A carboxyl methylation facilitates the assembly and enhances the activity of microtubule-associated PP2A in neuronal cells.[48,49] In addition to carboxyl methylation at L309, the catalytic subunit of PP2A activity is inhibited by tyrosine phosphorylation at Y307.[50] The C subunit can also be phosphorylated at T304.

## B. Protein Phosphatase 2A Linkage to Alzheimer's Disease

Postmortem studies have closely linked PP2A at multiple levels with AD. Ultimately, the activity of PP2A is reduced in the disease owing to reduced levels, increased inhibition, and alterations in its specificity and subcellular localization.[51–54] Decreased expression of mRNA encoding the PP2A catalytic subunit has been reported,[55] which may underlie the decreased levels of the protein that are observed.[53] A protein that binds and inhibits PP2A, termed SET or inhibitor 2 (I2), is highly expressed in AD brains.[56] Thus, in AD, increased levels of SET probably contribute to the general downregulation of PP2A associated with disease progression. Note, however, that the level of ARPP-19, which has recently been found to be an inhibitor of PP2A,[57,58] has also been found to be reduced in AD brain.[59] Importantly, a substantial decrease in PP2A methylation has been observed in postmortem brains of AD patients.[52,60] This is associated with decreased levels of a PP2A targeting subunit Bα/PR55, which binds preferentially to the methylated form of the PP2A catalytic core.[52,61]

Other recent studies have identified potential genetic links between AD and PP2A. The number of CAG repeats in the gene encoding the Bβ subunit, which had previously been linked to spinocerebellar ataxia 12, was found to be reduced in a Taiwanese AD cohort, and this might be associated with lower levels of expression of PP2A.[62] A follow-up study of a Japanese AD cohort found a similar result, particularly in a subgroup of samples who expressed the APOE4 allele.[63] Therefore, there are multiple mechanistic links between PP2A and AD.

## C. Pathophysiological Roles of Protein Phosphatase 2A in Alzheimer's Disease

Based on a number of biochemical studies, where the ability of different PPase preparations immunoprecipitated from human brain was used to dephosphorylate hyperphosphorylated tau, it appears that PP2A is the major phospho-tau phosphatase[28,30,33,34,64] (Table I). Thus the reduced PP2A activity mentioned above appears likely to be a major factor in increased tau phosphorylation and NFT pathology.[65] As tau is phosphorylated at so many sites, studies have been limited to only a subset of sites where specific phospho-antibodies are available. In addition, only qualitative assessment of the rates of dephosphorylation and affinity for any given site is possible. With these limitations in mind, T205, T212, S262, and S409 appear to be the preferred sites for PP2A, with S262 being the site that is most favored by PP2A compared to other PPases. However, as for other sites, these individual sites can be dephosphorylated by more than one PPase (see Table I).

As mentioned above, PP2A exists as a core enzyme consisting of the AC dimer together with various B subunits that influence substrate specificity. Only the Bα-containing isoform has been demonstrated to effectively bind to and dephosphorylate tau,[53] with the mechanism of selective interaction of the Bα subunit and phospho-tau revealed by X-ray crystallography studies.[66] Thus, although other forms of PP2A may be relevant for additional pathways that contribute to the disease, the data to date suggest that it is the deficiency in methylated ABαC that is the primary PP2A defect in AD pathogenesis.[52,53] Recent studies have also suggested that the ability of PP2A to dephosphorylate tau is regulated by the activity of the peptidyl-prolyl isomerase Pin1.[54,67,68] However, differing roles for Pin1 in promoting or inhibiting the effects of PP2A have been observed.

Regulation of PP2A has also been connected to APP. Phosphorylation of APP influences the ability of β-secretase to cleave APP to produce the most amyloidogenic and toxic peptide $A\beta_{1-42}$.[18] PP2A appears to inhibit the generation of this peptide,[69] presumably by dephosphorylating APP at Thr668 where it is phosphorylated by Cdk5[14] and/or c-jun N-terminal kinase.[70] Interestingly, a recent RNA Seq study using cells expressing the APPswe mutant form of APP showed that the expression levels of the PP2A catalytic subunit as well as several PP2A regulatory subunits were decreased.[71] The mechanism involved was not elucidated, but this preparation might be useful for understanding the effects of mutant APP expression on transcriptional events. The catalytic subunit of PP2A is known to be inhibited by phosphorylation of Y307.[50] Increased levels of Y307 were found in cells expressing the APPswe form of APP and in transgenic mice expressing APPswe and presenilin 1.[72] Increased phosphorylation of Y307 was also observed in sections of hippocampus and entorhinal cortex from human AD patients, especially in neurons containing NFTs. These results suggest that inhibition of PP2A may be caused by Aβ, and that this is linked to hyperphosphorylation of tau.

PP2A is also actively involved in apoptotic cell death and therefore can directly contribute to neurodegeneration, the final hallmark pathology of AD.[73] Generally, PP2A's proapoptotic functions involve dephosphorylation of Bcl family proteins.[74] In the progression of AD, the relevance of PP2A in early apoptotic processes may be relevant, such as in synaptic and axonal pruning, which ultimately lead to the programmed, controlled process of cell death.[75]

PP2A has diverse cellular roles beyond these pathological hallmarks, as it regulates multiple cellular signaling pathways. Neuroinflammatory processes are closely linked to a cascade of serine/threonine phosphorylation events, and the action of PP2A on the kinases in these pathways reduces their activity and can thus decrease inflammation.[76] Neuroinflammation is a key component of AD, with abnormally accumulated peptides such as Aβ being highly immunogenic.[77] Amyloid plaques are generally surrounded by microglia, and the associated inflammatory responses are thought to play a major role in neuronal cell death.[78] Decreased PP2A levels and activity may therefore be associated with increased inflammation. Anti-inflammatory agents are effective in reducing cognitive deficits in rodent models of AD[78–80] and have potential therapeutic utility in humans.[81,82]

PP2A is also intimately linked with cell cycle progression.[38] In AD, stress-induced re-entry/cell cycle re-entry has been proposed as a key potential initiating event.[83] Normally, quiescent neurons are seen to display markers of cell division,[84] and in an animal model in which cell cycle re-entry is driven by overexpression of the SV40 large T antigen, an inhibitor of PP2A, AD-like pathology is observed.[85] PP2A controls the G2/M transition,[86] and PP2A demethylation has been associated with cell division.[87] Thus, reduced PP2A methylation may be causally associated with the observed changes in cell cycle status.

## D. *In Vivo* Models Linking PP2A to Alzheimer's Disease

The central role that PP2A dysregulation appears to play in most of the pathological hallmarks of AD, and the mounting evidence of significant deficits in PP2A regulation in brains from individuals with AD, has led to the study of this mechanism in animal models. PP2A plays such critical roles in physiological functioning throughout the body that it has not been possible to generate knockout mouse models.[88] Knockdown of PP2A phosphatase activity has been achieved *in vivo* by overexpressing dominant negative forms of the catalytic subunit. This has been shown to lead to the tau hyperphosphorylation characteristic of the human disorder.[89,90] Pharmacological inhibition of PP2A with okadaic acid has yielded similar phenotypes. When okadaic acid was stereotactically injected into the brains of rats, many of the hallmark features of AD were recapitulated including Aβ deposition, tau hyperphosphorylation, and neurodegeneration.[91,92] But note that the effects of PP2A inhibitors such as okadaic acid may not be selective, and other PPases including PP1, PP4–6 may also be inhibited at the concentrations needed in these *in vivo* studies.

A chronic model of PP2A inhibition has been developed using I2 (SET) overexpression.[93] I2 is overexpressed in AD, and in the diseased state, it is cleaved into N- and C-terminal fragments that redistribute from the nucleus to the cytoplasm where they can act upon PP2A. Wang *et al.* have developed a model in which the C-terminal fragment of I2 ($I2_{CTF}$) is delivered to rat brains via adeno-associated virus (AAV), leading to its overexpression in the hippocampus.[93] AAV-$I2_{CTF}$ infected rats display Aβ deposition, tau hyperphosphorylation, neurodegeneration, and cognitive deficits linked to inhibition of PP2A activity. Another approach to test the effects of PP2A dysregulation involves introducing deficiencies in methylation through generation of a transgenic mouse with a mutated PP2A C subunit (L309A) that is unable to undergo C-terminal methylation. This mouse model displays tau hyperphosphorylation and microtubule dysfunction, consistent with the importance of PP2A methylation in tau regulation.[94] Although not directly linked to AD, a recent study has examined the effects of knockout of the B'δ subunit and found that this resulted in spatially restricted tauopathy in the brain stem and spinal cord.[95] There were no obvious cognitive impairments in this mouse model, and the effects of the B'δ knockout were likely explained through the ability of this subunit to influence the activities of GSK3 and/or Cdk5, which phosphorylated tau.

Additional models relevant to AD have further linked PP2A to the disorder. For example, anesthesia induces tau hyperphosphorylation,[96] which has been demonstrated to be due to decreases in PP2A activity associated with anesthesia-induced hypothermia. Anesthesia-induced hypothermia differentially affects activity of kinases and phosphatases, although even in normothermic conditions some anesthetics can inhibit PP2A to induce tau hyperphosphorylation.[97]

# III. Protein Phosphatase 2B (Calcineurin)

## A. General Properties

Protein phosphatase 2B (PP2B or calcineurin) is a $Ca^{2+}$/calmodulin-dependent protein phosphatase that is highly enriched in the central nervous system where it plays key roles in diverse aspects of signaling.[98] PP2B is comprised of a dimer of the catalytic A subunit and a $Ca^{2+}$-binding B subunit. Following increased intracellular $Ca^{2+}$, $Ca^{2+}$/calmodulin binds to the A subunit, relieving autoinhibition. PP2B is the sole target for the immunosuppressant drugs cyclosporin and FK506 which bind to cyclophilin or FKBP, respectively, and then bind to and inhibit PP2B activity. PP2B is required for long-term depression (LTD), where it mediates the actions of $Ca^{2+}$ downstream of NMDA glutamate receptors.[99–101] In this respect, it is involved in control of synaptic plasticity and learning and memory. PP2B also plays an important role in the striatum, where it regulates dopaminergic signaling.[42] PP2B acts in different cellular compartments, with substrates located at the pre- and postsynaptic side of the synapse.[102,103] An important family of substrates for PP2B is the NFAT transcription factors, to which PP2B interacts with in the cytosol, leading to dephosphorylation of NFATs at multiple sites allowing the import of NFAT into the nucleus and activation of specific transcriptional programs.[104]

## B. Protein Phosphatase 2B Linkage to Alzheimer's Disease

Decreases in PP2B levels have been associated with normal aging.[105] While earlier studies found either no change in PP2B protein levels or activity in AD brain[106] or reduction in activity,[107–109] more recent studies have consistently indicated that PP2B activity is increased.[28,110,111] The increase in activity appears to result from the formation of a truncated A subunit.[110,111] The position of the proteolytic cleavage is C-terminal to the autoinhibitory domain, and PP2B remains $Ca^{2+}$/calmodulin dependent.[110] However, the resulting activity of the truncated PP2B is higher than that of the full-length A subunit form of the enzyme. Proteolysis of the A subunit may be mediated by calpain I, which was also found to be more active in AD brain. Interestingly, proteolysis of PP2B and calpain was also found in hippocampus from subjects with mild cognitive impairment.[112] In this latter study, a shorter active form of the A subunit was detected, and it was also found that oligomeric forms of Aβ could increase PP2B proteolysis in cultured hippocampal neurons. Related to these results, a recent study of the Tg2576-APPswe mouse model of AD that expresses the human APPswe mutant linked to FAD found that caspase-3 activity was enhanced in hippocampus and that this was associated with truncation and activation of PP2B[113] (see further discussion below).

Although PP2B has been found to interact with fewer regulatory proteins compared to PP1 and PP2A, PP2B is known to be inhibited by calcipressin1 (also known as DSCR1, Adapt78, or RCAN1), the product of a gene encoded in the Down's syndrome critical region 1. Both mRNA and protein levels of calcipressin1 were found to be increased in AD brain, and the level of expression of calcipressin1 was correlated with the number of NFTs in the temporal cortex.[114–116]

The catalytic A subunit of PP2B dimerizes with the regulatory B subunit, a $Ca^{2+}$-binding protein related to calmodulin and other EF-hand $Ca^{2+}$-binding proteins. A recent genetic study, that aimed to identify associations between single nucleotide polymorphisms (SNPs) in AD subjects and levels of phospho-tau measured in patient CSF samples, has identified an SNP located in intron 5 of the B subunit (PPP3R1) of PP2B.[117] The SNP was associated with the rate of decline in disease progression but was not associated with the risk for AD, or the age of onset. Notably, the B subunit allele was associated with lower levels of PPP3R1 and higher levels of NFT pathology.

There are three genes that encode the catalytic A subunit (PPP3CA–C) and two genes that encode the regulatory B subunit (PPP3R1 and R2). A detailed analysis has identified a variety of novel PPP3CA variants that can generate a number of splicing isoforms of PP2B that exhibit differential expression in brain and nonbrain tissues.[118] The expression of several of these PPP3CA isoforms, but not all, that encode functional phosphatase domains was found to be decreased in the medial temporal gyrus from AD patient brains.

In summary, contradictory results have been obtained from studies of PP2B expression and/or activity. Both decreases and increases have been identified, with the increases being associated with proteolysis of the regulatory A subunit. Conceivably, proteolysis and increased activity might be related to postmortem conditions and not directly related to AD phenotype. It is notable that some genetic studies have found both decreased expression of A subunit isoforms and decreased levels of the B subunit that are associated with the degree of AD progression. Increased levels of a PP2B inhibitor, namely, calcipressin1, have also been detected. It remains possible that there may be local changes in PP2B subunit expression or expression of calcipressin1 in specific brain regions, or in specific subcellular compartments that lead to restricted inhibition of phosphatase activity which could, in turn, influence the phosphorylation of substrates such as tau (see below for further discussion). In this respect, one study of a double APP/PS1 AD mouse model has reported that there was increased PP2B immunoreactivity found in activated astrocytes, but not neurons, surrounding amyloid deposits.[119]

## C. Pathophysiological Roles of Protein Phosphatase 2B in Alzheimer's Disease

PP2B activity has been implicated in the generation of both Aβ and hyperphosphorylated tau, the two hallmarks of AD pathology. Early studies suggested a role for PP2B in the production of Aβ, but the molecular basis for the effects observed using PP2B inhibitors was never identified.[120] Several studies have implicated PP2B in the regulation of tau phosphorylation. As for PP2A (discussed above), PP2B can directly dephosphorylate specific sites in hyperphosphorylated tau (Table I).[28,30,32] These *in vitro* studies suggested that PP2B may not be a very active tau phosphatase, and consistent with this conclusion, other studies in AD brain have indicated that the increased proteolysis and activation of PP2B were not associated with dephosphorylation of tau but was correlated with hyperphosphorylation of tau.[110,111] In contrast, downregulation of PP2B using antisense oligonucleotides or the inhibitors cyclosporin or FK506 resulted in increased phosphorylation of a number of sites in tau that are implicated in formation of NFTs.[35–37] Given that PP2B can directly dephosphorylate some of the same sites *in vitro* that are altered by PP2B inhibitors, the simplest explanation of these results is that PP2B directly can dephosphorylate certain sites in tau *in vivo*. However, PP2B, like other PPases, may also act indirectly to control the activities of the kinases that phosphorylate tau. In this regard, PP2B was shown to dephosphorylate S9 of GSK3β in neuroblastoma cells, leading to activation of GSK3β and phosphorylation of tau.[121]

## D. Role of Protein Phosphatase 2B in the Synaptopathology of Alzheimer's Disease

As mentioned above, there is a growing recognition that the earliest cognitive impairments seen in AD may result from loss of functional synaptic transmission. Studies that argued against the original amyloid hypothesis of AD found that the patterns of synapse loss, rather than amyloid deposits, correlated best with the cognitive deficits in affected patients.[122–124] Subsequent studies in AD model mice and through the use of Aβ oligomeric preparations in *in vitro* studies indicated that synaptic plasticity at excitatory synapses was impaired.[125–127] Further studies indicated that application of Aβ to brain slices and into rodent brain induced synaptic loss, blocked LTP, and impaired cognitive function.[8,9,128–135] Aβ is produced in neurons in an activity-dependent manner and may normally be part of a negative feedback process that controls excitatory synaptic transmission.[127] However, higher levels of soluble Aβ oligomers lead to defective AMPA and NMDA glutamate receptor trafficking and ultimately synaptic loss.[133,136–138] The precise mechanisms involved in the actions of Aβ oligomers at the cell surface are not clear but may include both pre- and postsynaptic signaling processes and α7-nicotinic acetylcholine receptors, glutamate receptors, and the cellular prion protein.[133,136,139–141] Glutamate re-uptake at excitatory synapses may also be impaired, leading to altered synaptic and extrasynaptic signaling.[139] There is now a significant body of evidence that supports a key role for PP2B in mediating the disruptive effects of Aβ on synaptic structure and function.

The first suggestion for a role for PP2B in the actions of Aβ came from studies of LTP in dentate gyrus.[142] Both induction and maintenance of early and late phases of NMDA receptor-dependent LTP were found to be inhibited by application of Aβ1–42. Based on previous data which indicated that Aβ could increase intracellular $Ca^{2+}$ levels[143,144] and the knowledge that PP2B plays a key role in regulation of synaptic plasticity,[101] the effects of the immunosuppressant inhibitors of PP2B, namely, cyclosporin A and FK506, were studied. Both cyclosporin A and FK506, but not rapamycin (a control for FK506), prevented the effects of Aβ1–42 on LTP. The mechanism(s) involved in the effects of Aβ were investigated in detail by Snyder *et al.*,[136] who found that application of Aβ to cortical cultures resulted in decreased surface expression of NMDA glutamate receptors at synapses through promotion of endocytosis. Moreover, there was reduced surface expression of NMDA receptors in cultures obtained from the APPswe mouse model of AD. Further work established a role for α-7 nicotine receptor activation by Aβ and a requirement for PP2B, based on the ability of cyclosporin to block the effects of Aβ on NMDA receptor surface expression. As discussed in more detail below, a likely substrate for PP2B is the tyrosine phosphatase, STEP61, which is known to regulate NMDA receptor endocytosis through dephosphorylation of Tyr1472 of NR2B. Aβ was also able to block downstream signaling via NMDA receptors as demonstrated by reduced phosphorylation of the key transcription factor CREB in cultured neurons. Additional studies also showed that Aβ exposure resulted in increased endocytosis of AMPA receptors[137,138,145] and that PP2B was required for the effect of Aβ.[138,145] The effects of Aβ on disruption of synaptic signaling were also accompanied by loss of dendritic spines, and although the precise mechanisms have not been clearly resolved, PP2B has been shown to be required for the effects of Aβ on spine loss.[133,138]

PP2B is required for LTD, and studies of overexpression and inhibition of PP2B in mice show that it plays an important negative role in learning and memory.[101] Consistent with this role, intraperitoneal injection of FK506 was found to rescue deficits in contextual fear conditioning and novel object recognition in relatively young (5-month-old) mice expressing human APPswe (the Tg2576 AD mouse model).[146,147] Taken together, these studies strongly support a role for PP2B in mediating early effects of Aβ on synaptic plasticity and synaptic structure that is likely linked to impairment of learning and memory as well as cognition. Given the similarities observed, it has been suggested that the effects of Aβ on synaptic plasticity share common mechanisms with those involved in LTD.[133,138]

## E. Mechanisms of Action of Protein Phosphatase 2B in Mediating the Effects of Aβ

Despite the clear evidence that PP2B is involved in the actions of Aβ on synaptic function, a number of important questions remain to be answered. Several mechanisms may be involved in the activation of PP2B. Consistent with the growing appreciation that soluble oligomeric forms of Aβ are critical for the early synaptic dysfunction, activation of PP2B also requires Aβ oligomers.[133,145,148,149] The target for Aβ oligomers is not, however, clear and several possible binding partners have been suggested.[140,145] Snyder *et al.* implicated α7 nicotinic receptors in the effects of Aβ on NMDA receptor endocytosis,[136] but no role for α7 receptors were found in studies of Aβ on spine loss.[133] Rather, impairment in NMDA receptor-dependent $Ca^{2+}$ influx was suggested. This may seem paradoxical since PP2B requires $Ca^{2+}$ for activation. However, PP2B exhibits high affinity for $Ca^{2+}$/calmodulin and is preferentially responsive to small increases in $Ca^{2+}$, which is the basis for its selective activation in LTD.[100] Alternatively, as discussed above, PP2B may be proteolytically cleaved to produce a more active form.[110,111] While most of the studies implicating PP2B in Aβ actions have focused on soluble oligomeric forms, and processes that occur prior to amyloid plaque deposition, a recent study in AD model mice expressing human APP and presenilin 1 shows elevated $Ca^{2+}$ overload in neurites and spines.[150] Notably, the changes in $Ca^{2+}$ were located in the proximity of amyloid plaques, where it was shown that spino-dendritic $Ca^{2+}$ compartmentalization was perturbed. Intraperitoneal injection of FK506 prevented $Ca^{2+}$ overload and the associated structural changes, further complicating the identification of the source of $Ca^{2+}$ that might activate PP2B.

PP2B is a multifunctional phosphatase that dephosphorylates a wide variety of substrates in different cellular compartments.[101,103,151] Key phosphorylation sites in either NMDA or AMPA receptors that are associated with receptor trafficking may be direct substrates for PP2B.[103,152,153] In support of this, studies of the Tg2576 AD mouse model suggested that S845 of GluR1 was a direct target for PP2B.[113] Alternatively, PP2B may act to control glutamate receptor trafficking indirectly through regulation of substrates that include the tyrosine phosphatase STEP (see further discussion below). PP2B plays an important role in the regulation of transcription, and transcription factors, including NFAT1-4 and MEF2, are important neuronal substrates. MEF2 is activated by PP2B and this is known to be coupled to loss of dendritic spines,[154] although to date there are no studies of MEF2 in AD. NFAT proteins have been studied intensively in immune signaling, and the immunosuppressants cyclosporin A and FK506 work through inhibition of the actions of PP2B dephosphorylation of NFATs and nuclear translocation. Notably, two recent studies have implicated NFAT transcription factors in the actions of PP2B, acting downstream of Aβ.[149,155] While the general conclusions reached by the two studies were similar, the details were distinct. One study found increased nuclear translocation of NFAT1 and NFAT3 which was associated with mild cognitive decline or AD, respectively,[155] while the second study found increased NFAT4 in nuclear fractions from the cortex of AD patients.[149] Oligomeric Aβ was found to stimulate NFAT in astrocyte cultures and to influence glutamate-induced neuronal degeneration.[155] In contrast, in the second study, NFAT signaling was required for dendritic simplification and spine loss in neurons, a process dependent on PP2B.[149] Finally, it has also been suggested that proapoptotic proteins may be activated through their ability to be dephosphorylated and activated by PP2B, and that this is involved in the neurodegenerative actions of Aβ.[156,157]

# IV. Protein Phosphatase 1

## A. General Properties

PP1 is known to have an important role in several aspects of neuronal function.[100,158] It plays a key role in synaptic signaling, where it is required for LTD.[99,159] Behavioral studies have also elegantly shown that PP1 controls aspects of learning and memory.[160] There are four isoforms of PP1 that are the products of three genes, with PP1α and γ being highly enriched in dendritic spines where they are positioned to regulate early stages of postsynaptic signaling.[158] A notable feature of PP1 is that the catalytic subunits of the phosphatase can interact in a mutually exclusive way, with as many as 200 distinct regulatory proteins that target PP1 to specific subcellular locations where they influence substrate specificity.[161] For example, the F-actin-binding proteins spinophilin and neurabin are localized to actin-rich dendritic spines where they recruit PP1 to selectively dephosphorylate glutamate receptors.[162–164] PP1 also plays an important role in other cellular compartments such as the nucleus where it is a major phosphatase that dephosphorylates S133 in CREB[165,166] and is also targeted to histone modification via interactions with HDACs.[167] Extensive studies have shown that PP1 is a major target for dopamine signaling in striatal neurons where it is regulated by DARPP-32, a protein highly expressed in dopamine-innervated medium spiny neurons.[42]

## B. Pathophysiological Roles of Protein Phosphatase 1 in Alzheimer's Disease

Although most of the focus has been on the role of PP2A, PP1 has also been implicated in the regulation of tau dephosphorylation. For example, PP1 prepared from human postmortem brain was found to exhibit some site selectivity toward hyperphosphorylated tau.[28,29] T212, T217, S262, S396, and S422 were found to be preferentially dephosphorylated by the PP1 preparations, while T181, S199, S202, T205, S214, and S404 were not dephosphorylated. Of the selected sites, in one study, it was suggested that T212 might be a specific site for PP1,[29] as this site was apparently not a good substrate for PP2A or PP2B. However, this observation was not confirmed in another study.[28] It was also not clear what the status of the PP1 preparation was in terms of potential PP1 regulatory proteins, which might be expected to significantly influence substrate specificity.

In other recent studies, PP1 and hyperphosphorylated tau have been connected to deficits in axonal transport via a mechanism involving PP1. Using a squid axoplasm preparation, a reduced system in which various aspects of axonal transport can be studied, earlier studies found that filamentous human tau could inhibit fast anterograde axonal transport.[168] The effect of human tau was found to act via a process that involved activation of PP1, dephosphorylation and activation of GSK3, and subsequent phosphorylation of kinesin light chains by GSK3. This process depended on an 18-amno acid domain at the N-terminus of tau, which the authors suggested was a PP1-activating motif. In a follow-up study, it has been shown that pathogenic AD forms of tau enable greater exposure of the PP1-activating motif, which is normally sequestered by protein–protein interactions.[169] Moreover, there is a large increase in the accessibility of the PP1-activating motif in postmortem samples from AD patients. The mechanism by which PP1 might be activated is not known. However, it seems possible that this might involve recruitment of a form of PP1 that is targeted to GSK3 via a specific PP1-targeting protein. Alternatively, the N-terminal region of tau might perturb the interaction of a specific targeting subunit with PP1.

There is no obvious data related directly to studies of PP1 activity or levels in AD brain. However, PP1 activity may be required for mediating the effects of Aβ on synaptic plasticity. In a study that used transgenic mice expressing human APP with both Swedish and Arctic mutations, hippocampal slice LTP was found to be inhibited.[170] A selective PP1 inhibitor blocked the effects of Aβ, and a similar result was found in hippocampal slices isolated from APPswe/PS1 mice. Notably, the effects on LTP were not observed in mice that over-expressed a PP1 inhibitor, a result that is consistent with the known role for PP1 in LTD. Early studies of LTD in hippocampal neurons indicated that PP1 is required downstream of PP2B,[171] although the targets for PP1 have not been clearly identified. As for PP2B, PP1 may act directly to dephosphorylate NMDA and AMPA glutamate receptors and regulate their trafficking to and from synapses. Alternatively, PP1 may act at the level of regulation of substrates such as CaM kinase II, or transcription factors such as CREB. The tyrosine protein phosphatase, STEP, may also be a target since it can be regulated by PP1 (see below). Interestingly, PP1 has been suggested to be inhibited by Aβ,[172] but it is not clear how this would relate to known synaptic roles of PP1 in LTD where PP1 activity would be required for the disruptive effects of Aβ on synaptic plasticity.

A more indirect role for PP1 in AD has been suggested by studies of the translational control of BACE1 protein levels.[173] The level of BACE1 protein, which is the rate-limiting protease involved in Aβ formation, is known to be upregulated in AD.[174,175] In attempts to address the possible mechanisms involved, studies carried out in HEK cells and primary neuronal cultures indicated that cell stress leads to regulation of the unfolded protein response system and increased phosphorylation of the translational initiation factor eIF2α. Notably, while phosphorylation of eIF2α leads to a general inhibition of translation, certain mRNAs including that for BACE1 can be preferentially translated, leading to increased BACE1 protein synthesis. A large body of work has found that eIF2α dephosphorylation is regulated by PP1 in complexes with the specific PP1 targeting protein GADD34.[176,177] In primary neurons, the small molecule, salubrinal, which specifically inhibits PP1 in this complex,[178] selectively increased eIF2α phosphorylation leading to increased levels of BACE1 and Aβ. In contrast, blocking eIF2α phosphorylation had the opposite effect. Analysis of the 5×FAD AD mouse model found parallel increases in levels of BACE1 and eIF2α phosphorylation, and a similar effect on eIF2α was observed in samples from AD brains. As for PP2A (discussed earlier), these studies highlight the potential importance of studying specific forms of PP1 in complexes with unique targeting subunits that may play important roles in the regulation of neuronal processes related to AD.

# V. Protein Phosphatase 5

## A. General Properties

Protein phosphatase 5 (PP5) is a serine/threonine PPase related to PP2A, which is ubiquitously expressed but present in high levels in neurons.[179] It is unique in terms of the domain structure of the catalytic subunit, in that it contains three so-called tetratricopeptide repeat domains at the N-terminus which may play a role in autoinhibition of PPase activity. While PP5 is less studied than PP2A, a few reports have suggested a role for PP5 in regulation of tau dephosphorylation and also in the toxic effects of Aβ.

## B. Pathophysiological Roles of Protein Phosphatase 5 in Alzheimer's Disease

Using either recombinant PP5[33] or enzyme immunoprecipitated from rat or human brain,[28,34] PP5 has been shown to dephosphorylate a number of sites in hyperphosphorylated tau. T205, T212, and S409 were relatively good substrates for PP5, while S199, T202, S214, S396, and S404 were less efficiently phosphorylated (Table I). Interestingly, while S199 was not the best site for PP5 compared to PP1, PP2A, and PP2B, PP5 was the most effective PPase for this site. Analysis of postmortem samples from human AD brains indicated that PP5 levels and activity were reduced by ~20% compared to control samples, suggesting a contributing role for PP5 in the increased phosphorylation of tau seen in AD brain.[28,34]

PP5 may also be able to play a neuroprotective role to attenuate the effects of Aβ toxicity.[180] Previous studies have indicated that Aβ impairs mitochondrial function and can increase the levels of reactive oxygen species that may be causally involved in neuronal toxicity.[181] In cortical neurons in culture, PP5 downregulation was associated with increased Aβ-induced cell death, while overexpression of PP5 had the opposite effect. PP5 may prevent the actions of Aβ through its ability to suppress MAP kinase pathways involved in apoptosis.

# VI. Striatal-Enriched Tyrosine Phosphatase

## A. General Properties

STEP is an intracellular tyrosine phosphatase expressed in the striatum, hippocampus, neocortex, and other brain regions that are involved in learning and memory.[182–184] STEP isoforms include a cytosolic STEP46 and a membrane-associated STEP61, the latter being localized, in part, to the postsynaptic density.[185] A large body of work supports a model in which STEP opposes the development of synaptic strengthening through its ability to dephosphorylate various cellular substrates involved in this process (reviewed in Refs.[184,186]). These include key neuronal signaling molecules such as the MAP kinases, ERK1/2, and p38,[187,188] the Src family kinase Fyn,[189] and Pyk2.[190] STEP also regulates NMDA and AMPA receptor trafficking. STEP61 dephosphorylates the NR2B subunit of the NMDA receptor, leading to internalization of NR1/NR2B. Increased STEP61 activity was associated with increased internalization of GluR1/GluR2 receptor complexes and STEP KO mice have increased levels of these AMPA receptors on neuronal synaptic membranes, although whether STEP directly dephosphorylates the GluR2 subunit of the AMPA receptor is under investigation.[136,191–194]

STEP is also subject to regulation by various mechanisms that control its level of expression and activity. An important feature of STEP is the presence of a kinase-interacting motif (KIM domain) that is essential for substrate binding.[187,195] Notably, phosphorylation by protein kinase A at a serine residue within the KIM acts to inhibit the interaction of STEP with substrates,[196] and phosphorylation of the KIM domain can be reversed by the actions of either PP1 or PP2B.[188,196,197] The action of PP1 on STEP is subject to regulation in striatal neurons by dopamine through the ability of DARPP-32 to inhibit PP1, a process that contributes to synergistic activation of ERK1/2 by glutamate and dopamine and which plays an important role in the actions of drugs of abuse.[197] In addition to regulation of activity, STEP protein levels can be controlled by local translation in neurons[198] or through degradation by calpain-dependent cleavage or through ubiquitination and targeting to the proteasome.[199–201]

Given its critical role in regulation of synaptic function, as well as its regulation by various signaling pathways, it is not surprising that STEP has been implicated in various neuropsychiatric and neurological disorders, including drug addiction, schizophrenia, fragile X syndrome, and stroke.[186] Several recent studies have also revealed a key role for STEP in AD.

## B. Pathophysiological Roles of STEP in Alzheimer's Disease

As discussed above, there is a growing appreciation that the effects of Aβ on synaptic plasticity are responsible, at least in part, for the cognitive decline observed in AD. An important series of studies that contributed to this hypothesis has found that Aβ reduces surface expression of both NMDA and AMPA glutamate receptors at synapses through increased endocytosis and that STEP plays a central role in the actions of Aβ.[136,137,194,202] In the case of NMDA receptors, activation of α-7 nicotinic receptors is coupled to activation of PP2B, which promotes dephosphorylation of STEP at the regulatory site within the KIM domain.[136] PP2B may act directly on STEP as a substrate or act via inhibitor-1/DARPP-32 and PP1.[197] In either case, dephosphorylation of STEP would promote its interaction with substrates. As mentioned earlier, STEP is known to regulate NMDA receptor endocytosis through its ability to regulate the dephosphorylation the NR2B subunit.[193,194] STEP likely modulates the phosphorylation of the NR2B subunit of the NMDA receptor by two parallel pathways. STEP can directly dephosphorylate Y1472 of the NR2B subunit.[136,201] STEP may also act indirectly via dephosphorylation and inactivation of the Src family kinase Fyn, which is a kinase implicated in phosphorylation of Y1472.[189] Phosphorylation of Y1472 resides within a conserved tyrosine-dependent endocytic motif.[203] When dephosphorylated by STEP, the tyrosine residue in this motif binds to clathrin adapter proteins via strong hydrophobic interactions and promotes endocytosis of the NMDA receptor.

Further direct support for a role of STEP in mediating the actions of Aβ on glutamate receptor endocytosis has come from studies of STEP knockout mice.[194] STEP knockout mice[204] were crossed with 3×Tg-AD mice[205] to produce progeny null for STEP but with elevated Aβ levels (3×Tg-AD/STEP$^{-/-}$; double mutant).[194] Genetic reduction of STEP attenuated the loss of NR1 and NR2B NMDA receptor subunits from synaptosomal membranes observed in the 3×Tg-AD mice. Similar biochemical results were observed in Tg2576 mice in which STEP was knocked out. Importantly, the recovery in levels of NMDA receptors was accompanied by attenuation of the cognitive deficits normally seen in the 3×Tg-AD mice. Specifically, 6-month-old double mutant mice were improved, relative to 3×Tg-AD littermates, when tested for spatial reference memory in the Morris water maze, spontaneous alteration performance in the

Y maze, and nonspatial hippocampus-dependent memory in an object recognition task. Notably, all groups tested displayed similar locomotor activity and exploratory behavior in the open field task. An important aspect of this study was that the attenuation of AD-like cognitive deficits in double mutant mice took place despite unchanged levels of both Aβ and phospho-tau, which are measurable at 6 months of age in 3×Tg-AD mice. These findings suggest that the improved cognitive function was due to the decrease in STEP levels and that these improvements could be achieved without diminishing Aβ and phospho-tau levels at least at the earlier age tested.

STEP is also likely to be involved in regulation of AMPA receptor endocytosis by Aβ. Biochemical studies have implicated STEP in regulation of dephosphorylation of the GluR2 AMPA receptor subunit,[192] although the precise mechanism remains to be clearly elucidated. The GluR2 subunit of the AMPA receptor appears to be phosphorylated by Src family kinases.[206] A recent study has found that phosphorylation of Y876 in GluR2 controls its endocytosis via a mechanism involving the guanine-exchange factor BRAG2, the GTPAse Arf6, and the adaptor protein AP2.[207] As for NMDA receptors, STEP may therefore act either directly to dephosphorylate the GluR2 subunit at Y876 or alternatively to regulate the activity of Src family kinases or other tyrosine kinases that phosphorylate this site. In support of a direct role for STEP, recent studies have shown that the decrease in surface expression of GluR1 and GluR2 observed in Tg2576 mice is reversed when STEP is knocked out.[202]

As mentioned above, the Src family kinase Fyn is a substrate for STEP that may be involved in regulation of glutamate receptor endocytosis. For many years, Fyn has been implicated in various aspects of AD.[208] Fyn phosphorylates tau and may be involved in regulation of tau hyperphosphorylation.[209,210] Fyn has also been implicated in the synaptic and cognitive impairments caused by Aβ.[211,212] Interestingly, recent studies suggest that tau may act as a scaffold for Fyn that is needed for the effects of Aβ.[213,214] Activation of Fyn is achieved by intermolecular autophosphorylation of Y420 in its catalytic domain, and STEP can inactivate Fyn through dephosphorylation of Y420.[189] It is not clear how the actions of STEP might influence these various effects of active Fyn in terms of mediating the effects of Aβ, but it is possible that activation of Fyn and activation of STEP occur with different kinetics or in different postsynaptic locations in neurons affected by Aβ.

## C. Levels of STEP Protein Expression Are Elevated in Alzheimer's Disease

In addition to regulation of activity, the levels of expression of STEP protein have been implicated in AD. Elevated levels of the STEP61 isoform are present in Tg2576[201] and 3×Tg-AD mice.[194] Other studies have found that

STEP61 levels are elevated in the J20 AD mouse line.[212] Moreover, elevated levels of STEP61 have been found in prefrontal cortex from AD patients.[201] The mechanism for the increase in STEP61 levels involves Aβ-mediated inhibition of the ubiquitin proteasome system (UPS).[201] Increasing evidence suggests that UPS dysfunction plays an important role in the pathogenesis of AD.[215–217] In human AD brains, ubiquitin immunoreactive inclusion bodies accumulate and proteasomal activity is decreased.[218,219] Proteasomal inhibition results in the accumulation of ubiquitinated proteins, a decrease in free ubiquitin, and increased levels of several proteins involved in AD pathology, including tau and BACE1.[220–225] Notably, proteasome activity decreases with age in the brains of Tg2576 mice,[226] while restoring ubiquitin-recycling enzymes rescues memory deficits and dendritic spine alterations in AD mouse models.[224,227] STEP degradation is controlled by the UPS, and impairment of the UPS seen in response to Aβ in cell-based assays, or in the Tg2576 AD mouse model, has been shown to lead to increased levels of ubiquitinated STEP that remains active.[201] Thus Aβ modulates STEP via two parallel pathways which are not mutually exclusive. Aβ-induced activation of PP2B leads to dephosphorylation of STEP61 and activation, while Aβ-mediated inhibition of the UPS leads to reduced degradation of STEP61. In either case, elevated levels of Aβ result in upregulation of STEP activity and consequently lead to decreased phosphorylation and surface expression of NMDA and AMPA glutamate receptors, leading to reduced cognitive ability.

## VII. Other Protein Phosphatases

### A. Role of Additional Protein Phosphatases in the Pathophysiology of Alzheimer's Disease

Given the likely role of diverse signaling pathways being involved in the actions of Aβ in the synaptopathology of AD, and the complexity of the numerous sites being hyperphosphorylated in tau by various kinases, it is to be expected that roles for other serine/threonine and tyrosine protein phosphatases will be elucidated in the near future. In support of this, recent studies have begun to reveal unexpected roles for a number of protein phosphatases in AD-related contexts. For example, a recent study has shown that the leukocyte common antigen CD45, a receptor tyrosine phosphatase that plays an important role in the immune response, contributes to microglial-mediated clearance of Aβ oligomers.[228] Promotion of CD45 action might therefore be of help in treatment of AD. In another recent study, the PTPase PTP1B has been implicated in the susceptibility to diet-induced obesity and glucose intolerance in an APPswe/PSEN1 transgenic mouse line.[229] Increased PTP1B activity is likely associated with insulin resistance, a possible risk factor associated with AD. A number of previous studies in AD mouse models have indicated that there is aberrant upregulation of expression in postmitotic neurons of proteins normally associated with the cell cycle.[230] Included in these cell cycle proteins is the tyrosine phosphatase Cdc25, which plays a critical role at the G2/M transition in the cell cycle through its ability to activate Cdc2/cyclin B.[231,232]

## VIII. Protein Phosphatase-Directed Therapeutics for the Prevention and Treatment of Alzheimer's Disease

According to the National Institute on Aging, there are more than 5 million Americans who suffer from AD. Yet, despite a desperate need, development of an effective treatment for AD has been a major challenge. Four drugs are currently approved by the FDA to treat cognitive deficits in AD. Three of them are acetylcholinesterase inhibitors, which combat the loss of acetylcholine caused by the death of cholinergic neurons, while the other is a noncompetitive NMDA receptor antagonist, which inhibits overactivation of NMDA receptors by glutamate.[1] None of these drugs halts progression of the disease. Significant effort has been put into development of inhibitors of Aβ production.[233] However, targeting γ-secretase with nonselective γ-secretase inhibitors has deleterious effects on health, likely because γ-secretase also cleaves other substrates, such as Notch, which are essential for normal biological function.[234] This limitation is highlighted by the recent failure of the γ-secretase inhibitor semagacestat in Phase III clinical trials.[235] Given the lack of available drugs to treat AD, it is imperative to identify novel molecular processes that might be amenable to targeting through new drugs. Protein phosphatases, including PP2A, and STEP may be potential therapeutic targets for AD. It may also be possible to use a strategy that would combine one or more regulator of a protein phosphatase with kinase inhibitors that would provide a complementary approach to control of, for example, hyperphosphorylation of tau.

### A. PP2A Regulators

Postmortem analysis and *in vivo* models give strong indications of a central role for PP2A in AD. Understanding neuronal mechanisms of PP2A regulation could therefore provide useful pharmaceutical targets for rational therapeutic interventions to treat or prevent AD. Convincing data exist that reductions in PP2A activity can contribute to AD progression so that enhanced activity would likely be beneficial. Although generally more difficult to identify than inhibitors, it may be possible to either generate direct PP2A activators or take advantage of the endogenous biochemical regulatory mechanisms to achieve this goal. A number of compounds have been reported that activate PP2A (reviewed in Ref. 236). Sodium selenate has recently been tested in a model of AD.[237,238] This anionic PP2A activator was administered to tau transgenic animals where it was found to reduce tau hyperphosphorylation and enhance cognitive performance. The commonly used AD therapeutic memantine has been demonstrated to inhibit I2's activity toward PP2A, which may explain some of its therapeutic efficacy.[239] Metformin, which enhances PP2A levels by inhibiting its degradation, has been shown to be beneficial in models of AD.[240] Finally, it has recently been shown that an inhibitor of PP2A demethylation, namely, eiconsanoyl-5-hydroxytryptamide, provides enhanced PP2A activity, reduced protein phosphorylation, and cognitive benefits in rodent models for neurodegeneration.[241]

The encouraging data from multiple PP2A enhancing strategies suggest efficacy. Questions remain as to whether PP2A can be safely modulated in the therapeutic environment, particularly due to its ubiquitous nature and broad range of substrates. Some of the mechanisms of PP2A activation or stabilization described above may provide a level of selectivity: for example, modulation of only certain regulatory subunits. Given the compelling data linking PP2A to AD and the favorable outcomes from preliminary studies with rodent models, it is apparent that PP2A-targeted therapeutic approaches have significant potential for groundbreaking disease modifying pharmaceutical development.

## B. STEP Inhibitors

Based on studies which show that STEP normally acts to oppose synaptic strengthening, that knockout of STEP can attenuate behavioral and biochemical deficits in 3×Tg-AD mice, and that STEP is a tyrosine phosphatase with a known structure and enzymatic mechanism, STEP would appear to be a viable drug target, inhibition of which may potentially alleviate some of the synaptopathology of AD. However, past and current efforts to develop drugs targeting specific PTPs have been plagued by issues related to bioavailability and selectivity. This is due to the fact that the majority of PTP inhibitors carry a tyrosine phosphate-mimicking group that provides most of the binding energy through interaction with a highly conserved phosphate-binding loop in the catalytic center of every PTP. Based on emerging evidence, selectivity of inhibitors may be more readily achieved by targeting and stabilizing an inactive open-state PTP conformation, which dramatically differs from the active closed state.[242–244] In the open state, the flexible WPD-loop, which contains the catalytic acid/base aspartate, is distant from the catalytic center. Interestingly, the 3D structure of STEP exhibits an atypical open conformation that differs from most PTPs.[245] In this conformation, the WPD-loop is farther retracted, resulting in a large binding pocket that is not dominated by the conserved phosphate-binding loop. Hence, small molecules that bind STEP in its open state are likely to be very selective inhibitors. Moreover, STEP has a unique glutamine residue in the flexible WPD-loop, which may be utilized for specific interactions with a small molecule. In studies with the closely related phosphatase HePTP, a selective small-molecule inhibitor was found to interact specifically with the corresponding histidine residue in HePTP.[246] Several pharmaceutical companies have begun drug discovery programs to identify STEP inhibitors that will hopefully be available for preclinical studies in the near future.

## C. Other Protein Phosphatase Drug Targets

As for PP2A, PP5 appears to be downregulated in AD, and therefore activators of this enzyme would presumably be desirable. Conceivably, identification of drugs that could relieve autoinhibition of the enzyme might be useful. With respect to development of drugs that target PP1, a major limitation is the fact that there are potentially 200 or more forms of PP1 in complexes with the large number of regulatory/targeting subunits that have been identified. As discussed above, recent studies have identified the drug salubrinal that targets one specific form of PP1 in a complex with GADD34 and is used to increase the phosphorylation of initiation factor eIF2$\alpha$.[178] This demonstrates the feasibility of designing selective PP1 drugs. However, the specific example of salubrinal acts to paradoxically increase the synthesis of BACE1, which would not presumably be useful in the treatment of AD. Highly specific inhibitors of PP2B (calcineurin), that act as immunosuppressants, have been identified and used for several decades following transplantation surgery.[98,247,248] While commonly used, long-term cyclosporin and FK506 use is associated with nephrotoxicity, and in some cases, neurotoxicity.[249,250] However, given the recent results indicating that active fragments of PP2B are generated in response to A$\beta$, and that PP2B mediates many of the effects of A$\beta$, it may be worth considering the use of these immunosuppressant drugs in AD, or alternatively, using different types of drugs that might inhibit PP2B without unwanted side effects.[251]

## References

1. Castellani RJ, Rolston RK, Smith MA. Alzheimer disease. *Dis Mon* 2010;**56**:484–546.
2. Probst A, Langui D, Ulrich J. Alzheimer's disease: a description of the structural lesions. *Brain Pathol* 1991;**1**:229–39.
3. Haass C, Selkoe DJ. Soluble protein oligomers in neurodegeneration: lessons from the Alzheimer's amyloid beta-peptide. *Nat Rev Mol Cell Biol* 2007;**8**:101–12.
4. Goate A, Chartier-Harlin MC, Mullan M, Brown J, Crawford F, Fidani L, et al. Segregation of a missense mutation in the amyloid precursor protein gene with familial Alzheimer's disease. *Nature* 1991;**349**:704–6.

5. Pastor P, Goate AM. Molecular genetics of Alzheimer's disease. *Curr Psychiatry Rep* 2004;**6**:125–33.
6. Crews L, Rockenstein E, Masliah E. APP transgenic modeling of Alzheimer's disease: mechanisms of neurodegeneration and aberrant neurogenesis. *Brain Struct Funct* 2010; **214**:111–26.
7. Wolfe MS. Structure, mechanism and inhibition of gamma-secretase and presenilin-like proteases. *Biol Chem* 2010;**391**:839–47.
8. Venkitaramani DV, Chin J, Netzer WJ, Gouras GK, Lesne S, Malinow R, et al. Beta-amyloid modulation of synaptic transmission and plasticity. *J Neurosci* 2007;**27**:11832–7.
9. Palop JJ, Mucke L. Amyloid-beta-induced neuronal dysfunction in Alzheimer's disease: from synapses toward neural networks. *Nat Neurosci* 2010;**13**:812–8.
10. Suzuki T, Oishi M, Marshak DR, Czernik AJ, Nairn AC, Greengard P. Cell cycle-dependent regulation of the phosphorylation and metabolism of the Alzheimer amyloid precursor protein. *EMBO J* 1994;**13**:1114–22.
11. Suzuki T, Ando K, Isohara T, Oishi M, Lim GS, Satoh Y, et al. Phosphorylation of Alzheimer beta-amyloid precursor-like proteins. *Biochemistry* 1997;**36**:4643–9.
12. Oishi M, Nairn AC, Czernik AJ, Lim GS, Isohara T, Gandy SE, et al. The cytoplasmic domain of Alzheimer's amyloid precursor protein is phosphorylated at Thr654, Ser655, and Thr668 in adult rat brain and cultured cells. *Mol Med* 1997;**3**:111–23.
13. Isohara T, Horiuchi A, Watanabe T, Ando K, Czernik AJ, Uno I, et al. Phosphorylation of the cytoplasmic domain of Alzheimer's beta-amyloid precursor protein at Ser655 by a novel protein kinase. *Biochem Biophys Res Commun* 1999;**258**:300–5.
14. Iijima K, Ando K, Takeda S, Satoh Y, Seki T, Itohara S, et al. Neuron-specific phosphorylation of Alzheimer's beta-amyloid precursor protein by cyclin-dependent kinase 5. *J Neurochem* 2000;**75**:1085–91.
15. Suzuki T, Ando K, Iijima K, Oguchi S, Takeda S. Phosphorylation of amyloid precursor protein (APP) family proteins. *Methods Mol Med* 2000;**32**:271–82.
16. Sano Y, Nakaya T, Pedrini S, Takeda S, Iijima-Ando K, Iijima K, et al. Physiological mouse brain Abeta levels are not related to the phosphorylation state of threonine-668 of Alzheimer's APP. *PLoS One* 2006;**1**:e51.
17. da Cruz E, Silva EF, da Cruz E, Silva OA. Protein phosphorylation and APP metabolism. *Neurochem Res* 2003;**28**:1553–61.
18. Ando K, Iijima KI, Elliott JI, Kirino Y, Suzuki T. Phosphorylation-dependent regulation of the interaction of amyloid precursor protein with Fe65 affects the production of beta-amyloid. *J Biol Chem* 2001;**276**:40353–61.
19. Seeger M, Nordstedt C, Petanceska S, Kovacs DM, Gouras GK, Hahne S, et al. Evidence for phosphorylation and oligomeric assembly of presenilin 1. *Proc Natl Acad Sci USA* 1997;**94**:5090–4.
20. Walter J, Schindzielorz A, Grunberg J, Haass C. Phosphorylation of presenilin-2 regulates its cleavage by caspases and retards progression of apoptosis. *Proc Natl Acad Sci USA* 1999;**96**:1391–6.
21. Fluhrer R, Friedlein A, Haass C, Walter J. Phosphorylation of presenilin 1 at the caspase recognition site regulates its proteolytic processing and the progression of apoptosis. *J Biol Chem* 2004;**279**:1585–93.
22. Pastorino L, Ikin AF, Nairn AC, Pursnani A, Buxbaum JD. The carboxyl-terminus of BACE contains a sorting signal that regulates BACE trafficking but not the formation of total A(beta). *Mol Cell Neurosci* 2002;**19**:175–85.
23. von Arnim CA, Tangredi MM, Peltan ID, Lee BM, Irizarry MC, Kinoshita A, et al. Demonstration of BACE (beta-secretase) phosphorylation and its interaction with GGA1 in cells by fluorescence-lifetime imaging microscopy. *J Cell Sci* 2004;**117**:5437–45.

24. Lee VM, Goedert M, Trojanowski JQ. Neurodegenerative tauopathies. *Annu Rev Neurosci* 2001;**24**:1121–59.
25. Wray S, Saxton M, Anderton BH, Hanger DP. Direct analysis of tau from PSP brain identifies new phosphorylation sites and a major fragment of N-terminally cleaved tau containing four microtubule-binding repeats. *J Neurochem* 2008;**105**:2343–52.
26. Hanger DP, Anderton BH, Noble W. Tau phosphorylation: the therapeutic challenge for neurodegenerative disease. *Trends Mol Med* 2009;**15**:112–9.
27. Chung SH. Aberrant phosphorylation in the pathogenesis of Alzheimer's disease. *BMB Rep* 2009;**42**:467–74.
28. Liu F, Grundke-Iqbal I, Iqbal K, Gong CX. Contributions of protein phosphatases PP1, PP2A, PP2B and PP5 to the regulation of tau phosphorylation. *Eur J Neurosci* 2005;**22**:1942–50.
29. Rahman A, Grundke-Iqbal I, Iqbal K. Phosphothreonine-212 of Alzheimer abnormally hyperphosphorylated tau is a preferred substrate of protein phosphatase-1. *Neurochem Res* 2005;**30**:277–87.
30. Wang JZ, Grundke-Iqbal I, Iqbal K. Kinases and phosphatases and tau sites involved in Alzheimer neurofibrillary degeneration. *Eur J Neurosci* 2007;**25**:59–68.
31. Qian W, Shi J, Yin X, Iqbal K, Grundke-Iqbal I, Gong CX, et al. PP2A regulates tau phosphorylation directly and also indirectly via activating GSK-3beta. *J Alzheimers Dis* 2010;**19**:1221–9.
32. Rahman A, Grundke-Iqbal I, Iqbal K. PP2B isolated from human brain preferentially dephosphorylates Ser-262 and Ser-396 of the Alzheimer disease abnormally hyperphosphorylated tau. *J Neural Transm* 2006;**113**:219–30.
33. Gong CX, Liu F, Wu G, Rossie S, Wegiel J, Li L, et al. Dephosphorylation of microtubule-associated protein tau by protein phosphatase 5. *J Neurochem* 2004;**88**:298–310.
34. Liu F, Iqbal K, Grundke-Iqbal I, Rossie S, Gong CX. Dephosphorylation of tau by protein phosphatase 5: impairment in Alzheimer's disease. *J Biol Chem* 2005;**280**:1790–6.
35. Garver TD, Kincaid RL, Conn RA, Billingsley ML. Reduction of calcineurin activity in brain by antisense oligonucleotides leads to persistent phosphorylation of tau protein at Thr181 and Thr231. *Mol Pharmacol* 1999;**55**:632–41.
36. Yu DY, Luo J, Bu F, Song GJ, Zhang LQ, Wei Q. Inhibition of calcineurin by infusion of CsA causes hyperphosphorylation of tau and is accompanied by abnormal behavior in mice. *Biol Chem* 2006;**387**:977–83.
37. Luo J, Ma J, Yu DY, Bu F, Zhang W, Tu LH, et al. Infusion of FK506, a specific inhibitor of calcineurin, induces potent tau hyperphosphorylation in mouse brain. *Brain Res Bull* 2008;**76**:464–8.
38. Janssens V, Goris J. Protein phosphatase 2A: a highly regulated family of serine/threonine phosphatases implicated in cell growth and signalling. *Biochem J* 2001;**353**:417–39.
39. Strack S, Zaucha JA, Ebner FF, Colbran RJ, Wadzinski BE. Brain protein phosphatase 2A: developmental regulation and distinct cellular and subcellular localization by B subunits. *J Comp Neurol* 1998;**392**:515–27.
40. Saraf A, Virshup DM, Strack S. Differential expression of the B'beta regulatory subunit of protein phosphatase 2A modulates tyrosine hydroxylase phosphorylation and catecholamine synthesis. *J Biol Chem* 2007;**282**:573–80.
41. Van Kanegan MJ, Strack S. The protein phosphatase 2A regulatory subunits B'beta and B'delta mediate sustained TrkA neurotrophin receptor autophosphorylation and neuronal differentiation. *Mol Cell Biol* 2009;**29**:662–74.
42. Walaas SI, Hemmings Jr. HC, Greengard P, Nairn AC. Beyond the dopamine receptor: regulation and roles of serine/threonine protein phosphatases. *Front Neuroanat* 2011;**5**:50.
43. Hunter T. The age of crosstalk: phosphorylation, ubiquitination, and beyond. *Mol Cell* 2007;**28**:730–8.

44. Choudhary C, Mann M. Decoding signalling networks by mass spectrometry-based proteomics. *Nat Rev Mol Cell Biol* 2010;**11**:427–39.
45. Zolnierowicz S. Type 2A protein phosphatase, the complex regulator of numerous signaling pathways. *Biochem Pharmacol* 2000;**60**:1225–35.
46. Lee J, Stock J. Protein phosphatase 2A catalytic subunit is methyl-esterified at its carboxyl terminus by a novel methyltransferase. *J Biol Chem* 1993;**268**:19192–5.
47. Xie H, Clarke S. An enzymatic activity in bovine brain that catalyzes the reversal of the C-terminal methyl esterification of protein phosphatase 2A. *Biochem Biophys Res Commun* 1994;**203**:1710–5.
48. Tolstykh T, Lee J, Vafai S, Stock JB. Carboxyl methylation regulates phosphoprotein phosphatase 2A by controlling the association of regulatory B subunits. *EMBO J* 2000;**19**:5682–91.
49. Wu J, Tolstykh T, Lee J, Boyd K, Stock JB, Broach JR. Carboxyl methylation of the phosphoprotein phosphatase 2A catalytic subunit promotes its functional association with regulatory subunits in vivo. *EMBO J* 2000;**19**:5672–81.
50. Chen J, Martin BL, Brautigan DL. Regulation of protein serine-threonine phosphatase type-2A by tyrosine phosphorylation. *Science* 1992;**257**:1261–4.
51. Gong CX, Singh TJ, Grundke-Iqbal I, Iqbal K. Phosphoprotein phosphatase activities in Alzheimer disease brain. *J Neurochem* 1993;**61**:921–7.
52. Sontag E, Hladik C, Montgomery L, Luangpirom A, Mudrak I, Ogris E, et al. Downregulation of protein phosphatase 2A carboxyl methylation and methyltransferase may contribute to Alzheimer disease pathogenesis. *J Neuropathol Exp Neurol* 2004;**63**:1080–91.
53. Sontag E, Luangpirom A, Hladik C, Mudrak I, Ogris E, Speciale S, et al. Altered expression levels of the protein phosphatase 2A ABalphaC enzyme are associated with Alzheimer disease pathology. *J Neuropathol Exp Neurol* 2004;**63**:287–301.
54. Rudrabhatla P, Pant HC. Role of protein phosphatase 2A in Alzheimer's disease. *Curr Alzheimer Res* 2001. [Epub ahead of print].
55. Vogelsberg-Ragaglia V, Schuck T, Trojanowski JQ, Lee VM. PP2A mRNA expression is quantitatively decreased in Alzheimer's disease hippocampus. *Exp Neurol* 2001;**168**:402–12.
56. Tanimukai H, Grundke-Iqbal I, Iqbal K. Up-regulation of inhibitors of protein phosphatase-2A in Alzheimer's disease. *Am J Pathol* 2005;**166**:1761–71.
57. Gharbi-Ayachi A, Labbe JC, Burgess A, Vigneron S, Strub JM, Brioudes E, et al. The substrate of Greatwall kinase, Arpp 19, controls mitosis by inhibiting protein phosphatase 2A. *Science* 2010;**330**:1673–7.
58. Mochida S, Maslen SL, Skehel M, Hunt T. Greatwall phosphorylates an inhibitor of protein phosphatase 2A that is essential for mitosis. *Science* 2010;**330**:1670–3.
59. Kim SH, Nairn AC, Cairns N, Lubec G. Decreased levels of ARPP-19 and PKA in brains of Down syndrome and Alzheimer's disease. *J Neural Transm* 2001;**61**:263–72. Suppl.
60. Zhou XW, Gustafsson JA, Tanila H, Bjorkdahl C, Liu R, Winblad B, et al. Tau hyperphosphorylation correlates with reduced methylation of protein phosphatase 2A. *Neurobiol Dis* 2008;**31**:386–94.
61. Bryant JC, Westphal RS, Wadzinski BE. Methylated C-terminal leucine residue of PP2A catalytic subunit is important for binding of regulatory Balpha subunit. *Biochem J* 1999;**339** (Pt. 2):241–6.
62. Chen CM, Hou YT, Liu JY, Wu YR, Lin CH, Fung HC, et al. PPP2R2B CAG repeat length in the Han Chinese in Taiwan: association analyses in neurological and psychiatric disorders and potential functional implications. *Am J Med Genet B Neuropsychiatr Genet* 2009;**150B**:124–9.
63. Kimura R, Morihara T, Kudo T, Kamino K, Takeda M. Association between CAG repeat length in the PPP2R2B gene and Alzheimer disease in the Japanese population. *Neurosci Lett* 2011;**487**:354–7.

64. Liu F, Liang Z, Gong CX. Hyperphosphorylation of tau and protein phosphatases in Alzheimer disease. *Panminerva Med* 2006;**48**:97–108.
65. Iqbal K, Alonso AC, Chen S, Chohan MO, El-Akkad E, Gong CX, et al. Tau pathology in Alzheimer disease and other tauopathies. *Biochim Biophys Acta* 2005;**1739**:198–210.
66. Xu Y, Chen Y, Zhang P, Jeffrey PD, Shi Y. Structure of a protein phosphatase 2A holoenzyme: insights into B55-mediated Tau dephosphorylation. *Mol Cell* 2008;**31**:873–85.
67. Bulbarelli A, Lonati E, Cazzaniga E, Gregori M, Masserini M. Pin1 affects Tau phosphorylation in response to Abeta oligomers. *Mol Cell Neurosci* 2009;**42**:75–80.
68. Landrieu I, Smet-Nocca C, Amniai L, Louis JV, Wieruszeski JM, Goris J, et al. Molecular implication of PP2A and Pin1 in the Alzheimer's disease specific hyperphosphorylation of tau. *PLoS One* 2011;**6**:e21521.
69. Sontag E, Nunbhakdi-Craig V, Sontag JM, Diaz-Arrastia R, Ogris E, Dayal S, et al. Protein phosphatase 2A methyltransferase links homocysteine metabolism with tau and amyloid precursor protein regulation. *J Neurosci* 2007;**27**:2751–9.
70. Colombo A, Bastone A, Ploia C, Sclip A, Salmona M, Forloni G, et al. JNK regulates APP cleavage and degradation in a model of Alzheimer's disease. *Neurobiol Dis* 2009;**33**:518–25.
71. Shin J, Yu SB, Yu UY, Jo SA, Ahn JH. Swedish mutation within amyloid precursor protein modulates global gene expression towards the pathogenesis of Alzheimer's disease. *BMB Rep* 2010;**43**:704–9.
72. Liu R, Zhou XW, Tanila H, Bjorkdahl C, Wang JZ, Guan ZZ, et al. Phosphorylated PP2A (tyrosine 307) is associated with Alzheimer neurofibrillary pathology. *J Cell Mol Med* 2008;**12**:241–57.
73. Klumpp S, Krieglstein J. Serine/threonine protein phosphatases in apoptosis. *Curr Opin Pharmacol* 2002;**2**:458–62.
74. Ruvolo PP, Deng X, Ito T, Carr BK, May WS. Ceramide induces Bcl2 dephosphorylation via a mechanism involving mitochondrial PP2A. *J Biol Chem* 1999;**274**:20296–300.
75. Culmsee C, Landshamer S. Molecular insights into mechanisms of the cell death program: role in the progression of neurodegenerative disorders. *Curr Alzheimer Res* 2006;**3**:269–83.
76. Shanley TP, Vasi N, Denenberg A, Wong HR. The serine/threonine phosphatase, PP2A: endogenous regulator of inflammatory cell signaling. *J Immunol* 2001;**166**:966–72.
77. Tuppo EE, Arias HR. The role of inflammation in Alzheimer's disease. *Int J Biochem Cell Biol* 2005;**37**:289–305.
78. Lee YJ, Han SB, Nam SY, Oh KW, Hong JT. Inflammation and Alzheimer's disease. *Arch Pharm Res* 2010;**33**:1539–56.
79. Lim GP, Yang F, Chu T, Chen P, Beech W, Teter B, et al. Ibuprofen suppresses plaque pathology and inflammation in a mouse model for Alzheimer's disease. *J Neurosci* 2000;**20**:5709–14.
80. Cole GM, Morihara T, Lim GP, Yang F, Begum A, Frautschy SA. NSAID and antioxidant prevention of Alzheimer's disease: lessons from in vitro and animal models. *Ann N Y Acad Sci* 2004;**1035**:68–84.
81. Breitner JC. The role of anti-inflammatory drugs in the prevention and treatment of Alzheimer's disease. *Annu Rev Med* 1996;**47**:401–11.
82. Szekely CA, Breitner JC, Zandi PP. Prevention of Alzheimer's disease. *Int Rev Psychiatry* 2007;**19**:693–706.
83. Woods J, Snape M, Smith MA. The cell cycle hypothesis of Alzheimer's disease: suggestions for drug development. *Biochim Biophys Acta* 2007;**1772**:503–8.
84. Nagy Z, Esiri MM, Smith AD. Expression of cell division markers in the hippocampus in Alzheimer's disease and other neurodegenerative conditions. *Acta Neuropathol* 1997;**93**:294–300.

85. Park KH, Hallows JL, Chakrabarty P, Davies P, Vincent I. Conditional neuronal simian virus 40 T antigen expression induces Alzheimer-like tau and amyloid pathology in mice. *J Neurosci* 2007;**27**:2969–78.
86. Wurzenberger C, Gerlich DW. Phosphatases: providing safe passage through mitotic exit. *Nat Rev Mol Cell Biol* 2011;**12**:469–82.
87. Turowski P, Fernandez A, Favre B, Lamb NJ, Hemmings BA. Differential methylation and altered conformation of cytoplasmic and nuclear forms of protein phosphatase 2A during cell cycle progression. *J Cell Biol* 1995;**129**:397–410.
88. Gotz J, Probst A, Ehler E, Hemmings B, Kues W. Delayed embryonic lethality in mice lacking protein phosphatase 2A catalytic subunit Calpha. *Proc Natl Acad Sci USA* 1998;**95**:12370–5.
89. Kins S, Crameri A, Evans DR, Hemmings BA, Nitsch RM, Gotz J. Reduced protein phosphatase 2A activity induces hyperphosphorylation and altered compartmentalization of tau in transgenic mice. *J Biol Chem* 2001;**276**:38193–200.
90. Deters N, Ittner LM, Gotz J. Substrate-specific reduction of PP2A activity exaggerates tau pathology. *Biochem Biophys Res Commun* 2009;**379**:400–5.
91. Arendt T, Holzer M, Fruth R, Bruckner MK, Gartner U. Paired helical filament-like phosphorylation of tau, deposition of beta/A4-amyloid and memory impairment in rat induced by chronic inhibition of phosphatase 1 and 2A. *Neuroscience* 1995;**69**:691–8.
92. Arendt T, Holzer M, Fruth R, Bruckner MK, Gartner U. Phosphorylation of tau, Abeta-formation, and apoptosis after in vivo inhibition of PP-1 and PP-2A. *Neurobiol Aging* 1998;**19**:3–13.
93. Wang X, Blanchard J, Kohlbrenner E, Clement N, Linden RM, Radu A, et al. The carboxy-terminal fragment of inhibitor-2 of protein phosphatase-2A induces Alzheimer disease pathology and cognitive impairment. *FASEB J* 2010;**24**:4420–32.
94. Schild A, Schmidt K, Lim YA, Ke Y, Ittner LM, Hemmings BA, et al. Altered levels of PP2A regulatory B/PR55 isoforms indicate role in neuronal differentiation. *Int J Dev Neurosci* 2006;**24**:437–43.
95. Louis JV, Martens E, Borghgraef P, Lambrecht C, Sents W, Longin S, et al. Mice lacking phosphatase PP2A subunit PR61/B'delta (Ppp 2r5d) develop spatially restricted tauopathy by deregulation of CDK5 and GSK3beta. *Proc Natl Acad Sci USA* 2011;**108**:6957–62.
96. Planel E, Richter KE, Nolan CE, Finley JE, Liu L, Wen Y, et al. Anesthesia leads to tau hyperphosphorylation through inhibition of phosphatase activity by hypothermia. *J Neurosci* 2007;**27**:3090–7.
97. Whittington RA, Virag L, Marcouiller F, Papon MA, El Khoury NB, Julien C, et al. Propofol directly increases tau phosphorylation. *PLoS One* 2011;**6**:e16648.
98. Aramburu J, Rao A, Klee CB. Calcineurin: from structure to function. *Curr Top Cell Regul* 2000;**36**:237–95.
99. Mulkey RM, Endo S, Shenolikar S, Malenka RC. Involvement of a calcineurin/inhibitor-1 phosphatase cascade in hippocampal long-term depression. *Nature* 1994;**369**:486–8.
100. Winder DG, Sweatt JD. Roles of serine/threonine phosphatases in hippocampal synaptic plasticity. *Nat Rev Neurosci* 2001;**2**:461–74.
101. Mansuy IM. Calcineurin in memory and bidirectional plasticity. *Biochem Biophys Res Commun* 2003;**311**:1195–208.
102. Cousin MA, Robinson PJ. The dephosphins: dephosphorylation by calcineurin triggers synaptic vesicle endocytosis. *Trends Neurosci* 2001;**24**:659–65.
103. Sanderson JL, Dell'Acqua ML. AKAP signaling complexes in regulation of excitatory synaptic plasticity. *Neuroscientist* 2011;**17**:321–36.
104. Oliveira AM, Bading H. Calcium signaling in cognition and aging-dependent cognitive decline. *Biofactors* 2011;**37**:168–74.

105. Agbas A, Zaidi A, Michaelis EK. Decreased activity and increased aggregation of brain calcineurin during aging. *Brain Res* 2005;**1059**:59–71.
106. Billingsley ML, Ellis C, Kincaid RL, Martin J, Schmidt ML, Lee VM, et al. Calcineurin immunoreactivity in Alzheimer's disease. *Exp Neurol* 1994;**126**:178–84.
107. Ladner CJ, Czech J, Maurice J, Lorens SA, Lee JM. Reduction of calcineurin enzymatic activity in Alzheimer's disease: correlation with neuropathologic changes. *J Neuropathol Exp Neurol* 1996;**55**:924–31.
108. Lian Q, Ladner CJ, Magnuson D, Lee JM. Selective changes of calcineurin (protein phosphatase 2B) activity in Alzheimer's disease cerebral cortex. *Exp Neurol* 2001;**167**:158–65.
109. Celsi F, Svedberg M, Unger C, Cotman CW, Carri MT, Ottersen OP, et al. Beta-amyloid causes downregulation of calcineurin in neurons through induction of oxidative stress. *Neurobiol Dis* 2007;**26**:342–52.
110. Liu F, Grundke-Iqbal I, Iqbal K, Oda Y, Tomizawa K, Gong CX. Truncation and activation of calcineurin A by calpain I in Alzheimer disease brain. *J Biol Chem* 2005;**280**:37755–62.
111. Qian W, Yin X, Hu W, Shi J, Gu J, Grundke-Iqbal I, et al. Activation of protein phosphatase 2B and hyperphosphorylation of Tau in Alzheimer's disease. *J Alzheimers Dis* 2011;**23**:617–27.
112. Mohmmad AH, Baig I, Levine III H, Guttmann RP, Norris CM. Proteolysis of calcineurin is increased in human hippocampus during mild cognitive impairment and is stimulated by oligomeric Abeta in primary cell culture. *Aging Cell* 2011;**10**:103–13.
113. D'Amelio M, Cavallucci V, Middei S, Marchetti C, Pacioni S, Ferri A, et al. Caspase-3 triggers early synaptic dysfunction in a mouse model of Alzheimer's disease. *Nat Neurosci* 2011;**14**:69–76.
114. Ermak G, Morgan TE, Davies KJ. Chronic overexpression of the calcineurin inhibitory gene DSCR1 (Adapt78) is associated with Alzheimer's disease. *J Biol Chem* 2001;**276**:38787–94.
115. Ermak G, Davies KJ. Gene expression in Alzheimer's disease. *Drugs Today (Barc)* 2002;**38**:509–16.
116. Cook CN, Hejna MJ, Magnuson DJ, Lee JM. Expression of calcipressin1, an inhibitor of the phosphatase calcineurin, is altered with aging and Alzheimer's disease. *J Alzheimers Dis* 2005;**8**:63–73.
117. Cruchaga C, Kauwe JS, Mayo K, Spiegel N, Bertelsen S, Nowotny P, et al. SNPs associated with cerebrospinal fluid phospho-tau levels influence rate of decline in Alzheimer's disease. *PLoS Genet* 2010;**6**:e1001101.
118. Chiocco MJ, Zhu X, Walther D, Pletnikova O, Troncoso JC, Uhl GR, et al. Fine mapping of calcineurin (PPP3CA) gene reveals novel alternative splicing patterns, association of 5'UTR trinucleotide repeat with addiction vulnerability, and differential isoform expression in Alzheimer's disease. *Subst Use Misuse* 2010;**45**:1809–26.
119. Norris CM, Kadish I, Blalock EM, Chen KC, Thibault V, Porter NM, et al. Calcineurin triggers reactive/inflammatory processes in astrocytes and is upregulated in aging and Alzheimer's models. *J Neurosci* 2005;**25**:4649–58.
120. Desdouits F, Buxbaum JD, Desdouits-Magnen J, Nairn AC, Greengard P. Amyloid beta peptide formation in cell-free preparations. Regulation by protein kinase C, calmodulin, and calcineurin. *J Biol Chem* 1996;**271**:24670–4.
121. Kim Y, Lee YI, Seo M, Kim SY, Lee JE, Youn HD, et al. Calcineurin dephosphorylates glycogen synthase kinase-3 beta at serine-9 in neuroblast-derived cells. *J Neurochem* 2009;**111**:344–54.
122. Katzman R, Terry R, DeTeresa R, Brown T, Davies P, Fuld P, et al. Clinical, pathological, and neurochemical changes in dementia: a subgroup with preserved mental status and numerous neocortical plaques. *Ann Neurol* 1988;**23**:138–44.
123. DeKosky ST, Scheff SW. Synapse loss in frontal cortex biopsies in Alzheimer's disease: correlation with cognitive severity. *Ann Neurol* 1990;**27**:457–64.

124. Terry RD, Masliah E, Salmon DP, Butters N, DeTeresa R, Hill R, et al. Physical basis of cognitive alterations in Alzheimer's disease: synapse loss is the major correlate of cognitive impairment. *Ann Neurol* 1991;**30**:572–80.
125. Hsia AY, Masliah E, McConlogue L, Yu GQ, Tatsuno G, Hu K, et al. Plaque-independent disruption of neural circuits in Alzheimer's disease mouse models. *Proc Natl Acad Sci USA* 1999;**96**:3228–33.
126. Larson J, Lynch G, Games D, Seubert P. Alterations in synaptic transmission and long-term potentiation in hippocampal slices from young and aged PDAPP mice. *Brain Res* 1999;**840**:23–35.
127. Kamenetz F, Tomita T, Hsieh H, Seabrook G, Borchelt D, Iwatsubo T, et al. APP processing and synaptic function. *Neuron* 2003;**37**:925–37.
128. Mucke L, Masliah E, Yu GQ, Mallory M, Rockenstein EM, Tatsuno G, et al. High-level neuronal expression of abeta 1–42 in wild-type human amyloid protein precursor transgenic mice: synaptotoxicity without plaque formation. *J Neurosci* 2000;**20**:4050–8.
129. Walsh DM, Klyubin I, Fadeeva JV, Cullen WK, Anwyl R, Wolfe MS, et al. Naturally secreted oligomers of amyloid beta protein potently inhibit hippocampal long-term potentiation in vivo. *Nature* 2002;**416**:535–9.
130. Cleary JP, Walsh DM, Hofmeister JJ, Shankar GM, Kuskowski MA, Selkoe DJ, et al. Natural oligomers of the amyloid-beta protein specifically disrupt cognitive function. *Nat Neurosci* 2005;**8**:79–84.
131. Palop JJ, Chin J, Roberson ED, Wang J, Thwin MT, Bien-Ly N, et al. Aberrant excitatory neuronal activity and compensatory remodeling of inhibitory hippocampal circuits in mouse models of Alzheimer's disease. *Neuron* 2007;**55**:697–711.
132. Lacor PN, Buniel MC, Furlow PW, Clemente AS, Velasco PT, Wood M, et al. Abeta oligomer-induced aberrations in synapse composition, shape, and density provide a molecular basis for loss of connectivity in Alzheimer's disease. *J Neurosci* 2007;**27**:796–807.
133. Shankar GM, Bloodgood BL, Townsend M, Walsh DM, Selkoe DJ, Sabatini BL. Natural oligomers of the Alzheimer amyloid-beta protein induce reversible synapse loss by modulating an NMDA-type glutamate receptor-dependent signaling pathway. *J Neurosci* 2007;**27**:2866–75.
134. Selkoe DJ. Alzheimer's disease is a synaptic failure. *Science* 2002;**298**:789–91.
135. Selkoe DJ. Soluble oligomers of the amyloid beta-protein impair synaptic plasticity and behavior. *Behav Brain Res* 2008;**192**:106–13.
136. Snyder EM, Nong Y, Almeida CG, Paul S, Moran T, Choi EY, et al. Regulation of NMDA receptor trafficking by amyloid-beta. *Nat Neurosci* 2005;**8**:1051–8.
137. Almeida CG, Tampellini D, Takahashi RH, Greengard P, Lin MT, Snyder EM, et al. Beta-amyloid accumulation in APP mutant neurons reduces PSD-95 and GluR1 in synapses. *Neurobiol Dis* 2005;**20**:187–98.
138. Hsieh H, Boehm J, Sato C, Iwatsubo T, Tomita T, Sisodia S, et al. AMPAR removal underlies Abeta-induced synaptic depression and dendritic spine loss. *Neuron* 2006;**52**:831–43.
139. Li S, Hong S, Shepardson NE, Walsh DM, Shankar GM, Selkoe D. Soluble oligomers of amyloid Beta protein facilitate hippocampal long-term depression by disrupting neuronal glutamate uptake. *Neuron* 2009;**62**:788–801.
140. Lauren J, Gimbel DA, Nygaard HB, Gilbert JW, Strittmatter SM. Cellular prion protein mediates impairment of synaptic plasticity by amyloid-beta oligomers. *Nature* 2009;**457**:1128–32.
141. Gimbel DA, Nygaard HB, Coffey EE, Gunther EC, Lauren J, Gimbel ZA, et al. Memory impairment in transgenic Alzheimer mice requires cellular prion protein. *J Neurosci* 2010;**30**:6367–74.
142. Chen QS, Wei WZ, Shimahara T, Xie CW. Alzheimer amyloid beta-peptide inhibits the late phase of long-term potentiation through calcineurin-dependent mechanisms in the hippocampal dentate gyrus. *Neurobiol Learn Mem* 2002;**77**:354–71.

143. Mattson MP, Chan SL. Dysregulation of cellular calcium homeostasis in Alzheimer's disease: bad genes and bad habits. *J Mol Neurosci* 2001;**17**:205–24.
144. Camandola S, Mattson MP. Aberrant subcellular neuronal calcium regulation in aging and Alzheimer's disease. *Biochim Biophys Acta* 2011;**1813**:965–73.
145. Zhao WQ, Santini F, Breese R, Ross D, Zhang XD, Stone DJ, et al. Inhibition of calcineurin-mediated endocytosis and alpha-amino-3-hydroxy-5-methyl-4-isoxazolepropionic acid (AMPA) receptors prevents amyloid beta oligomer-induced synaptic disruption. *J Biol Chem* 2010;**285**:7619–32.
146. Dineley KT, Hogan D, Zhang WR, Taglialatela G. Acute inhibition of calcineurin restores associative learning and memory in Tg2576 APP transgenic mice. *Neurobiol Learn Mem* 2007;**88**:217–24.
147. Taglialatela G, Hogan D, Zhang WR, Dineley KT. Intermediate- and long-term recognition memory deficits in Tg2576 mice are reversed with acute calcineurin inhibition. *Behav Brain Res* 2009;**200**:95–9.
148. Dineley KT, Kayed R, Neugebauer V, Fu Y, Zhang W, Reese LC, et al. Amyloid-beta oligomers impair fear conditioned memory in a calcineurin-dependent fashion in mice. *J Neurosci Res* 2010;**88**:2923–32.
149. Wu HY, Hudry E, Hashimoto T, Kuchibhotla K, Rozkalne A, Fan Z, et al. Amyloid beta induces the morphological neurodegenerative triad of spine loss, dendritic simplification, and neuritic dystrophies through calcineurin activation. *J Neurosci* 2010;**30**:2636–49.
150. Kuchibhotla KV, Goldman ST, Lattarulo CR, Wu HY, Hyman BT, Bacskai BJ. Abeta plaques lead to aberrant regulation of calcium homeostasis in vivo resulting in structural and functional disruption of neuronal networks. *Neuron* 2008;**59**:214–25.
151. Robinson RA, Lange MB, Sultana R, Galvan V, Fombonne J, Gorostiza O, et al. Differential expression and redox proteomics analyses of an Alzheimer disease transgenic mouse model: effects of the amyloid-beta peptide of amyloid precursor protein(Xi). *Neuroscience* 2011;**177**:207–22.
152. Ehlers MD. Reinsertion or degradation of AMPA receptors determined by activity-dependent endocytic sorting. *Neuron* 2000;**28**:511–25.
153. Snyder GL, Galdi S, Fienberg AA, Allen P, Nairn AC, Greengard P. Regulation of AMPA receptor dephosphorylation by glutamate receptor agonists. *Neuropharmacology* 2003;**45**:703–13.
154. Flavell SW, Cowan CW, Kim TK, Greer PL, Lin Y, Paradis S, et al. Activity-dependent regulation of MEF2 transcription factors suppresses excitatory synapse number. *Science* 2006;**311**:1008–12.
155. Abdul HM, Sama MA, Furman JL, Mathis DM, Beckett TL, Weidner AM, et al. Cognitive decline in Alzheimer's disease is associated with selective changes in calcineurin/NFAT signaling. *J Neurosci* 2009;**29**:12957–69.
156. Agostinho P, Lopes JP, Velez Z, Oliveira CR. Overactivation of calcineurin induced by amyloid-beta and prion proteins. *Neurochem Int* 2008;**52**:1226–33.
157. Reese LC, Zhang W, Dineley KT, Kayed R, Taglialatela G. Selective induction of calcineurin activity and signaling by oligomeric amyloid beta. *Aging Cell* 2008;**7**:824–35.
158. Mansuy IM, Shenolikar S. Protein serine/threonine phosphatases in neuronal plasticity and disorders of learning and memory. *Trends Neurosci* 2006;**29**:679–86.
159. Morishita W, Connor JH, Xia H, Quinlan EM, Shenolikar S, Malenka RC. Regulation of synaptic strength by protein phosphatase 1. *Neuron* 2001;**32**:1133–48.
160. Genoux D, Haditsch U, Knobloch M, Michalon A, Storm D, Mansuy IM. Protein phosphatase 1 is a molecular constraint on learning and memory. *Nature* 2002;**418**:970–5.
161. Bollen M. Combinatorial control of protein phosphatase-1. *Trends Biochem Sci* 2001;**26**:426–31.

162. Hsieh-Wilson LC, Benfenati F, Snyder GL, Allen PB, Nairn AC, Greengard P. Phosphorylation of spinophilin modulates its interaction with actin filaments. *J Biol Chem* 2003;**278**:1186–94.
163. Sarrouilhe D, di TA, Metaye T, Ladeveze V. Spinophilin: from partners to functions. *Biochimie* 2006;**88**:1099–113.
164. Ragusa MJ, Dancheck B, Critton DA, Nairn AC, Page R, Peti W. Spinophilin directs protein phosphatase 1 specificity by blocking substrate binding sites. *Nat Struct Mol Biol* 2010;**17**:459–64.
165. Hagiwara M, Alberts A, Brindle P, Meinkoth J, Feramisco J, Deng T, et al. Transcriptional attenuation following cAMP induction requires PP-1-mediated dephosphorylation of CREB. *Cell* 1992;**70**:105–13.
166. Koshibu K, Graff J, Beullens M, Heitz FD, Berchtold D, Russig H, et al. Protein phosphatase 1 regulates the histone code for long-term memory. *J Neurosci* 2009;**29**:13079–89.
167. Canettieri G, Morantte I, Guzman E, Asahara H, Herzig S, Anderson SD, et al. Attenuation of a phosphorylation-dependent activator by an HDAC-PP1 complex. *Nat Struct Biol* 2003;**10**:175–81.
168. Lapointe NE, Morfini G, Pigino G, Gaisina IN, Kozikowski AP, Binder LI, et al. The amino terminus of tau inhibits kinesin-dependent axonal transport: implications for filament toxicity. *J Neurosci Res* 2009;**87**:440–51.
169. Kanaan NM, Morfini GA, Lapointe NE, Pigino GF, Patterson KR, Song Y, et al. Pathogenic forms of tau inhibit kinesin-dependent axonal transport through a mechanism involving activation of axonal phosphotransferases. *J Neurosci* 2011;**31**:9858–68.
170. Knobloch M, Farinelli M, Konietzko U, Nitsch RM, Mansuy IM. Abeta oligomer-mediated long-term potentiation impairment involves protein phosphatase 1-dependent mechanisms. *J Neurosci* 2007;**27**:7648–53.
171. Mulkey RM, Herron CE, Malenka RC. An essential role for protein phosphatases in hippocampal long-term depression. *Science* 1993;**261**:1051–5.
172. Vintem AP, Henriques AG, da Cruz E, Silva OA, da Cruz E, Silva EF. PP1 inhibition by Abeta peptide as a potential pathological mechanism in Alzheimer's disease. *Neurotoxicol Teratol* 2009;**31**:85–8.
173. O'Connor T, Sadleir KR, Maus E, Velliquette RA, Zhao J, Cole SL, et al. Phosphorylation of the translation initiation factor eIF2alpha increases BACE1 levels and promotes amyloidogenesis. *Neuron* 2008;**60**:988–1009.
174. Yang LB, Lindholm K, Yan R, Citron M, Xia W, Yang XL, et al. Elevated beta-secretase expression and enzymatic activity detected in sporadic Alzheimer disease. *Nat Med* 2003;**9**:3–4.
175. Zhao J, Fu Y, Yasvoina M, Shao P, Hitt B, O'Connor T, et al. Beta-site amyloid precursor protein cleaving enzyme 1 levels become elevated in neurons around amyloid plaques: implications for Alzheimer's disease pathogenesis. *J Neurosci* 2007;**27**:3639–49.
176. Brush MH, Weiser DC, Shenolikar S. Growth arrest and DNA damage-inducible protein GADD34 targets protein phosphatase 1 alpha to the endoplasmic reticulum and promotes dephosphorylation of the alpha subunit of eukaryotic translation initiation factor 2. *Mol Cell Biol* 2003;**23**:1292–303.
177. Marciniak SJ, Ron D. Endoplasmic reticulum stress signaling in disease. *Physiol Rev* 2006;**86**:1133–49.
178. Boyce M, Bryant KF, Jousse C, Long K, Harding HP, Scheuner D, et al. A selective inhibitor of eIF2alpha dephosphorylation protects cells from ER stress. *Science* 2005;**307**:935–9.
179. Hinds TD, Jr., Sanchez ER. Protein phosphatase 5. *Int J Biochem Cell Biol* 2008;**40**:2358–62.
180. Sanchez-Ortiz E, Hahm BK, Armstrong DL, Rossie S. Protein phosphatase 5 protects neurons against amyloid-beta toxicity. *J Neurochem* 2009;**111**:391–402.

181. Reddy PH, Beal MF. Amyloid beta, mitochondrial dysfunction and synaptic damage: implications for cognitive decline in aging and Alzheimer's disease. *Trends Mol Med* 2008;**14**:45–53.
182. Lombroso PJ, Naegele JR, Sharma E, Lerner M. A protein tyrosine phosphatase expressed within dopaminoceptive neurons of the basal ganglia and related structures. *J Neurosci* 1993;**13**:3064–74.
183. Boulanger LM, Lombroso PJ, Raghunathan A, During MJ, Wahle P, Naegele JR. Cellular and molecular characterization of a brain-enriched protein tyrosine phosphatase. *J Neurosci* 1995;**15**:1532–44.
184. Braithwaite SP, Paul S, Nairn AC, Lombroso PJ. Synaptic plasticity: one STEP at a time. *Trends Neurosci* 2006;**29**:452–8.
185. Oyama T, Goto S, Nishi T, Sato K, Yamada K, Yoshikawa M, et al. Immunocytochemical localization of the striatal enriched protein tyrosine phosphatase in the rat striatum: a light and electron microscopic study with a complementary DNA-generated polyclonal antibody. *Neuroscience* 1995;**69**:869–80.
186. Goebel-Goody SM, Baum M, Paspalas C, Carty NC, Fernandez S, Kurup P, et al. Therapeutic implications for striatal-enriched protein tyrosine phosphatase (STEP) in neuropsychiatric disorders. *Pharmacol Rev* 2011. [Epub ahead of print].
187. Pulido R, Zuniga A, Ullrich A. PTP-SL and STEP protein tyrosine phosphatases regulate the activation of the extracellular signal-regulated kinases ERK1 and ERK2 by association through a kinase interaction motif. *EMBO J* 1998;**17**:7337–50.
188. Paul S, Nairn AC, Wang P, Lombroso PJ. NMDA-mediated activation of the tyrosine phosphatase STEP regulates the duration of ERK signaling. *Nat Neurosci* 2003;**6**:34–42.
189. Nguyen TH, Liu J, Lombroso PJ. Striatal enriched phosphatase 61 dephosphorylates Fyn at phosphotyrosine 420. *J Biol Chem* 2002;**277**:24274–9.
190. Venkitaramani DV, Moura PJ, Picciotto MR, Lombroso PJ. Striatal-enriched protein tyrosine phosphatase (STEP) knockout mice have enhanced hippocampal memory. *Eur J Neurosci* 2011;**33**:2288–98.
191. Pelkey KA, Askalan R, Paul S, Kalia LV, Nguyen TH, Pitcher GM, et al. Tyrosine phosphatase STEP is a tonic brake on induction of long-term potentiation. *Neuron* 2002;**34**:127–38.
192. Zhang Y, Venkitaramani DV, Gladding CM, Zhang Y, Kurup P, Molnar E, et al. The tyrosine phosphatase STEP mediates AMPA receptor endocytosis after metabotropic glutamate receptor stimulation. *J Neurosci* 2008;**28**:10561–6.
193. Braithwaite SP, Adkisson M, Leung J, Nava A, Masterson B, Urfer R, et al. Regulation of NMDA receptor trafficking and function by striatal-enriched tyrosine phosphatase (STEP). *Eur J Neurosci* 2006;**23**:2847–56.
194. Zhang Y, Kurup P, Xu J, Carty N, Fernandez SM, Nygaard HB, et al. Genetic reduction of striatal-enriched tyrosine phosphatase (STEP) reverses cognitive and cellular deficits in an Alzheimer's disease mouse model. *Proc Natl Acad Sci USA* 2010;**107**:19014–9.
195. Munoz JJ, Tarrega C, Blanco-Aparicio C, Pulido R. Differential interaction of the tyrosine phosphatases PTP-SL, STEP and HePTP with the mitogen-activated protein kinases ERK1/2 and p38alpha is determined by a kinase specificity sequence and influenced by reducing agents. *Biochem J* 2003;**372**:193–201.
196. Paul S, Snyder GL, Yokakura H, Picciotto MR, Nairn AC, Lombroso PJ. The Dopamine/D1 receptor mediates the phosphorylation and inactivation of the protein tyrosine phosphatase STEP via a PKA-dependent pathway. *J Neurosci* 2000;**20**:5630–8.
197. Valjent E, Pascoli V, Svenningsson P, Paul S, Enslen H, Corvol JC, et al. From The Cover: regulation of a protein phosphatase cascade allows convergent dopamine and glutamate signals to activate ERK in the striatum. *Proc Natl Acad Sci USA* 2005;**102**:491–6.

198. Hu Y, Zhang Y, Venkitaramani DV, Lombroso PJ. Translation of striatal-enriched protein tyrosine phosphatase (STEP) after beta1-adrenergic receptor stimulation. *J Neurochem* 2007;**103**:531–41.
199. Nguyen TH, Paul S, Xu Y, Gurd JW, Lombroso PJ. Calcium-dependent cleavage of striatal enriched tyrosine phosphatase (STEP). *J Neurochem* 1999;**73**:1995–2001.
200. Xu J, Kurup P, Zhang Y, Goebel-Goody SM, Wu PH, Hawasli AH, et al. Extrasynaptic NMDA receptors couple preferentially to excitotoxicity via calpain-mediated cleavage of STEP. *J Neurosci* 2009;**29**:9330–43.
201. Kurup P, Zhang Y, Xu J, Venkitaramani DV, Haroutunian V, Greengard P, et al. Abeta-mediated NMDA receptor endocytosis in Alzheimer's disease involves ubiquitination of the tyrosine phosphatase STEP61. *J Neurosci* 2010;**30**:5948–57.
202. Zhang Y, Kurup P, Xu J, Anderson GM, Greengard P, Nairn AC, et al. Reduced levels of the tyrosine phosphatase STEP block beta amyloid-mediated GluA1/GluA2 receptor internalization. *J Neurochem* 2011;**119**:664–72.
203. Roche KW, Standley S, McCallum J, Dune LC, Ehlers MD, Wenthold RJ. Molecular determinants of NMDA receptor internalization. *Nat Neurosci* 2001;**4**:794–802.
204. Venkitaramani DV, Paul S, Zhang Y, Kurup P, Ding L, Tressler L, et al. Knockout of striatal enriched protein tyrosine phosphatase in mice results in increased ERK1/2 phosphorylation. *Synapse* 2009;**63**:69–81.
205. Oddo S, Caccamo A, Shepherd JD, Murphy MP, Golde TE, Kayed R, et al. Triple-transgenic model of Alzheimer's disease with plaques and tangles: intracellular Abeta and synaptic dysfunction. *Neuron* 2003;**39**:409–21.
206. Hayashi T, Huganir RL. Tyrosine phosphorylation and regulation of the AMPA receptor by SRC family tyrosine kinases. *J Neurosci* 2004;**24**:6152–60.
207. Scholz R, Berberich S, Rathgeber L, Kolleker A, Kohr G, Kornau HC. AMPA receptor signaling through BRAG2 and Arf6 critical for long-term synaptic depression. *Neuron* 2010;**66**:768–80.
208. Shirazi SK, Wood JG. The protein tyrosine kinase, fyn, in Alzheimer's disease pathology. *Neuroreport* 1993;**4**:435–7.
209. Lee G, Thangavel R, Sharma VM, Litersky JM, Bhaskar K, Fang SM, et al. Phosphorylation of tau by fyn: implications for Alzheimer's disease. *J Neurosci* 2004:**24**:2304–12.
210. Derkinderen P, Scales TM, Hanger DP, Leung KY, Byers HL, Ward MA, et al. Tyrosine 394 is phosphorylated in Alzheimer's paired helical filament tau and in fetal tau with c-Abl as the candidate tyrosine kinase. *J Neurosci* 2005;**25**:6584–93.
211. Chin J, Palop JJ, Yu GQ, Kojima N, Masliah E, Mucke L. Fyn kinase modulates synaptotoxicity, but not aberrant sprouting, in human amyloid precursor protein transgenic mice. *J Neurosci* 2004;**24**:4692–7.
212. Chin J, Palop JJ, Puolivali J, Massaro C, Bien-Ly N, Gerstein H, et al. Fyn kinase induces synaptic and cognitive impairments in a transgenic mouse model of Alzheimer's disease. *J Neurosci* 2005;**25**:9694–703.
213. Ittner LM, Ke YD, Delerue F, Bi M, Gladbach A, van EJ, et al. Dendritic function of tau mediates amyloid-beta toxicity in Alzheimer's disease mouse models. *Cell* 2010;**142**:387–97.
214. Roberson ED, Halabisky B, Yoo JW, Yao J, Chin J, Yan F, et al. Amyloid-beta/Fyn-induced synaptic, network, and cognitive impairments depend on tau levels in multiple mouse models of Alzheimer's disease. *J Neurosci* 2011;**31**:700–11.
215. de Vrij FM, Fischer DF, van Leeuwen FW, Hol EM. Protein quality control in Alzheimer's disease by the ubiquitin proteasome system. *Prog Neurobiol* 2004;**74**:249–70.
216. Upadhya SC, Hegde AN. Role of the ubiquitin proteasome system in Alzheimer's disease. *BMC Biochem* 2007;**8**(Suppl. 1):S12.

217. Oddo S. The ubiquitin-proteasome system in Alzheimer's disease. *J Cell Mol Med* 2008;**12**:363–73.
218. Mori H, Kondo J, Ihara Y. Ubiquitin is a component of paired helical filaments in Alzheimer's disease. *Science* 1987;**235**:1641–4.
219. Lam YA, Pickart CM, Alban A, Landon M, Jamieson C, Ramage R, et al. Inhibition of the ubiquitin-proteasome system in Alzheimer's disease. *Proc Natl Acad Sci USA* 2000;**97**:9902–6.
220. David DC, Layfield R, Serpell L, Narain Y, Goedert M, Spillantini MG. Proteasomal degradation of tau protein. *J Neurochem* 2002;**83**:176–85.
221. Qing H, Zhou W, Christensen MA, Sun X, Tong Y, Song W. Degradation of BACE by the ubiquitin-proteasome pathway. *FASEB J* 2004;**18**:1571–3.
222. He G, Qing H, Cai F, Kwok C, Xu H, Yu G, et al. Ubiquitin-proteasome pathway mediates degradation of APH-1. *J Neurochem* 2006;**99**:1403–12.
223. He G, Qing H, Tong Y, Cai F, Ishiura S, Song W. Degradation of nicastrin involves both proteasome and lysosome. *J Neurochem* 2007;**101**:982–92.
224. Gong B, Cao Z, Zheng P, Vitolo OV, Liu S, Staniszewski A, et al. Ubiquitin hydrolase Uch-L1 rescues beta-amyloid-induced decreases in synaptic function and contextual memory. *Cell* 2006;**126**:775–88.
225. Tseng BP, Green KN, Chan JL, Blurton-Jones M, LaFerla FM. Abeta inhibits the proteasome and enhances amyloid and tau accumulation. *Neurobiol Aging* 2008;**29**:1607–18.
226. Oh S, Hong HS, Hwang E, Sim HJ, Lee W, Shin SJ, et al. Amyloid peptide attenuates the proteasome activity in neuronal cells. *Mech Ageing Dev* 2005;**126**:1292–9.
227. Smith DL, Pozueta J, Gong B, Arancio O, Shelanski M. Reversal of long-term dendritic spine alterations in Alzheimer disease models. *Proc Natl Acad Sci USA* 2009;**106**:16877–82.
228. Zhu Y, Hou H, Rezai-Zadeh K, Giunta B, Ruscin A, Gemma C, et al. CD45 deficiency drives amyloid-beta peptide oligomers and neuronal loss in Alzheimer's disease mice. *J Neurosci* 2011;**31**:1355–65.
229. Mody N, Agouni A, McIlroy GD, Platt B, Delibegovic M. Susceptibility to diet-induced obesity and glucose intolerance in the APP (SWE)/PSEN1 (A246E) mouse model of Alzheimer's disease is associated with increased brain levels of protein tyrosine phosphatase 1B (PTP1B) and retinol-binding protein 4 (RBP4), and basal phosphorylation of S6 ribosomal protein. *Diabetologia* 2011;**54**:2143–51.
230. Nagy Z, Esiri MM, Cato AM, Smith AD. Cell cycle markers in the hippocampus in Alzheimer's disease. *Acta Neuropathol* 1997;**94**:6–15.
231. Ding XL, Husseman J, Tomashevski A, Nochlin D, Jin LW, Vincent I. The cell cycle Cdc25A tyrosine phosphatase is activated in degenerating postmitotic neurons in Alzheimer's disease. *Am J Pathol* 2000;**157**:1983–90.
232. Vincent I, Bu B, Hudson K, Husseman J, Nochlin D, Jin L. Constitutive Cdc25B tyrosine phosphatase activity in adult brain neurons with M phase-type alterations in Alzheimer's disease. *Neuroscience* 2001;**105**:639–50.
233. De SB, Vassar R, Golde T. The secretases: enzymes with therapeutic potential in Alzheimer disease. *Nat Rev Neurol* 2010;**6**:99–107.
234. Cummings J. What can be inferred from the interruption of the semagacestat trial for treatment of Alzheimer's disease? *Biol Psychiatry* 2010;**68**:876–8.
235. Extance A. Alzheimer's failure raises questions about disease-modifying strategies. *Nat Rev Drug Discov* 2010;**9**:749–51.
236. Voronkov M, Braithwaite SP, Stock JB. Phosphoprotein phosphatase 2A: a novel druggable target for Alzheimer's disease. *Future Med Chem* 2011;**3**:821–33.
237. Corcoran NM, Martin D, Hutter-Paier B, Windisch M, Nguyen T, Nheu L, et al. Sodium selenate specifically activates PP2A phosphatase, dephosphorylates tau and reverses memory deficits in an Alzheimer's disease model. *J Clin Neurosci* 2010;**17**:1025–33.

238. van EJ, Ke YD, Liu X, Delerue F, Kril JJ, Gotz J, et al. Sodium selenate mitigates tau pathology, neurodegeneration, and functional deficits in Alzheimer's disease models. *Proc Natl Acad Sci USA* 2010;**107**:13888–93.
239. Chohan MO, Khatoon S, Iqbal IG, Iqbal K. Involvement of I2PP2A in the abnormal hyperphosphorylation of tau and its reversal by Memantine. *FEBS Lett* 2006;**580**:3973–9.
240. Kickstein E, Krauss S, Thornhill P, Rutschow D, Zeller R, Sharkey J, et al. Biguanide metformin acts on tau phosphorylation via mTOR/protein phosphatase 2A (PP2A) signaling. *Proc Natl Acad Sci USA* 2010;**107**:21830–5.
241. Lee KW, Chen W, Junn E, Im JY, Grosso H, Sonsalla PK, et al. Enhanced phosphatase activity attenuates alpha-Synucleinopathy in a mouse model. *J Neurosci* 2011;**31**:6963–71.
242. Liu S, Zeng LF, Wu L, Yu X, Xue T, Gunawan AM, et al. Targeting inactive enzyme conformation: aryl diketoacid derivatives as a new class of PTP1B inhibitors. *J Am Chem Soc* 2008;**130**:17075–84.
243. Wu S, Bottini M, Rickert RC, Mustelin T, Tautz L. In silico screening for PTPN22 inhibitors: active hits from an inactive phosphatase conformation. *ChemMedChem* 2009;**4**:440–4.
244. Zhang X, He Y, Liu S, Yu Z, Jiang ZX, Yang Z, et al. Salicylic acid based small molecule inhibitor for the oncogenic Src homology-2 domain containing protein tyrosine phosphatase-2 (SHP2). *J Med Chem* 2010;**53**:2482–93.
245. Barr AJ, Ugochukwu E, Lee WH, King ON, Filippakopoulos P, Alfano I, et al. Large-scale structural analysis of the classical human protein tyrosine phosphatome. *Cell* 2009;**136**:352–63.
246. Sergienko E, Xu J, Liu We, Dahl R, Critton DA, Su Y, et al. A specific inhibitor of hematopoietic protein tyrosine phosphatase augments ERK and p38 activation *in vivo*. *ACS Chem Biol* 2011. [In press].
247. Borel JF, Feurer C, Gubler HU, Stahelin H. Biological effects of cyclosporin A: a new antilymphocytic agent. *Agents Actions* 1976;**6**:468–75.
248. Calne RY, Rolles K, White DJ, Thiru S, Evans DB, McMaster P, et al. Cyclosporin A initially as the only immunosuppressant in 34 recipients of cadaveric organs: 32 kidneys, 2 pancreases, and 2 livers. *Lancet* 1979;**2**:1033–6.
249. Naesens M, Kuypers DR, Sarwal M. Calcineurin inhibitor nephrotoxicity. *Clin J Am Soc Nephrol* 2009;**4**:481–508.
250. Bechstein WO. Neurotoxicity of calcineurin inhibitors: impact and clinical management. *Transpl Int* 2000;**13**:313–26.
251. Sieber M, Baumgrass R. Novel inhibitors of the calcineurin/NFATc hub—alternatives to CsA and FK506? *Cell Commun Signal* 2009;**7**:25.

# G-substrate: The Cerebellum and Beyond

Shogo Endo

*Aging Regulation Research Team, Tokyo Metropolitan Institute of Gerontology, Tokyo, Japan*



The discovery of nitric oxide (NO) as an activator of soluble guanylate cyclase (sGC) has stimulated extensive research on the NO–sGC–3′:5′-cyclic guanosine monophosphate (cGMP)–cGMP-dependent protein kinase (PKG) pathway. However, the restricted localization of pathway components and the lack

Progress in Molecular Biology
and Translational Science, Vol. 106
DOI: 10.1016/B978-0-12-396456-4.00004-3

of information on PKG substrates have hindered research seeking to examine the physiological roles of the NO–sGC–cGMP–PKG pathway. An excellent substrate for PKG is the G-substrate, which was originally discovered in the cerebellum. The role of G-substrate in the cerebellum and other brain structures has been revealed in recent years. This review discusses the relationship between the G-substrate and other components of the NO–sGC–cGMP–PKG pathway and describes the characteristics of the G-substrate gene and protein related to diseases. Finally, we discuss the physiological role of G-substrate in the cerebellum, where it regulates cerebellum-dependent long-term memory, and its role in the ventral tegmental area and retina, where it acts as an effective neuroprotectant.

**Abbreviations:** 6-OHDA, 6-hydroxydopamine; cAMP, 3′:5′-cyclic adenosine monophosphate; cGMP, 3′:5′-cyclic guanosine monophosphate; DA, dopaminergic; DARPP-32, dopamine- and cAMP-regulated phosphoprotein of 32kDa; DR, diabetic retinopathy; I-1, protein phosphatase inhibitor-1; KO, knockout; LTD, long-term depression; NMDA, *N*-methyl-D-aspartic acid; NO, nitric oxide; NOS, nitric oxide synthase; OKR, optokinetic response; PD, Parkinson's disease; PDE, phosphodiesterase; PKA, cAMP-dependent protein kinase; PKG, cGMP-dependent protein kinase; PP, protein phosphatase; RGC, retinal ganglion cell; sGC, soluble guanylate cyclase; SNP, single-nucleotide polymorphism; TH, tyrosine hydroxylase; VOR, vestibulo-ocular reflex.

## I. Introduction

Cyclic nucleotides constitute an essential part of signal transduction systems involved in cell regulation, development, growth, neuronal plasticity, and brain function. The 3′:5′-cyclic adenosine monophosphate (cAMP) pathway is a cyclic nucleotide pathway that has been studied intensively in many tissues because it is ubiquitous and also because its key component, cAMP-dependent protein kinase (PKA), is found at similar concentrations in most tissues. By contrast, the 3′:5′-cyclic guanosine monophosphate (cGMP) pathway has received little attention because it and most of its components are present in very few tissues. Furthermore, the mechanistic details underlying increased intracellular cGMP concentrations were also lacking until recent years. Although its key component, namely, cGMP-dependent protein kinase (PKG), is expressed at relatively high levels in lung, heart, smooth muscle, and intestine, it is expressed at low levels in other tissues.[1–5] For these reasons, the identification of substrates for downstream components of the cGMP pathway and the determination of their physiological roles have progressed slowly.[6–8] Major targets of cGMP research have been the cardiovascular and gastrointestinal systems.[9,10] Only more recently has research focused on the physiological role of the cGMP pathway in the central nervous system.

The family of natriuretic peptides and their receptors are observed in the brain[11–16] (for a recent review, see Ref. 17). Natriuretic peptides activate membrane-bound guanylate cyclase to increase intracellular cGMP concentrations.[12,18] The physiological function of natriuretic peptide is not well characterized in the brain.

The components of the nitric oxide–soluble guanylate cyclase–cGMP–PKG (NO–sGC–cGMP–PKG) pathway (Fig. 1) are observed in a few restricted areas in the central nervous system.[7,19–21] Among these areas, the cerebellum is very rich in PKG and its preferred, high-affinity substrate, namely, the G-substrate.[22–31] The isolation of the G-substrate, its almost exclusive localization to Purkinje cells, and the presence of high PKG levels in Purkinje cells prompted us to study the physiological functions of the cGMP pathway and the G-substrate in the cerebellum. In addition, we also found that the G-substrate plays essential neuroprotective roles in the midbrain and the retina, where a limited number of cells contain the G-substrate.[32–34] In this review, we focus on the structure, regulation, and physiological roles of the G-substrate in the cerebellum and in other areas of the nervous system.

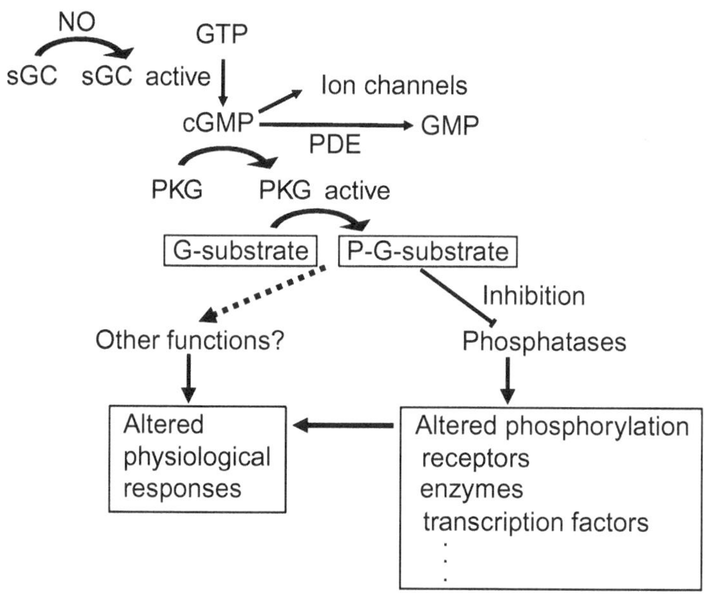


FIG. 1. NO–sGC–cGMP–PKG pathway scheme and G-substrate. NO, nitric oxide; sGC, soluble guanylate cyclase; PDE, phosphodiesterase; PKG, cGMP-dependent protein kinase.

## II. Components of the NO–sGC–cGMP–PKG Pathway

In the 1970s, PKG was isolated first and characterized in invertebrates (lobster muscle),[35] shown to be present in mammalian tissues,[3] and then purified and characterized.[36,37] However, research progressed slowly until the 1990s, when the major mechanisms underlying the elevation of intracellular cGMP concentrations were shown to involve activators of PKG: the NO–sGC pathway (Fig. 1)[38,39] and the natriuretic peptide-membrane-bound guanylate cyclase pathway.[12,18] In particular, the finding that NO directly activates sGC intracellularly and intercelllularly[39] brought about a large amount of research examining the physiological role of the NO–sGC–cGMP–PKG pathway in many tissues, including the central nervous system.[38,40,41] Cellular targets of cGMP (Fig. 1) in the nervous system are classified into three categories: (1) ion channels, (2) phosphodiesterases (PDEs), and (3) PKGs.

### A. Ion Channels as Targets of cGMP

Cyclic GMP can activate two types of ion channels: (1) cyclic nucleotide-dependent channels, and (2) hyperpolarization-activated cyclic nucleotide-dependent (HCN) channels.[42–44] Cyclic nucleotide-dependent channels are ligand-gated ion channels hypothesized to have homo- or heterotetrameric structures composed of a combination of six subunits (CNGA1, CNGA2, CNGA3, CNGA4, CNGB1a/b, and CNGB3). Each of these subunits has a cyclic nucleotide-binding domain in the C-terminus tail that senses the intracellular concentration of cAMP and/or cGMP. Cyclic GMP-activated channels have been extensively characterized electrophysiologically (for reviews, see Refs. 45,46).

Although HCN channels are also ion channels, they are regulated primarily by voltage (hyperpolarization), which is modulated by cAMP and/or cGMP (for a recent review, see Ref. 47). HCN channels are hypothesized to contain four subunits, and their intracellular C-terminal domain has a cyclic nucleotide-binding site. In mammals, four HCN channel subunits (HCN 1–4) exist, and homo- and heterotetrameric channels with distinct electrophysiological features have been reported (for review, see Ref. 47). Recently, hyperpolarization-activated channels have been extensively characterized in the deep cerebellar nucleus of the cerebellum,[44] where the channels are abundant and other components of NO–sGC–cGMP–PKG pathway are observed. Further study of how these channels contribute to the physiological functions of the cerebellum may reveal novel roles for these channels in cerebellum-dependent behaviors.

### B. PDEs: The Controllers of Intracellular cGMP

PDEs are enzymes that convert cyclic nucleotides (cAMP and cGMP) to their corresponding 5′-nucleotides (5′-AMP and 5′-GMP) in order to inactivate the cyclic nucleotides. There are currently 11 members in the PDE superfamily (for recent reviews, see Refs. 20,48). PDE1–3, 5, 6, and 9–11 can hydrolyze cGMP. Cyclic GMP stimulates PDE2 and 5 activities but inhibits PDE3 activity. PDEs have similar structures, and, in some cases, more than one PDE subtype can be expressed in a single cell. Each PDE subtype has a unique subcellular localization; thus, PDEs have recently attracted much attention in terms of the roles they play in the local regulation of cyclic nucleotide concentration.[9,49] PDEs can be activated or inhibited by cGMP and cAMP,[20,48] leading to cross-talk between cGMP and cAMP. This is achieved through a PDE-mediated degradation step that spatially and temporally regulates cyclic nucleotides in subcellular compartments.[50] In addition, PDE2, 5, 6, 10, and 11 have a GAF domain, suggesting that these PDEs potentially interact with and are regulated by other molecules.[51] The GAF domain is a domain found in numerous proteins in various species and stands for cGMP-binding PDEs (mammals), cyanobacterial adenylate cyclases CyaB1 and B2 (*Anabaena* sp.), and formate hydrogenlyase transcription activator (*Escherichia coli*).[52] A variety of subtype-specific PDE inhibitors have been designed, including cilostazol for PDE3, rolipram for PDE4, and sildenafil for PDE5. Such signal-transduction-targeted inhibitors, which are effective intracellularly, may lead to new approaches to treating diseases such as cognitive deficits (for recent reviews, see Refs. 49,53). This approach contrasts with those that use the strategy of targeting extracellular receptors for developing treatments for diseases. The latter is the major approach used today.

### C. PKGs: Ultimate Targets of cGMP

Another important intracellular target of cGMP is PKGs. PKGs are Ser/Thr kinases that are present in eukaryotes, ranging from paramecium[54,55] to humans.[19,56,57] In mammals, PKGI and PKGII arise from different genes.[19] Moreover, PKGI has two splicing variants, PKGIα and PKGIβ. PKGIα activation requires cGMP concentrations lower than that needed for PKGIβ activation. PKGIα localizes mainly to the cytosol, whereas PKGIβ mainly localizes to plasma membranes due to its N-terminal modification by myristoylation.[58] PKG contains both a catalytic domain and regulatory domains within a single peptide. Unlike PKA, which releases its catalytic domain following cAMP-mediated activation, PKG does not release its catalytic domain upon activation. Research in mice lacking PKGs has demonstrated that PKGs play a variety of physiological roles. Hofmann *et al.*[19] and Kleppisch and Feil[21] provide excellent reviews of recent progress in PKG research involving the central nervous system.

## III. G-Substrate Genes

Even though the partial amino acid sequence of the G-substrate was identified in the 1980s,[26] it was not until the late 1990s that its full-length cDNA was isolated and characterized.[28,29,31] The isolation of the G-substrate cDNA ignited a cascade of studies, leading to the extensive characterization of the substrate through molecular biological methods. In addition, the Genome Project has identified G-substrate genes in a variety of animals, including humans and mice (see Figs. 2 and 3). These rapid and intense methodological developments have made it possible for us to generate G-substrate knockout (KO) mice, with the aim of deciphering the physiological functions of the G-substrate. Moreover, in humans, single nucleotide polymorphisms (SNPs) of the G-substrate gene and their potential clinical relevance have been identified through high-throughput DNA sequencing technology.[59]

## A. Structure of the G-Substrate Gene

Figure 2 shows the structure of the G-substrate gene in humans (GSBS, PPP1R17, C7orf16; 21.3 kbp), rats (Gsbs, Ppp1r17; 16.7 kbp), and mice (Gsbs, Ppp1R17; 15.2kbp). Interestingly, the G-substrate gene of these animals has similar exon–intron structures: (1) They consist of five exons; (2) exon 2 contains an initial methionine; (3) PKG-phosphorylation sites 1 and 2 are in separate exons; and (4) exon 5, the longest exon of the five exons, contains a stop codon. In addition, the exon–intron junctions are well conserved across species, and the amino acid residues corresponding to the donor–acceptor site are identical in human, mouse, and rat G-substrate genes. Recent genome sequencing of various species has revealed that many species have G-substrate orthologs. Characteristics of human, rat, and mouse G-substrate genes are well conserved

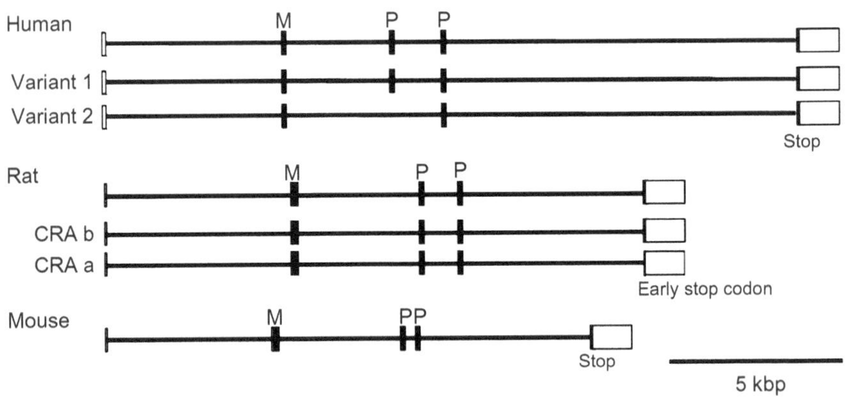


FIG. 2. Structures and splicing variants of G-substrate gene in humans, mice, and rats. Human (NC 000007.13 Chromosome 7 (31726799–31748069), 21.3kbp); mouse (*Mus musculus*, C57BL/6, NC 000072.5 Chromosome 6 (5597509–55982682), 15.2kbp); rat (*Rattus norvegius*, NC 005103.2 Chromosome 4 (84846563–84863277), 16.7kbp). Black boxes indicate the coding sequences, and white boxes indicate the noncoding sequences. M, initial methionine; P, PKG phosphorylation site.

in various vertebrate species such as chimpanzee (*Pan traglodytes*), monkey (*Macaca mulatta, Macaca fascicularis*), rabbit (*Oryctolagus cuniculus*), bovine (*Bos taurus*), xenopus (*Xenopus tropicalis*), chicken (*Gallus gallus*), and zebra finch (*Taeniopygia guttata*). The conservation of the G-substrate gene structure and characteristics suggests that G-substrate has played important physiological roles throughout evolution.

## B. SNPs in G-Substrate Gene

The identification of SNPs are being carried out extensively in order to broaden our understanding about the links between common disorders and genes, to customize medical treatments based on a patient's genetic make-up, and to maximize effectiveness and minimize side effects of therapeutic treatments.

Our laboratory was the first to clone the human G-substrate cDNA sequence[28] based on an EST clone obtained from a female infant (Soares cDNA library).[60] The human G-substrate gene (GSBS, PPP1R17, C7orf16) is located in the 7p15 region of chromosome 7.[28] The gene encoding the coiled-coil domain containing 129 (CCDC129) and the gene encoding PDE1C are positioned to the 5′ and 3′ sides, respectively, of the G-substrate gene. The same alignment of these three genes is conserved in rat (chromosome 4) and mouse (chromosome 6).

Recently, the Human Genome Project has identified and characterized SNPs in detail. Among the more than 300 human G-substrate gene SNPs registered in the National Center for Biotechnology Information (NCBI) database, thirteen SNPs are found in the G-substrate coding exons. Some of them are observed around PKG phosphorylation sites, which are important for the phosphatase inhibitory activity (Fig. 3), including Pro-to-Leu (rs145437298), Ala-to-Thr (rs146525582), Ala-to-Val (rs161112620) for the phosphorylation site 1, and Ala-to-Val (rs113903903) for phosphorylation site 2. In mouse (C57BL/6), two silent SNPs reside in the coding region of exons 2 and 5, whereas in rat (*Rattus norvegicus*) one SNP resides in the noncoding region of exon 5. However, no SNPs have been identified thus far in the coding regions of the rat G-substrate gene. Further examination of other human individuals may be necessary in order to identify all G-substrate SNPs and to determine the physiological influences of these SNPs.

The Genome Project also identified specific regions of the genome termed "recombination hotspots," regions with relatively high HapMap recombination rates.[61,62] Hotspots are observed around the 5′-end of the human G-substrate gene. High recombination rates are hypothesized to occur in important genes that have undergone positive selection throughout evolution.[63,64] This suggests that the presence of the G-substrate in the cerebellum and other areas in the nervous system may have been advantageous for the functioning of these structures during evolution.[63,64] Further analyses of other species may reveal the relationship between recombinations in the G-substrate gene that have occurred during evolution and may clarify the functional roles of the G-substrate in cerebellum and other neuronal tissues.

### C. G-Substrate SNPs and Disease in Humans

In 2003, Ono *et al.*[59] described a G-substrate SNP that was associated with hypercholesterolemia in certain individuals from the east-central part of Japan. They observed an association between an SNP in the 5′-promoter region of the G-substrate gene, -1323T-C (604088.0001), and total cholesterol levels in adults from this area. Individuals bearing the T allele had 14% higher total cholesterol levels than those lacking the T allele. The probability of having hypercholesterolemia was approximately two-fold higher in individuals bearing the T allele compared to those lacking it. However, the influence of SNP in the promoter region on the G-substrate expression was not reported. Currently, although the G-substrate-associated mechanisms responsible for controlling plasma cholesterol levels are unknown, the cGMP pathway has been implicated in hypercholesteromia.[65–67] Further study on the tissue distribution of G-substrate in systems other than the nervous system may be necessary in order to assess the physiological role of the G-substrate and cGMP pathway in the regulation of plasma cholesterol levels.

## IV. G-Substrate Gene Transcripts

### A. Splicing Variants of G-Substrate mRNA

In humans and rats, splicing variants of the G-substrate are reported in the NCBI database (Figs. 2 and 3). In humans, transcript variant 1 (longer isoform; NP_006658) contains all five exons, yielding the G-substrate isoform 1 (AAD13030), which is composed of 155 amino acid residues. On the other hand, transcript variant 2 (shorter variant; NM_001145123) lacks exon 3, yielding G-substrate isoform 2 (NM_006658), which is composed of 104 amino acid residues. Although both human variants 1 and 2 adhere to typical exon–intron junctions and obey normal donor–acceptor sequence rules, variant 1 seems to be the major isoform in the brain. Polymerase chain reaction (PCR) using a human brain cDNA library as a template failed to detect variant 2 (Endo and Suzuki, unpublished observations), suggesting that variant 2 represents a considerably minor mRNA specie in the brain. Further analysis of splicing variant 2 using tissues other than neural tissues is necessary.

Isoforms 1 and 2 of the human G-substrate share a common N-terminus and C-terminus (Fig. 3). The major difference between the two is that isoform 1 contains two PKG phosphorylation sites (site 1 and 2), whereas

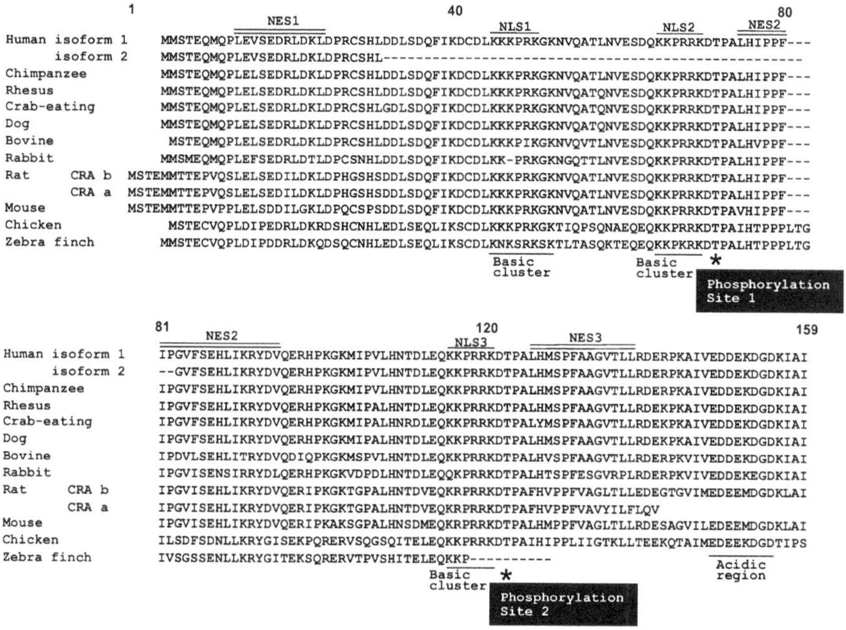


FIG. 3. Primary structure of G-substrate in various species. Amino acid (AA) sequences were deduced from isolated cDNA or from G-substrate gene. The information on the cDNAs and genes used in this figure is as follows: humans (*Homo sapiens*), DNA sequence AF097730 (cDNA), AA sequence isoform 1 (AAD13030) and isoform 2 (NP001138595); chimpanzee (*Pan traglodytes*), DNA sequence XM001166815 (genomic), AA sequence XP001166815 (predicted); rhesus monkey (*Macaca mulatta*), DNA sequence XM001083835 (genomic), AA sequence XP001083835 (predicted); crab-eating macaque (*Macaca fascicularis*), DNA Sequence AB170873 (cDNA), AA sequence BAE87936; dog (*Canine familiaris*), DNA sequence XM539505 (genomic), AA sequence XP5395059 (predicted); bovine (*Bos taurus*), DNA sequence XM868976 (genomic), AA sequence XP874069 (predicted); rabbit (*Oryctolagus cuniculus*), DNA sequence AY250082 (cDNA), AA sequence AAP76222; rat (*Rattus norvegicus*), DNA sequence AF294688 (cDNA), AA sequence AAN27914; rat CRA_a, AA sequence EDL88067 (predicted); rat CRA_b, EDL88068 (predicted); mouse (*Mus musculus*), DNA sequence AF500906 (cDNA), AA sequence AAM21706; chicken (*Gallus gallus*), DNA sequence XM418851 (genomic), AA sequence XP418851 (predicted); zebra finch (*Taeniopygia guttata*), DNA sequence (cDNA, partial) DV579352, AA sequence (predicted from partial cDNA). NES, potential nuclear export signal; NLS, potential nuclear localization signal; *, phosphorylation sites by PKG.

isoform 2 contains only one of these sites (site 2) as shown in Fig. 3. The physiological differences between G-substrate isoforms 1 and 2 are currently unknown. As we will see later in this review, however, phosphorylation of site 2 confers to it potent protein phosphatase (PP) inhibitory activity.[29] Experiments using G-substrate harboring a mutation in site 1 showed, however, that G-substrate isoform 2—the isoform that contains only phosphorylation site

2—still may have PP inhibitory activity.[29] Further examination of the expression and physiological role of the G-substrate is necessary. Future studies will have to determine whether a shorter isoform exists in the tissues other than the brain.

Rat (*Rattus norvegius*) G-substrate mRNA also has splicing variants (Figs. 2 and 3). A transcript containing all exons yields a longer isoform (NP_703197 or CRA_b isoform, EDL88068), which is composed of 159 amino acid residues. The other transcript is a shorter isoform (CRA_a isoform, EDL88067), which is composed of 141 amino acid residues. The shorter CRA_a isoform uses an unusual exon–intron junction between exon 4 and 5 of the G-substrate gene. This leads to an early stop codon in exon 5 to yield the shorter G-substrate isoform, which has 18 amino acid residues fewer than the longer CRA_b isoform. Due to the frame shift caused by the unusual exon–intron conjunction, the shorter isoform has C-terminal amino acid residues that are distinct from those of the longer isoform. Both isoforms, however, contain exactly the same N-terminal amino acid sequences. We have been unable to detect the shorter isoform (CRA_a) in a rat brain cDNA library or in reverse-transcription PCR (RT-PCR) products from rat cerebellum (Endo and Suzuki, unpublished observations). Both isoforms contain PKG phosphorylation sites 1 and 2 and are thus expected to be potent PP inhibitors. Further studies are necessary to reveal the physiological functions of isoforms that have different C-terminal sequences and to determine whether the shorter CRA_a isoform also exists in non-neural tissues.

So far, no splicing variants have been reported for the mouse G-substrate. Considering the existence of G-substrate isoforms in human and rat, further examination is necessary in order to determine whether mouse G-substrate variants exist in non-neural tissues.

## B. Expression of G-Substrate mRNA During Development, Aging, and Disease

Mouse G-substrate mRNA can be observed as early as the notochord stage (EST clone MM3UTR C36).[68] However, details of the G-substrate expression pattern in notochord remain unknown. It would be interesting to trace the expression of the G-substrate throughout brain development, especially with regard to the developing cerebellum[69–73] which contains higher amounts of G-substrate than any other structure in the nervous system. Using a fluorescence differential display PCR method, Iida *et al.*[74] identified five genes that are expressed in an age-dependent manner in the brains of 3-day-old to 15-month-old mice. Quantitative PCR analyses revealed that the expression of G-substrate mRNA was quite low in the brains of 3-day-old and 1-week-old mice, but was modest in 2-week-old to 9-month-old mice. Highest expression was observed in 15-month-old mice, with mRNA levels being approximately two times that in 6- and 9-month-old mice.

The sudden increase in G-substrate expression in the brains of 2-week-old mice corresponds to the maturation of cerebellar neuronal circuitry,[69–73,75–77] including the maturation of cerebellar Purkinje cells where the G-substrate is concentrated. The age-dependent increase in G-substrate expression—that is, the increased expression observed in 15-month-old mice—suggests that the G-substrate may be playing a physiological role in compensating for the age-dependent decline in motor function,[78–80] a function in which the cerebellum plays important roles. Indeed, strong PP inhibitory activity accompanied by elevated G-substrate levels may assist feeble neuronal signal transduction in the brains of aged individuals. For instance, the PP inhibitory activity of the G-substrate may maintain phosphorylation induced by various stimuli for longer durations and at higher levels. Similar to the potential role of the G-substrate in aged animals, the enhanced signal transduction by higher concentrations of cAMP and cGMP through PDE inhibition is a target for development of treatments for a variety of diseases, including certain neurological diseases.[49,81–83] A variety of subtype-specific PDE inhibitors have been designed to improve the activation and maintenance of the cyclic nucleotide system. One striking example is the discovery of a specific inhibitor of PDE5, which is a cGMP-hydrolyzing PDE, for the treatment of erectile dysfunction.[84]

Decreased expression of G-substrate mRNA was observed in the gene profiling of mice with spinocerebellar ataxia type 1 (SCA1). SCA1 is a neurodegenerative disease caused by the expansion of a glutamine repeat in the protein called ataxin-1.[85] This autosomal-dominant disorder typically has its clinical onset in adulthood, with major symptoms of loss of coordination of voluntary movements. SCA1 is a member of a group of neurological disorders designated the polyglutamine diseases, which include Huntington disease, spinobulbar muscular atrophy, and the spinocerebellar ataxias SCA1, SCA2, SCA3, SCA6, SCA7, and SCA17.[86] Although ataxin-1 is ubiquitously expressed,[87] the disease affects specific subsets of neurons such as cerebellar Purkinje cells and neurons within the brainstem. Similar to what is observed in SCA1 patients, progressive ataxia and Purkinje cell degeneration occur in transgenic mice overexpressing a mutant SCA1 allele with 82 CAG repeats.[88–90] The translocation of overexpressed mutant protein into nucleus of cerebellar Purkinje cell is necessary to develop Purkinje cell pathology or neurological dysfunction.[91]

Serra *et al.*[92] compared the gene expression patterns in the SCA1 "ataxic" transgenic mouse line (B05-ataxin-1[82Q]) with those in two "non-ataxic" lines (A02-ataxin-1[30Q] and K772T-[82Q]). Nine genes were identified whose expression was consistently altered in the cerebellum of ataxic mice

(B05-ataxin-1[82Q]) at 5 and 12 weeks of age; *Homer-3, carbonic anhydrase-like sequence, EAAT4, IP3R1, Purkinje cell protein 1, SERCA3, EST (AI842002), homeotic protein Gtx, and G-substrate*. G-substrate mRNA was significantly downregulated in the ataxic animals. The results may simply mean the loss of G-substrate mRNA associated with loss of cerebellar Purkinje cells in the ataxic SCA1 animals. In these animals, a variety of pathological changes including loss of Purkinje cells are observed before the onset of ataxia. As early as age of 3 weeks, large vacuoles are detected in Purkinje cells[88,89]; at age of 4 weeks, intracellular aggregates of mutant ataxin-1 are observed.[90] The number of Purkinje cells carrying the aggregates increases to 90% at age of 12 weeks.[90] Accompanying the intracellular accumulation of ataxin-1 aggregates, the loss of Purkinje cell dendrites as well as shrinkage of molecular layer is observed by age of 5 weeks.[89] By 12-week-old, when ataxia is detectable, considerable number of Purkinje cells are lost and the loss continues.[89,90]

In ataxic SCA1 mice, the transcription of G-substrate and other genes might have been suppressed by the nuclear translocation of ataxin-1 aggregates. Considering the strong neuroprotective role of G-substrate in the midbrain and in the retina (see Section VII), the downregulation of G-substrate expression might lead to further degeneration of Purkinje cells in SCA1 mice. It would be of interest to examine whether the overexpression of G-substrate in cerebellum of SCA1 mice is able to rescue the neurodegenerative phenotype.

## V. Cellular Functions of the G-Substrate

The G-substrate was originally isolated and characterized in rabbit cerebellum. Molecular cloning of the G-substrate further revealed its unique features through recombinant proteins. In this section, we look at the unique features of the primary structures of the G-substrate and the role these structures play, such as in PP inhibitory activity.

### A. Primary Structure of the G-Substrate

Figure 3 shows the primary structures of the G-substrate in several vertebrates. G-substrate is a protein with 155–159 amino acid residues. The N-terminal end of the rat and mouse G-substrate has an extra MSTE sequence in addition to the M(M)STE commonly found in the G-substrate N-terminus of most species. G-substrate contains three clusters of basic amino acids (Fig. 3): –KKKPR/IKGK–KKPRRKDT(Pi)PA–KKPRRKDT(Pi)PA–. The amino acid sequences of the second and third clusters are identical and were originally determined by Aitken *et al.*[26] through direct protein sequencing of the G-substrate purified from rabbit cerebellum. In addition, these two clusters are located next to the Thr residues phosphorylated by PKG. The amino acid sequence of the two phosphorylation sites is identical in the species listed in Fig. 3. This reflects the potential functional importance of these sequences for phosphorylation and for phosphorylated G-substrate.

In the C-terminus, the G-substrate contains a short stretch of acidic amino acids, D (Asp) and E (Glu), with potential RNA binding and modulating capabilities that are observed in other thermostable PP proteins, such as $I_1^{PP2A}$ (also known as LANP, MAPM, ANP32A, pp32, or PHAP-1)[93–96]; $I_2^{PP2A}$ (also known as IGAAD, TAF-I(β), SET, PHAP-II)[93–95]; and a RNA-binding modulating protein APRIL (also known as PHAPI2, SSP29, ANP32B).[96,97] The G-substrate might confer the ability for binding modulation and transport of mRNA through the Hu–CRM1–mRNA complex. Further investigation of the interaction between the G-substrate and the Hu–CRM1–mRNA complex is needed.

Human, rat, and mouse G-substrate have a molecular weight of approximately 18 kDa, as deduced from cDNA.[28–31] G-substrate purified from rabbit cerebellum behaves like a molecule with a Stokes radius of 31 Å when assessed with Sephadex G-100, indicating that its apparent molecular weight is 54 kDa; this contrasts with SDS-PAGE analysis indicating that its apparent molecular weight is 23 kDa.[24] Further sedimentation coefficient analysis indicated that its molecular weight is 21.7 kDa, which is more consistent with the apparent molecular weight determined by SDS-PAGE.[24,27] These results indicate that G-substrate is a protein with an asymmetric and unfolded structure.[24,27] Recombinant preparations from human and rat G-substrate in *E. coli* and in established mammalian cell lines also have characteristics similar to those of the G-substrate purified from cerebellum, such as thermostability, acid stability,[28,29,31] and a higher apparent molecular weight, as determined through gel filtration (Endo and Suzuki, unpublished observations).

## B. Comparison of the Sequence of G-Substrate Phosphorylation Sites and the Sequences of Other Protein Phosphatase Inhibitors Activated by Phosphorylation

The amino acid sequences surrounding the two phosphorylation sites in the G-substrate are strikingly similar to those of PP inhibitors, inhibitor-1 (I-1), and dopamine- and cAMP-regulated phosphoprotein of 32 kDa (DARPP-32) (Fig. 4). In the cases of I-1 and DARPP-32, the Thr residue phosphorylated by PKA is surrounded by two Pro (-P-T-P-). On the other hand, in the G-substrate, the Thr residues phosphorylated by PKG have a single Pro (-D-T-P-) and an additional Pro residue in the N-terminal side of the phosphorylation sites (-“P”-R-R-K-D-T-P-). The presence of Pro residues

```
                                                      *
G-substrate   Site 1 ...CDLKKKPRKGKNVQATLNVESDQKKPRRKDTPALHI...
              Site 2 ...YDVQERHPKGKMIPVLHNTDLEQKKPRRKDTPALHM...
DARPP-32             ...KDRKKIQFSVPAPPSQLDPRQVEMIRRRRPTPAMLF...
Inhibitor-1          ...NSPRKIQFTVPLLEPHLDPEAAEQIRRRRPTPPATL...
```

FIG. 4. Comparison of the amino acid sequences surrounding the phosphorylation sites of G-substrate, DARPP-32, and inhibitor-1 (I-1). The Thr (T) residues marked with an asterisk are phosphorylated by PKG for G-substrate and by PKA for DARPP-32 and I-1. The underlined K/R-K-I-Q-F in DARPP-32 and underlined RKIQF in I-1 are necessary for efficient binding to and inhibition of PP1 (for details see Section V.C). G-substrate lacks the K/R-K-I-Q-F sequence for both sites 1 and 2. Amino acid sequences are adapted from Endo *et al.*[28,29,31] for G-substrate; Kurihara *et al.*[98] and Ehrlich *et al.*[99] for DARPP-32; and Elbrecht *et al.*[100] and Endo *et al.*[101] for inhibitor-1 (I-1).

in a protein sequence often markedly influences the protein architecture. The positioning of Pro residues around the phosphorylation sites of G-substrate, DARPP-32, and I-1 may confer unique substrate specificity toward PKA and PKG phosphorylation[7,25,28] and may contribute to their inhibitory activity against specific types of PP.[28–31]

## C. G-Substrate as a Protein Phosphatase Inhibitor

The sequence similarities between G-substrate, DARPP-32, and I-1 around their phosphorylation sites suggest that G-substrate functions as an endogenous protein phosphatase (PP) inhibitor (Fig. 4). In fact, the PKG-phosphorylated G-substrate strongly inhibits PPs.[28,29,31] However, the G-substrate lacks the additional component required for specific PP1 inhibition: the K/R-K-I/V-Q-F motif observed in DARPP-32, I-1, and other PP1 regulators.[101–105] In G-substrate phosphorylation sites 1 and 2, the corresponding amino acid residues similar to the K/R-K-I/V-Q-F motif are K-K-K-P-R and Q-E-R-H-P, respectively (Figs. 3 and 4). Lack of the K/R-I/V-Q-F sequence, together with lack of Pro in the G-substrate phosphorylation sites, might contribute to the PP inhibitory specificities of the G-substrate. The G-substrate inhibits the catalytic subunit of PP1 with an $IC_{50}$ of 14–131 nM.[28,29,31] Since the estimated concentration of the G-substrate is approximately 10 μM in cerebellar Purkinje cells,[106] phosphorylation of 0.1–1% of G-substrate is necessary in order to inhibit 50% of PP1 activity. Locally, a higher level of G-substrate phosphorylation and stronger inhibition of PP can be potentially achieved in cells.

Upon phosphorylation by PKG, the human and rat G-substrate inhibits PP2A more strongly than PP1,[28,29] whereas the mouse G-substrate inhibits PP1 more strongly than PP2A.[31] These specificity differences cannot be explained completely by the species differences in the G-substrate, since G-substrate and its phosphorylation sites have significant sequence similarities across different species (Fig. 3). The G-substrate inhibition selectivity for PP1 and PP2A might have been influenced by the PP preparations used by Hall *et al.*[31] and Endo *et al.*[28,29] For example, Hall *et al.*[31] used $PP2A_1$ to assay the PP inhibitory activity of mouse G-substrate, whereas Endo *et al.*[28,29] used PP2Ac to assay PP inhibitory activity of human and rat G-substrate. Alternatively, the way PP was stored might have contributed to these differences.[102] Further examination is necessary to determine the mechanisms underlying the G-substrate-mediated inhibition of PP, as cellular PP1 and PP2A exist as complexes with other regulatory subunits that determine the location, substrate specificity, and physiological functions of the G-substrate. At least in A10 cells in the midbrain and retinal amacrine cells, G-substrate works as a strong PP2A inhibitor *in vivo* to protect these cells (see Section VII).

## D. Phosphorylation of G-Substrate by PKG and PKA

In general, PKG substrates are phosphorylated by other kinases, including PKA and PKC.[7] Like other PKG substrates, G-substrate can be phosphorylated by PKA.[27–29] The $K_m$ of G-substrate is approximately 0.2 μM for PKG and approximately 6.0 μM for PKA, and its $V_{max}$ is two-fold higher for PKG compared to PKA.[25,27] Both PKG and PKA phosphorylate the same Thr residues on the G-substrate.[26–29] Mutation of PKG phosphorylation sites (Thr to Ala) completely eliminates phosphorylation by both PKA and PKG, indicating that phosphorylation by PKA and PKG occurs at the same Thr residues.[29]

## E. Nuclear Localization Signal (NLS) and Nuclear Export Signal (NES) of G-substrate

The G-substrate has potential NLSs and NESs (Fig. 3). Classical NLSs are characterized by a cluster of basic amino acids,[107] whereas NESs are characterized by hydrophobic amino acid residues interspersed in a short sequence.[108] The presence of NLSs and NESs in the G-substrate suggests that the G-substrate may shuttle between the cytosol and nucleus using these signals. Supporting this hypothesis is the observation that the G-substrate is located not only in the cytosol but also in the nucleus of cerebellar Purkinje cells in brain slices (Fig. 5). Two NLSs of the G-substrate are located next to phosphorylation sites by PKG. Addition of negative charges by phosphorylation may disrupt the cluster of positive charges of basic amino acids in NLS. Then, this may contribute to controlling the location and shuttling of the G-substrate in the cells. Nuclear export can be mediated by Leu-rich NES via a CRM1-dependent pathway.[108] CRM1, which binds to the NES of proteins and mediates transport to the cytosol, is also found in Purkinje cells (Fig. 6). Examining the role of NLS and NES in the G-substrate will clarify the physiological roles of

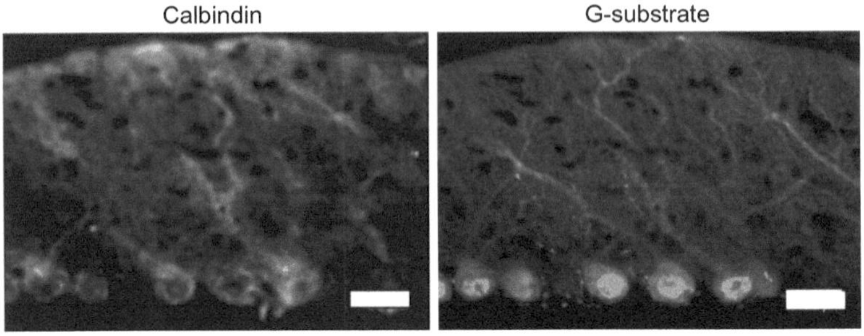


FIG. 5. G-substrate localizing to the cytosol and nucleus of mouse cerebellum. Sagittal section of mouse cerebellum was stained with antibodies against calbindin (left panel) and G-substrate (right panel). Calbindin was specifically observed in Purkinje cells in the cerebellum. G-substrate was observed not only in somas and dendrites but also in the nuclei of cerebellar Purkinje cells. Scale bars: 30 μm.

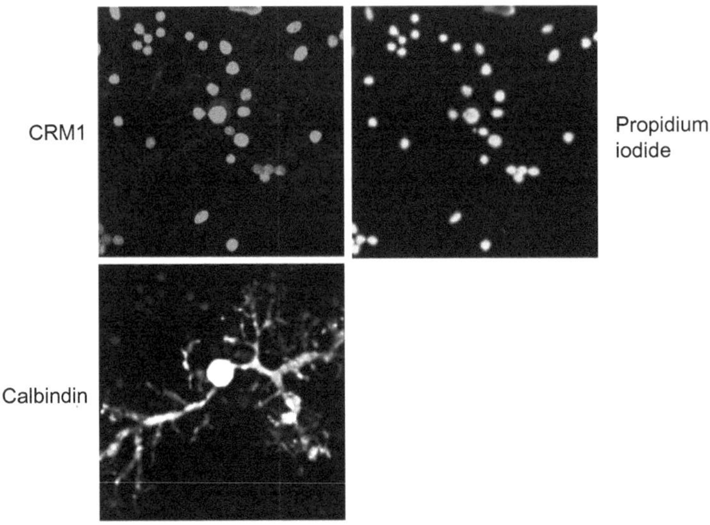


FIG. 6. Expression of CRM1 and calbindin in primary cultures of rat cerebellum. Rat cerebellar primary cultures were double-immunostained with rabbit anti-CRM1 and mouse anti-calbindin antibodies. The cultures were then treated with Alexa Fluor 488 anti-rabbit IgG and Alexa Fluor 633 anti-mouse IgG, and then counterstained with propidium iodide. Intense nuclear staining was observed with the CRM1 antibody. CRM1 was also expressed in Purkinje cells immunostained for calbindin.

the substrate in the cytosol and nucleus of Purkinje cells. The translocation of the G-substrate in cells of the retina and ventral tegmental area is yet to be investigated.

# VI. G-Substrate and Brain Function

## A. Distribution of G-Substrate in the Nervous System

Immunohistochemical analyses have revealed the distribution of G-substrate in rat brain[22,23,27,29,31,106,109] and mouse brain.[30,31] G-substrate is expressed in very limited subsets of cells in the brain. Strong immunohistochemical staining occurs in cerebellar Purkinje cells (Fig. 5).[22,23,27,29–31] Other cerebellar cells were not stained by G-substrate antibodies.[22,23,27,29–31] In addition to the cerebellum, we also observed G-substrate immunoreactivity in the suprachiasmatic nucleus (SCN), in the paraventricular area, in relatively large neurons in parts of the hypothalamus near the third ventricle, and in the amygdala and in the olfactory bulb (Fig. 7). G-substrate is observed also in the A10 area of the ventral tegmental area.[34] We also characterized G-substrate in the retina in detail.[32,33] Although only a small number of cells in the ventral tegmental area and the retina express G-substrate, the G-substrate in these cells plays a significant neuroprotective role against chemically induced injury[32–34]. Details of the protective role of G-substrate will be discussed later in this review (see Section VII). Although the importance of the NO–sGC–cGMP–PKG system in fear memory mediated by the amygdala[110–113] and in the circadian rhythm mediated by the SCN[114,115] has been demonstrated, substrates for PKG in these tissues have not been well characterized. G-substrate might play a role as a downstream component of PKG in these areas. Further examination is necessary to determine the precise role of G-substrate in these areas.

## B. G-Substrate in the Cerebellum

In the 1970s, the presence of a PKG substrate in mammalian tissues, including brain, was demonstrated, with a high content in the cerebellum.[1,2,22,23] However, it was not until the 1980s that G-substrate was purified and characterized from rabbit cerebellum.[24–27] Seven hundred rabbit cerebella (950 g) were used to obtain 330 μg of purified G-substrate through several steps of chromatography.[24] Then purified rabbit G-substrate was used for the biochemical characterization and for protein sequencing.[25,26] Further characterization demonstrated that G-substrate was observed in the somas, axons, and dendrites of cerebellar Purkinje cells.[27–31,106,109] In Purkinje cell somas, G-substrate is observed in both the cytosol and nucleus[30,106,109] (Fig. 5) and the concentration of G-substrate in Purkinje cell is estimated at approximately 10μM.[106]

These lines of evidence prompted us to study the physiological functions of G-substrate in cerebellum in detail.[28–30] G-substrate is a PKG substrate that is positioned downstream of the NO–sGC–cGMP–PKG pathway (Fig. 1). Even though Purkinje cells do not express NO synthase,[109,116] cells surrounding

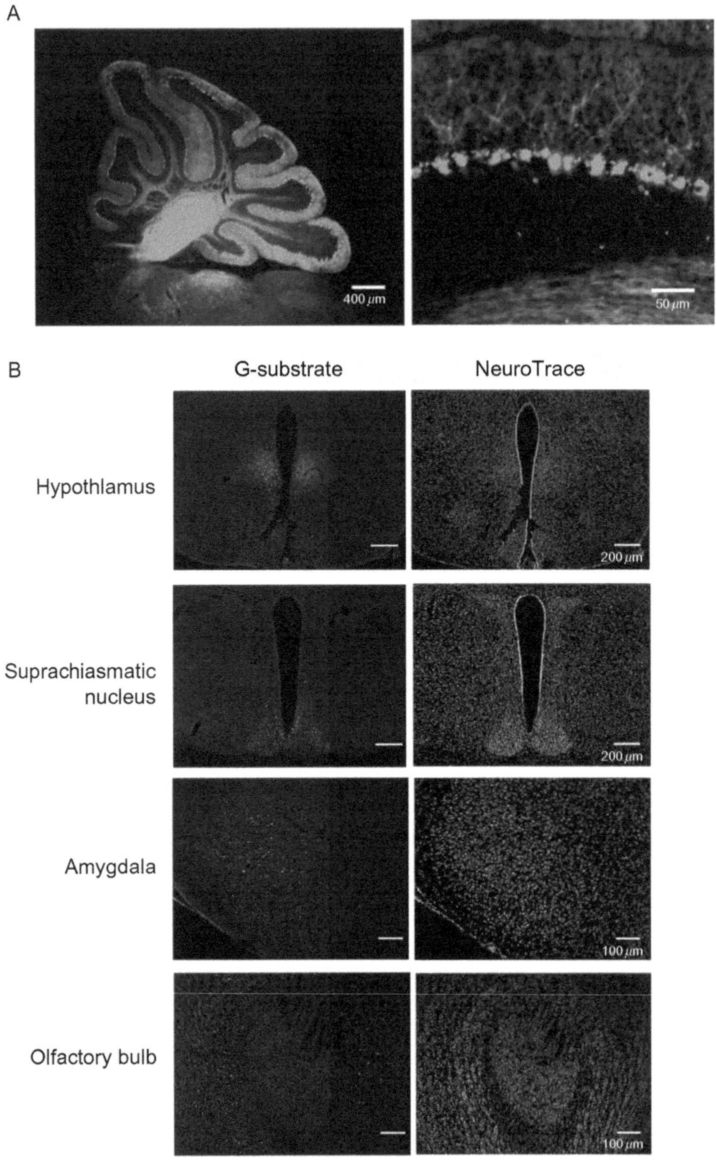


FIG. 7. Distribution of G-substrate in the central nervous system. A. Sagittal sections of mouse brain stained with anti-G-substrate antibody. G-substrate localized not only to somas but also to dendrites and axons of Purkinje cells. Scales bars: left panel, 400 μm; right panel, 50 μm. B. Coronal section of mouse brain stained with anti-G-substrate antibody. G-substrate was observed in restricted areas in hypothalamus, amygdala, suprachiasmatic nucleus, and olfactory bulb. Neurons are stained with NeuroTrace. Cells around Purkinje cells are granule cells which are the most abundant neuronal cells in the cerebellum. Scale bars: 200 μm for hypothalamus and suprachismatic nucleus; 100 μm for amygdala and olfactory bulb.

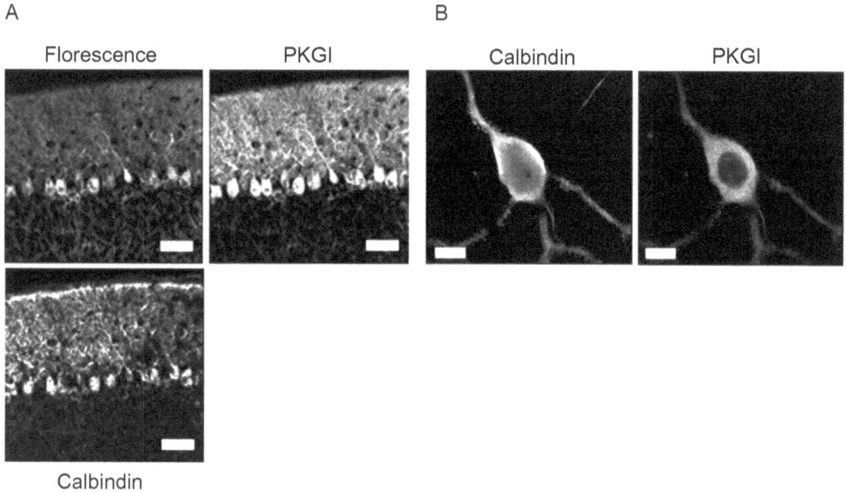


FIG. 8. Distribution of PKGI in cerebellum and in cerebellar Purkinje cells. A. Sagittal sections of cerebellum from G-substrate KO mice stained with antibodies against PKGI (right) and calbindin (bottom). G-substrate KO mice expressed AFAP (GFP derivative) in Purkinje cells under the control of G-substrate promoter (Florescence). Scale bars: 50 μm. B. Purkinje cell from a mouse cerebellar primary culture stained with antibodies against calbindin (left) and PKGI (right). PKGI was observed in dendrites and somas of cerebellar Purkinje cells. In primary cell cultures, PKGI was mainly observed in the cytosol. Scale bars: 10 μm.

them do. Thus, NO that diffuses to Purkinje cells can activate soluble guanylate cyclase,[39] which in turn enhances PKG activity[117] by increasing the concentration of cGMP. Purkinje cells are rich in PKGI (Fig. 8). In cerebellar slices, G-substrate in Purkinje cells is effectively phosphorylated in response to a membrane-permeable analog of cGMP that activates PKG.[29]

Thus far, each component of the NO–sGC–cGMP–PKG pathway has been shown to be necessary for the induction of cerebellar long-term depression (LTD),[30,118–125,125a] a characteristic form of synaptic plasticity displayed by cerebellar Purkinje cells (for review, see Refs. 75–77,126, 126a). LTD is hypothesized to be an important cellular basis for motor learning and other cognitive functions.[75–77,126,126a] In LTD, synaptic transmission from parallel fibers (PFs) to Purkinje cells is persistently depressed after simultaneous stimulation of the PFs and climbing fibers (CFs). CF stimuli can be replaced by delivering depolarizing pulses to the Purkinje cell membrane. Peculiarly, NO is not required for LTD induction in young, cultured Purkinje cells,[127] but is required for LTD induction in cerebellar slices.[128] These results suggest that the NO–sGC–cGMP–PKG pathway acts as a modulator, whose requirement for LTD may depend on various circumstances.[76]

## C. Motor Learning in G-Substrate KO Mice

Cerebellar LTD has been considered to be a cellular mechanism that underlies motor learning.[75–77,126,126a] To determine whether G-substrate, LTD, and motor behaviors are correlated, we established G-substrate KO mice and analyzed them in detail (Table I).[30] Homozygous G-substrate KO ( −/−) mice survived, and their life span was comparable to that of wild-type mice (Endo and Suzuki, unpublished observations). Homozygous G-substrate KO mice grew and reproduced normally, and bore pups normally. Overall, they performed similarly to wild-type mice in behavioral tests, including open field test, elevated plus maze, light–dark box, prepulse inhibition, social interaction test, and fear conditioning test (tone and context dependent).

With regard to the induction of cerebellar LTD, we discovered a distinct age-dependent deficit in G-substrate-deficient Purkinje cells: LTD in acute slice preparation occurs during postnatal 4–5 weeks but diminishes at 6 weeks and then recovers at 10 weeks and thereafter.[30] This may be related to the development-dependent expression of PP which we and other groups observed previously in cerebellar Purkinje cells.[129,130] Thus, PP1 is required for the induction of cerebellar LTD during early states of cerebellar maturation, whereas PP2A is required for the induction of cerebellar LTD during mature stages, such as that observed in mature primary cultures.[129,130] Details of molecular mechanisms underlying PP subtype-dependent cerebellar LTD are unknown.

With regard to cerebellum-dependent behaviors, G-substrate of KO mice are mostly free of deficits (Table I). For instance, they perform normally in motor learning tasks such as eyeblink conditioning,[131] and adaptive

TABLE I
CHARACTERISTICS OF G-SUBSTRATE KNOCKOUT MICE[a]

| | Comparison to wild-type mice |
|---|---|
| General behaviors | No difference |
| Life span and reproduction | No difference |
| Ataxia | No apparent ataxia |
| Cerebellar long-term depression | Significant reduction at around 6 weeks of age |
| Cerebellum-dependent behaviors | |
| Eye blink conditioning | No difference |
| Motor coordination | No difference |
| Adaptive locomotion | No difference |
| Optokinetic response | No difference in short-term adaptation<br>Reduced long-term adaptation |

[a]Adapted from: Endo *et al.* (2009).[30]

locomotion.[132,168] It is possible that we did not observe deficits during eyeblink conditioning, a task generally believed to require the cerebellum, because the task does not necessarily depend completely on cerebellar function.[133,134]

We also examined how the age-dependent reduction of LTD is reflected in the age profile of adaptation in the optokinetic eye movement response (OKR), a simple form of motor learning.[30] OKR and vestibulo-ocular reflex (VOR) function are cooperative so that an image on the retina remains motionless even though the body or landscape may move.[75,126,126a] The NO–sGC–cGMP–PKG pathway has been demonstrated to play an important role in OKR and VOR adaptation[135,136] (Table II) in mice. In 6-week-old G-substrate KO mice, we detected a significant difference in short-term OKR adaptation between homozygous G-substrate KO mice and wild-type mice. The magnitude of short-term OKR adaptation was smaller in G-substrate KO mice than in wild-type mice. These results are consistent with our findings of attenuated LTD in 6-week-old mice but recovery in 12-week-old mice.[30]

Next, we examined OKR in adult G-substrate KO mice (more than 12-week old).[30] The only deficit exhibited by adult G-substrate KO mice was a significant reduction in long-term OKR adaptation. Thus, the functional role of the G-substrate gene is defined by its differential involvement in LTD-driven short-term OKR adaptation and non-LTD-driven long-term OKR adaptation. This long-term adaptation is caused by a slowly developed potentiation of synaptic transmission or an intrinsic excitability in vestibular relay neurons.[137] The impairment of long-term VOR adaptation with normal short-term adaptation was also reported in mice deficient in PKGI,[125] another component of the NO–sGC–cGMP–PKG pathway, which phosphorylates the G-substrate. Long-term VOR and OKR adaptation may share a common synaptic mechanism at the level of vestibular relay neurons (see Refs. 77,137). PKGI is concentrated and colocalizes with G-substrate in Purkinje cells. PKGI phosphorylates G-substrate efficiently and effectively to endow it with PP inhibitory activity.[28–31] Therefore, the NO–sGC–cGMP–PKG–G-substrate cascade appears to play an essential role in the induction of long-term VOR and OKR adaptation, a role that cannot be compensated by another pathway. Interestingly, genetically modified mice for nNOS, PKGI, and G-substrate demonstrate quite similar behavioral phenotype: normal in motor coordination and eyeblink conditioning, but deficits in eye movement (OKR and/or VOR) (Table II).[30,125,136]

Currently, it is not known how the lack of G-substrate in Purkinje cells impairs the adaptive mechanism in the vestibular nucleus. However, G-substrate KO mice and PKGI KO mice potentially provide excellent models for examining the cellular and molecular mechanisms underlying long-term OKR and VOR adaptation and "long-term" memory. G-substrate and the

TABLE II

COMPARISON OF CEREBELLAR DYSFUNCTION IN MICE HAVING GENETIC MODIFICATIONS IN THE NO–sGC–cGMP–PKG PATHWAY[a]

| | Electrophysiological properties of Purkinje cells | | | Behaviors | | | |
|---|---|---|---|---|---|---|---|
| | | | | VOR/OKR adaptation | | | |
| Knockout mice | Parallel fiber response | Climbing fiber response | Cerebellar LTD | Short-term | Long-term | Motor coordination | Eyeblink conditioning |
| nNOS (null knockout) | n.d. | n.d. | Impaired | Impaired | n.d. | Normal* | Normal |
| PKGI (Purkinje cell-specific conditional knockout) | n.d. | n.d. | Impaired | Normal | Impaired | Normal | n.d. |
| G-substrate (null knockout) | Normal | Normal | Impaired (young age) | Normal | Impaired | Normal | Normal |

*Nocturnal deficit. Motor coordination was examined by the rotor rod test. n.d., not determined.

[a]Adapted from: Katoh *et al.*[136] and Kreigsfeld *et al.*[168] for nNOS; Feil *et al.*[125] for PKGI; Endo *et al.*[30] for G-substrate.

NO–sGC–cGMP–PKG pathway may be involved in the location shift of memory engram associated with the conversion of memory from short-term to long-term in OKR and/or VOR.[137,138]

## D. Use of G-Substrate KO Mice

G-substrate KO mice express AFAP (a GFP derivative) under the control of G-substrate promoter.[30,32,33,139] Not only Purkinje cell somas but also dendrites and axons originating from Purkinje cells are clearly visible under a fluorescence microscope[30] (Fig. 8). The expression of G-substrate can be used as a marker to trace differentiation of Purkinje cells in cerebellum, a particular type of amacrine cells in the retina,[32] and A10 cells in ventral tegmental area.[34] In an experiment in which isolated mesenchymal stem cells from G-substrate KO mice was transplanted into cerebellum *in vivo,* some transplanted cells expressed AFAP, had Purkinje cell-like morphology, and expressed a Purkinje cell marker.[139] Thus, G-substrate KO mice can be used to assess the activity of the G-substrate promoter *in vivo*, such as in the implantation of stem cells.

## E. G-Substrate in Zebra Finch

Zebra finch (*T. guttata*) has been used in studies characterizing the mechanisms underlying behaviors such as singing and learning songs.[140–142] Jarvis' group has characterized the genes and gene expression related to song learning in zebra finch.[140] They constructed subtracted cDNA libraries and analyzed these libraries using zebra finch cDNA expression arrays comparing (1) the cDNA of two male singers minus the cDNA of one male silent control and (2) the cDNA of one juvenile learner minus the cDNA of one male silent control.[140] In either case, G-substrate was enriched in these subtracted libraries. Even though cloned zebra finch G-substrate cDNA is not the full-length cDNA (Fig. 3)—it seems to lack the part of the C-terminal region that includes PKG-phosphorylation site 2—the N-terminal portion of zebra finch G-substrate overall shares similarities to that of other vertebrates. Zebra finch G-substrate, however, has three extra amino acids (L-T-G) inserted into the amino acid sequence at positions 81–83. This unique insertion is also observed in chicken G-substrate (Fig. 3).

Singing requires intensive learning and training. Minute and precise control of muscles is required for singing and pronunciation.[141] The expression of G-substrate in mature songbirds and juvenile birds learning to sing may play an important role in these behaviors. Further analysis, including the identification of full-length G-substrate cDNA and the distribution of G-substrate in zebra finch brain, will help to reveal the potential physiological role of G-substrate in the song learning behaviors in birds.

## VII. Neuroprotection by G-Substrate

We observed that G-substrate has a neuroprotection effect on A10 dopaminergic (DA) cells in the midbrain[34] and on amacrine cells in the retina.[32,33] Interestingly, in both cases, Akt activity seems to be regulated by the strong PP inhibitory activity of G-substrate. The neuroprotection experiments were carried out using a DA cell injury model for Parkinson's disease (PD) and a retina excitotoxicity model for diabetic retinopathy (DR). Treatment with G-substrate might prove to be a good method for protecting vulnerable DA cells in the substantia nigra and vulnerable cells in the retina.

### A. Dopaminergic Cells in Ventral Tegmental Area

In PD, there is a striking difference between A9 and A10 DA neurons in the midbrain with regard to susceptibility to degeneration. DA neurons in the A9 region (substantia nigra) are considerably more vulnerable than those in the adjacent A10 region (ventral tegmental area).[143] Animal models of PD also show distinct differences in vulnerability.[144,145] Even though A9 and A10 DA cells share many similarities, they have quite different gene expression profiles.[146–148] One of the major differences in gene expression is that A10 DA cells express greater levels of G-substrate mRNA than A9 DA cells.[146,147] Intriguingly, vulnerable A9 DA neurons express higher levels of two isoforms of the PP2A regulatory subunit B (B56),[147] indicating that PP2A activity toward certain targets may be elevated in A9 neurons. These results implicate the G-substrate-PP2A system in the vulnerability difference between A9 and A10 DA cells.

Overexpression of G-substrate protects DA BE(2)-M17 cells against injuries caused by 6-hydroxydopamine (6-OHDA) and carbobenzoxy-Leu-Leu-Leu-al (MG-132).[34] On the other hand, the knockdown of endogenous G-substrate by RNAi increases vulnerability of cells to these toxins.[34] The activation of PP2A induced by 6-OHDA *in vitro* is inhibited by overexpression of G-substrate and causes a significant increase in the basal phosphorylation levels of Akt, GSK3, and ERK2. However, G-substrate overexpression does not alter the phosphorylation level of proapoptotic pp38.[34] Reducing Akt by RNAi suppresses the protective effect of G-substrate against 6-OHDA, indicating that Akt may act as a downstream component of the G-substrate–PP2A pathway for cell protection.

At the cellular level, G-substrate seems to modulate a subset of PP2A substrates known to promote cell survival rather than work as a general PP2A inhibitor, such as okadaic acid and calycrin A, which reduce the basal level of PP2A activity. Hence, G-substrate may work as a PP2A inhibitor but more selectively in cells. One possibility is that G-substrate may modulate substrate specificity by binding to PP2A subunits. PP2A exists as a complex of AC or

ABC; each subunit has many isoforms in cells. Each complex has distinct characteristics in catalytic activity, substrate specificity, and cellular location.[149] Searching for and examination of G-substrate-interacting proteins, including PP2A subunits, will reveal whether the notion above is correct.

In addition to the *in vitro* experiments mentioned above, we conducted *in vivo* experiments to characterize the neuroprotective qualities of G-substrate. *In vivo* lentiviral delivery of G-substrate into rat substantia nigra protected A9 DA neurons from 6-OHDA-induced toxicity.[34] The 6-OHDA protocol causes severe injury, leading to 80% loss of tyrosine hydroxylase (TH) positive DA neurons in the midbrain and complete loss of TH-positive striatal terminals. Approximately 50% of G-substrate-transduced A9 DA neurons were protected against 6-OHDA toxicity *in vivo*. These results suggest that enhancing G-substrate levels may be a neuroprotective strategy for protecting vulnerable A9 DA neurons in PD. Protection of PC12 cells from 6-OHDA-induced apoptosis through the NO–sGC–cGMP–PKG pathway also supports the role of G-substrate as a neuroprotectant.[150] The subcellular distribution of G-substrate, however, has not been established in PC 12 cells.

## B. Amacrine Cells in Retina

G-substrates partially purified from mouse and rat retina have a similar molecular weight as that purified from the cerebellum. Like cerebellar G-substrate, retinal G-substrate is a heat-stable protein (at 95°C for 5min). Antibodies against the N- and C-termini of cerebellar G-substrate cross-react with retinal G-substrate.[32] The G-substrate is located in a restricted subset of cells in the retina. In adult rats and mice, retinal G-substrate is located in a subpopulation of amacrine cells and in C38-positive retinal ganglion cells (RGCs) but not in alpha RGCs.[32] In addition, G-substrate is co-expressed with other upstream and downstream signaling components of the NO–sGC–cGMP–PKG–G-substrate–protein phosphatase pathway in the adult retina.[32]

Under conditions of stress, contradictory findings exist regarding the action of NO, with findings that retinal cell protection depends on the concentration of NO.[151] In addition, there are some clinical reports on the actions of NO in patients with glaucoma and DR.[152–154] However, an intracellular signal transduction cascade influenced by NO is unknown in these diseases.[33] DR is the most common complication of diabetes and is the leading cause of severe visual impairment,[155] which leads to the severe restriction of the patient's behavior and to a significant reduction in the quality of life. DR is clinically characterized by the presence of microangiopathy.[156,157] One of the causes of DR is thought to be increased glutamate concentration,[158,159] where extracellular glutamate induces amacrine cell death through enhanced intracellular $Ca^{2+}$ concentration through the activation of *N*-methyl-D-aspartic acid receptor (NMDA-R)s.[160–163]

In the retina, NMDA induces Akt dephosphorylation through PP2A activation.[164] On the other hand, BDNF (brain-derived neurotrophic factor) suppresses Akt dephosphorylation and reduces decreased retinal excitotoxicity.[164] Colocalization of G-substrate and other NO–sGC–cGMP–PKG pathway components in amacrine cells prompted us to investigate the role of G-substrate in Akt regulation in retinal cell death. Amacrine cells are classified into more than 20 classes based on their shape and staining with different cellular markers.[165] Approximately 30% of amacrine cells in mouse retina are G-substrate positive[32]; this population of cells is functionally classified as extremely susceptible to excitotoxicity and they are the target of adverse effects caused by NMDA-induced increases in glutamate concentrations.

G-substrate immunoreactivity in amacrine cells decreases rapidly as the result of NMDA-associated injury. The calpain inhibitors ALLN (*N*-acetyl-Leu-Leu-Nle-al) and calpain inhibitor III (carbobenzoxy-Val-Phe-al) block the loss of G-substrate immunoreactivity.[33] G-substrate is efficiently degraded by calpain I and II *in vitro*.[33] These results clearly demonstrate that G-substrate is a good substrate for calpain and may be degraded in cells through calpain activated by increased intracellular $Ca^{2+}$ concentrations. In addition, amacrine cells in homozygous G-substrate KO mice are more susceptible to NMDA injury than those of wild-type mice. Moreover, 60% of apoptotic amacrine cells did not express G-substrate due to the G-substrate gene knockout.[33] Because most amacrine cells express the NMDA NR1 receptor,[166] NMDA would be expected to cause similar damage to these cells. However, susceptibility to NMDA injury varies across different populations of amacrine cells, perhaps in part due to the presence of G-substrate in different subpopulations of amacrine cells.[32] This idea is consistent with the role of G-substrate in contributing to the differential vulnerability of DA neurons in models of PD[34], as mentioned in the previous section.

Based on these results, G-substrate-positive amacrine cells are primary targets for the treatment of retinal excitotoxicity. Considering the involvement of G-substrate in the PP2A-mediated dephosphorylation of Akt, G-substrate plays an essential role in the pro-apoptotic activity of Akt in retinal amacrine cells.

These findings have important implications for treating retinal diseases associated with excitotoxicity, such as vessel occlusion and DR. Moreover, elucidating the system responsible for regulating calcium signaling and PP activation may also shed light on the general mechanism underlying excitotoxicity-induced neuronal degeneration.

## VIII. Aging and G-Substrate

Aging causes a variety of cellular and molecular changes in most body tissues. For example, in the NO–sGC–cGMP–PKG system of aged animals, it is hypothesized that basal NOS activity increases and, potentially, NO modifies and activates NMDA-Rs. On the other hand, with aging, the activities of cGMP-hydrolyzing PDEs increase, thereby decreasing intracellular concentrations of cGMP produced through the NO–sGC pathway (for a recent review, see Refs. 167). This may lead to the impairment of hippocampus-dependent memory and/or other cognitive functions. In mice, the expression of G-substrate mRNA significantly increases in the cerebellum of aged mice.[74] The increased expression of G-substrate may compensate for reduced cGMP–PKG function caused by the low cellular concentrations of cGMP in aged animals. Age-related decline of cognitive functions, including memory, might be caused by the reduced concentration of cGMP and its resulting downstream activation of secondary messenger systems. Further research on how aging affects the cGMP system may reveal potential targets for treatments aimed at improving the decline of cognitive functions. In this regard, some researchers are currently testing cAMP-hydrolyzing PDEs[49] and a variety of PDE inhibitors for their ability to improve or enhance cognitive functions. Further research on the cGMP pathway and G-substrate may lead us to future treatments for improving cognitive deficits related to injury, diseases, and aging.

## IX. Conclusions

In this review, we have focused on G-substrate, a component of the NO–sGC–cGMP–PKG pathway in cells. The pathway has received less attention compared to the cAMP system. However, recent progress in cGMP-associated research, such as the discovery of the mechanism mediating cGMP production, has led to tremendous interest in this area. G-substrate is expressed primarily in cerebellar Purkinje cells and in a few restricted areas of the nervous system. The role of G-substrate in cerebellum has been extensively examined by using G-substrate KO mice. Furthermore, the important role of G-substrate in cerebellum-dependent long-term memory has been demonstrated. The shuttling of G-substrate between the cytosol and nucleus in Purkinje cells may support the gene regulation necessary for long-term memory.

In the retina and the midbrain, G-substrate plays a strong neuroprotective role. Further investigation may lead to future clinical applications for G-substrate. Even though G-substrate itself does not have enzyme activity, it efficiently regulates the enzymatic activity of PP in cells. As mentioned in this review, the modulatory effects of phosphorylation have tremendous outcomes, ranging from those that affect long-term memory to those that confer neuroprotection. The NO–sGC–cGMP–PKG pathway and G-substrate may be a potential target of novel drugs and treatments for various diseases, such as learning impairment and neurodegenerative diseases caused by a variety of traumas and pathophysiological processes.

## Acknowledgments

The author wishes to thank Ms. Masako Suzuki for tremendous technical help throughout the G-substrate research project. Without her help, this project would not have progressed successfully. The author would also like thank the many other collaborators who have also contributed to G-substrate research. Especially, the generous support by Drs. Angus Nairn and Paul Greengard from the initial stage of G-substrate project is gratefully acknowledged. This work was supported in part by the Naito Foundation, MEXT, Japan, and the Japan Foundation for Aging and Health.

## References

1. Casnellie JE, Greengard P. Guanosine 3′:5′-cyclic monophosphate-dependent phosphorylation of endogenous substrate proteins in membranes of mammalian smooth muscle. *Proc Natl Acad Sci USA* 1974;**71**:1891–5.
2. de Jonge HR. Cyclic nucleotide-dependent phosphorylation of intestinal epithelium proteins. *Nature* 1976;**262**:591–3.
3. Lincoln TM, Hall CL, Park CR, Corbin JD. Guanosine 3′:5′-cyclic monophosphate binding proteins in rat tissues. *Proc Natl Acad Sci USA* 1976;**73**:2559–63.
4. Walter U. Distribution of cyclic-GMP-dependent protein kinase in various rat tissues and cell lines determined by a sensitive and specific radioimmunoassay. *Eur J Biochem* 1981;**118**:339–46.
5. Walter U, Lohmann SM, Reiser G, Schwoch G, Miller P, DeCamilli P. Studies on regulation and cellular localization of cAMP- and cGMP-dependent protein kinases. *Prog Clin Biol Res* 1982;**102**:173–82.
6. Pearson RB, Kemp BE. Protein kinase phosphorylation site sequences and consensus specificity motifs: tabulations. *Methods Enzymol* 1991;**200**:62–81.
7. Wang X, Robinson PJ. Cyclic GMP-dependent protein kinase and cellular signaling in the nervous system. *J Neurochem* 1997;**68**:443–56.
8. Schlossmann J, Desch M. cGK substrates. *Handb Exp Pharmacol* 2009;**191**:163–93.
9. Beavo JA, Brunton LL. Cyclic nucleotide research—still expanding after half a century. *Nat Rev Mol Cell Biol* 2002;**3**:710–8.
10. Rybalkin SD, Yan C, Bornfeldt KE, Beavo JA. Cyclic GMP phosphodiesterases and regulation of smooth muscle function. *Circ Res* 2003;**93**:280–91.
11. DiCicco-Bloom E, Lelièvre V, Zhou X, Rodriguez W, Tam J, Waschek JA. Embryonic expression and multifunctional actions of the natriuretic peptides and receptors in the developing nervous system. *Dev Biol* 2004;**271**:161–75.
12. Kuhn M. Structure, regulation, and function of mammalian membrane guanylyl cyclase receptors, with a focus on guanylyl cyclase-A. *Circ Res* 2003;**93**:700–9.
13. Komatsu Y, Nakao K, Suga S, Ogawa Y, Mukoyama M, Arai H, et al. C-type natriuretic peptide (CNP) in rats and humans. *Endocrinology* 1991;**129**:1104–6.
14. Sudoh T, Minamino N, Kangawa K, Matsuo H. C-type natriuretic peptide (CNP): a new member of natriuretic peptide family identified in porcine brain. *Biochem Biophys Res Commun* 1990;**168**:863–70.
15. Sudoh T, Kangawa K, Minamino N, Matsuo H. A new natriuretic peptide in porcine brain. *Nature* 1988;**332**:78–81.
16. Herman JP, Langub Jr. MC, Watson Jr. RE. Localization of C-type natriuretic peptide mRNA in rat hypothalamus. *Endocrinology* 1993;**133**:1903–6.

17. Potter LR, Yoder AR, Flora DR, Antos LK, Dickey DM. Natriuretic peptides: their structures, receptors, physiologic functions and therapeutic applications. *Handb Exp Pharmacol* 2009;**191**:341–66.
18. Kuhn M. Function and dysfunction of mammalian membrane guanylyl cyclase receptors: lessons from genetic mouse models and implications for human diseases. *Handb Exp Pharmacol* 2009;**191**:47–69.
19. Hofmann F, Bernhard D, Lukowski R, Weinmeister P. cGMP regulated protein kinases (cGK). *Handb Exp Pharmacol* 2009;**191**:137–62.
20. Kleppisch T. Phosphodiesterases in the central nervous system. *Handb Exp Pharmacol* 2009;**191**:71–92.
21. Kleppisch T, Feil R. cGMP signalling in the mammalian brain: role in synaptic plasticity and behaviour. *Handb Exp Pharmacol* 2009;**191**:549–759.
22. Schlichter DJ, Casnellie JE, Greengard P. An endogenous substrate for cGMP-dependent protein kinase in mammalian cerebellum. *Nature* 1978;**273**:61–2.
23. Schlichter DJ, Detre JA, Aswad DW, Chehrazi B, Greengard P. Localization of cyclic GMP-dependent protein kinase and substrate in mammalian cerebellum. *Proc Natl Acad Sci USA* 1980;**77**:5537–41.
24. Aswad DW, Greengard P. A specific substrate from rabbit cerebellum for guanosine 3′:5′-monophosphate-dependent protein kinase. I. Purification and characterization. *J Biol Chem* 1981;**256**:3487–93.
25. Aswad DW, Greengard P. A specific substrate from rabbit cerebellum for guanosine 3′:5′-monophosphate-dependent protein kinase. II. Kinetic studies on its phosphorylation by guanosine 3′:5′-monophosphate-dependent and adenosine 3′:5′-monophosphate-dependent protein kinases. *J Biol Chem* 1981;**256**:3494–500.
26. Aitken A, Bilham T, Cohen P, Aswad D, Greengard P. A specific substrate from rabbit cerebellum for guanosine-3′:5′- monophosphate-dependent protein kinase. III. Amino acid sequences at the two phosphorylation sites. *J Biol Chem* 1981;**256**:3501–6.
27. Nairn AC, Greengard P. Cyclic GMP-dependent protein phosphorylation in mammalian brain. *Fed Proc* 1983;**42**:3107–13.
28. Endo S, Suzuki M, Sumi M, Nairn AC, Morita R, Yamakawa K, et al. Molecular identification of human G-substrate, a possible downstream component of the cGMP-dependent protein kinase cascade in cerebellar Purkinje cells. *Proc Natl Acad Sci USA* 1999;**96**:2467–72.
29. Endo S, Nairn AC, Greengard P, Ito M. Thr123 of rat G-substrate contributes to its action as a protein phosphatase inhibitor. *Neurosci Res* 2003;**45**:79–89.
30. Endo S, Shutoh F, Dinh TL, Okamoto T, Ikeda T, Suzuki M, et al. Dual involvement of G-substrate in motor learning revealed by gene deletion. *Proc Natl Acad Sci USA* 2009;**106**:3525–30.
31. Hall KU, Collins SP, Gamm DM, Massa E, DePaoli-Roach AA, Uhler MD. Phosphorylation-dependent inhibition of protein phosphatase-1 by G-substrate. A Purkinje cell substrate of the cyclic GMP-dependent protein kinase. *J Biol Chem* 1999;**274**:3485–95.
32. Nakazawa T, Endo S, Shimura M, Kondo M, Ueno S, Tamai M. Retinal G-substrate, potential downstream component of NO/cGMP/PKG pathway, is located in subtype of retinal ganglion cells and amacrine cells with protein phosphatases. *Brain Res Mol Brain Res* 2005;**135**:58–68.
33. Nakazawa T, Shimura M, Mourin R, Kondo M, Yokokura S, Saido TC, et al. Calpain-mediated degradation of G-substrate plays a critical role in retinal excitotoxicity for amacrine cells. *J Neurosci Res* 2009;**87**:1412–23.
34. Chung CY, Koprich JB, Endo S, Isacson O. An endogenous serine/threonine protein phosphatase inhibitor, G-substrate, reduces vulnerability in models of Parkinson's disease. *J Neurosci* 2007;**27**:8314–23.

35. Kuo JF, Greengard P. Cyclic nucleotide-dependent protein kinases. VI. Isolation and partial purification of a protein kinase activated by guanosine 3′,5′-monophosphate. *J Biol Chem* 1970;**245**:2493–8.
36. Gill GN, Walton GM, Sperry PJ. Guanosine 3′:5′-monophosphate-dependent protein kinase from bovine lung. Subunit structure and characterization of the purified enzyme. *J Biol Chem* 1997;**252**:6443–9.
37. Lincoln TM, Dills WL, Jr., Corbin JD. Purification and subunit composition of guanosine 3′:5′-monophosphate-dependent protein kinase from bovine lung. *J Biol Chem* 1977;**252**: 4269–75.
38. Snyder SH. Nitric oxide: first in a new class of neurotransmitters. *Science* 1992;**257**:494–6.
39. Stone JR, Marletta MA. Soluble guanylate cyclase from bovine lung: activation with nitric oxide and carbon monoxide and spectral characterization of the ferrous and ferric states. *Biochemistry* 1994;**33**:5636–40.
40. Jaffrey SR, Snyder SH. Nitric oxide: a neural messenger. *Annu Rev Cell Dev Biol* 1995;**11**:417–40.
41. Kots AY, Martin E, Sharina IG, Murad F. A short history of cGMP, guanylyl cyclases, and cGMP-dependent protein kinases. *Handb Exp Pharmacol* 2009;**191**:1–14.
42. Tsay D, Dudman JT, Siegelbaum SA. HCN1 channels constrain synaptically evoked $Ca^{2+}$ spikes in distal dendrites of CA1 pyramidal neurons. *Neuron* 2007;**56**:1076–89.
43. Nolan MF, Malleret G, Dudman JT, Buhl DL, Santoro B, Gibbs E, et al. A behavioral role for dendritic integration: HCN1 channels constrain spatial memory and plasticity at inputs to distal dendrites of CA1 pyramidal neurons. *Cell* 2004;**119**:719–32.
44. Wilson GW, Garthwaite J. Hyperpolarization-activated ion channels as targets for nitric oxide signalling in deep cerebellar nuclei. *Eur J Neurosci* 2010;**31**:1935–45.
45. Barnstable CJ, Wei JY, Han MH. Modulation of synaptic function by cGMP and cGMP-gated cation channels. *Neurochem Int* 2004;**45**:875–84.
46. Biel M, Michalakis S. Cyclic nucleotide-gated channels. *Handb Exp Pharmacol* 2009;**191**:111–36.
47. Wahl-Schott C, Biel M. HCN channels: structure, cellular regulation and physiological function. *Cell Mol Life Sci* 2009;**66**:470–94.
48. Francis SH, Corbin JD, Bischoff E. Cyclic GMP-hydrolyzing phosphodiesterases. *Handb Exp Pharmacol* 2009;**191**:367–408.
49. Reneerkens OA, Rutten K, Steinbusch HW, Blokland A, Prickaerts J. Selective phosphodiesterase inhibitors: a promising target for cognition enhancement. *Psychopharmacology (Berl)* 2009;**202**:419–43.
50. Baillie GS. Compartmentalized signalling: spatial regulation of cAMP by the action of compartmentalized phosphodiesterases. *FEBS J* 2009;**276**:1790–9.
51. Schultz JE. Structural and biochemical aspects of tandem GAF domains. *Handb Exp Pharmacol* 2009;**191**:93–109.
52. Zoraghi R, Corbin JD, Francis SH. Properties and functions of GAF domains in cyclic nucleotide phosphodiesterases and other proteins. *Mol Pharmacol* 2004;**65**:267–78.
53. Blokland A, Schreiber R, Prickaerts J. Improving memory: a role for phosphodiesterases. *Curr Pharm Des* 2006;**12**:2511–23.
54. Miglietta LA, Nelson DL. A novel cGMP-dependent protein kinase from Paramecium. *J Biol Chem* 1988;**263**:16096–161605.
55. Klumpp S, Steiner AL, Schultz JE. Immunocytochemical localization of cyclic GMP, cGMP-dependent protein kinase, calmodulin and calcineurin in Paramecium tetraurelia. *Eur J Cell Biol* 1983;**32**:164–70.
56. Francis SH, Corbin JD. Cyclic nucleotide-dependent protein kinases: intracellular receptors for cAMP and cGMP action. *Crit Rev Clin Lab Sci* 1999;**36**:275–328.

57. Pfeifer A, Ruth P, Dostmann W, Sausbier M, Klatt P, Hofmann F. Structure and function of cGMP-dependent protein kinases. *Rev Physiol Biochem Pharmacol* 1999;**135**:105–49.
58. Vaandrager AB, Ehlert EM, Jarchau T, Lohmann SM, de Jonge HR. N-terminal myristoylation is required for membrane localization of cGMP-dependent protein kinase type II. *J Biol Chem* 1996;**271**:7025–9.
59. Ono S, Ezura Y, Emi M, Fujita Y, Takada D, Sato K, et al. A promoter SNP (-1323T>C) in G-substrate gene (GSBS) correlates with hypercholesterolemia. *J Hum Genet* 2003;**48**:447–50.
60. Soares MB, Bonaldo MF, Jelene P, Su L, Lawton L, Efstratiadis A. Construction and characterization of a normalized cDNA library. *Proc Natl Acad Sci USA* 1994;**91**:9228–32.
61. Jeffreys AJ, Kauppi L, Neumann R. Intensely punctate meiotic recombination in the class II region of the major histocompatibility complex. *Nat Genet* 2001;**29**:217–22.
62. Myers S, Bottolo L, Freeman C, McVean G, Donnelly P. A fine-scale map of recombination rates and hotspots across the human genome. *Science* 2005;**310**:321–4.
63. Freudenberg J, Fu YH, Ptácek LJ. Enrichment of HapMap recombination hotspot predictions around human nervous system genes: evidence for positive selection? *Eur J Hum Genet* 2007;**15**:1071–8.
64. Freudenberg J, Fu YH, Ptácek LJ. Human recombination rates are increased around accelerated conserved regions—evidence for continued selection? *Bioinformatics* 2007;**23**:1441–3.
65. Landmesser U, Hornig B, Drexler H. Endothelial dysfunction in hypercholesterolemia: mechanisms, pathophysiological importance, and therapeutic interventions. *Semin Thromb Hemost* 2000;**26**:529–37.
66. Francois M, Kojda G. Effect of hypercholesterolemia and of oxidative stress on the nitric oxide-cGMP pathway. *Neurochem Int* 2004;**45**:955–61.
67. Münzel T, Daiber A, Ullrich V, Mülsch A. Vascular consequences of endothelial nitric oxide synthase uncoupling for the activity and expression of the soluble guanylyl cyclase and the cGMP-dependent protein kinase. *Arterioscler Thromb Vasc Biol* 2005;**25**:1551–7.
68. Sowden JC, Morrison K, Putt W, Beddington R, Edwards YH. The identification of novel sequences expressed in the mouse notochord. *Mamm Genome* 1997;**8**:42–4.
69. Altman J, Bayer SA. Time of origin and distribution of a new cell type in the rat cerebellar cortex. *Exp Brain Res* 1977;**29**:265–74.
70. Altman J, Bayer SA. Prenatal development of the cerebellar system in the rat. II. Cytogenesis and histogenesis of the inferior olive, pontine gray, and the precerebellar reticular nuclei. *J Comp Neurol* 1978;**179**:49–75.
71. Altman J, Bayer SA. Prenatal development of the cerebellar system in the rat. I. Cytogenesis and histogenesis of the deep nuclei and the cortex of the cerebellum. *J Comp Neurol* 1978;**179**:23–48.
72. Altman J, Bayer SA. Embryonic development of the rat cerebellum. II. Translocation and regional distribution of the deep neurons. *J Comp Neurol* 1985;**231**:27–41.
73. Altman J, Bayer SA. Embryonic development of the rat cerebellum. I. Delineation of the cerebellar primordium and early cell movements. *J Comp Neurol* 1985;**231**:1–26.
74. Iida R, Yasuda T, Tsubota E, Takatsuka H, Masuyama M, Matsuki T, et al. Five age-dependently expressed genes in mouse brain revealed by the fluorescence differential display-PCR technique. *Exp Gerontol* 2002;**37**:1121–6.
75. Ito M. Cerebellar long-term depression: characterization, signal transduction, and functional roles. *Physiol Rev* 2001;**81**:1143–95.
76. Ito M. The molecular organization of cerebellar long-term depression. *Nat Rev Neurosci* 2002;**3**:896–902.
77. Ito M. Cerebellar circuitry as a neuronal machine. *Progr Neurobiol* 2006;**78**:272–303.
78. Duong JK, Gardner K, Rucker LM. Development and retention of fine psychomotor skills: implications for the aging dentist. *J Can Dent Assoc* 2010;**76**:a25.

79. Seidler RD, Bernard JA, Burutolu TB, Fling BW, Gordon MT, Gwin JT, et al. Motor control and aging: links to age-related brain structural, functional, and biochemical effects. *Neurosci Biobehav Rev* 2010;**34**:721–33.
80. Scherder E, Dekker W, Eggermont L. Higher-level hand motor function in aging and (preclinical) dementia: its relationship with (instrumental) activities of daily life—a mini-review. *Gerontology* 2008;**54**:333–41.
81. Menniti FS, Faraci WS, Schmidt CJ. Phosphodiesterases in the CNS: targets for drug development. *Nat Rev Drug Discov* 2006;**5**:660–70.
82. Siuciak JA. The role of phosphodiesterases in schizophrenia: therapeutic implications. *CNS Drugs* 2008;**22**:983–93.
83. Montoliu C, Rodrigo R, Monfort P, Llansola M, Cauli O, Boix J, et al. Cyclic GMP pathways in hepatic encephalopathy. Neurological and therapeutic implications. *Metab Brain Dis* 2010;**25**:39–48.
84. Palit V, Eardley I. An update on new oral PDE5 inhibitors for the treatment of erectile dysfunction. *Nat Rev Urol* 2010;**7**:603–9.
85. Orr HT, Chung MY, Banfi S, Kwiatkowski Jr. TJ, Servadio A, Beaudet AL, et al. Expansion of an unstable trinucleotide CAG repeat in spinocerebellar ataxia type 1. *Nat Genet* 1993;**4**:221–6.
86. Zoghbi HY, Orr HT. Glutamine repeats and neurodegeneration. *Ann Rev Neurosci* 2000;**23**:217–47.
87. Servadio A, Koshy B, Armstrong D, Antalffy B, Orr HT, Zoghbi HY. Expression analysis of the ataxin-1 protein in tissues from normal and spinocerebellar ataxia type 1 individuals. *Nat Genet* 1995;**10**:94–8.
88. Burright EN, Clark HB, Servadio A, Matilla T, Feddersen RM, Yunis WS, et al. SCA1 transgenic mice: a model for neurodegeneration caused by an expanded CAG trinucleotide repeat. *Cell* 1995;**82**:937–48.
89. Clark HB, Burright EN, Yunis WS, Larson S, Wilcox C, Hartman B, et al. Purkinje cell expression of a mutant allele of SCA1 in transgenic mice leads to disparate effects on motor behaviors, followed by a progressive cerebellar dysfunction and histological alterations. *J Neurosci* 1997;**17**:7385–95.
90. Skinner PJ, Vierra-Green CA, Clark HB, Zoghbi HY, Orr HT. Altered trafficking of membrane proteins in purkinje cells of SCA1 transgenic mice. *Am J Pathol* 2001;**159**:905–13.
91. Klement IA, Skinner PJ, Kaytor MD, Yi H, Hersch SM, Clark HB, et al. Ataxin-1 nuclear localization and aggregation: role in polyglutamine-induced disease in SCA1 transgenic mice. *Cell* 1998;**95**:41–53.
92. Serra HG, Byam CE, Lande JD, Tousey SK, Zoghbi HY, Orr HT. Gene profiling links SCA1 pathophysiology to glutamate signaling in Purkinje cells of transgenic mice. *Hum Mol Genet* 2004;**13**:2535–43.
93. Li M, Damuni Z. $I1^{PP2A}$ and $I2^{PP2A}$. Two potent protein phosphatase 2A-specific inhibitor proteins. *Methods Mol Biol* 1998;**93**:59–66.
94. Li M, Guo H, Damuni Z. Purification and characterization of two potent heat-stable protein inhibitors of protein phosphatase 2A from bovine kidney. *Biochemistry* 1995;**34**:1988–96.
95. Matilla A, Radrizzani M. The Anp32 family of proteins containing leucine-rich repeats. *Cerebellum* 2005;**4**:7–18.
96. Gallouzi IE, Steitz JA. Delineation of mRNA export pathways by the use of cell-permeable peptides. *Science* 2001;**294**:1895–901.
97. Brennan CM, Gallouzi IE, Steitz JA. Protein ligands to HuR modulate its interaction with target mRNAs in vivo. *J Cell Biol* 2000;**151**:1–14.
98. Kurihara T, Lewis RM, Eisler J, Greengard P. Cloning of cDNA for DARPP-32, a dopamine- and cyclic AMP-regulated neuronal phosphoprotein. *J Neurosci* 1988;**8**:508–17.

99. Ehrlich ME, Kurihara T, Greengard P. Rat DARPP-32: cloning, sequencing, and characterization of the cDNA. *J Mol Neurosci* 1990;**2**:1–10.
100. Elbrecht A, DiRenzo J, Smith RG, Shenolikar S. Molecular cloning of protein phosphatase inhibitor-1 and its expression in rat and rabbit tissues. *J Biol Chem* 1990;**265**:13415–8.
101. Endo S, Zhou X, Connor J, Wang B, Shenolikar S. Multiple structural elements define the specificity of recombinant human inhibitor-1 as a protein phosphatase-1 inhibitor. *Biochemistry* 1996;**35**:5220–8.
102. Endo S, Connor JH, Forney B, Zhang L, Ingebritsen TS, Lee EY, et al. Conversion of protein phosphatase 1 catalytic subunit to a $Mn^{2+}$-dependent enzyme impairs its regulation by inhibitor 1. *Biochemistry* 1997;**36**:6986–92.
103. Huang HB, Horiuchi A, Watanabe T, Shih SR, Tsay HJ, Li HC, et al. Characterization of the inhibition of protein phosphatase-1 by DARPP-32 and inhibitor-2. *J Biol Chem* 1999;**274**:7870–8.
104. Kwon YG, Huang HB, Desdouits F, Girault JA, Greengard P, Nairn AC. Characterization of the interaction between DARPP-32 and protein phosphatase 1 (PP-1): DARPP-32 peptides antagonize the interaction of PP-1 with binding proteins. *Proc Natl Acad Sci USA* 1997;**94**:3536–41.
105. Egloff MP, Johnson DF, Moorhead G, Cohen PT, Cohen P, Barford D. Structural basis for the recognition of regulatory subunits by the catalytic subunit of protein phosphatase 1. *EMBO J* 1997;**16**:1876–87.
106. Detre JA, Nairn AC, Aswad DW, Greengard P. Localization in mammalian brain of G-substrate, a specific substrate for guanosine 3′, 5′-cyclic monophosphate-dependent protein kinase. *J Neurosci* 1984;**4**:2843–9.
107. Lange A, Mills RE, Lange CJ, Stewart M, Devine SE, Corbett AH. Classical nuclear localization signals: definition, function, and interaction with importin alpha. *J Biol Chem* 2007;**282**:5101–5.
108. Fornerod M, Ohno M, Yoshida M, Mattaj IW. CRM1 is an export receptor for leucine-rich nuclear export signals. *Cell* 1997;**90**:1051–60.
109. Qian Y, Chao DS, Santillano DR, Cornwell TL, Nairn AC, Greengard P, et al. cGMP-dependent protein kinase in dorsal root ganglion: relationship with nitric oxide synthase and nociceptive neurons. *J Neurosci* 1996;**16**:3130–8.
110. Kelley JB, Anderson KL, Itzhak Y. Pharmacological modulators of nitric oxide signaling and contextual fear conditioning in mice. *Psychopharmacology (Berl)* 2010;**210**:65–74.
111. Ota KT, Pierre VJ, Ploski JE, Queen K, Schafe GE. The NO-cGMP-PKG signaling pathway regulates synaptic plasticity and fear memory consolidation in the lateral amygdala via activation of ERK/MAP kinase. *Learn Mem* 2008;**15**:792–805.
112. Ota KT, Monsey MS, Wu MS, Young GJ, Schafe GE. Synaptic plasticity and NO-cGMP-PKG signaling coordinately regulate ERK-driven gene expression in the lateral amygdala and in the auditory thalamus following Pavlovian fear conditioning. *Learn Mem* 2010;**17**:221–35.
113. Paul C, Schöberl F, Weinmeister P, Micale V, Wotjak CT, Hofmann F, et al. Signaling through cGMP-dependent protein kinase I in the amygdala is critical for auditory-cued fear memory and long-term potentiation. *J Neurosci* 2008;**28**:14202–12.
114. Eskin A, Takahashi JS, Zatz M, Block GD. Cyclic guanosine 3′:5′-monophosphate mimics the effects of light on a circadian pacemaker in the eye of aplysia. *J Neurosci* 1984;**4**:2466–71.
115. Golombek DA, Agostino PV, Plano SA, Ferreyra GA. Signaling in the mammalian circadian clock: the NO/cGMP pathway. *Neurochem Int* 2004;**45**:929–36.
116. Bredt DS, Hwang PM, Snyder SH. Localization of nitric oxide synthase indicating a neural role for nitric oxide. *Nature* 1990;**347**:768–70.
117. Hartell NA, Furuya S, Jacoby S, Okada D. Intercellular action of nitric oxide increases cGMP in cerebellar Purkinje cells. *Neuroreport* 2001;**12**:25–8.

118. Crepel F, Jaillard D. Protein kinases, nitric oxide and long-term depression of synapses in the cerebellum. *Neuroreport* 1990;**1**:133–6.
119. Ito M, Karachot L. Messengers mediating long-term desensitization in cerebellar Purkinje cells. *Neuroreport* 1990;**1**:129–32.
120. Shibuki K, Okada D. Endogenous nitric oxide release required for long-term synaptic depression in the cerebellum. *Nature* 1991;**349**:326–8.
121. Daniel H, Hemart N, Jaillard D, Crepel F. Long-term depression requires nitric oxide and guanosine 3′:5′ cyclic monophosphate production in rat cerebellar Purkinje cells. *Eur J Neurosci* 1993;**5**:1079–82.
122. Lev-Ram V, Nebyelul Z, Ellisman MH, Huang PL, Tsien RY. Absence of cerebellar long-term depression in mice lacking neuronal nitric oxide synthase. *Learn Mem* 1997;**4**:169–77.
123. Boxall AR, Garthwaite J. Long-term depression in rat cerebellum requires both NO synthase and NO-sensitive guanylyl cyclase. *Eur J Neurosci* 1996;**8**:2209–12.
124. Jacoby S, Sims RE, Hartell NA. Nitric oxide is required for the induction and heterosynaptic spread of long-term potentiation in rat cerebellar slices. *J Physiol (Lond)* 2001;**535**:825–39.
125. Feil R, Hartmann J, Luo C, Wolfsgruber W, Schilling K, Feil S, et al. Impairment of LTD and cerebellar learning by Purkinje cell-specific ablation of cGMP-dependent protein kinase I. *J Cell Biol* 2003;**163**:295–302.
125a. Shin JH, Linden DJ. An NMDA receptor/nitric oxide cascade is involved in cerebellar LTD but is not localized to the parallel fiber terminal. *J Neurophysiol* 2005;**94**:4281–9.
126. Ito M. *The cerebellum and neural control.* New York: Raven Press; 1984.
126a. Ito M. *The cerebellum: Brain for an implicit self.* New Jersey: FT Press; 2011.
127. Linden DJ, Connor JA. Long-term depression of glutamate currents in cultured cerebellar Purkinje neurons does not require nitric oxide signaling. *Eur J Neurosci* 1992;**4**:10–5.
128. Shin JH, Linden DJ. An NMDA receptor/nitric oxide cascade is involved in cerebellar LTD but is not localized to the parallel fiber terminal. *J Neurophysiol* 2005;**94**:4281–9.
129. Launey T, Endo S, Sakai R, Harano J, Ito M. Protein phosphatase 2A inhibition induces cerebellar long-term depression and declustering of synaptic AMPA receptor. *Proc Natl Acad Sci USA* 2004;**101**:676–81.
130. Eto M, Bock R, Brautigan DL, Linden DJ. Cerebellar long-term synaptic depression requires PKC-mediated activation of CPI-17, a myosin/moesin phosphatase inhibitor. *Neuron* 2002;**36**:1145–58.
131. Chapman PF, Atkins CM, Allen MT, Haley JE, Steinmetz JE. Inhibition of nitric oxide synthesis impairs two different forms of learning. *Neuroreport* 1992;**3**:567–70.
132. Yanagihara D, Kondo I. Nitric oxide plays a key role in adaptive control of locomotion in cat. *Proc Natl Acad Sci USA* 1996;**93**:13292–7.
133. Sakamoto T, Endo S. Amygdala, deep cerebellar nuclei and red nucleus contribute to delay eyeblink conditioning in C57BL/6 mice. *Eur J Neurosci* 2010;**32**:1537–51.
134. Boele HJ, Koekkoek SK, De Zeeuw CI. Cerebellar and extracerebellar involvement in mouse eyeblink conditioning: the ACDC model. *Front Cell Neurosci* 2010;**3**:19.
135. Nagao S, Ito M. Subdural application of hemoglobin to the cerebellum blocks vestibuloocular reflex adaptation. *Neuroreport* 1991;**2**:193–6.
136. Katoh A, Kitazawa H, Itohara S, Nagao S. Inhibition of nitric oxide synthesis and gene knockout of neuronal nitric oxide synthase impaired adaptation of mouse optokinetic response eye movements. *Learn Mem* 2000;**7**:220–6.
137. Shutoh F, Ohki M, Kitazawa H, Itohara S, Nagao S. Memory trace of motor learning shifts transsynaptically from cerebellar cortex to nuclei for consolidation. *Neuroscience* 2006;**139**:767–77.
138. Kassardjian CD, Tan YF, Chung JY, Heskin R, Peterson MJ, Broussard DM. The site of a motor memory shifts with consolidation. *J Neurosci* 2005;**25**:7979–85.

139. Bae JS, Furuya S, Shinoda Y, Endo S, Schuchman EH, Hirabayashi Y, et al. Neurodegeneration augments the ability of bone marrow-derived mesenchymal stem cells to fuse with Purkinje neurons in Niemann-Pick type C mice. *Hum Gene Ther* 2005;**16**:1006–11.
140. Wada K, Howard JT, McConnell P, Whitney O, Lints T, Rivas MV, et al. A molecular neuroethological approach for identifying and characterizing a cascade of behaviorally regulated genes. *Proc Natl Acad Sci USA* 2006;**103**:15212–7.
141. Jarvis ED. Learned birdsong and the neurobiology of human language. *Ann N Y Acad Sci* 2004;**1016**:749–77.
142. Konishi M. Birdsong: from behavior to neuron. *Annu Rev Neurosci* 1985;**8**:125–70.
143. Hirsch E, Graybiel AM, Agid YA. Melanized dopaminergic neurons are differentially susceptible to degeneration in Parkinson's disease. *Nature* 1988;**334**:345–8.
144. German DC, Dubach M, Askari S, Speciale SG, Bowden DM. 1-Methyl-4-phenyl-1,2,3,6-tetrahydropyridine-induced parkinsonian syndrome in *Macaca fascicularis*: which midbrain dopaminergic neurons are lost? *Neuroscience* 1988;**24**:161–74.
145. Rodriguez M, Barroso-Chinea P, Abdala P, Obeso J, González-Hernández T. Dopamine cell degeneration induced by intraventricular administration of 6-hydroxydopamine in the rat: similarities with cell loss in Parkinson's disease. *Exp Neurol* 2001;**169**:163–81.
146. Grimm J, Mueller A, Hefti F, Rosenthal A. Molecular basis for catecholaminergic neuron diversity. *Proc Natl Acad Sci USA* 2004;**101**:13891–6.
147. Chung CY, Seo H, Sonntag KC, Brooks A, Lin L, Isacson O. Cell type-specific gene expression of midbrain dopaminergic neurons reveals molecules involved in their vulnerability and protection. *Hum Mol Genet* 2005;**14**:1709–25.
148. Greene JG, Dingledine R, Greenamyre JT. Gene expression profiling of rat midbrain dopamine neurons: implications for selective vulnerability in parkinsonism. *Neurobiol Dis* 2005;**18**:19–31.
149. Lechward K, Awotunde OS, Swiatek W, Muszyńska G. Protein phosphatase2A: variety of forms and diversity of functions. *Acta Biochim Pol* 2001;**48**:921–33.
150. Ha KS, Kim KM, Kwon YG, Bai SK, Nam WD, Yoo YM, et al. Nitric oxide prevents 6-hydroxydopamine-induced apoptosis in PC12 cells through cGMP-dependent PI3 kinase/Akt activation. *FASEB J* 2003;**17**:1036–47.
151. Kashii S, Mandai M, Kikuchi M, Honda Y, Tamura Y, Kaneda K, et al. Dual actions of nitric oxide in N-methyl-D-aspartate receptor-mediated neurotoxicity in cultured retinal neurons. *Brain Res* 1996;**711**:93–101.
152. Neufeld AH, Sawada A, Becker B. Inhibition of nitric-oxide synthase 2 by aminoguanidine provides neuroprotection of retinal ganglion cells in a rat model of chronic glaucoma. *Proc Natl Acad Sci USA* 1999;**96**:9944–8.
153. Sennlaub F, Courtois Y, Goureau O. Inducible nitric oxide synthase mediates retinal apoptosis in ischemic proliferative retinopathy. *J Neurosci* 2002;**22**:3987–93.
154. Vorwerk CK, Hyman BT, Miller JW, Husain D, Zurakowski D, Huang PL, et al. The role of neuronal and endothelial nitric oxide synthase in retinal excitotoxicity. *Invest Ophthalmol Vis Sci* 1997;**38**:2038–44.
155. Resnikoff S, Pascolini D, Etya'ale D, Kocur I, Pararajasegaram R, Pokharel GP, et al. Global data on visual impairment in the year 2002. *Bull World Health Organ* 2004;**82**:844–51.
156. Campochiaro PA. Ocular neovascularisation and excessive vascular permeability. *Expert Opin Biol Ther* 2004;**4**:1395–402.
157. Skondra D, Noda K, Almulki L, Tayyari F, Frimmel S, Nakazawa T, et al. Characterization of azurocidin as a permeability factor in the retina: involvement in VEGF-induced and early diabetic blood–retinal barrier breakdown. *Invest Ophthalmol Vis Sci* 2008;**49**:726–31.

158. Ambati J, Chalam KV, Chawla DK, D'Angio CT, Guillet EG, Rose SJ, et al. Elevated gamma-aminobutyric acid, glutamate, and vascular endothelial growth factor levels in the vitreous of patients with proliferative diabetic retinopathy. *Arch Ophthalmol* 1997;**115**:1161–6.
159. Deng J, Wu DZ, Gao R. Detection of glutamate and gamma-aminobutyric acid in vitreous of patients with proliferative diabetic retinopathy. *Yan Ke Xue Bao* 2000;**16**:199–202.
160. Lam TT, Abler AS, Kwong JM, Tso MO. N-methyl-D-aspartate (NMDA)-induced apoptosis in rat retina. *Invest Ophthalmol Vis Sci* 1999;**40**:2391–7.
161. Kido N, Tanihara H, Honjo M, Inatani M, Tatsuno T, Nakayama C, et al. Neuroprotective effects of brain-derived neurotrophic factor in eyes with NMDA-induced neuronal death. *Brain Res* 2000;**884**:59–67.
162. Kwong JM, Lam TT. N-methyl-D-aspartate (NMDA) induced apoptosis in adult rabbit retinas. *Exp Eye Res* 2000;**71**:437–44.
163. Kitaoka Y, Kitaoka Y, Kumai T, Lam TT, Kuribayashi K, Isenoumi K, et al. Involvement of RhoA and possible neuroprotective effect of fasudil, a Rho kinase inhibitor, in NMDA-induced neurotoxicity in the rat retina. *Brain Res* 2004;**1018**:111–8.
164. Nakazawa T, Shimura M, Endo S, Takahashi H, Mori N, Tamai M. N-Methyl-D-Aspartic acid suppresses Akt activity through protein phosphatase in retinal ganglion cells. *Mol Vis* 2005;**11**:1173–82.
165. MacNeil MA, Masland RH. Extreme diversity among amacrine cells: implications for function. *Neuron* 1998;**20**:971–82.
166. Araki CM, Hamassaki-Britto DE. Calretinin co-localizes with the NMDA receptor subunit NR1 in cholinergic amacrine cells of the rat retina. *Brain Res* 2000;**869**:220–4.
167. Domek-Łopacińska KU, Strosznajder JB. Cyclic GMP and nitric oxide synthase in aging and Alzheimer's disease. *Mol Neurobiol* 2010;**41**:129–37.
168. Kriegsfeld LJ, Eliasson MJ, Demas GE, Blackshaw S, Dawson TM, Nelson RJ, et al. Nocturnal motor coordination deficits in neuronal nitric oxide synthase knock-out mice. *Neuroscience* 1999;**89**:311–5.

# Index

Note: Page numbers followed by "*f*" indicate figures, and "*t*" indicate tables.

## D

## H



## I

## N



## O

## R



## S

## T



## U

## V



## W

Kolibri 360 GmbH
Ferdinand-Jühlke-Straße 7
99095 Erfurt, Deutschland
produktsicherheit@kolibri360.de